Lecture Notes in Mechanical Engineering

Lecture Notes in Mechanical Engineering (LNME) publishes the latest developments in Mechanical Engineering—quickly, informally and with high quality. Original research reported in proceedings and post-proceedings represents the core of LNME. Volumes published in LNME embrace all aspects, subfields and new challenges of mechanical engineering. Topics in the series include:

- Engineering Design
- Machinery and Machine Elements
- Mechanical Structures and Stress Analysis
- Automotive Engineering
- Engine Technology
- Aerospace Technology and Astronautics
- Nanotechnology and Microengineering
- Control, Robotics, Mechatronics
- MEMS
- Theoretical and Applied Mechanics
- Dynamical Systems, Control
- Fluid Mechanics
- Engineering Thermodynamics, Heat and Mass Transfer
- Manufacturing
- Precision Engineering, Instrumentation, Measurement
- Materials Engineering
- Tribology and Surface Technology

To submit a proposal or request further information, please contact the Springer Editor of your location:

China: Dr. Mengchu Huang at mengchu.huang@springer.com
India: Priya Vyas at priya.vyas@springer.com
Rest of Asia, Australia, New Zealand: Swati Meherishi at swati.meherishi@springer.com
All other countries: Dr. Leontina Di Cecco at Leontina.dicecco@springer.com

To submit a proposal for a monograph, please check our Springer Tracts in Mechanical Engineering at http://www.springer.com/series/11693 or contact Leontina.dicecco@springer.com

Indexed by SCOPUS. All books published in the series are submitted for consideration in Web of Science.

More information about this series at http://www.springer.com/series/11236

S. R. Mishra · T. N. Dhamala · O. D. Makinde
Editors

Recent Trends in Applied Mathematics

Select Proceedings of AMSE 2019

Springer

Editors
S. R. Mishra
Siksha 'O' Anusandhan (Deemed to be University)
Bhubaneswar, Odisha, India

T. N. Dhamala
Tribhuvan University
Kathmandu, Nepal

O. D. Makinde
Faculty of Military Science
Stellenbosch University
Stellenbosch, South Africa

ISSN 2195-4356 ISSN 2195-4364 (electronic)
Lecture Notes in Mechanical Engineering
ISBN 978-981-15-9816-6 ISBN 978-981-15-9817-3 (eBook)
https://doi.org/10.1007/978-981-15-9817-3

This Springer imprint is published by the registered company Springer Nature Singapore Pte Ltd.
The registered company address is: 152 Beach Road, #21-01/04 Gateway East, Singapore 189721, Singapore

Contents

About the Editors

Dr. S. R. Mishra is an Associate Professor in the Department of Mathematics, Siksha 'O' Anusandhan (Deemed to be University), Bhubaneswar, India. He has completed his Ph.D. degree in the year of 2013 and published more than 84 papers in the national and international journals of repute. He has also guided 4 research scholars and currently guiding 5 scholars. His broad areas of research in the field are heat and mass transfer of various nanofluids, statistical analysis of various parameters using artificial neural network, etc.

Prof. Dr. T. N. Dhamala is working as the Full Professor in the Central Department of Mathematics, Tribhuvan University, Kathmandu, Nepal, and leading the research group of Optimization, Graph Theory, Scheduling Theory and Algorithms. He has supervised a number of Ph.D., M.Phil. and Master's degree students. He also held the position of the Head of Department in the Central Department of Computer Science and Technology, Tribhuvan University, in the years 2007–2013. He coordinated the MPhil Program in Mathematics at the Central Department of Mathematics, Tribhuvan University, from 2016 to 2019.

Prof. O. D. Makinde is presently a Distinguished Professor of Computational and Applied Mathematics at the Faculty of Military Science, Stellenbosch University, South Africa. He is also a visiting professor to several other universities, including the Vellore University of Technology in India; Nelson Mandela African Institute of Science and Technology in Tanzania; the Pan African University Institute for Basic Sciences, Technology and Innovation in Kenya; the African University of Science and Technology in Nigeria; the Adama Science and Technology University in Ethiopia, etc. He was a Full Professor & Head of Applied Mathematics Department at the University of Limpopo, South Africa (1998–2008), and Senior Professor & Director of Postgraduate Studies at the Cape Peninsula University of Technology, South Africa (2008–2013). He obtained his B.Sc. (Hons.) degree – First Class with Faculty Prize M.Sc. degree qualifications in Mathematics from Obafemi Awolowo University in Nigeria and Ph.D. degree in Computational Applied Mathematics from the University of Bristol in UK under the prestigious Commonwealth Scholarship. Prof. Makinde's research work covers three broad areas which include fluid mechanics, mathematical biology and computational mathematics. He authored 4 Applied Mathematics textbooks & monographs, edited 6 advanced research textbooks on Heat Transfer in Fluids & Solids and published over 450 research papers in many reputable international journals worldwide. Prof. Makinde's scientific metrics according to Google Scholar show H-Index = 51, Citations Index = 10390 and i10-index = 229. This bibliometric statistic continues to increase due to high quality and global impact of his research work. He has supervised over 30 Ph.D.s.

A Multi-item Deteriorating Inventory Model Under Stock Level-Dependent, Time-Varying, and Price-Sensitive Demand

Abhijit Barman and P. K. De

Abstract This paper advocates a multi-item deteriorating inventory model where shortages are not allowed. Here, we have proposed a single-stage EOQ model for deteriorating items where the demand function is depending on nonlinear selling price, nonlinear time, and inventory stock. The model is developed under a known initial inventory. The main objective of this model is to determine the selling price and time length until the inventory reaches zero for each item. To demonstrate our model, one numerical example has been given which is followed by a sensitivity analysis of the major parameters involved in this model.

Keywords Multi-item inventory · Deteriorating items · Selling price · Order quantity · Hessian matrix

1 Introduction

In real-life situations, it is observed that demand for an inventory model changes for the number of items increases in the stocks. That is why companies or any firm owners deal with the multi-item inventory system. The present paper presents a multi-item inventory system over a single period with a finite time horizon. The product deteriorates with the passes of time under the different deteriorating rates. Most of the items that undergo decay over time are medicine, blood banks, volatile liquids, vegetables, etc. Demand for the items is deterministic which depends on inventory label, selling price, and time-varying. The main goal of this model is to determine the unit selling price of a product and the length of the period up to zero inventory that maximizes the overall profit of a retailer or any inventory warehouse.

A. Barman (✉) · P. K. De
Department of Mathematics, National Institute of Technology Silchar, Silchar 788010, Assam, India
e-mail: abhijitmath93@gmail.com

P. K. De
e-mail: pijusde@gmail.com

S. R. Mishra et al. (eds.), *Recent Trends in Applied Mathematics*, Lecture Notes in Mechanical Engineering, https://doi.org/10.1007/978-981-15-9817-3_1

Even though sufficient literature is available in the area of deteriorating items, but still very less literature is available on multi-item inventory system with deterioration.

The first effort to illustrate the optimum order policies for deteriorating items was established by Chare and Schratures [1]. They introduced an EOQ model for an exponentially decaying inventory system. Later, Covert and Philip [2] extended this model by incorporating variable deterioration rate with two parameters Weibull distribution. Bhattacharya [3] proposed a new method for deteriorating items with linear stock-dependent demand rate in a two items inventory system. Dye et al. [4] discussed pricing and ordering policy for deteriorating items with shortages where the deterioration and demand rate are continuous as well as a differentiable function of time and price, respectively. Pal et al. [5] established a multi-item EOQ model with nonlinear price-dependent and price break-sensitive demand. In the case of non-instantaneous deteriorating items, a joint pricing and inventory model has been established by Maihami and Kamalabadi [6]. Linear price-sensitive and nonlinear time-dependent demand functions have been considered to develop this model with partially backlogging. Sarkar et al. [7] established an inventory model for deteriorating items considering time-sensitive demand with a finite production rate. The selling price and component cost are considered at a continuous rate of time. Yang [8] studied an EOQ model where the holding cost is stock-dependent and the demand rate is also stock-dependent with relaxed terminal environments under shortages. The prime goal of this model is profit maximization by determining optimum order quantity and level of ending inventory. Janssen et al. [9] reviewed 393 articles that are published from January 2012 to December 2015 and categorized the articles based on the different demand characteristics and the deterioration of the items. Feng et al. [10] used the demand as a multivariate function of stock, price, and freshness in an EOQ model. Chen et al. [11] discovered an inventory model for time elapse deteriorating items with a short lifecycle. This model is designed for the stock label, time-varying, and price-sensitive deterministic demand in a finite horizon multi-period setting.

This paper address an EOQ model for n numbers of different items in a finite time horizon. For each item, an initial inventory stock depending on store capacity has been taken separately. The deterministic demand function is taken in a pattern of the nonlinear selling price, exponential time-varying, and linear stock-dependent. Shortages of products are not allowed in this multi-item inventory system. Thus, this paper determines the optimum selling price, time length for which the inventory reaches zero for each item and the overall profit.

The rest of the paper is organized as follows. In Sect. 2, we describe the notations and assumptions used throughout the model. We inaugurate the mathematical model with necessary and sufficient conditions in Sect. 3. In Sect. 4, a numerical example has been provided to illustrate the solution procedure. In Sect. 5, a sensitivity analysis of the optimum solutions concerning different parameters has also been provided. Finally, the summarized findings and some future research suggestions are discussed in Sect. 6.

2 Notations and Assumptions

The following notations and assumption are adopted to develop the model:

Notations

M_i	Manufacturing cost per item for *i*th product
O_i	Ordering cost for *i*th product
h_i	Holding cost per unit time for *i*th product
Q_i	Initial order quantity for *i*th product
p_i	Unit selling price for *i*th item
R_i	Demand rate for each product
θ_i	Deterioration rate for *i*th product
$a_i, b_i, c_i, \mu_i, \alpha_i, \lambda_i$	Demand and stock elasticity parameters
$I(p_i, t)$	Inventory level for *i*th product at time t
T_i	Time length up to zero inventory
TP	Total profit

Assumptions

- The model is considered for n number of different types of products for deteriorating items in a single stage.
- Shortages are not considered in this inventory model i.e. $I_i(p_i, t) \geq 0$ for $i = 1, 2, 3 \ldots n$.
- The replenishment rate is infinite and lead time is negligible.
- Deterioration rate θ_i is constant for *i*th product.
- Demand rate R_i is deterministic in nature and a function of inventory level $I_i(p_i, t)$ with nonlinear selling price $(a_i - b_i p_i - c_i p_i^2)$ and exponentially time varying. For $i = 1, 2, 3 \ldots n$ with considering $a_i \gg b_i \gg c_i$, R_i is represented by

$$R_i = (a_i - b_i p_i - c_i p_i^2)\alpha_i e^{\lambda_i t} + \mu_i I(p_i, t).$$

3 Mathematical Model Formulation and Solution Methodology

At the beginning of the cycle, the system starts with inventory Q_i for the ith product. Over the course of the period, the inventory level down due to both demand and deterioration until it reaches zero at time T_i. During the time interval $[0, T_i]$, the following differential equation represents the inventory status for the ith product

$$\frac{dI_i(p_i, t)}{dt} + \theta_i I(p_i, t) = -R_i \tag{1}$$

with two boundary conditions, $I_i(p_i, 0) = Q_i$ and $I_i(p_i, T_i) = 0$ for $i = 1, 2, \ldots n$. Solving the inventory system and using the boundary conditions, we get the level of inventory of ith item at time t is

$$I_i(p_i, t) = Q_i e^{-(\theta_i+\mu_i)t} + \frac{(a_i - b_i p_i - c_i p_i^2)\alpha_i}{(\theta_i + \mu_i + \lambda_i)}[e^{-(\theta_i+\mu_i)t} - e^{\lambda_i t}] \tag{2}$$

From the second boundary condition, we have

$$T_i = \frac{1}{(\theta_i + \mu_i + \lambda_i)} Log\left[\frac{Q_i(\theta_i + \mu_i + \lambda_i)}{(a_i - b_i p_i - c_i p_i^2)\alpha_i} + 1\right] \tag{3}$$

Next, for $i = 1, 2, 3 \ldots n$, the total profit in the whole cycle consists of the following five elements:

- Total ordering cost for the ith product is given by

$$OC_i = O_i \tag{4}$$

- Inventory holding cost for the ith product is given by

$$HC_i = h_i \int_0^{T_i} I(p_i, t)dt = h_i K_i \tag{5}$$

- Total manufacturing cost for the ith product is

$$MC_i = M_i Q_i \tag{6}$$

- Deteriorating cost for the ith product is given by

$$DC_i = M_i \int_0^{T_i} \theta_i I(p_i, t)dt = M_i \theta_i K_i \tag{7}$$

- Sales revenue for the *i*th product is written by:

$$\begin{aligned} SR_i &= p_i \int_0^{T_i} R_i(p_i, t) \\ &= \frac{1}{\lambda_i} p_i \alpha_i (a_i - b_i p_i - c_i p_i^2)(e^{\lambda_i T_i} - 1) + p_i \mu_i K_i \end{aligned} \tag{8}$$

where

$$\begin{aligned} K_i &= \int_0^{T_I} I_i(p_i, t)dt = \frac{Q_i}{(\theta_i + \mu_i)}[1 - e^{-(\theta_i + \mu_i)T_i}] \\ &+ \left\{ \frac{(a_i - b_i p_i - c_i p_i^2)\alpha_i}{(\theta_i + \mu_i + \lambda_i)} \left[\frac{1}{\lambda_i} + \frac{1}{(\theta_i + \mu_i)} - \frac{e^{\lambda_i T_i}}{\lambda_i} - \frac{e^{-(\theta_i + \mu_i)}}{(\theta_i + \mu_i)} \right] \right\} \end{aligned} \tag{9}$$

Therefore, the total profit of the retailer for all the items in the whole cycle is

$$\begin{aligned} TP(p_1, p_2, \ldots p_n) &= \sum_{i=1}^{n} [SR_i - (OC_i + HC_i + MC_i + DC_i)] \\ &= \sum_{1}^{n} \left[\frac{1}{\lambda_i} p_i \alpha_i (a_i - b_i p_i - c_i p_i^2)(e^{\lambda_i T_i} - 1) + (p_i \mu_i - h_i - M_i \theta_i) K_i - O_i - M_i Q \right]_i \end{aligned} \tag{10}$$

TP $(p_1, p_2, \ldots p_n)$ is function of $p_1, p_2, \ldots p_n$. So, for some T_i (*from Eq.* 3), the necessary conditions for the overall profit function come from $\frac{\partial TP(p_1, p_2, \ldots p_n)}{\partial p_i} = 0$ for $i = 1, 2, 3 \ldots n$.

$$
\begin{aligned}
\frac{\partial TP(p_1, p_2, \dots p_n)}{\partial p_i} = & -\frac{p_i Q_i(-b_i - 2c_i p_i)\left[1 + \frac{Q_i(\theta_i+\mu_i+\lambda_i)}{(a_i - b_i p_i - c_i p_i^2)\alpha_i}\right]^{-1+\frac{\lambda_i}{\theta_i+\mu_i+\lambda_i}}}{(a_i - b_i p_i - c_i p_i^2)} + \frac{\mu_i Q_i\left[-1 + \left(1 + \frac{Q_i(\theta_i+\mu_i+\lambda_i)}{(a_i - b_i p_i - c_i p_i^2)\alpha_i}\right)^{\frac{-(\theta_i+\mu_i)}{(\theta_i+\mu_i+\lambda_i)}}\right]}{(\theta_i + \mu_i)} \\
& + \frac{\alpha_i[\mu_i(a_i - b_i p_i - c_i p_i^2) + (p_i \mu_i - h_i - M_i \theta_i)(-b_i - 2c_i p_i)]\left[\frac{1}{\lambda_i} + \frac{1}{\theta_i+\mu_i} - \frac{\left(1+\frac{Q_i(\theta_i+\mu_i+\lambda_i)}{(a_i-b_i p_i - c_i p_i^2)\alpha_i}\right)^{\frac{\lambda_i}{\theta_i+\mu_i+\lambda_i}}}{\lambda_i} - \frac{\left(1+\frac{Q_i(\theta_i+\mu_i+\lambda_i)}{(a_i-b_i p_i - c_i p_i^2)}\right)^{\frac{-(\theta_i+\mu_i)}{\theta_i+\mu_i+\lambda_i}}}{\theta_i+\mu_i}\right]}{\theta_i + \mu_i + \lambda_i} \\
& + \frac{(p_i \mu_i - h_i - M_i \theta_i)\alpha_i(a_i - b_i p_i - c p_i^2)\left[\frac{Q_i(-b_i - 2c_i p_i)\left(1+\frac{Q_i(\theta_i+\mu_i+\lambda_i)}{(a_i-b_i p_i-c_i p_i^2)\alpha_i}\right)^{-1+\frac{\lambda_i}{\theta_i+\mu_i+\lambda_i}}}{(a_i - b_i p_i - c_i p_i^2)^2\alpha_i} + \frac{Q_i(-b_i-2c_i p_i)(-\theta_i-\mu_i)\left(1+\frac{Q_i(\theta_i+\mu_i+\lambda_i)}{(a_i-b_i p_i-c_i p_i^2)\alpha_i}\right)^{-1+\frac{-(\theta_i+\mu_i)}{(\theta_i+\mu_i+\lambda_i)}}}{(a_i-b_i p_i-c_i p_i^2)\alpha_i(\theta_i+\mu_i)}\right]}{(\theta_i + \mu_i + \lambda_i)} \\
& + \frac{(p_i \mu_i - h_i - M_i \theta_i) Q_i^2(-b_i - 2c_i p_i)(-\theta_i - \mu_i)\left(1 + \frac{Q_i(\theta_i+\mu_i+\lambda_i)}{(a_i - b_i p_i - c_i p_i^2)\alpha_i}\right)^{-1+\frac{-\theta_i-\mu_i}{(\theta_i+\mu_i+\lambda_i)}}}{(a_i - b_i p_i - c_i p_i^2)^2 \alpha_i(\theta_i + \mu_i)} \\
& + \frac{\alpha_i[p_i(-b_i - 2c_i p_i) + (a_i - b_i p_i - c_i p_i^2)]\left[-1 + \left(1 + \frac{Q_i(\theta_i+\mu_i+\lambda_i)}{(a_i - b_i p_i - c_i p_i^2)\alpha_i}\right)^{\frac{\lambda_i}{(\theta_i+\mu_i+\lambda_i)}}\right]}{\lambda_i}
\end{aligned}
\tag{11}
$$

This gives a system of nonlinear equations with n number of unknowns p_i for i = *1,2...n*.

Concavity test by the Hessian matrix

To check whether the profit function (10) is concave, we have to determine the hessian matrix

$$H = \begin{bmatrix} \frac{\partial^2 TP}{\partial p_1^2} & \frac{\partial^2 TP}{\partial^2 p_1 \partial p_2} & \cdots & \frac{\partial^2 TP}{\partial p_1 \partial p_n} \\ \frac{\partial TP}{\partial p_2 \partial p_1} & \frac{\partial^2 TP}{\partial p_2^2} & \cdots & \frac{\partial^2 TP}{\partial p_2 \partial p_n} \\ \cdots & \cdots & \cdots & \cdots \\ \frac{\partial^2 TP}{\partial p_n \partial p_1} & \frac{\partial^2 TP}{\partial p_n \partial p_2} & \cdots & \frac{\partial^2 TP}{\partial p_n^2} \end{bmatrix}$$

We have to show all the principal minors of the hessian matrix will alternate their sign starting with a negative sign. Since the expression of the second-order derivatives is highly nonlinear, we will check the result numerically with some graphical representation. We have shown numerically as well as graphically the concavity of profit function $TP(p_1, p_2, \ldots p_n)$ in the given numerical example.

4 Numerical Investigation

Let us consider a storehouse problem with two different items with different demand rates and deterioration rates are listed by the following parametric values in Table 1.

Table 1 Different parametric values

Value of the parameter	Item 1	Item 2
Demand elasticity parameter	$a_1 = 120$; $b_1 = 3.0$; $c_1 = 0.005$; $\lambda_1 = 0.1$; $\alpha_1 = 1$	$a_2 = 140$; $b_2 = 3.2$; $c_2 = 0.005$; $\lambda_2 = 0.15$; $\alpha_2 = 1.5$
Manufacturing cost ($)	$M_1 = 10$	$M_2 = 12$
Ordering cost ($)	$O_1 = 70$	$O_2 = 80$
Holding cost ($)	$h_1 = 3$	$h_2 = 4$
Deterioration rate	$\theta_1 = 0.08$	$\theta_2 = 0.05$
Stock elasticity parameter	$\mu_1 = 0.6$	$\mu_2 = 0.8$
Initial inventory level	160 units	200 units

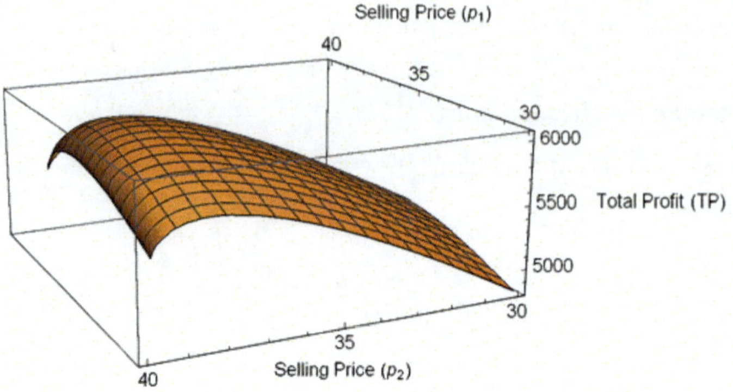

Fig. 1 Variation of optimal profit with respect to selling prices p_1 and p_2

Using Mathematica software, we determine the optimum selling price, time length for the different products, and also the net profit. Unit selling price of first item (p_1^*) = \$35.284, Unit selling price of second item (p_2^*) = \$38.292, Cycle length for first item (T_1^*) = 4.3837 days, Cycle length for second item (T_2^*) = 3.3807 days, and the total profit (TP^*) = \$6049.46. We get $\Delta_1 = -40.55 \leq 0$ and $\Delta_2 = 3174.3878 \geq 0$. The sign of principle minor of the hessian matrix alternate starting with a negative sign. So, the condition of sufficiency is also satisfied.

The following figure (Fig. 1) represents the nature of concavity of the profit function (10). For selling price $p_1^* = \$35.284$ and $p_2^* = \$38.292$, the corresponding total profit $TP^* = \$6049.46$ gives the global maximum in the concave Fig. 1.

5 Sensitivity Analysis

A post-optimality analysis is carried out to analyze the outcome of the changes of different parameters on the optimal solutions. The results of the post-optimality analysis are listed in Table 2 and Table 3, respectively. The changes to the optimum values p_1^*, p_2^*, T_1^*, T_2,* TP^* have been done by decreasing/increasing the values of the major parameters $a_i, b_i, c_i, \mu_i, \alpha_i, \theta_i, \lambda_i$ for $i = 1, 2 \ldots n$. The post-optimality analysis is accomplished from -20% to 20% by changing one parameter at a time and keeping remain parametric values unchanged.

The sensitive analysis which is explored in Tables 2 and 3 indicates the following observations:

- It is visible in Table 2 that, with the increase of purchasing cost (M_i), the selling price (p_i) for both the products as well as the time interval (T_i) will decrease. So, the total profit (TC) will decrease with increasing the purchasing cost (M_i).
- It also observed that with the increasing of holding cost (h_i), the selling price (p_i), time length (T_i), and the overall profit (TC) decrease rapidly. From a financial

Table 2 Sensitivity analysis with respect to different parameters

Parameter	Percentage change in parameter	p_1	p_2	T_1	T_2	TP
M_i	−20%	35.367	38.336	4.4410	3.4156	6916.08
	−10%	35.326	38.314	4.4121	3.3981	6482.71
	$M_1 = 10$; $M_2 = 12$	**35.284**	**38.292**	**4.3837**	**3.3807**	**6049.46**
	10%	35.242	38.269	4.3558	3.3636	5616.31
	20%	35.200	38.247	4.3284	3.3468	5183.28
O_i	−20%	35.284	38.292	4.3837	3.3807	6079.46
	−10%	35.284	38.292	4.3837	3.3807	6064.46
	O_i	**35.284**	**38.292**	**4.3837**	**3.3807**	**6049.46**
	10%	35.284	38.292	4.3837	3.3807	6034.46
	20%	35.284	38.292	4.3837	3.3807	6019.46
h_i	−20%	35.597	38.584	4.6106	3.6346	6384.55
	−10%	35.441	38.439	4.4930	3.5012	6215.58
	h_i	**35.284**	**38.292**	**4.3837**	**3.3807**	**6049.46**
	10%	35.127	38.143	4.282	3.2713	5886.05
	20%	34.969	37.994	4.1863	3.1714	5725.22
Q_i	−20%	35.036	38.096	3.8780	2.9830	4998.59
	−10%	35.160	38.195	4.1369	3.1864	5561.73
	$Q_1 = 160$; $Q_2 = 200$	**35.284**	**38.292**	**4.3837**	**3.3807**	**6049.46**
	10%	35.405	38.384	4.6199	3.5669	6678.74
	20%	35.522	38.472	4.8470	3.7459	7234.29

lookout, it is clear that increasing inventory holding cost directly affects the total profit.

- From Table 2, it is clear that ordering cost (O_i) does not affect the selling price (p_i). So, the change of overall profit (TC) is negligible for that case.
- From Table 2, it is found that increasing with initial inventory stock level (Q_i) the selling price and time interval for both the items will increase. For this case, the net profit will also increase rapidly.
- When the parametric value a_1, a_2 increase the value of selling price (p_i) for both the product will increase but the time length for each item (T_i) of the inventory cycle will decrease. In this case, the overall profit (TC) also increases very rapidly which is shown in Table 3.
- With increasing the value of the other parameter b_i, c_i the value of the selling price (p_i) decreases. So, the value of the total profit (TC) also decreases. For that case, b_i is more effective other than the parameter c_i.
- The effect of change in overall profit (TC) with respect to the change in parameters λ_i and α_i are shown in Table 3 separately. It is observed that, as λ_i or α_i increases, the increment of total profit (TC) is not significant.

Table 3 Sensitivity analysis with respect to different parameters

Parameter	Percentage change in parameter	p_1	p_2	T_1	T_2	TP
a_i	−20%	28.425	31.045	4.6555	3.5545	3630.42
	−10%	31.873	34.697	4.5140	3.4647	4847.11
	$a_1 = 120; a_2 = 140$	**35.284**	**38.292**	**4.3837**	**3.3807**	**6049.46**
	10%	38.658	41.832	4.2626	3.3018	7238.05
	20%	41.998	45.3201	4.1501	3.2276	8413.38
b_i	−20%	43.192	46.108	4.5789	3.5519	8548.57
	−10%	38.885	41.887	4.4817	3.4666	7191.25
	$b_1 = 3.0; b_2 = 3.2$	**35.284**	**38.292**	**4.3837**	**3.3807**	**6049.46**
	10%	32.237	35.204	4.2869	3.2962	5079.12
	20%	29.630	32.529	4.1929	3.2140	4246.66
c_i	−20%	35.718	38.989	4.4207	3.4289	6222.68
	−10%	35.499	38.635	4.4019	3.4044	6134.79
	$c_1 = 0.005; c_2 = 0.008$	**35.284**	**38.292**	**4.3837**	**3.3807**	**6049.46**
	10%	35.074	37.959	4.3659	3.3579	5966.55
	20%	34.868	37.638	4.3486	3.336	5885.94
λ_i	−20%	35.275	38.291	4.4768	3.4654	6025.37
	−10%	35.279	38.291	4.4296	3.4224	6037.45
	$\lambda_1 = 0.1; \lambda_2 = 0.15$	**35.284**	**38.292**	**4.3837**	**3.3807**	**6049.46**
	10%	35.288	38.293	4.3391	3.3405	6061.39
	20%	35.293	38.294	4.2957	3.3015	6073.23
μ_i	−20%	34.323	37.545	4.2037	3.2193	5608.43
	−10%	34.830	37.944	4.2915	3.2984	5835.33
	$\mu_1 = 0.4; \mu_2 = 0.6$	**35.284**	**38.292**	**4.3837**	**3.3807**	**6049.46**
	10%	35.685	38.590	4.4802	3.4663	6250.93
	20%	36.036	38.843	4.5810	3.5549	6439.94
θ_i	−20%	35.594	38.454	4.6934	3.5441	6362.40
	−10%	35.437	38.373	4.5311	3.4597	6239.93
	$\theta_1 = 0.08; \theta_2 = 0.05$	**35.284**	**38.292**	**4.3837**	**3.3807**	**6049.46**
	10%	35.134	38.212	4.2491	3.3066	6006.71
	20%	34.987	38.131	4.1255	3.2369	5895.45
α_i	−20%	35.579	38.514	4.9575	3.8329	5921.67
	−10%	35.418	38.394	4.6456	3.5871	5986.68
	$\alpha_1 = 1; \alpha_2 = 2$	**35.284**	**38.292**	**4.3837**	**3.3807**	**6049.46**
	10%	35.172	38.204	4.1598	3.2045	6109.82
	20%	35.077	38.129	3.9658	3.0518	6167.71

- When the value of parameter μ_i increases then the selling price (p_i), time length (T_i), and the total profit (TC) also increase. The physical phenomena of this parameter suggest that demand proportional to the inventory of the firm.
- Increasing of parameter θ_i the overall profit (TC) will decrease. The economic viewpoint of this observation shows that as increasing the deterioration rate the profit will be minimized.

6 Conclusion

In this paper, we explored a short life period multi-item EOQ model where deterioration is considered. A stock level-dependent, time-varying, and price-sensitive deterministic demand have been considered to develop the model under a known primary stock. To design the model, the effect of nonlinear selling price and nonlinear time-varying demand functions has been estimated. Our model is demonstrated and illustrated with one numerical example with a graphical explanation. Sensitivity analysis is shown to see the changes in overall profit with respect to the variant of several parameters involved in this model. The contribution of this paper helps decision-makers to increase the overall profit by understanding the market demand situation. As a result, retailers may change their earlier selling price of the items to earn the maximum profit.

This paper can be extended by incorporating various other concepts like inflation, reliability, or some other fuzzy environments.

Acknowledgements First, the author is gratefully thankful to MHRD, Govt. of India for giving support for carrying out research at the National Institute of Technology Silchar and also thankful to TEQIP III for supporting financially to present this paper in AMSE-2019 at Bhubaneswar.

References

1. P. Chare, G. Schrader, A model for exponentially decaying inventories. J. Ind. Eng. **15**, 238–243 (1963)
2. R.P. Covert, G.C. Philip, An EOQ model for items with Weibull distribution deterioration. AIIE Trans. **5**(4), 323–326 (1973)
3. D.K. Bhattacharya, On multi-item inventory. Eur. J. Oper. Res. **162**(3), 786–791 (2005)
4. C.Y. Dye, Joint pricing and ordering policy for a deteriorating inventory with partial backlogging. Omega **35**(2), 184–189 (2007)
5. B. Pal, S.S. Sana, K. Chaudhuri, Multi-item EOQ model while demand is sales price and price break sensitive. Econ. Model. **29**(6), 2283–2288 (2012)
6. R. Maihami, I.N. Kamalabadi, Joint pricing and inventory control for non-instantaneous deteriorating items with partial backlogging and time and price dependent demand. Int. J. Prod. Econ. **136**(1), 116–122 (2012)
7. B. Sarkar, S. Saren, H.M. Wee, An inventory model with variable demand, component cost and selling price for deteriorating items. Econ. Model. **30**, 306–310 (2013)

8. C.T. Yang, An inventory model with both stock-dependent demand rate and stock-dependent holding cost rate. Int. J. Prod. Econ. **155**, 214–221 (2014)
9. L. Janssen, T. Claus, J. Sauer, Literature review of deteriorating inventory models by key topics from 2012 to 2015. Int. J. Prod. Econ. **182**, 86–112 (2016)
10. L. Feng, Y.L. Chan, L.E. Cárdenas-Barrón, Pricing and lot-sizing polices for perishable goods when the demand depends on selling price, displayed stocks, and expiration date. Int. J. Prod. Econ. **185**, 11–20 (2017)
11. L. Chen, X. Chen, M.F. Keblis, G. Li, Optimal pricing and replenishment policy for deteriorating inventory under stock-level-dependent, time-varying and price-dependent demand. Comput. Ind. Eng. **135**, 1294–1299 (2019)

On Estimation of Reliability Following Selection from Pareto Populations

Ajaya Kumar Mahapatra, Brijesh Kumar Jha, and Chiranjibi Mahapatra

Abstract Let $\prod_1, \prod_2, \ldots, \prod_k$ be k populations, where $\prod_i$ follows a Pareto distribution with unknown scale parameters α_i and common known shape parameters $\beta_i; i = 1, \ldots, k$. Independent random samples are drawn from each of these populations. Let T_i be the smallest observation in the ith sample. The natural selection rule is to select the population with the largest T_i. Then, we consider the estimation of the reliability function of the selected population. The uniform minimum variance unbiased estimator is derived. A class of scale equivariant estimators have been proposed. An inadmissibility result in regards to the class of scale equivariant estimators is established generally

Keywords Selection rule · UMVUE · MLE · Scale equivariant estimators · Brewster–Zidek technique

1 Introduction

Let $\prod_1, \prod_2, \ldots, \prod_k$ be as defined above with each of them corresponding to a probability density function/ probability mass function $f(x|\theta_i), i = 1, \ldots, k$. A common problem is to choose the population or a subset of populations having the best. The population may be regarded as the best according to some attributes such as the largest mean, smallest variance, etc. An important practical problem is to estimate the parameters of the selected population or an attribute of the selected subset. These problems are in general mentioned as "Estimation after selection". Such problems have been at first constructed and explored by Rubinstein [16]. Estimation of the quantile of a

A. K. Mahapatra (✉) · C. Mahapatra
Centre for Applied Mathematics and Computing, Siksha O Anusandhan Deemed To Be University, Bhubaneswar 751030, India
e-mail: ajayamohapatra@soa.ac.in

B. K. Jha
Department of Mathematics, Siksha O Anusandhan Deemed To Be University, Bhubaneswar 751030, India

S. R. Mishra et al. (eds.), *Recent Trends in Applied Mathematics*, Lecture Notes in Mechanical Engineering, https://doi.org/10.1007/978-981-15-9817-3_2

selected population has been considered by Sharma and Vellaisamy [17], Kumar and Kar ([10–12]) and Vellaisamy [18]. Mishra and van der Meulen [15] have studied the estimation after selection in general truncation distributions. The estimation of the reliability function of a selected subset was studied by Kumar et al. [13] for the case of two-parameter exponential distribution. It was assumed that the scale parameters are known and the location parameters are unknown and unequal. They derived the Uniform Minimum Variance Unbiased Estimator(UMVUE) for the survival function and proposed some natural estimators. These estimators are compared in terms of their risks using Brewster and Zidek technique. Further, this estimator is also improved by solving a differential inequality in the light of Vellaisamy and Punen [19]. They have considered the estimation of the location parameter from a selected subset of exponential distribution.

Income distributions were studied initially using Pareto distribution. Later on, it was applied to reliability and life testing, industrial and engineering studies. Johnson and Katz [8], Harris [7], Davis and Feldstein [4], Freiling [6], Berger and Mandelbrot [2], etc., have described several situations where the Pareto model is very useful. This model has been found suitable to describe the allotment of service times in regard to city maintenance, allocation of fallout of nuclear particles, etc. Kumar and Gangopadhyaya [9] have taken up the case to estimate the scale parameter of the chosen Pareto population. In this paper, we study the estimation of the reliability function in the following selection. In Sect. 4, we have derived the UMVUE for the reliability function of the selected Pareto population. In Sect. 5, an inadmissibility result has been established generally for the estimators in the scale equivariant class.

2 Deriving the UMVUE

Independent random samples $X_{i1}, X_{i2}, \ldots, X_{in}, i = 1, \ldots, k$ are drawn from k populations $\prod_1, \prod_2, \ldots, \prod_k$, respectively. Let these observations from the respective populations have an associated probability density function $f_i(.)$, given by the Pareto model.

$$f_i(x) = \begin{cases} \frac{\beta\alpha_i^\beta}{x^{\beta+1}}, & \text{if } \alpha_i \leq x < \infty, \alpha_i > 0, \beta > 0, \\ 0, & \text{elsewhere, for } i = 1, 2, \ldots, k. \end{cases} \tag{1}$$

Let us assume overall that the scale parameters $\alpha_1, \alpha_2, \ldots, \alpha_k$ are completely unknown and the common shape parameter β is known. Let $T_i = \min(X_{i1}, X_{i2}, \ldots, X_{in})$. Then the statistic $\underline{T} = (T_1, T_2, \ldots, T_k)$ is complete and sufficient. It is also the maximum likelihood estimator (MLE) of $\underline{\alpha} = (\alpha_1, \alpha_2, \ldots, \alpha_k), i = 1, \ldots, k$. It may be seen that T_i follows a Pareto distribution with shape parameter $n\beta$ and scale parameter $\alpha_i, i = 1, \ldots, k$. It is given by

$$g_i(y) = \begin{cases} \frac{n\beta\alpha_i^{n\beta}}{y^{n\beta+1}}, \text{ if } \alpha_i \leq y < \infty, \\ 0, \quad \text{elsewhere, for } i = 1, 2, \ldots, k. \end{cases} \tag{2}$$

The reliability function of the population $\prod_i$ is given by

$$R_i(t) = P(X_{ij} > t) = \left(\frac{\alpha_i}{t}\right)^{\beta}, \ \alpha_i < t. \tag{3}$$

Our goal is to choose the population associated with the highest reliability $R_i(t)$, $i = 1, \ldots, k$. The probability density of T_i has monotone likelihood ratio property in (α_i, T_i), $i = 1, \ldots, k$. A logical selection rule is to select the population $\prod_i$ if $T_i = max(T_1, \ldots, T_k)$, $i = 1, \ldots, k$. Optimality properties in this regard have been scrutinized by Bahadur and Goodman [1], Lehmann [14] and Eaton [5]. Let $T_{(1)} \leq T_{(2)}, \leq T_{(k)}$ stand for the ordered values of $T_i's$. We want to estimate

$$R_J(t) = \sum_{j=1}^{k} \left(\frac{\alpha_j}{t}\right)^{\beta} I_j, \tag{4}$$

where

$$I_j = \begin{cases} 1, \text{ if } T_j = T_{(k)}, j = 1, \ldots, k; \\ 0, \text{ otherwise.} \end{cases} \tag{5}$$

An unbiased estimator δ of $R_J(t)$ satisfies $E(\delta - R_J(t)) = 0$. To derive the UMVUE of $R_J(t)$, we need the following lemmas.

Lemma 2.1 *Let X be a random variable with pdf $g_i(.)$, given by*

$$g_i(x) = \begin{cases} \frac{n\beta\alpha_i^{n\beta}}{x^{n\beta+1}} \ \textit{if } \alpha_i \leq x < \infty \\ 0 \quad \textit{otherwise, for } i = 1, \ldots, k. \end{cases} \tag{6}$$

Suppose that U(x) be a real-valued function defined on R, such that
(a) $E_\alpha(U(x)) < \infty \quad \forall \alpha \in \Omega$,
(b) The integral $\int_x^\infty U(t)P(t, \beta)dt$ exists and is finite,
where $P(x, \beta) = \frac{n}{\beta^{n-1}x^n\beta+1}$ for $0 < x < \infty$,
(c) $\lim_{x\to\infty}[x^\beta \int U(t)P(t, \beta)dt] = 0$.
Then the function

$$V(x) = x^\beta U(x) - \frac{\beta x^{\beta-1}}{P(x, \beta)} \int_x^\infty U(t)P(t, \beta)dt$$

satisfies

$$E_\alpha V(x) = \alpha^\beta E_\alpha U(x) \tag{7}$$

Proof The proof follows using integration by parts to the second expression. The following lemma is a generalization of the above lemma.

Lemma 2.2 *Let $T_1, T_2, \ldots, T_k$ be k independent random variables with pdf $g_i(.)$ as defined in* (4.2).
Suppose that $U_1(\underline{t}), U_2(\underline{t}), \ldots, U_k(\underline{t})$ be k real-valued function defined on R, such that
(a) $E_{\underline{\alpha}}(U_i(\underline{T})) < \infty \ \ \forall \alpha_i > 0,\ i = 1, \ldots, k.$
(b) The integral $\int_{t_i}^{\infty} U(t_1, t_2, \ldots, t_{i-1}, x, t_{i+1}, \ldots, t_k) P(x, \beta) dx$ exists and is finite, where $P(x, \beta) = \frac{n}{\beta^{n-1} x^n \beta + 1}$ for $0 < t_i < \infty$,
(c) Then the function

$$V_i(\underline{T}) = t_i^{\beta} U_i(\underline{T}) - \frac{\beta t_i^{\beta-1}}{P(t_i, \beta)} \int_{t_i}^{\infty} U(t_1, t_2, \ldots, t_{i-1}, x, t_{i+1}, \ldots, t_k) P(x, \beta) dx$$

satisfies

$$E_{\alpha} V_i(\underline{T}) = {\alpha_i}^{\beta} E_{\alpha} U_i(\underline{T})$$

Next, since $R_J(t) = \sum_{j=1}^{k} \frac{{\alpha_j}^{\beta}}{t^{\beta}} I_j$, define

$$U_i(\underline{t}) = \begin{cases} \frac{1}{t^{\alpha}} \text{ if } T_j = T_{(k)} \\ 0 \ \text{ otherwise} \end{cases} \tag{8}$$

then we can write $E(R_J(t)) = \sum_{i=1}^{k} {\alpha_i}^{\beta} E[U_i(\underline{t})]$, from lemma (4.2) we have

$$E[\sum_{i=1}^{k} V_i(\underline{T})] = E[\sum_{i=1}^{k} {\alpha_i}^{\beta} U_i(\underline{T})]$$

which is the unbiased estimator of $R_J(t)$.

Theorem 2.1 *The UMVUE of $R_J(t)$ is given by* $\hat{R_J^U}(t) = \frac{1}{n}\left(\frac{T_{(k)}}{t}\right)^{\beta} \left[n - 1 - \sum_{i=1}^{k-1} \left(\frac{T_i}{T_{(k)}}\right)^{n\beta+\beta}\right].$

Proof We have

$$\begin{aligned} V_i(\underline{t}) &= t_i^{\beta} U_i(\underline{t}) - \frac{\beta t_i^{\beta-1}}{P(t_i, \beta)} \int_{t_i}^{\infty} U_i(t_1, t_2, \ldots, t_{i-1}, x, t_{i+1}, \ldots, t_k) P(x, \beta) dx \\ &= \frac{t_i^{\beta}}{t^{\beta}} I(t_i \geq \max_{i \neq j} t_j) - \frac{\beta t_i^{\beta-1}}{P(t_i, \beta)} \int_{t_i}^{\infty} I(t_i \geq \max_{i \neq j} t_j) \frac{P(x, \beta)}{t^{\beta}} dx \\ &= \left[t_i^{\beta} - \frac{\beta t_i^{\beta-1}}{P(t_i, \beta)} \int_{t_i}^{\infty} P(x, \beta) dx \right] \frac{I(t_i \geq \max_{i \neq j} t_j)}{t^{\beta}} \end{aligned}$$

$$\Rightarrow E\Big(\sum_{i=1}^{k} V_i(\underline{T})\Big) = \frac{1}{t^\beta}\Big[T_{(k)}^\beta - \Big(\sum_{i=1}^{k-1}\frac{\beta T_i^\beta}{P(T_i,\beta)}\Big)\int_{T_{(k)}}^{\infty} P(x,\beta)dx\Big]$$

$$= \frac{1}{n}\Big(\frac{T_{(k)}}{t}\Big)^\beta\Big[n-1-\sum_{i=1}^{k-1}\Big(\frac{T_i}{T_{(k)}}\Big)^{n\beta+\beta}\Big],$$

which is the UMVUE of $R_J(t)$.

3 An Inadmissibility Result

In this section, we will try to find out the form of an equivariant estimator of $R_J(t)$. For this, let us consider the scale group of transformations $G = \{g_c : g_c(x) = cx, c > 0\}$. Under this transformation $\alpha \to c\alpha$, $R_J \to c^\beta R_J$. Hence, the decision problem is invariant under this transformation in regards to the quadratic loss, given by

$$L(\hat{R}_J(t), R_J(t)) = \left(\frac{\hat{R}_J(t) - R_J(t)}{R_J(t)}\right)^2,$$

where $\hat{R}_J(t)$ is any estimator of $R_J(t)$. An estimator $h(\underline{T})$ is said to be equivariant if

$$h(cT_1, cT_2, \ldots, cT_k) = c^\beta h(T_1, T_2, \ldots, T_k).$$

Let $c = \frac{1}{T_k}$, we have

$$h\Big(\frac{T_1}{T_k}, \frac{T_2}{T_k}, \ldots, \frac{T_{k-1}}{T_k}, 1\Big) = \frac{1}{T_k^\beta}h(\underline{T})$$

$$\Rightarrow h(\underline{T}) = T_k^\beta h(\underline{Z}), \tag{9}$$

where $\underline{Z} = (Z_1, Z_2, \ldots, Z_{k-1})$, $Z_i = \frac{T_i}{T_k}$, for $i = 1, \ldots, k-1$ and let $\underline{z} = (z_1, z_2, \ldots, z_{k-1})$ be any observed value of $\underline{Z}$.

It can be easily seen that the UMVUE is a scale equivariant estimator. We now use the Brewster–Zidek technique for improving upon the equivariant estimators.

The risk of $h(\underline{T})$ for estimating $R_i(t)$, for $i = 1, 2, \ldots, k$ is given by

$$R(T_k^\beta h(\underline{Z}), R_i(t)) = E[T_k^\beta h(\underline{Z}) - R_i(t)]^2 = E^{\underline{Z}}[E^{\underline{T}|\underline{Z}}(T_k^\beta h(\underline{Z}) - R_i(t))^2|\underline{Z}]$$

Differentiating the above equation with respect to $h(\underline{Z})$. We see that the inner conditional expectation is minimized by

$$h_{\beta}^{*}(\underline{z}) = R_i(t)\frac{E(T_k^{\beta}|\underline{Z} = \underline{z})}{E(T_k^{2\beta}|\underline{Z} = \underline{z})} \tag{10}$$

In order to find out the expectation above, we need the conditional density of T_k given $\underline{Z} = \underline{z}$. It is given by

$$\frac{kn\beta\, \alpha_i^{kn\beta}}{t_k^{kn\beta+1} {z_i}^{kn\beta}}, \quad \frac{\alpha_i}{z_i} \leq t_k < \infty, \ \text{ and } \frac{\alpha_{i+1}}{\alpha_i} \leq z_i < \infty. \tag{11}$$

(See also Kumar and Kar [9].
Hence

$$h_{\beta}^{*}(\underline{Z}) = \frac{kn-2}{kn-1}\left(\frac{\alpha_j}{t\alpha_i}\right)^{\beta} {z_i}^{\beta}, \quad i \neq j.$$

If we fix j and vary i such that $i \neq j$, we have

$$\hat{h_{\beta}}(\underline{z}) = \sup\ h_{\beta}^{*}(\underline{Z}) = \frac{kn-2}{kn-1}\left(\frac{\max(z_1, \ldots, z_{k-1}, 1)}{t}\right)^{\beta} \quad \text{also } \inf h_{\beta}^{*}(\underline{z}) = 0. \tag{12}$$

Summarizing the above results the following theorem is concluded immediately.

Theorem 3.1 *Let* $\Psi(\underline{Z})$ *be an estimator of* R *as defined in* (5.1)*, then define an estimator* $\Psi^{*}(\underline{Z})$ *by*

$$\Psi^{*}(\underline{Z}) = \begin{cases} \Psi(\underline{Z}), & \text{if } \Psi(\underline{Z}) < \hat{h_{\beta}}(\underline{Z}), \\ \hat{h_{\beta}}(\underline{Z}), & \text{otherwise.} \end{cases} \tag{13}$$

Then, $\Psi^{*}(\underline{Z})$ *is an improved estimator of* $\Psi(\underline{Z})$ *provided* $P\{\Psi(\underline{Z}) \geq \hat{h_{\beta}}(\underline{Z})\} > 0$.

Remark 3.1 It can be seen that Theorem 3.1 will also hold good even for the usual squared error loss function. This is because the proof of Brewster–Zidek [3] technique was established on the orbits of $\underline{Z} = \underline{z}$.

Remark 3.2 For $n > 2$, then the estimator $\hat{R_J^c} = \frac{1}{n}\left(\frac{T_{(k)}}{t}\right)^{\beta}\left[c - \sum_{i=1}^{k-1}\left(\frac{T_i}{T_{(k)}}\right)^{n\beta+\beta}\right]$ uniformly dominates $\hat{R_J^U}(t) = \frac{1}{n}\left(\frac{T_{(k)}}{t}\right)^{\beta}\left[n - 1 - \sum_{i=1}^{k-1}\left(\frac{T_i}{T_{(k)}}\right)^{n\beta+\beta}\right]$ for $\frac{n^2-2n-1}{n-1} \leq c < n-1$.

Proof Consider the risk difference RD_1 of the above two estimators. So,

$$\begin{aligned} RD_1 &= E[(\hat{R_J^U}(t), R_J)^2 - (\hat{R_J^c}(t) - R_J(t))^2] \\ &= E[(\hat{R_J^U}(t) + \hat{R_J^c}(t))((\hat{R_J^U}(t) - \hat{R_J^c}(t)))] \\ &- E[2R_J(t)((\hat{R_J^U}(t) - \hat{R_J^c}(t)))]. \end{aligned}$$

We have $\hat{R_J^U}(t) + \hat{R_J^c}(t) = \frac{1}{n}\Big(\frac{T_{(k)}}{t}\Big)^{\beta}\Big[n-1+c-2\sum_{i=1}^{k-1}\Big(\frac{T_i}{T_{(k)}}\Big)^{n\beta+\beta}\Big],$

$$\hat{R_J^U}(t) - \hat{R_J^c}(t) = \frac{n-1+c}{n}\Big(\frac{T_{(k)}}{t}\Big)^{\beta}$$

and $E[R_J(t)(\hat{R}_J(t) - \hat{R_J^c}(t))] = \frac{1}{t^{\beta}}E\Big[\frac{n-1-c}{n}\sum_{i=1}^{k}\Big(\frac{\alpha_i}{t}\Big)^{\beta}T_i{}^{\beta}I(T_i \geq \max_{i\neq j} T_j)\Big]$

Taking $U_i(\underline{T}) = T_i{}^{\beta}I(T_i \geq \max_{i\neq j} T_j)$, from Lemma 2.2, we have

$$\begin{aligned} E\Big[\sum_{i=1}^{k}\Big(\frac{\alpha_i}{t}\Big)^{\beta}T_i{}^{\beta}I(T_i \geq \max_{i\neq j} T_j)\Big] &= \sum_{i=1}^{k}E\Big[T_i{}^{2\beta}I(T_i \geq \max_{i\neq j} T_j) \\ &- \beta T_i{}^{n\beta+\beta}\int_{T_i}^{\infty}\frac{I(x \geq \max_{i\neq j} T_j)}{x^{n\beta-\beta+1}}dx\Big] \\ &= E\Big[\frac{T_{(k)}^{2\beta}}{n-1}\Big(n-2-\sum_{i=1}^{k-1}\Big(\frac{T_i}{T_{(k)}}\Big)^{n\beta+\beta}\Big)\Big]. \end{aligned} \tag{14}$$

With the help of (14), we are in a position to compute RD_1.

$$\begin{aligned} RD_1 &= \frac{1}{t^{\beta}}\Big[\frac{n-1-c}{n^2}E\Big\{T_{(k)}^{2\beta}\Big(n-1+c-2\sum_{i=1}^{k-1}\Big(\frac{T_i}{T_{(k)}}\Big)^{n\beta+\beta}\Big)\Big\} \\ &- 2\frac{n-1-c}{n(n-1)}E\Big\{T_{(k)}^{2\beta}\Big(n-2-2\sum_{i=1}^{k-1}\Big(\frac{T_i}{T_{(k)}}\Big)^{n\beta+\beta}\Big)\Big\}\Big] \\ &= \frac{1}{t^{\beta}}\frac{n-1-c}{n^2(n-1)}E\Big[T_{(k)}^{2\beta}\Big(1+2n-n^2+c(n-1)+2\sum_{i=1}^{k-1}\Big(\frac{T_i}{T_{(k)}}\Big)^{n\beta+\beta}\Big)\Big]. \end{aligned} \tag{15}$$

We see that for $\frac{n^2-2n-1}{n-1} \leq c < n-1$ and $n > 2$, then $RD_1 > 0$. Hence the conclusion follows immediately.

Remark 3.3 The natural estimator $\hat{R_J^*}(t) = (\frac{T_{(k)}}{t})^{\beta}$ is inadmissible.

Proof Consider the counterpart of the UMVUE for the component problem $\hat{R_J^{AU}}(t) = \frac{n-1}{n}(\frac{T_{(k)}}{t})^{\beta}$. We claim that this estimator dominates uniformly the natural estimator for $n > 2$. The risk difference between these two estimators RD_2 is given by

$$
\begin{aligned}
RD_2 &= E[(\hat{R}_J^*(t), R_J(t))^2 - (\hat{R}_J^{AU}(t) - R_J(t))^2] \\
&= E[(\hat{R}_J^*(t) + \hat{R}_J^{AU}(t))((\hat{R}_J^*(t) - \hat{R}_J^{AU}(t))) \\
&\quad - E[2R_J(t)((\hat{R}_J^*(t) - \hat{R}_J^{AU}(t)))]
\end{aligned}
$$

$$
\begin{aligned}
\text{We have}\quad \hat{R}_J^*(t) + \hat{R}_J^{AU}(t) &= \frac{2n-1}{n}\left(\frac{T_{(k)}}{t}\right)^{\beta}, \\
\hat{R}_J^*(t) - \hat{R}_J^{AU}(t) &= \frac{1}{n}\left(\frac{T_{(k)}}{t}\right)^{\beta} \\
\text{and}\quad E[R_J(t)(\hat{R}_J^*(t) - \hat{R}_J^{AU}(t))] &= \frac{1}{t^{\beta}}E\left[\frac{1}{n}\sum_{i=1}^{k}\left(\frac{\alpha_i}{t}\right)^{\beta} T_i{}^{\beta} I(T_i \geq \max_{i \neq j} T_j)\right]
\end{aligned}
$$

Proceeding similarly as above, we have

$$
RD_2 = \frac{1}{t^{\beta}}\frac{2n-1}{n^2(n-1)}E\left[T_{(k)}^{2\beta}\left(n+1+2n\sum_{i=1}^{k-1}\left(\frac{T_i}{T_{(k)}}\right)^{n\beta+\beta}\right)\right] > 0. \tag{16}
$$

Hence the conclusion follows.

Remark 3.4 The estimator $\hat{R}_J^{AU}(t) = \frac{n-1}{n}(\frac{T_{(k)}}{t})^{\beta}$ is inadmissible.

Proof For the component problem, let us consider the counterpart of the best scale equivariant estimator, given by, $\hat{R}_J^S(t) = \frac{n-2}{n-1}(\frac{T_{(k)}}{t})^{\beta}$. Here also we see that the estimator $\hat{R}_J^S(t)$ dominates uniformly $\hat{R}_J^{AU}$ for $n > 2$. We compute the risk difference RD_3, which is given by

$$
\begin{aligned}
RD_3 &= E[(\hat{R}_J^U(t), R_J(t))^2 - (\hat{R}_J^S(t) - R_J(t))^2] \\
&= \frac{1}{t^{\beta}}\frac{1}{n^2(n-1)^2}E\left[T_{(k)}^{2\beta}\left(1+2n(n-1)\sum_{i=1}^{k-1}\left(\frac{T_i}{T_{(k)}}\right)^{n\beta+\beta}\right)\right] > 0.
\end{aligned} \tag{17}
$$

Hence the conclusion follows.

Conclusion: Under the mean squared error criterion the estimator $\hat{R}_J^S$ dominates $\hat{R}_J^U$, which once again improves upon the natural estimator $\hat{R}_J^*$. Hence $\hat{R}_J^U$ is preferred. One should not prefer $\hat{R}_J$ unless one is not interested in the class of unbiased estimators. Also we have $\hat{R}_J^c$ which dominates $\hat{R}_J$.

References

1. R.R. Bahadur, A.L. Goodman, Impartial decision rules and sufficient statistics. Ann. Math. Stat. **23**, 553–562 (1952)
2. J.O. Berger, B. Mandelbrot, A new model for error clustering in telephone circuits. IBM J. Res. Dev. **7**, 224–236 (1963)
3. J.F. Brewster, J.V. Zidek, Improving on equivariant estimators. Ann. Stat. **2**, 21–38 (1974)
4. H.T. Davis, M.L. Feldstein, The generalized Pareto law as a model for progressively censored survival data. Biometrika **66**, 299–306 (1979)
5. M.L. Eaton, Some optimum properties of ranking procedures. Ann. Math. Stat. **38**, 124–137 (1967)
6. E.C. Freiling, *A Comparison of the Fallout Mass-Size Distributions Calculated by Lognormal and Power Law Models* (U.S. Naval Radiological Defence Laboratory, San Francisco (AD-646019), 1966)
7. C.M. Harris, The Pareto distributions as a queue service discipline. Oper. Res. **16**, 307–313 (1968)
8. N.L. Johnson, S. Kotz, *Continuous Univariate Distributions-I* (John Wiley and Sons, New York, 1970)
9. S. Kumar, A. Kar, Estimating quantiles of a selected exponential population. Stat. Prob. Lett. **52**, 9–29 (2005)
10. S. Kumar, A. Kar, Quantile estimation for a selected normal population. Commun. Stat. Theory Methods **29**(2), 437–444 (2000)
11. S. Kumar, A. Kar, Estimating quantiles of a selected exponential population. Stat. Prob. Lett. **52**, 9–19 (2001a)
12. S. Kumar, A. Kar, Minimum variance unbiased estimation of quantile of a selected exponential population. Am. J. Math. Manage. Sci. **21**(1–2), 183–191 (2001b)
13. S. Kumar, A.K. Mahapatra, P. Vellaisamy, Reliability estimation of the selected exponential populations. Stat. Prob. Lett. **79**, 1372–1377 (2009)
14. E.L. Lehmann, On a theorem of Bahadur and Goodman. Ann. Math. Stat. **37**, 1–6 (1966)
15. N. Misra, E.C. van der Meulen, On estimation following selection from nonregular distributions. Commun. Stat. Theory Methods **30**(12), 2543–2561 (2001)
16. Rubinstein, D. (1961). Estimation of the reliability of a system in development. Rep. R-61-ELC-1, Tech. Info. Service, GE Co
17. D. Sharma, P. Vellaisamy, Quantile estimation of the selected normal population. Sankhya, Ser. B **51**, 311–322 (1989)
18. P. Vellaisamy, Quantile estimation of a selected exponential population. J. Stat. Plann. Infer. **115**, 461–470 (2003)
19. P. Vellaisamy, A.P. Punnen, Improved estimators for the selected location parameters. Stat. Pap. **43**(2), 291–299 (2002)
20. J.H. Venter, Estimation of the mean of the selected population. Commun Stat. Theory Methods **17**(3), 797–805 (1988)

Inventory Model with Partial Backordering and Single Deteriorating Item for a Two-Warehouse System

S. K. Indrajitsingha, A. K. Sahoo, P. N. Samanta, and U. K. Misra

Abstract Present paper explores a model of inventories having deteriorating items with kept in two warehouses. We also considered the shortages which are fulfilled by partial backlogging in which the demand is depending upon the selling price. If the ordering quantity exceeds the capacity of the owned warehouse, the excess amount of stock will store in a rented warehouse. To reduce the storage cost, the items in the rented warehouse are released first. Due to demand and deterioration in the first interval the items in rented warehouse decreases to zero. On the other hand, the items in the owned warehouse decreases only due to deterioration. After a certain period, the inventory level in owned warehouse reaches zero and hence shortage starts. It is assumed that the rate of backlogging and the demand are exponential functions. In order to establish the model, we discussed with numerical examples and to study the behaviour of the model, sensitivity analysis has been carried out.

Keywords Demand · Deterioration · Shortages · Partially backlogging · Two-warehouse

AMS Classification No 90B05

S. K. Indrajitsingha
Department of Mathematics, Saraswati Degree Vidya Mandir, Berhampur 760002, Odisha, India
e-mail: skismath86@gmail.com

A. K. Sahoo (✉) · P. N. Samanta
Department of Mathematics, Berhampur University, Berhampur, Bhanja Bihar 760007, Odisha, India
e-mail: ajit2ksahoo@gmail.com

P. N. Samanta
e-mail: pns.math@buodisha.edu.in

U. K. Misra
NIST, Palur Hills, Berhampur 761008, Odisha, India
e-mail: umakanta_misra@yahoo.com

S. R. Mishra et al. (eds.), *Recent Trends in Applied Mathematics*, Lecture Notes in Mechanical Engineering, https://doi.org/10.1007/978-981-15-9817-3_3

1 Introduction

The most successful operation research techniques that are used by the modern industries or business is inventory theory. The economic value of a company always depends upon the inventory and thus it plays an extremely important role in the economy of any country. Since the primary function of inventory is to provide the customer service in right time at right cost and at right place, the management part of inventory should be taken very carefully. Though it depends upon various aspects having different cost, our ultimate goal is to maximise the profit in the business. Among them the main factor is to maintain more stock by minimizing the investment cost. Here, the main objective is to resolve, where the inventories be stored. Most of the researchers worked by taking a single warehouse with unlimited capacity. But in actual practice, a warehouse has limited capacity. Hartely [1] is the first person to explore the model having different warehouses (owned and rented) in 1976. The inventory are put away first in the owned warehouse having fixed limit 'W' and the main overabundance units are put away in rented warehouse. Clearly, the holding cost of a rented warehouse is higher than that of owned warehouse. In 1987, a deterministic model with two storage facilities have been developed by Sharma [2].

In 1992, Goswami and Chaudhuri [3] designed an inventory model, which includes both with and without shortages having linearly increasing demand. Subsequently in the same year Pakkala and Achary [4] planned a two-warehouse inventory model for deteriorating items with finite replenishment rate. In 1997, Bomkherouf [5] extended the work of Palkkala by considering two shortage facilities. However, in 2004, Yang [6] gave an effective result of the Bomkherouf"s model under inflation. Other prominent papers along this path have been established by Lee and Hsu [7], Yang [8], Jaggi and Verma [9].

A recent work towards this direction are by Cheng and Zhang [10], Cardenas-Barron [11], Liang and Zhou [12] and Sett et al. [13]. It is found that numerous analysts have been working on inventory management taking two-warehouse for keeping more stock with exponentially growing demand. In 2012, Yadav et al. [14] considered a fuzzy inventory model for deteriorating items with two-warehouse and demand depending upon stock by the help of genetic algorithm. Yang and Chang [15] developed a two-warehouse inventory model with partially backlogging in 2013. In 2015, Shabani et al. [16] designed an inventory model with fuzzy deterioration and demand rate in which stocks are stored in two warehouses. In 2016, Tiwari et al. [17] did an effective work on impact of trade credit and inflation on retailers ordering policies for non-instantaneous deteriorating items in a two-ware-house environment. Mandal and Giri [18] developed an inventory model under stock-dependent demand quantity with discount offer in two stockrooms. Most recently, Indrajitsingha et al. [19] considered a two-warehouse inventory model with a linear demand depending upon selling price and partially backlogging shortages. But in this model, the demand is exponential function. In the present article, we extended the above model by developing a two-warehouse inventory model in which demand is exponentially increasing and affecting the total average cost of the model in a short period of time.

2 Notations

OW	Owned warehouse.
RW	Rented warehouse.
$I_r(t)$	Amount of stock in RW at time t.
$I_o(t)$	Amount of stock in OW at time t.
θ	Rate of deterioration.
α	Initial demand rate.
β	Positive demand parameter.
t_r	Time at which the inventory level in rented warehouse depletes to zero.
t_o	Time at which the inventory level in owned warehouse depletes to zero.
W	Storage capacity of OW.
p	Selling price ($/unit/year).
$D(p)$	Demand rate depending upon selling price.
q	Order quantity.
PC	Purchasing cost.
HC	Holding cost.
DC	Deterioration cost.
SC	Shortage cost.
LC	Lost sale cost.
S	Initial stock level.
q_1	Backorder quantity during stock out.
T	Length of the replenishment cycle.
c	Purchasing cost ($/unit/day).
k	Backlogging rate.
h_r	Holding cost ($/unit/year) in RW.
h_o	Holding cost ($/unit/year) in OW.
d	Unit deterioration cost ($/unit/day).
C_1	Shortage cost per unit ($/unit/day).
C_2	Unit lost sale cost ($/unit/day).
$TAC(t_r, t_o)$	Total average cost ($/unit/day).

3 Assumptions

Through out the paper we assume that:

(i) The model consists of a finite planning horizon.
(ii) The demand rate is exponential and depending upon selling price, i.e. $D(p) = \alpha p^{-\beta},\ \alpha, \beta > 0$.
(iii) Negligible lead time.
(iv) The shortages are allowed and backlogged partially.
(v) The limited capacity owned warehouse.
(vi) The unlimited capacity rented warehouse.

(vii) The unit holding cost of RW is more than that of OW.
(viii) Those products have deteriorating nature are considered.
(ix) Higher powers of θ are neglected.
(x) Items are kept in OW first.
(xi) Items are stored in RW will be consumed first.

4 Mathematical Formulation

The issue which we have discussed here is the means by which retailers know whether or not to take a rented warehouse to hold the things. If the order quantity $q \leq W$, then it is not necessary to take rented warehouse. But, if $q > W$, then W units are stored in owned warehouse and the remaining is dispatched into the rented warehouse. So during the interval (o, t_r), the items in rented warehouse decrease due to demand and deterioration until it reaches level zero. On the other hand, a portion on the inventory in owned warehouse is depleted due to deterioration only. During the interval (t_r, t_o) the inventory in owned warehouse decreases due to demand and deterioration. After $t = t_o$, shortages start and backlogging rate is negative exponential function of time. The pictorial representation of the above system can be represented by Fig. 1 and the behaviour of the problem is instructed by the following differential equations.

$$\frac{dI_r(t)}{dt} = -\theta I_r(t) - \alpha p^{-\beta}, 0 \leq t \leq t_r \tag{1}$$

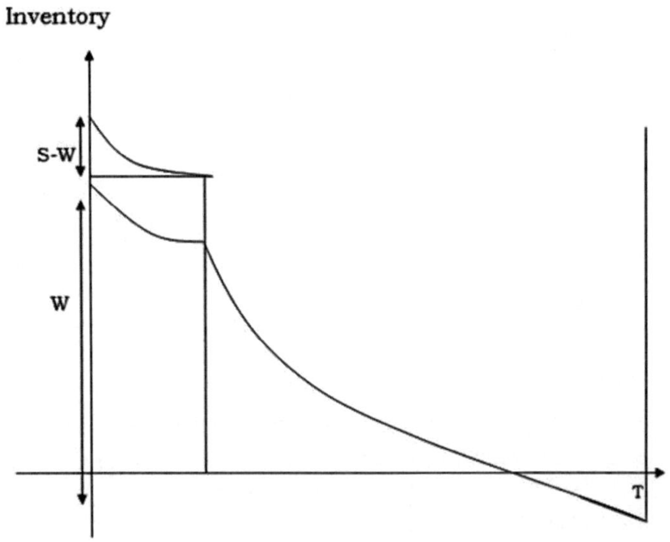

Fig. 1 Inventory time graph for two warehouses

with $I_r(t_r) = 0$.

$$\frac{dI_o(t)}{dt} = -\theta I_o(t), 0 \le t \le t_r \tag{2}$$

with $I_o(0) = W$.

$$\frac{dI_o(t)}{dt} = -\theta I_o(t) - \alpha p^{-\beta}, t_r \le t \le t_o \tag{3}$$

with $I_o(t_o) = 0$.
The solutions of (1), (2) and (3) are as follows:

$$I_r(t) = \frac{\alpha p^{-\beta}}{\theta}\left\{e^{\theta(t_r - t)}\right\} \quad 0 \le t \le t_r, \tag{4}$$

$$I_o(t) = We^{-\theta t}, \; 0 \le t \le t_r, \tag{5}$$

$$I_o(t) = \frac{-\alpha p^{-\beta}}{\theta} + \frac{\alpha e^{\theta t_o}.p^{-\beta}}{\theta} e^{-\theta t} \quad t_r \le t \le t_o. \tag{6}$$

From (5), we have

$$I_r(0) = S - W$$

$$S = \frac{\alpha p^{-\beta}}{\theta}\left(e^{\theta t_r} - 1\right) \tag{7}$$

At $t = t_r$, Eqs. (5) and (6) yield.

$$W = \frac{\alpha p^{-\beta}}{\theta} e^{\theta t_r}\left\{e^{\theta(t_o - t_r)} - 1\right\} \tag{8}$$

$$S = \frac{\alpha p^{-\beta}}{\theta}\left\{e^{\theta t_o} - 1\right\} \tag{9}$$

With the above data following parameters are calculated:

$$PC = (S + q_1)c,$$

where

$$q_1 = \int_{t_o}^{T} k\alpha p^{-\beta} dt.$$

Then,

$$PC = \left\{\alpha p^{-\beta}\left\{\frac{e^{\theta t_o} - 1}{\theta} + k(T - t_o)\right\}\right\}c. \tag{10}$$

$$HC = HC_r + HC_o,$$

where

$$HC_r = h_r \int_0^{t_r} I_r(t)dt = \frac{h_r \alpha p^{-\beta}}{\theta}\left\{\left(\frac{e^{\theta t_r} - 1}{\theta}\right) - t_r\right\}. \tag{11}$$

And

$$\begin{aligned} HC_o &= h_o\left\{\int_0^{t_r} I_o(t)dt + \int_{t_r}^{t_o} I_o(t)dt\right\} \\ &= \frac{h_o}{\theta}\left\{W\left(1 - e^{-\theta t_r}\right) - \frac{\alpha p^{-\beta}}{\theta}\left(1 - e^{\theta(t_o - t_r)}\right) - \frac{\alpha p^{-\beta}}{\theta}(t_o - t_r)\right\} \end{aligned} \tag{12}$$

$$DC = DC_r + DC_o,$$

where

$$DC_r = d\left\{\alpha p^{-\beta}\left\{\frac{e^{\theta t_r} - 1}{\theta} - t_r\right\}\right\}. \tag{13}$$

And

$$DC_o = d\left\{W - \alpha p^{-\beta}(t_o - t_r)\right\}. \tag{14}$$

$$SC = C_2 \int_{t_o}^{T} \alpha p^{-\beta} dt = C_2 \alpha p^{-\beta}(T - t_o) \tag{15}$$

$$LC = C_3 \int_{t_o}^{T} (1 - k)\alpha p^{-\beta} dt = C_3(1 - k)\alpha p^{-\beta}(T - t_o). \tag{16}$$

Total average cost $TAC(t_r, t_o)$ for this model during a cycle is given by

$$\begin{aligned} TAC(t_r, t_o) &= \frac{1}{T}[PC + HC + DC + SC + LC] \\ &= \frac{1}{T}\left[\begin{array}{l} \left\{\alpha p^{-\beta}\left\{\frac{e^{\theta t_o} - 1}{\theta} + k(T - t_o)\right\}\right\}c + \frac{h_r \alpha p^{-\beta}}{\theta}\left\{\left(\frac{e^{\theta t_r} - 1}{\theta}\right) - t_r\right\} \\ + \frac{h_o}{\theta}\left\{W\left(1 - e^{-\theta t_r}\right) - \frac{\alpha p^{-\beta}}{\theta}\left(1 - e^{\theta(t_o - t_r)}\right) - \frac{\alpha p^{-\beta}}{\theta}(t_o - t_r)\right\} \\ + d\left\{\alpha p^{-\beta}\left\{\frac{e^{\theta t_r} - 1}{\theta} - t_r\right\}\right\} + d\left\{W - \alpha p^{-\beta}(t_o - t_r)\right\} + C_1 \alpha p^{-\beta}(T - t_o) \\ + C_2(1 - k)\alpha p^{-\beta}(T - t_o) \end{array}\right]. \end{aligned} \tag{17}$$

To minimize the total average cost function $TAC(t_1, t_2)$ per unit time, the values of t_r and t_o can be obtained by solving the equations:

$$\frac{\partial TAC(t_r, t_o)}{\partial t_r} = 0 \text{ and } \frac{\partial TAC(t_r, t_o)}{\partial t_o} = 0. \tag{18}$$

Thus, values of t_r and t_o obtained from the above equations will minimize the total cost function.

5 Numerical Examples

Let us consider the realistic values of the parameters in these two examples which are not considered from any case study.

Example 1 Suppose $\alpha = 85$ units, $\beta = 0.001$, $p =$ \$23/unit/day, $k = 0.7$ unit, $C_1 =$ \$12/unit/day, $C_2 =$\$11/unit/day, $c =$ \$15/unit/day, $\theta = 0.006$, $W = 100$ units, $d =$ 0.5 unit, $\tilde{h}_r =$ \$. 0.07/unit/day, $\tilde{h}_o =$ \$ 0.06/unit/day, T = 365 days. Corresponding to these input values, $t_r = 30.1824$ days, $t_o = 73.5827$ days will minimize *TAC* and the minimum value is $TAC(t_r, t_o) =$ \$2085.64.

Example 2 Suppose $\alpha = 80$ units, $\beta = 0.01$, $p =$ \$23/unit/day, $k = 0.7$ unit, $C_1 =$ \$12/unit/day, $C_2 =$ \$11/unit/day, $c =$ \$15/unit/day, $\theta = 0.006$, $W = 100$ units, $d =$ 0.5 unit, $\tilde{h}_r =$ \$. 0.07/unit/day, $\tilde{h}_o =$ \$ 0.06/unit/day, T = 365 days. Corresponding to these input values, $t_r = 30.1424$ days, $t_o = 73.5684$ days will minimize *TAC* and the minimum value is $TAC(t_r, t_o) =$ \$1908.38.

6 Sensitivity Analysis

The following observations are carried out by changing the parameters θ, β, α and T using Mathematica 11.1 software.

- Table 1 indicates the value of θ increases, total average cost increases with a decrease in t_r and t_o.

Table 1 Sensitivity analysis of Deterioration rate (θ)

θ	t_r	t_o	TAC
0.006	30.1555	73.5731	1962.99
0.008	22.9873	58.0709	1982.51
0.010	18.3142	47.954	1995.35
0.012	15.0313	40.8352	2004.44
0.014	12.6015	35.551	2011.21
0.016	10.7324	31.4834	2016.44

(continued)

Table 1 (continued)

θ	t_r	t_o	TAC
0.018	9.25135	28.2481	2020.61
0.020	8.05005	25.6156	2024.01
0.022	7.05693	23.432	2026.84
0.024	6.2229	21.5913	2029.22

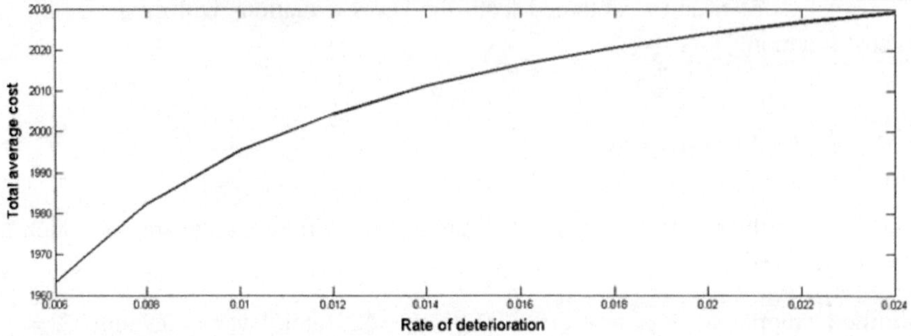

- Table 2 indicates as the value of β increases, total average cost decreases with a decrease in t_r and t_o very slowly.

Table 2 Sensitivity analysis of positive demand parameter (β)

β	t_r	t_o	TAC
0.001	30.1555	73.5731	1962.99
0.003	30.1526	73.5721	1950.72
0.005	30.1497	73.5711	1938.53
0.007	30.1468	73.57	1926.41
0.009	30.1439	73.569	1914.37
0.011	30.1409	73.5679	1902.41
0.013	30.138	73.5668	1890.52
0.015	30.135	73.5658	1878.71
0.017	30.132	73.5647	1866.97
0.019	30.1289	73.5636	1855.3

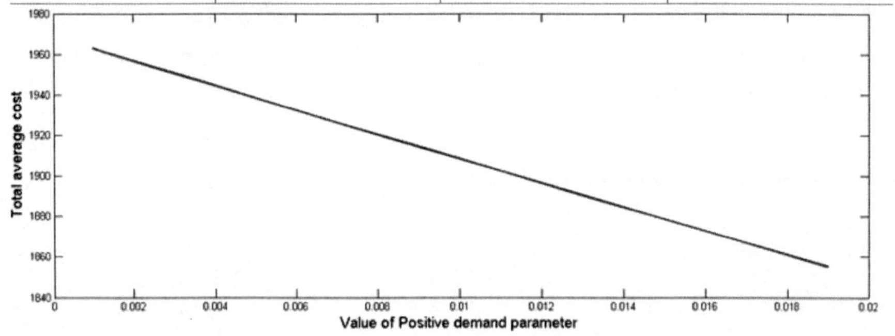

Table 3 Sensitivity analysis of initial demand rate (α)

α	t_r	t_o	TAC
70	30.0901	73.5497	1717.69
72	30.1046	73.5549	1766.75
74	30.1184	73.5598	1815.81
76	30.1314	73.5645	1864.87
78	30.1438	73.5689	1913.93
80	30.1555	73.5731	1962.99
82	30.1667	73.5771	2012.05
84	30.1773	73.5809	2061.11
86	30.1875	73.5845	2110.17
88	30.1971	73.588	2159.23

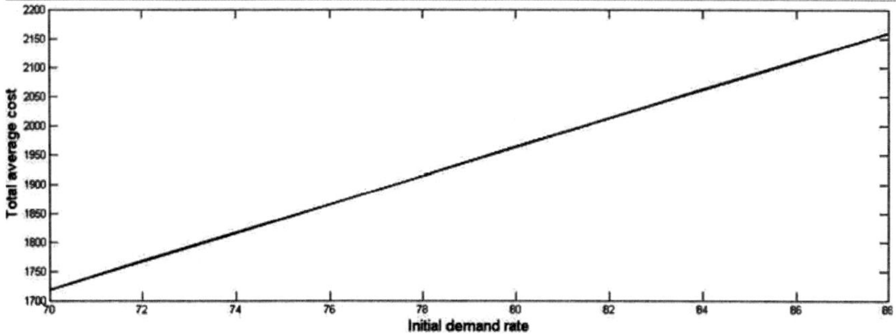

- Table 3 indicates as the value of α increases, total average cost increases with a very little increasing effect on t_r and t_o.
- Table 4 indicates as the value of T decreases, total average cost decreases and there is no effect on t_r and t_o.

Table 4 Sensitivity analysis of total time period (T)

T	t_r	t_o	TAC
365	30.1555	73.5731	1962.99
355	30.1555	73.5731	1960.32
345	30.1555	73.5731	1957.51
335	30.1555	73.5731	1954.52
325	30.1555	73.5731	1951.35
315	30.1555	73.5731	1947.98
305	30.1555	73.5731	1944.39

(continued)

Table 4 (continued)

T	t_r	t_o	TAC
295	30.1555	73.5731	1940.55
285	30.1555	73.5731	1936.45
275	30.1555	73.5731	1932.04

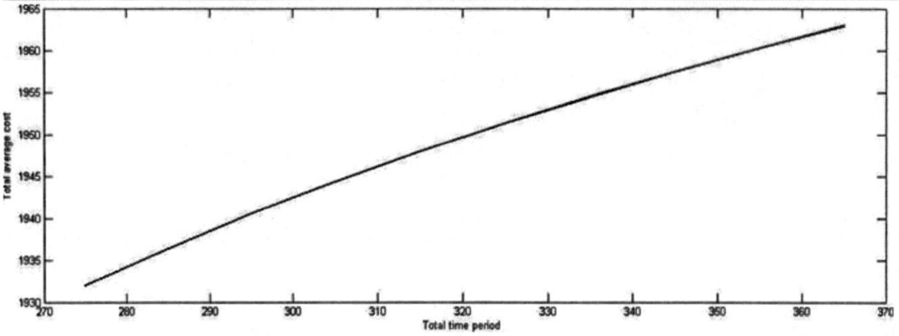

7 Conclusion

This article amalgaments the concepts of two-warehouse inventory model of single deteriorating items with shortages under partially backlogging, where demand is considered as an exponential function. From this model, we conclude that the time period of rented warehouse and owned warehouse has no effect, while the total time period is either going increase or decrease and it is directly proportional to total average cost. Therefore this model will be very much profitable for the business in which the stock period is less. For future research, this study can be extended by considering preservation technology in deterioration.

References

1. V.R. Hartely, *Operations Research—A Managerial Emphasis* (Santa Monica, CA, 1976), pp. 315–317
2. K.V.S. Sarma, A deterministic order-level inventory model for deteriorating items with two storage facilities. Eur. J. Oper. Res. **29**, 70–73 (1987)
3. A. Goswami, K.S. Chaudhuri, An economic order quantity model for items with two levels of storage for a linear trend in demand. J. Oper. Res. Soc. **43**, 157–167 (1992)
4. T.P.M. Pakkala, K.K. Achary, A deterministic inventory model for deteriorating items with two warehouses and finite replenishment rate. Eur. J. Oper. Res. **57**, 71–76 (1992)
5. L. Benkherouf, A deterministic inventory model for deteriorating items with two storage facilities. Int. J. Prod. Econ. **48**(1), 167–175 (1997)
6. H.H. Yang, Two-ware house inventory models for deteriorating items with shortages under inflation. Eur. J. Oper. Res. **157**, 344–356 (2004)
7. C.C. Lee, S.L. Hsu, A two-warehouse production model for deteriorating inventory items with time dependent demands. Euro. J. Oper. Res. **194**(3), 700–710 (2009)

8. H.L. Yang, Two-warehouse inventory models for deteriorating items with shortages under inflation. Euro. J. Oper. Res. **157**(2), 344–356 (2004)
9. C.K. Jaggi, P. Verma, Joint optimization of price and order quantity with shortages for a two-warehouse system. Top (Spain) **16**(1), 195–213 (2008)
10. M. Cheng, B. Zhang, G. Wang, Optimal policy for deteriorating items with trapezoidal type demand and partial backlogging. Appl. Math. Model. **35**, 3552–3560 (2011)
11. L.E. Cardenas-Barron, The economic production quantity (EPQ) with shortage derived algebraically. Int. J. Prod. Econ. **70**, 289–292 (2011)
12. Y. Liang, F. Zhou, A two-warehouse inventory model for deteriorating items under conditionally permissible delay in payments. Appl. Math. Model. **35**(1), 2221–2231 (2011)
13. B.K. Sett, B. Sarkar, A. Goswami, A two-ware house inventory model with increasing demand and time varying deterioration. Sci. Iran. Trans. E Ind. Eng. **19**, 306–310 (2012)
14. D. Yadav, S.R. Singh, R. Kumari, Inventory model of deteriorating items with two-warehouse and stock dependent demand using genetic algorithm in fuzzy environment. Yugoslav J. Operat. Res. **22**(1), 51–78 (2012)
15. H.L. Yang, C.T. Chang, A two-warehouse partial backlogging inventory model for deteriorating items with permissible delay in payment under inflation. Appl. Math. Model. **37**(1), 2717–2726 (2013)
16. S. Shabani, A. Mirzazadeh, E. Sharifi, A two-warehouse inventory model with fuzzy deterioration rate and fuzzy demand rate under conditionally permissible delay in payment. J. Indus. Prod. Eng. **33**(8), 516–532 (2015)
17. S. Tiwari, L.E. Cardenas-Barron, A. Khanna, C.K. Jaggi, Impact of trade credit and inflation on retailer's ordering policies for non-instantaneous deteriorating items in a two-warehouse environment. Int. J. Prod. Econ. **176**, 154–169 (2016)
18. P. Mandal, B.C. Giri, A two-warehouse integrated inventory model with imperfect production process under stock dependent demand quantity discount offer. Int. J. Syst. Sci. Operat. Logist. **4**(4), 1–12 (2017)
19. S.K. Indrajitsingha, P.N. Samanta and U.K. Misra, A fuzzy two-warehouse inventory model for single deteriorating item with selling price dependent demand and shortages under partially backlogged condition. Applicat. Appl. Math. **14**(1), 511–536, ISSN 1932-9466 (2019)

On Factorization of Sixth-Degree Polynomial of Type-(3,3)

Anjan Debnath

Abstract A special type of monic polynomial equations of degree 6 with real coefficients has been chosen, and it is shown how to obtain the complete factorization provided the given equation is factorisable in to two cubic equations satisfying some special conditions. Important examples and several results have been provided in order to establish the ideas and methodologies.

Keywords Computation of polynomials · Factorization of polynomials · Sixth-degree equation

2010 Mathematics Subject Classification 12-06 · 12Y05 · 12Y99 · 11-06 · 11Y40

1 Introduction

An expression of the form

$$p_n(x) := a_0x^n + a_1x^{n-1} + \cdots + a_{n-1}x + a_n$$

is called a univariate polynomial in x or simply a polynomial in x with coefficients $a_0, a_1, \ldots, a_n$, which we regard as reals only in this paper. $n \in \mathbb{N}$ is the degree of this polynomial $p_n(x)$. The coefficients a_0 and a_n are called the leading coefficient and the constant term, respectively. Clearly, the equation is of degree n implies the leading coefficient a_0 is non-zero. The expression $p_n(x) = 0$ is called polynomial equation in x.

In literature, solution of any univariate polynomial equation under radical sign is an interesting problem provided when the degree of the polynomial is at most 4. In case, the degree is 5 or higher, it is very well known to us, that the general method of determination of solution under radical sign is impossible. There are

A. Debnath (✉)
Department of Mathematics, The Heritage College, Kolkata 700107, India
e-mail: anjandebnath@rocketmail.com

S. R. Mishra et al. (eds.), *Recent Trends in Applied Mathematics*, Lecture Notes in Mechanical Engineering, https://doi.org/10.1007/978-981-15-9817-3_4

some special methods through which one can find the solution of some special polynomial equations. In [1], a special case of pentic polynomials has been shown how to factorize.

In the present paper, we have considered a different approach. We consider a very specific type of polynomials based on their factorisation. Let us first define them. This definition is provided in [1] already. We restate the same here for the present purpose. We shall use $\mathbb{S}[x]$ to denote the set of all polynomials with coefficients from the set $\mathbb{S}$ where $\mathbb{S}$ can be either $\mathbb{R}$ or $\mathbb{Q}$ or $\mathbb{Z}$.

Definition 1.1 A polynomial $p_n(x)$ with real coefficients is said to be of type $(l_1, l_2, \ldots, l_m) \in \mathbb{N} \times \cdots \times \mathbb{N}$ if $p_n(x)$ can be expressed as a product of m number of non-constant polynomials $g_{l_1}(x), \ldots, g_{l_m}(x)$ of degree $l_1, \ldots, l_m$, respectively.

For example, the polynomial $x^4 - 1$ is of type $(2, 2)$ because $x^4 - 1 = (x^2 - 1)(x^2 + 1)$. Similarly, the polynomial $x^5 + 6x^3 + x^2 + 8x + 4$ is of type $(3, 2)$ type because

$$x^5 + 6x^3 + x^2 + 8x + 4 = (x^3 + 2x + 1)(x^2 + 4).$$

Remark 1.1 It is important to note that, if a polynomial is not of type $(l_1, l_2, \ldots, l_m) \in \mathbb{N} \times \cdots \times \mathbb{N}$, the polynomial is not of our interest and we ignore them.

Remark 1.2 Secondly, a polynomial in $\mathbb{R}[x]$, if be reducible in to smaller degree non-constant polynomials in $\mathbb{R}[x]$, can be of several types. For example, $x^4 - 1$ is of type (2, 2) as well as (2, 1, 1) because

$$x^4 - 1 = (x^2 + 1)(x + 1)(x - 1).$$

Similarly, the polynomial $p_4(x) := x^4 + 5x^3 + 5x^2 - 5x - 6$ is of type (3, 1) as well as of type (2, 2), (2, 1, 1), (1, 1, 1, 1) because

$$p_4(x) = \begin{cases} (x^3 + 6x^2 + 11x + 6)(x - 1) \\ (x^2 - 1)(x^2 + 5x + 6) \\ (x^2 - 1)(x + 2)(x + 3) \\ (x - 1)(x + 1)(x + 2)(x + 3). \end{cases}$$

Remark 1.3 Thirdly, it is not important in which order the polynomial is being factored to determine its type. For example, in the above example $p_4(x)$ has been written as $(x^2 - 1)(x + 2)(x + 3)$. Note that $p_4(x)$ can also be written $(x^2 + 3x + 2)(x - 1)(x + 3)$ as well. In either case, $p_4(x)$ can be regarded of type (2, 1, 1). Hence, in general, we just need a particular factorisation of the given polynomial to be of some specific type and that's all.

Remark 1.4 A monic equation is a polynomial equation in which the leading coefficient is 1. Else the equation is referred as non-monic. Every non-monic polynomial equation of degree n say, $a_0x^n + a_1x^{n-1} + \cdots + a_n = 0$ can be transformed into a monic equation through the transformation $y = a_0x$. Therefore, without loss of generality, we consider monic equations only.

Remark 1.5 In this paper, we consider monic bicubic polynomials with real coefficients. Since factorization of any bicubic polynomials with real coefficients is itself a difficult task, we focus on a specific case where the coefficients are integers and the given monic bicubic polynomial is of type (3, 3) such that sum of three roots of the corresponding bicubic equation is zero.

Remark 1.6 Finally, the main aim of this paper is not to show how to solve a sixth-degree polynomial equation but, how to factorize in the desired form. Once such factorization is complete, by solving the equations formed by equating to zero each of the factors, one can derive the solutions. The solution part is left to the reader. We rather focus on factorization only.

2 Methodology

Let $f_6(x) := x^6 + ax^5 + bx^4 + cx^3 + dx^2 + ex + f \in \mathbb{R}[x]$ where e is merely a real coefficient, not the usual Euler's exponential real constant "e" and $f \neq 0$ be such that where $f \neq 0$ be such that the sum of three roots of $f_6(x) = 0$ is 0. We assume $f_6(x)$ can be written as product of two cubic polynomials $f_{31}(x)$, $f_{32}(x)$ say, $f_6(x) = f_{31}(x)f_{32}(x)$. We further assume the roots are $x = -\alpha, -\beta, -\gamma, -\delta, -\lambda, -\mu$ and $\delta + \lambda + \mu = 0$ where

$$f_{31}(x) = x^3 + (\alpha + \beta + \gamma)x^2 + (\alpha\beta + \alpha\gamma + \beta\gamma)x + \alpha\beta\gamma, \tag{2.1}$$

$$f_{32}(x) = x^3 + (\delta\lambda + \delta\mu + \lambda\mu)x + \delta\lambda\mu. \tag{2.2}$$

Since $f_6(x) = f_{31}(x)f_{32}(x)$, equating the coefficients of similar terms, we obtain the following list

$$a = \alpha + \beta + \gamma, \tag{2.3}$$

$$b = (\alpha\beta + \alpha\gamma + \beta\gamma) + (\delta\lambda + \delta\mu + \lambda\mu), \tag{2.4}$$

$$c = (\alpha + \beta + \gamma)(\delta\lambda + \delta\mu + \lambda\mu) + \alpha\beta\gamma + \delta\lambda\mu, \tag{2.5}$$

$$d = (\alpha + \beta + \gamma)\delta\lambda\mu + (\alpha\beta + \alpha\gamma + \beta\gamma)(\delta\lambda + \delta\mu + \lambda\mu), \tag{2.6}$$

$$e = (\alpha\beta + \alpha\gamma + \beta\gamma)\delta\lambda\mu + \alpha\beta\gamma(\delta\lambda + \delta\mu + \lambda\mu), \tag{2.7}$$

$$f = \alpha\beta\gamma \times \delta\lambda\mu. \tag{2.8}$$

For Eq. (2.3), the Eqs. (2.5) and (2.6) reduce to

$$c = a(\delta\lambda + \delta\mu + \lambda\mu) + \alpha\beta\gamma + \delta\lambda\mu, \tag{2.9}$$
$$d = a(\delta\lambda\mu) + (\alpha\beta + \alpha\gamma + \beta\gamma)(\delta\lambda + \delta\mu + \lambda\mu). \tag{2.10}$$

For Eq. (2.4) and Eq. (2.5), the Eq. (2.7) becomes

$$\begin{aligned} ae &= \alpha\beta\gamma\{a(\delta\lambda + \delta\mu + \lambda\mu)\} + \delta\lambda\mu\{a[b - (\delta\lambda + \delta\mu + \lambda\mu)]\} \\ \Rightarrow\ ae &= u(c - u - v) + v[ab - (c - u - v)] \\ \Rightarrow\ u^2 &- v^2 - uc - v(ab - c) + ae = 0, \end{aligned} \tag{2.11}$$

where $u = \alpha\beta\gamma$, $v = \delta\lambda\mu$. Clearly, $uv = f$ and as $f \neq 0$ so are u, v.

Again from Eq. (2.9), we get

$$a(\delta\lambda + \delta\mu + \lambda\mu) = c - u - v. \tag{2.12}$$

Using Eqs. (2.12), (2.4) becomes

$$\begin{aligned} &\alpha\beta + \alpha\gamma + \beta\gamma = b - (\delta\lambda + \delta\mu + \lambda\mu) \\ \Rightarrow\ &a(\alpha\beta + \alpha\gamma + \beta\gamma) = (ab - c) + u + v. \end{aligned} \tag{2.13}$$

Multiplying Eqs. (2.12) and (2.13), it follows that

$$\begin{aligned} &a^2(\alpha\beta + \alpha\gamma + \beta\gamma)(\delta\lambda + \delta\mu + \lambda\mu) = (c - u - v)(ab - c + u + v) \\ \Rightarrow\ &a^2(d - av) = c(ab - c) + cu + cv \\ &\qquad - (ab - c)u - u^2 - uv - (ab - c)v - uv - v^2 \\ \Rightarrow\ &u^2 + v^2 + (ab - 2c)u + (ab - 2c - a^3)v - c(ab - c) + 2f + a^2d = 0. \end{aligned} \tag{2.14}$$

Now adding Eqs. (2.11) and (2.14), we see that

$$\begin{aligned} &2u^2 + (ab - 3c)u + (ab - 2c - a^3 - ab + c)\frac{f}{u} \\ &\qquad + ae + a^2d + 2f - c(ab - c) = 0 \\ \Rightarrow\ &2u^3 + (ab - 3c)u^2 + [a^2d + ae + 2f - c(ab - c)]u - (a^3 + c)f = 0. \end{aligned} \tag{2.15}$$

Also, subtracting Eq. (2.11) from Eq. (2.14) and using $uv = f$, we get

$$\begin{aligned} &2v^2 + (ab - 2c + c)u + (ab - 2c - a^3 + ab - c)v \\ &\qquad + 2f + a^2d - c(ab - c) - ae = 0 \\ \Rightarrow\ &(ab - c)u^3 + [2f + a^2d - c(ab - c) - ae]u^2 \\ &\qquad + f(2ab - 3c - a^3)u + 2f^2 = 0. \end{aligned} \tag{2.16}$$

Here, Eq. (2.16) is a cubic equation in u provided $ab - c \neq 0$. As a consequence, we have obtain a system of two cubic equations in u:

- $2u^3 + (ab - 3c)u^2 + [a^2d + ae + 2f - c(ab - c)]u - (a^3 + c)f = 0,$ (2.17)
- $(ab - c)u^3 + [2f + a^2d - c(ab - c) - ae]u^2 + f(2ab - 3c - a^3)u + 2f^2 = 0.$ (2.18)

This system will have a common root if the resultant of this system is zero, which is regarded as required condition that $f_6(x)$ can be factored into the desired form stated in first page.

Example 2.1 Let us consider $f_6(x) = 0$ where

$$f_6(x) := x^6 + 5x^5 + 9x^4 + 26x^3 + 38x^2 + 45x + 28.$$

Here, $a = 5, b = 9, c = 26, d = 38, e = 45, f = 28$. Then $A_1 = 2$, $B_1 = ab - 3c = -33$, $C_1 = a^2d + ae + 2f - c(ab - c) = 737$, $D_1 = -f(a^3 + c) = -4228$.

Also, $A_2 = ab - c = 19$, $B_2 = 2f + a^2d - c(ab - c) - ae = 287$, $C_2 = f(2ab - 3c - a^3) = -3164$, $D_2 = 2f^2 = 1568$.

Since $ab - c \neq 0$, Eqs. (2.17) and (2.18) become

$$2u^3 - 33u^2 + 737u - 4228 = 0, \tag{2.19}$$

$$19u^3 + 287u^2 - 3164u + 1568 = 0. \tag{2.20}$$

Here, Eq. (2.19)×19− Eq. (2.20)×2 and Eq. (2.19)×19+ Eq. (2.20)×2 yield, respectively

$$-1201u^2 + 20331u - 83468 = 0, \tag{2.21}$$

$$2981u^2 + 41489u - 436492 = 0. \tag{2.22}$$

Since this system has at least one common solution, using technique of cross-multiplication, we get

$$\frac{u^2}{-5411315000} = \frac{u}{-773045000} = \frac{1}{-110435000}$$

$$\Rightarrow u = \frac{u^2}{u} = \frac{u}{1} = +\sqrt{\frac{u^2}{1}}$$

$$\Rightarrow u = 7, 7, +7.$$

Clearly $u = c_2 = 7$. Hence $uv = f$ gives $v = 4$. But $ab_2 = c - u - v$ where $b_2 = \delta\lambda + \delta\mu + \lambda\mu$. So $b_2 = 3$ and, therefore, $x^3 + b_2x + c_2$, i.e., $x^3 + 3x + 4$ is a factor of $f_6(x)$. Dividing $f_6(x)$ by $x^3 + 3x + 4$, the other factor $x^3 + 5x^2 + 6x + 7$ is obtained and so

$$f_6(x) = (x^3 + 5x^2 + 6x + 7)(x^3 + 3x + 4).$$

Clearly solving $x^3 + 3x + 4 = 0$ and $x^3 + 5x^2 + 6x + 7 = 0$, one can obtain the roots of $f_6(x) = 0$.

Remark 2.1 The above process is valid when $ab - c \neq 0$. If $ab = c$ occurs, the system of equations consisting Eqs. (2.17) and (2.18) reduces to a system in which Eq. (2.17) is cubic equation and Eq. (2.18) is either a quadratic equation or linear equation or merely a constant. The last case is impossible except the case $0 = 0$. For the other two cases, it is the best choice to solve Eq. (2.18) directly and check the valid value of u by substituting in Eq. (2.17).

For example, consider $F(x) := x^6 + 4x^5 + 9x^4 + 36x^3 + 44x^2 + 74x + 60$. Here, $a = 4, b = 9, c = 36, d = 44, e = 74, f = 60$, and $ab - c = 0$. Therefore, we have

$$2u^3 - 72u^2 + 1120u - 6000 = 0, \tag{2.23}$$

$$528u^2 - 6000u + 7200 = 0. \tag{2.24}$$

Equation (2.24) is not cubic equation but a quadratic equation. Solving we get, $u = 10, \frac{15}{11}$, of which $u = 10$ satisfies Eq. (2.23). So the required common root is $u = 10$ and consequently one factor $x^3 + 5x + 6$ of $F(x)$ is achieved. The other one will be $x^3 + 4x^2 + 4x + 10$ and so

$$F(x) = (x^3 + 5x + 6)(x^3 + 4x^2 + 4x + 10).$$

The roots of $F(x) = 0$ will be found after solving $x^3 + 5x + 6 = 0$ and $x^3 + 4x^2 + 4x + 10 = 0$.

Remark 2.2 We shall refer the system consisting the Eqs. (2.17) and (2.18) as the system of auxiliary equations and shall use $\mathcal{A}(f_6(x))$ to denote this system. At a later stage, we shall derive more other alternative auxiliary equations.

We have already known the resultant is necessary and sufficient condition of the auxiliary equations in the system $\mathcal{A}(f_6(x))$ which gives us permission to derive the common root u. Since the resultant of the system $\mathcal{A}(f_6(x))$ is a determinant of order 6×6 and might be gigantic from computational point of view, we proceed to search for alternative criteria in the next section. Let us denote this criteria as $\mathcal{C}(\mathcal{A})$ where $\mathcal{A}$ stands for $\mathcal{A}(f_6(x))$.

3 Determination of Criteria $\mathcal{C}(\mathcal{A})$

First of all, the system $\mathcal{A}(f_6(x))$ is of the form

$$A_1u^3 + B_1u^2 + C_1u + D_1 = 0, \tag{3.1}$$

$$A_2u^3 + B_2u^2 + C_2u + D_2 = 0. \tag{3.2}$$

where $A_1 \neq 0$ and Eq. (3.2) is at most a cubic equation. We assume this system possesses at least one common solution.

Now $A_2 \times$ Eq. (3.1) $-A_1 \times$ Eq. (3.2) gives

$$B_{12}u^2 + C_{12}u + D_{12} = 0,$$

where $B_{12} = A_2B_1 - A_1B_2, C_{12} = A_2C_1 - A_1C_2, D_{12} = A_2D_1 - A_1D_2$. Next, $D_2 \times$ Eq. (3.1) $-D_1 \times$ Eq. (3.2) gives

$$\bar{A}_{12}u^3 + \bar{B}_{12}u^2 + \bar{C}_{12}u = 0$$
$$\Rightarrow \bar{A}_{12}u^2 + \bar{B}_{12}u + \bar{C}_{12} = 0$$

as $uv = f$ and $f \neq 0$ imply $u \neq 0, v \neq 0$, where $\bar{A}_{12} = A_1D_2 - A_2D_1, \bar{B}_{12} = B_1D_2 - B_2D_1, \bar{C}_{12} = C_1D_2 - C_2D_1$.

Therefore, we now have got the following system

$$B_{12}u^2 + C_{12}u + D_{12} = 0, \tag{3.3}$$

$$\bar{A}_{12}u^2 + \bar{B}_{12}u + \bar{C}_{12} = 0. \tag{3.4}$$

Hence,

$$\frac{u^2}{C_{12}\bar{C}_{12} - D_{12}\bar{B}_{12}} = \frac{u}{D_{12}\bar{A}_{12} - B_{12}\bar{C}_{12}} = \frac{1}{B_{12}\bar{B}_{12} - C_{12}\bar{A}_{12}}. \tag{3.5}$$

Since $u = \frac{u^2}{u} = \frac{u}{1} = \pm\sqrt{\frac{u^2}{1}}$ where plus or minus sign will be taken according as both ratios $\frac{u^2}{u}, \frac{u}{1}$ are plus or minus, respectively, therefore

$$u = \frac{\Delta_2}{\Delta_1} = \frac{\Delta_1}{\Delta_0} = \pm\sqrt{\frac{\Delta_2}{\Delta_0}}$$

where

$$\Delta_2 = \begin{vmatrix} C_{12} & D_{12} \\ \bar{B}_{12} & \bar{C}_{12} \end{vmatrix}, \Delta_1 = \begin{vmatrix} D_{12} & A_{12} \\ \bar{C}_{12} & \bar{A}_{12} \end{vmatrix}, \Delta_0 = \begin{vmatrix} B_{12} & C_{12} \\ \bar{A}_{12} & \bar{B}_{12} \end{vmatrix}.$$

Equating the first two ratios, we get the desired condition $\mathcal{C}(\mathcal{A})$ as

$$\begin{vmatrix} B_{12} & C_{12} \\ \bar{A}_{12} & \bar{B}_{12} \end{vmatrix} \begin{vmatrix} C_{12} & D_{12} \\ \bar{B}_{12} & \bar{C}_{12} \end{vmatrix} = \begin{vmatrix} B_{12} & D_{12} \\ \bar{A}_{12} & \bar{C}_{12} \end{vmatrix}^2, \tag{3.6}$$

where

$$B_{12} = -\begin{vmatrix} A_1 & B_1 \\ A_2 & B_2 \end{vmatrix}, C_{12} = -\begin{vmatrix} A_1 & C_1 \\ A_2 & C_2 \end{vmatrix}, D_{12} = -\begin{vmatrix} A_1 & D_1 \\ A_2 & D_2 \end{vmatrix},$$

$$\bar{A}_{12} = \begin{vmatrix} A_1 & D_1 \\ A_2 & D_2 \end{vmatrix}, \bar{B}_{12} = \begin{vmatrix} B_1 & D_1 \\ B_2 & D_2 \end{vmatrix}, \bar{C}_{12} = \begin{vmatrix} C_1 & D_1 \\ C_2 & D_2 \end{vmatrix}$$

with

$$\begin{array}{ll} A_1 = 2, & A_2 = ab - c, \\ B_1 = ab - 3c, & B_2 = 2f + a^2d - ae - c(ab - c), \\ C_1 = a^2d + ae + 2f - c(ab - c), & C_2 = f(2ab - 3c - a^3), \\ D_1 = -f(a^3 + c), & D_2 = 2f^2. \end{array}$$

In Example 2.1, we found $A_1 = 2$, $B_1 = -33$, $C_1 = 737$, $D_1 = -4228$ and $A_2 = 19$, $B_2 = 287$, $C_2 = -3164$, $D_2 = 1568$. The LHS (3.6) of $\mathcal{C}(\mathcal{A})$ is

$$\begin{vmatrix} -1201 & 20331 \\ 83468 & 1161692 \end{vmatrix} \begin{vmatrix} 20331 & -83468 \\ 1161692 & -12221776 \end{vmatrix} = 2^{10}5^8 7^4 13^2 1699^2,$$

where RHS (3.6) is

$$\begin{vmatrix} -1201 & -83468 \\ 83468 & -12221776 \end{vmatrix}^2 = 468517280467600000000 = 2^{10}5^8 7^4 13^2 1699^2.$$

Thus, $\mathcal{C}(\mathcal{A})$ is satisfied. Required common root u is given by

$$u = \begin{cases} \frac{C_{12}\bar{C}_{12} - D_{12}\bar{B}_{12}}{D_{12}\bar{A}_{12} - B_{12}\bar{C}_{12}} = 7 \\ \frac{D_{12}\bar{A}_{12} - B_{12}\bar{C}_{12}}{B_{12}\bar{B}_{12} - C_{12}\bar{A}_{12}} = 7 \\ \pm\sqrt{\frac{C_{12}\bar{C}_{12} - D_{12}\bar{B}_{12}}{B_{12}\bar{B}_{12} - C_{12}\bar{A}_{12}}} = \pm 7, \end{cases}$$

where we chose plus sign.

4 More Auxiliary Equations

4.1. First, we recall the Eqs. (2.11) and (2.14) once again.

$$u^2 - v^2 - uc - v(ab - c) + ae = 0, \tag{4.1}$$

$$u^2 + v^2 + (ab - 2c)u + (ab - 2c - a^3)v - c(ab - c) + 2f + a^2d = 0. \tag{4.2}$$

Since $uv = f$, i.e., $v = \frac{f}{u}$, so Eq. (4.1) becomes

$$u^4 - cu^3 + aeu^2 - f(ab - c)u - f^2 = 0. \tag{4.3}$$

Similarly, Eq. (4.2) becomes

$$u^4 + (ab - 2c)u^3 + [a^2d + 2f - c(ab - c)]u^2 + f(ab - 2c - a^3)u + f^2 = 0. \tag{4.4}$$

4.2. Let us write

$$\sum \alpha = \alpha + \beta + \gamma,$$
$$\sum \alpha\beta = \alpha\beta + \alpha\gamma + \beta\gamma,$$
$$\sum \delta\lambda = \delta\lambda + \delta\mu + \lambda\mu.$$

Then all the equations from Eqs. (2.3) till (2.10) can be rewritten as

$$a = \sum \alpha, \tag{4.5}$$
$$b = \sum \alpha\beta + \sum \delta\lambda, \tag{4.6}$$
$$c = \sum \alpha \sum \delta\lambda + u + v, \tag{4.7}$$
$$d = v\sum \alpha + \sum \alpha\beta \sum \delta\lambda, \tag{4.8}$$
$$e = v\sum \alpha\beta + u\sum \delta\lambda, \tag{4.9}$$
$$f = u \times v, \tag{4.10}$$
$$c - u - v = a\sum \delta\lambda, \tag{4.11}$$
$$d - av = \sum \alpha\beta \sum \delta\lambda. \tag{4.12}$$

From Eq. (4.12), we get

$$a(d - av) = \sum \alpha\beta \times a\sum \delta\lambda$$
$$\Rightarrow ad - a^2v = (c - u - v)\sum \alpha\beta \tag{4.13}$$

by Eq. (4.11).

Again, from Eq. (4.9),

$$e(c - u - v) = v(c - u - v)\sum \alpha\beta + u(c - u - v)\sum \delta\lambda$$
$$\Rightarrow ae(c - u - v) = av(ad - a^2v) + u(c - u - v) \times a\sum \delta\lambda. \tag{4.14}$$

Let us simplify Eq. (4.14). Using Eq. (4.7), we see that

$$\begin{aligned} &av(ad - a^2v) + u(c - u - v) \times a \sum \delta\lambda \\ &= a^2v(d - av) + u(c - u - v) \times (c - u - v) \\ &= u^3 - 2cu^2 - a^3v^2 + (c^2 + 2f)u + (a^2d + f)v - 2cf. \end{aligned}$$

Hence Eq. (4.14) becomes

$$u^3 - 2cu^2 - a^3v^2 + (c^2 + 2f + ae)u + (a^2d + f + ae)v - (2f + ae)c = 0, \tag{4.15}$$

$$\begin{aligned} \Rightarrow\ & u^5 - (2c)u^4 + (c^2 + 2f + ae)u^3 \\ & - c(2f + ae)u^2 + f(a^2d + f + ae)u - a^3f^2 = 0 \end{aligned} \tag{4.16}$$

which is another auxiliary equation in u of degree 5.

4.3. Let us reconsider Eq. (4.9). Using Eq. (4.6), we obtain

$$e = ub + (v - u) \sum \alpha\beta$$

$$\Rightarrow e(c - u - v) = ub(c - u - v) + (v - u)(ad - a^2v), \text{ by Eq(4.13)}$$

$$\Rightarrow bu^2 + a^2v^2 + (ad - bc - e)u - (ad + e)v + (bf - a^2f + ce) = 0, \tag{4.17}$$

$$\Rightarrow bu^4 + (ad - bc - e)u^3 + (bf - a^2f + ce)u^2 - f(ad + e)u + a^2f^2 = 0. \tag{4.18}$$

4.4. Furthermore, adding Eqs. (4.8), (4.9) and using Eq. (4.6), we have

$$a^2(d + e) = a^2[av + v(b - \sum \delta\lambda) + (u + b - \sum \delta\lambda) \sum \delta\lambda]$$

$$\begin{aligned} \Rightarrow a^2(d + e) &= a^3v + av(ab - c + u + v) \\ &+ (au + ab - c + u + v)(c - u - v) \end{aligned}$$

$$\begin{aligned} \Rightarrow\ & (a + 1)u^2 - (a - 1)v^2 + (ab - 2c - ac)u - (a^3 + a^2b - ac - ab + 2c)v \\ & + (a^2(d + e) - abc + c^2 + 2f) = 0, \end{aligned} \tag{4.19}$$

$$\begin{aligned} \Rightarrow\ & (a + 1)u^4 + (ab - 2c - ac)u^3 + [a^2(d + e) - abc + c^2 + 2f]u^2 \\ & - f(a^3 + a^2b - ac - ab + 2c)u - f^2(a - 1) = 0. \end{aligned} \tag{4.20}$$

On the other hand, subtracting Eq. (4.9) from Eq. (4.8) and using Eq. (4.6) we get

$$d - e = av - v(b - \sum \delta\lambda) + (b - \sum \delta\lambda - u) \sum \delta\lambda,$$

$$\Rightarrow a^2(d-e) = a^3 v - av(ab - c + u + v) + (ab - c + u + v - au)(c - u - v)$$

$$\Rightarrow a^2(d-e) = (a-1)u^2 - (a+1)v^2 - (ab - 2c + ac)u$$
$$+ (a^3 - a^2 b + ac - ab + 2c)v + (abc - c^2 - 2f)$$

$$\Rightarrow (a-1)u^2 - (a+1)v^2 - (ab + ac - 2c)u + (a^3 - a^2 b + ac - ab + 2c)v$$
$$+ [abc - c^2 - 2f - a^2(d-e)] = 0, \tag{4.21}$$

$$\Rightarrow (a-1)u^4 - (ab + ac - 2c)u^3 + [abc - c^2 - 2f - a^2(d-e)]u^2$$
$$+ f(a^3 - a^2 b + ac - ab + 2c)u - f^2(a+1) = 0. \tag{4.22}$$

Compiling together the list of equations involving u and v is shown below.

- $u^2 - v^2 - uc - (ab - c)v + ae = 0,$ (4.23)
- $u^2 + v^2 + (ab - 2c)u + (ab - 2c - a^3)v + a^2 d + 2f - c(ab - c) = 0,$ (4.24)
- $u^3 - 2cu^2 - a^3 v^2 + (c^2 + 2f + ae)u + (a^2 d + f + ae)v$
  $- c(2f + ae) = 0,$ (4.25)
- $bu^2 + a^2 v^2 + (ad - bc - e)u - (ad + e)v + (bf - a^2 f + ce) = 0,$ (4.26)
- $(a+1)u^2 - (a-1)v^2 + (ab - 2c - ac)u - (a^3 + a^2 b - ac - ab + 2c)v$
  $+ (a^2(d+e) - abc + c^2 + 2f) =,$ (4.27)
- $(a-1)u^2 - (a+1)v^2 - (ab + ac - 2c)u + (a^3 - a^2 b + ac - ab + 2c)v$
  $+ [abc - c^2 - 2f - a^2(d-e)] = 0.$ (4.28)

Eliminating v from the equations enlisted above taken two at a time, one can obtain more auxiliary equations in u. Interested reader can derive such list.

5 Conclusion

Techniques on the factorization of given monic sixth-degree univariate polynomial with real coefficients have been shown as was promised. Auxiliary equations involving u and v have been prepared and shown with examples. Also, demonstration of factorization through example has been done. The reader can solve the equation involving the cubic factors and determine the roots of the given equation. In the end, a list of alternative auxiliary equations has also been included. For future work, the reader can work the same on $\mathbb{C}[x]$. Since there are several beautiful gems are available in solving equations of higher degree in literature, interested reader can take help from them as well. For example, [2–7] can be utilized for this purpose.

Acknowledgements The author wishes to express his gratitude and acknowledgement to Dr. Sushama Agarwal Narayandas, Director and HOD of Ramanujan Institute for Advanced Study in Mathematics, for her constant support and encouragement to complete the project in time.

References

1. A. Debnath, R. Sahadevan, On factorisation of certain 5th degree polynomials. J. Indian Acad. Math. **37** (2015)
2. N.D. Bagis, On a general sextic equation solved by the Rogers-Ramanujan continued fraction (2012). arXiv:1111.6023v2 [math.GM]
3. R.B. King, *Beyond the Quartic Equation* (Birkhauser, Boston, 1996)
4. M. Riaz, Geometric solutions of algebraic equations. Amer. Math. Monthly **69**(7), 654–658 (1962)
5. E.M. Radford, On the solution of certain equations of the seventh degree. Quart. J. Math. **30**, 263–306 (1898)
6. R.G. Kulkarni, Unified method for solving general polynomial equations of degree less than five. Alabama J. Math. **30**, 1–18 (2006)
7. H. Umemura, Resolution of algebraic equations by theta constants, in *Tata Lectures on Theta II*, vol. 3(261-3) (BirkhaÃijser, Boston, 1984), p. 272

On Determination of $\varphi^{-1}(2^a p^{a_1})$

Anjan Debnath and Avishek Adhikari

Abstract In this paper, we have considered the set $\varphi^{-1}(2^a p^{a_1})$ where p is an odd prime, $a, a_1 \in \mathbb{N}$ and have shown how to determine the elements of it. Several important results on behalf of this have been discussed.

Keywords Euler's phi function · Arithmetic function · Number theory

2010 Mathematics Subject Classification: 11A25 · 11A99

1 Introduction

The Euler's totient function $\varphi(n)$ is a multiplicative function that counts the total number of positive integers less than or equal to $n \in \mathbb{N}$ and relatively prime to n. In other words,

$$\varphi(n) = \#\{r \in \mathbb{N} : 1 \leqslant r \leqslant n, gcd(r, n) = 1\}.$$

For such given n, the value $\varphi(n)$ is called the totient value of n whereas n is referred as the pre-totient value of $\varphi(n)$.

In other words, if $\varphi(x) = n$ holds for some $x, n \in \mathbb{N}$ then x is called pre-totient of n and n is called the totient of x. When the canonical form of x is known, say $x = p_1^{a_1} p_2^{a_2} \dots p_r^{a_r}$ where $p_1 < p_2 < \cdots < p_r$ are primes and each $a_i \in \mathbb{N}$, determination of $\varphi(x)$ becomes easy because of the working formula $\varphi(x) = x \prod_{i=1}^{r}(1 - \frac{1}{p_i})$. For example, $\varphi(60) = 60(1 - \frac{1}{2})(1 - \frac{1}{3})(1 - \frac{1}{5}) = 16$.

However it is comparably a difficult task to move on in the reverse direction. In other words, when n is given say 16, the determination of x is not at all an easy job.

A. Debnath (✉)
Department of Mathematics, The Heritage College, Kolkata 700107, India
e-mail: anjandebnath@rocketmail.com

A. Adhikari
Department of Mathematics, Presidency University, Kolkata 700073, India
e-mail: avishek.adh@gmail.com

S. R. Mishra et al. (eds.), *Recent Trends in Applied Mathematics*, Lecture Notes in Mechanical Engineering, https://doi.org/10.1007/978-981-15-9817-3_5

In fact, there are multiple pre-images of 16 under φ. For example each of 32, 34, 40, 48, 60 has pre-totient value 16 and, $\varphi(17) = 16$ as well.

In [1, 2] some discoveries have been made regarding determination of pre-totients. However in [2], a predetermined list of smaller pre-totients is needed in order to deal with bigger cases and in [1], some particular cases have been discussed. However, in this current paper, we shall consider the case of determination of the pre-totient values of $2^a p^{a_1}$, where p is an odd prime and both $a, a_1 \in \mathbb{N}$. The set of all such pre-totients will be denoted by $\varphi^{-1}(2^a p^{a_1})$ i.e.,

$$\varphi^{-1}\left(2^a p^{a_1}\right) = \{x \in \mathbb{N} : \varphi(x) = 2^a p^{a_1}\}.$$

It is to be noted that, some of these sets might be empty as well. For example, $\varphi^{-1}(2.7)$ is an empty set. In fact, 14 is the first even number, of which there is no pre-totient x at all. This part will be established at a later stage. For now, let us assume, for a given $n = 2^a p^{a_1}$ where p is odd prime and $a, a_1 \in \mathbb{N}$, the set $\varphi^{-1}(n)$ is non-empty, containing at least one element say $x \in \mathbb{N}$. Then either $x \in E(2^a p^{a_1})$ or $x \in O(2^a p^{a_1})$ where, for $n \in \mathbb{N}$

$$E(n) = \{x \in \varphi^{-1}(n) : x \text{ is even}\}$$
$$O(n) = \{x \in \varphi^{-1}(n) : x \text{ is odd}\}$$

Evidently $\varphi^{-1}(n)$ is disjoint union of $E(n)$ and $O(n)$. In Theorem 5.1 of [1] a very particular case about set wise equality of $E(2s)$ and $O(2s)$ where s is odd positive integer greater than 1 is shown. However we shall show much general situation in the current paper. We start our journey with the first case $x \in E(n)$. Before going into details, let us get introduced with the following definitions which will be used frequently.

Definition 1.1 For any set $S \subseteq \mathbb{N}$, the cardinality of S is the number of elements of S and will be denoted by $|S|$ or $\#S$.

Definition 1.2 For $m, a, b \in \mathbb{N}$, we define

$$\begin{aligned}
\mathbb{N}_m &= \{1, 2, \ldots, m\},\\
\mathbb{W}_m &= \{0, 1, 2, \ldots, m\},\\
\mathbb{W} &= \{0, 1, 2, \ldots\},\\
{}_m\mathbb{N}_n &= \{m, m+1, m+2, \ldots, n-1, n\},\\
{}_a\mathbb{N} &= \{x \in \mathbb{N} : x \geqslant a\},\\
\mathbb{P} &= \{p \in \mathbb{N} : p \text{ is prime}\}
\end{aligned}$$

Definition 1.3 The greatest common divisor of two integers a, b (not both zero) is a common divisor $d \in \mathbb{N}$, which is divisible by any common divisor of a and b. It is denoted by $d = (a, b)$. In case, if $d = 1$, the integers a, b are called relatively prime integers.

2 When $x \in E(2^a P^{a_1})$

Let us first see what can be the upper limit of e if $2^e|x$ as x is assumed to be even, where e is just a variable, not the usual Euler's exponential constant e and where by the symbol $a|b$ we meant that the non-zero integer a divides the integer b.

Theorem 2.1 *If $x \in E(2^a p^{a_1})$ where p is odd prime and $a, a_1 \in \mathbb{N}$ then x cannot be a multiple of 2^{a+1}.*

Proof If possible let $x = 2^{a+1}m$ where $m \in \mathbb{N}$. If $(2, m) = 1$ then $\varphi(x) = 2^a p^{a_1}$ will give $\varphi(2^{a+1}m) = 2^a p^{a_1}$, i.e.,

$$\varphi(m) = p^{a_1} \tag{1}$$

The contradiction is already obtained for $m = 1$. If $m > 2$ then in Eq. (1), LHS is even but RHS is odd, contradiction again.

On the other hand, if $(2, m) \neq 1$ then $x = 2^{a+n}m_1$ where $n \geqslant 2$ and $(2, m_1) = 1$. Once again, $\varphi(x) = 2^a p^{a_1}$ will produce $2^{n-1}\varphi(m_1) = p^{a_1}$. For $n - 1 \neq 0$, absurdity is achieved and if $n = 1$, contradiction is still there as $n \geqslant 2$. ■

Deduction 2.1 If $x \in E(2^a p^{a_1})$ then x will never be a multiple of $2^{a+2}, 2^{a+3}, \ldots,$ $2^{a+n}, \ldots$ etc. where $n \geqslant 2$.

Deduction 2.2 Let $x \in E(2^a p^{a_1})$. For each $e \in \mathbb{N}_a$ there is at least one $m \in \mathbb{N} \setminus 2\mathbb{N}$ such that

$$x = 2^e m \in E(2^a p^{a_1}).$$

Deduction 2.3 If $m_0 \in O(2^{a+1-e} p^{a_1}), e \in \mathbb{N}_a$, then $2^e m_0 \in E(2^a p^{a_1})$ and vice-versa.

Theorem 2.2 *If $m \in O(2^a p^{a_1})$ then $2m \in E(2^a p^{a_1})$. Conversely, if $2m \in E(2^a p^{a_1})$ where m is odd, then $m \in O(2^a p^{a_1})$.*

Proof Trivial. ■

Therefore, whenever we determine the set $O(2^a p^{a_1})$ completely, the set $E(2^a p^{a_1})$ will enlist the elements which would be double of the elements of $O(2^a p^{a_1})$. We now see if any odd prime other than p appears in the canonical form of $x \in \varphi^{-1}(2^a p^{a_1})$, what its power could be.

Theorem 2.3 *Let $x \in \varphi^{-1}(2^a p^{a_1})$. If there be an odd prime q other than p such that $q^{\beta_1} \mid x$ then $\beta_1 = 1$.*

Proof Let $x = 2^e q^{\beta_1} m_1$ where $(2q, m_1) = 1$ and $e \in \mathbb{W}_a$, by deduction 2.2. Then

$$2^a p^{a_1} = \varphi(x) = 2^{e-1} q^{\beta_1 - 1}(q - 1)\varphi(m_1)$$

If $\beta_1 - 1 \in \mathbb{N}$, the above equality shows LHS$\not\equiv$RHS(mod q). Hence $\beta_1 - 1 = 0$, i.e., $\beta_1 = 1$. ■

Definition 2.1 For $\lambda \in \mathbb{N}$ and $A \subseteq \mathbb{N}$, we define the set

$$\lambda A = \{\lambda a : a \in A\}.$$

Theorem 2.4 $2E(2^a p^{a_1}) = E(2^{a+1} p^{a_1}) \setminus \{2m_0 : m_0 \in O(2^{a+1} p^{a_1})\}$.

Proof Let $x \in E(2^a p^{a_1})$. Then $x = 2^e m$ where m is odd and $e \in \mathbb{N}_a$. Therefore,

$$\begin{aligned} 2^a p^{a_1} &= \varphi(x) = \varphi(2^e m) \\ \Leftrightarrow 2^{a+1} p^{a_1} &= 2^e \varphi(m) = \varphi(2^{e+1} m) = \varphi(2x) \end{aligned}$$

Since $x \in E(2^a p^{a_1})$ and $1 \leqslant e \leqslant a$, $2x \in E(2^{a+1} p^{a_1})$ implies $2 \leqslant e \leqslant a + 1$. So, $x \in E(2^a p^{a_1}) \Leftrightarrow 2x \in E(2^{a+1} p^{a_1}) \setminus \{2m_0\}$ for each $m_0 \in O(2^{a+1} p^{a_1})$. ■

Corollary 2.1 $|E(2^{a+1} p^{a_1})| = |O(2^{a+1} p^{a_1})| + 2|E(2^a p^{a_1})|$ *for all* $a \in \mathbb{N}$.

Proof By using Theorem 2.2, $2m_0 \in \varphi^{-1}(2^a p^{a_1})$ iff $m_0 \in \varphi^{-1}(2^a p^{a_1})$. In other words, $2m_0 \in E(2^a p^{a_1})$ iff $m_0 \in O(2^a p^{a_1})$. Theorem 2.4 is now sufficient enough to finish the proof. ■

Remark 2.1 Let us denote $|E(2^\alpha p^\beta)|$ and $|O(2^\alpha p^\beta)|$ by $\varepsilon_{\alpha,\beta}$ and $\theta_{\alpha,\beta}$, respectively. Then Corollary 2.1 states that $\varepsilon_{a+1,a_1} = 2\varepsilon_{a,a_1} + \theta_{a+1,a_1}$.

Now $\varepsilon_{a,a_1} = \theta_{a,a_1} + 2\varepsilon_{a-1,a_1}$ provided $a - 1 \in \mathbb{N}$. Hence

$$\varepsilon_{a+1,a_1} = 2^2\varepsilon_{a-1,a_1} + 2\theta_{a,a_1} + \theta_{a+1,a_1}.$$

Proceeding similarly, we obtain

$$\varepsilon_{a+1,a_1} = 2^3\varepsilon_{a-2,a_1} + 2^2\theta_{a-1,a_1} + 2\theta_{a,a_1} + \theta_{a+1,a_1}$$

provided $a - 2 \in \mathbb{N}$ and in general, when $a - r \in \mathbb{N}$ for some $r \in \mathbb{N}$, we get the following deduction.

Deduction 2.4 For $a \in \mathbb{N}$, if $a - r \in \mathbb{N}$ for some $r \in \mathbb{N}$,

$$\begin{aligned} \varepsilon_{a+1} &= 2^{a-r}\varepsilon_{r+1} \\ &\quad + 2^{a-r-1}\theta_{r+2} + \cdots + 2^3\theta_{a-2} + 2^2\theta_{a-1} + 2\theta_a + \theta_{a+1} \end{aligned}$$

where ε_α and θ_α are same as $\varepsilon_{\alpha,\beta}$ and $\theta_{\alpha,\beta}$, by ignoring β in the indices, respectively.

Deduction 2.5 For $a \in \mathbb{N}$,

$$\varepsilon_{a+1} = 2^a\varepsilon_1 + 2^{a-1}\theta_2 + \cdots + 2^2\theta_{a-1} + 2\theta_a + \theta_{a+1}$$

After this, we now move on to the case when $x \in O(2^a p^{a_1})$ and see the configuration of the set of odd pre-totient elements of $2^a p^{a_1}$.

3 When $x \in O(2^a p^{a_1})$

We start this section with the following theorem which estimates the upper limit of the number of distinct prime divisor of x.

Theorem 3.1 *If $x \in O(2^a p^{a_1})$ then x has at most a number of distinct odd prime factors.*

Proof Let $x = m_0 \in O(2^a p^{a_1})$. Then $m_0 \in {}_2\mathbb{N} \setminus 2\mathbb{N}$.

Let $\omega(m_0) = r$ where $\omega(n)$ denotes the total number of prime divisors of $n \in {}_2\mathbb{N}$. Then we can write the canonical form as

$$m_0 = q_1^{a_1} q_2^{a_2} \cdots q_r^{a_r}.$$

Hence

$$2^a p_1^{a_1} = \varphi(m_0) = \prod_{i=1}^{r} q_i^{a_i - 1}(q_i - 1) \tag{2}$$

shows that each odd prime factor on right-hand side will give at least 2 as a factor and hence the right side must be divisible by 2^r at least. In other words, $2^r | 2^a$, i.e., $r \in \mathbb{N}_a$. ■

3.1 When $r = 1$

Since the total number of odd prime factor of $x = m_0 \in O(2^a p^{a_1})$ is 1, x must be one of the form either p^β or q^β where $\beta \in \mathbb{N}$ and $q \in \mathbb{P} \setminus \{2, p\}$.

Let us start with first possibility. When $m_0 = p^\beta$ then Eq. (2) becomes $p - 1 = 2^a p^{a_1 + 1 - \beta}$.

Here if $a_1 + 1 - \beta \neq 0$, LHS$\not\equiv$ RHS$[p]$. So $\beta = a_1 + 1$ and hence $p = 2^a + 1$. Which clearly indicates the Fermat's numbers are defined by p must be a Fermat's prime F_α for some $a = 2^\alpha \in \mathbb{N}$ where, for $n \in \mathbb{W}$,

$$F_n := 2^{2^n} + 1.$$

See [3–7] for latest updates on Fermat's primes in order to utilize them. sOn the other hand, if $x = m_0 = q^\beta$, by Theorem 2.3, $\beta = 1$ and then Eq. (2) will show

$$q = (2^a p^{a_1} + 1)_*$$

where the asterisk denotes we shall consider the element once it passes primality test. As a consequence, we now have the following theorem ready.

Theorem 3.2 *Let* $x = m_0 \in O(2^a p^{a_1})$ *and* $\omega(x) = 1$. *Then* $m_0 =$

(1) p^β *provided* $\beta = a_1 + 1$ *and* $p = F_\alpha$ *for some* $2^\alpha \in \mathbb{N}$;
(2) q^β *where* $\beta = 1$ *and* $q = (2^a p^{a_1} + 1)_*$

3.2 When $r = 2$

Next, let $\omega(x) = 2$. Since exactly two odd primes are to be present as prime divisors in the canonical form of x, we see that m_0 will be one of the form: $p^{\beta_1} q_2^{\beta_2}$ or $q_1^{\beta_1} q_2^{\beta_2}$, where once again each $\beta_i \in \mathbb{N}$ and q_1, q_2 are odd primes other than p.

3.2.1 $m_0 = p^{\beta_1} q_2^{\beta_2}$

In this case, first of all, by Theorem 2.3 $\beta_2 = 1$. Also Eq. (2) gives

$$2^a p^{a_1+1-\beta_1} = (p-1)(q_2-1) \tag{3}$$

Now let

$$p - 1 = 2^{\lambda_1} \tag{4}$$
$$q_2 - 1 = 2^{\lambda_2} p^{\mu_{21}} \tag{5}$$

Therefore, Eq. (3) yields $\lambda_1 + \lambda_2 = a$, $\mu_{21} = a_1 + 1 - \beta_1$, $\beta_1 \in \mathbb{N}_{a_1+1}$.

From Eq. (4), it is once again clear that λ_1 is a power of 2, say $\lambda = 2^\alpha \in \mathbb{N}_{a-1}$ for some $\alpha \in \mathbb{N}$ and p is nothing but a Fermat's prime F_α. Hence $\lambda_2 = a - 2^\alpha$ and then Eq. (5) gives $q_2 = (2^{a-2^\alpha} F_\alpha^{a_1+1-\beta_1} + 1)_*$.

3.2.2 $m_0 = q_1^{\beta_1} q_2^{\beta_2}$

By Theorem 2.3, $\beta_1 = \beta_2 = 1$ and so $m = q_1 q_2$/ For Eq. (3), we assume $q_1 - 1 = 2^{\lambda_1} p^{\mu_{11}}$, $q_2 - 1 = 2^{\lambda_2} p^{\mu_{21}}$ so that

$$a = \lambda_1 + \lambda_2 \text{ in } \mathbb{N} \tag{6}$$
$$a_1 = \mu_{11} + \mu_{21} \text{ in } \mathbb{W} \tag{7}$$

where the terms *in* $\mathbb{N}$ mean to solve the Eq. (6) in $\mathbb{N}$ and so do *in* $\mathbb{W}$ mean. Hence $m_0 = q_1 q_2 \in O(2^a p^{a_1})$ provided $q_1 = (2^{\lambda_1} p^{\mu_{11}} + 1)_*$, $q_2 = (2^{a-\lambda_1} p^{a_1-\mu_{11}} + 1)_*$.

As a consequence we now have established the following theorem.

Theorem 3.3 *Let $x = m_0 \in O(2^a p^{a_1})$ and $\omega(x) = 2$. Then*

(1) $m_0 = p^{\beta_1} q_2$ *provided*

  (a) $\beta_1 = a_1 + 1, a = 2^\alpha \in \mathbb{N}_{a-1}$
  (b) $p = F_\alpha$
  (c) $q_2 = (2^a p^{a_1} + 1)_*$

(2) $m_0 = q_1 q_2$ *provided*

  (a) $q_1 = (2^{\lambda_1} p^{\mu_{11}} + 1)_*, q_2 = (2^{a-\lambda_1} p^{a_1-\mu_{11}} + 1)_*$
  (b) $\lambda_1 \in \mathbb{N}_{a-1}, \mu_{11} \in \mathbb{W}_{a_1}$

3.3 When $r = 3$

This time $x = m_0$ will be one of the forms: $p^{\beta_1} q_2 q_3$ or $q_1 q_2 q_3$ where $\beta \in \mathbb{N}$ and q_1, q_2, q_3 are odd primes other than p, with 1 as exponent by the aid of Theorem 2.3.

3.3.1 $m_0 = p^{\beta_1} q_2 q_3$

Equation (2) will reduce now into

$$2^a p^{a_1+1-\beta_1} = (p-1)(q_2-1)(q_3-1) \tag{8}$$

We assume

$$p - 1 = 2^{\lambda_1} \tag{9}$$

$$q_2 - 1 = 2^{\lambda_2} p^{\mu_{21}} \tag{10}$$

$$q_3 - 1 = 2^{\lambda_3} p^{\mu_{31}} \tag{11}$$

so that

$$\lambda_1 + \lambda_2 + \lambda_3 = a \text{ in } \mathbb{N} \tag{12}$$

$$\mu_{21} + \mu_{31} = a_1 + 1 - \beta_1 \text{ in } \mathbb{W} \tag{13}$$

Once again, $\lambda_1 = 2^\alpha \in \mathbb{N}_{a-1}$ and $p = F_\alpha$.

3.3.2 $m_0 = q_1 q_2 q_3$

In this case, Eq. (2) reduces to

$$2^a p_1^{a_1} = (q_1 - 1)(q_2 - 1)(q_3 - 1)$$

If we let $q_1 - 1 = 2^{\lambda_1} p^{\mu_{11}}, q_2 - 1 = 2^{\lambda_2} p^{\mu_{21}}, q_3 - 1 = 2^{\lambda_3} p^{\mu_{31}}$ clearly then

$$\lambda_1 + \lambda_2 + \lambda_3 = a \text{ in } \mathbb{N}$$
$$\mu_{11} + \mu_{21} + \mu_{31} = a_1 \text{ in } \mathbb{W}$$

Therefore the proof of the next theorem is done.

Theorem 3.4 *Let $\omega(x) = 3$ where $x = m_0 \in O(2^a p^{a_1})$. Then*

(1) $m_0 = p^{\beta_1} q_2 q_3$ provided

- *(a) $p_1 = F_\alpha$ for some $2^\alpha \in \mathbb{N}_{a-2}, \beta_1 \in \mathbb{N}_{a_1+1}$*
- *(b) $q_2 = (2^{\lambda_2} p^{\mu_{21}} + 1)_*$*
- *(c) $q_3 = (2^{a-2^\alpha-\lambda_2} p^{(a_1+1-\beta_1)-\mu_{21}} + 1)_*$*
- *(d) $\lambda_2 \in \mathbb{N}_{a-2^\alpha-1}, \mu_{21} \in \mathbb{W}_{a_1+1-\beta_1}$*

(2) $m_0 = q - 1 q_2 q_3$ provided

- *(a) $q_i = (2^{\lambda_i} p^{\mu_{i1}} + 1)_*, i = 1, 2, 3$*
- *(b) $\lambda_1 + \lambda_2 + \lambda_3 = a$ in $\mathbb{N}_{a-2}$*
- *(c) $\mu_{11} + \mu_{21} + \mu_{31} = a_1$ in $\mathbb{W}$.*

Proceeding in a similar way the further general case when $r \geqslant 4$ occurs can also be shown. We just provided the statement of that.

Theorem 3.5 *Let $\omega(x) = r \geqslant 4$ where $x \in O(2^a p^{a_1})$. Then*

(1) $x = p^{\beta_1} q_2 q_3 whereq_r$ provided

- *(a) $p = F_\alpha 2^\alpha \in \mathbb{N}_a, \beta_1 \in \mathbb{N}_{a_1+1}$*
- *(b) $q_i = (2^{\lambda_i} p^{\mu_i} + 1)_*, i \in {}_2\mathbb{N}_r$*
- *(c) $\lambda_2 + \cdots + \lambda_r = a - 2^\alpha$ in $\mathbb{N}$*
- *(d) $\mu_2 + \cdots + \mu_r = a_1$ in $\mathbb{W}$*

(2) $x = q_1 q_2 q_3 \cdots q_r$ provided

- *(a) $q_i = (2^{\lambda_i} p^{\mu_i} + 1)_*, i \in \mathbb{N}_r$*
- *(b) $\lambda_1 + \lambda_2 + \cdots + \lambda_r = a - 2^\alpha$ in $\mathbb{N}$*
- *(c) $\mu_1 + \mu_2 + \cdots + \mu_r = a_1$ in $\mathbb{W}$*

Having acquainted with all these, let us look at few examples on construction of $\varphi^{-1}\left(2^a p_1^{a_1}\right)$ for different values of a.

4 Construction of $\varphi^{-1}\left(2^a p_1^{a_1}\right)$

4.1 When $a = 1$

To find $\varphi^{-1}(2p^{a_1})$ where $a_1 \in \mathbb{N}$, we see that any element x of it will be of the form $2m_0, m_0$ where $m_0 \in O(2p^{a_1})$. Now evidently $\omega(m)$ is only 1 and so by Theorem 3.2,

$$m_1 = \begin{cases} 3^{a_1+1} \text{ if } p = 3 \\ (2p^{a_1}+1)_* \end{cases} \tag{14}$$

Deduction 4.1 If $p \geqslant 5$ be odd prime and $\varphi^{-1}(2p^{a_1})$ be non-empty then $2p^{a_1}+1$ is a prime.

Proof Here $\varphi^{-1}(2p^{a_1}) = \{2m, m\}$ where $m = (2p^{a_1}+1)_*$, since $p \neq 3$. If $(2p^{a_1}+1)$ is not prime, $\varphi^{-1}(2p^{a_1})$ will become empty, a contradiction. ■

Remark 4.1 Here the following conjecture can be made. *If* $|\varphi^{-1}(n)| = 2$ *then* $n+1$ *is prime and* n *is not divisible by 4*. The interested reader can work on it for further development.

We have seen that $\varphi^{-1}(2p^{a_1}) = \{2m, m\}$ where $m = 3^{a_1+1}$ if $p = 3$ and $m = (2p^{a_1}+1)_*$. Let us discuss few interesting cases for different values of $a_1 \in \mathbb{N}$.

4.1.1 $a_1 = 1$

We consider $\varphi^{-1}(2p)$. Clearly

$$O(2p) = \begin{cases} 3^2;\ p = 3 \\ (2p+1)_* \end{cases}$$

In other words, $p = 3$ implies

$$\varphi^{-1}(2.3) = \varphi^{-1}(6) = \{2.3^2, 2.(2.3+1)_*, 3^2, (2.3+1)_*\} = \{18, 14, 9, 7\}.$$

On the other hand, if $p > 3$ then $\varphi^{-1}(2p) = \{2(2p+1)_*, (2p+1)_*\}$.

Deduction 4.2 If $p > 3$ be odd prime then $|\varphi^{-1}(2p)|$ is either 0 or 2.

What possible form does $2p+1$ take ? If it is divisible by 3, we must have $p \equiv 1[3]$. Also, $p \equiv 1[2]$. So $p \equiv 1[6]$. In other words, $\varphi^{-1}(2p)$ is empty whenever the odd prime p is of the form $6k+1$ for some positive integer k. The smallest such prime is 7, and therefore $\varphi^{-1}(14)$ is the first set in the list of $\varphi^{-1}(2p)$ which is empty because

$$\begin{aligned} \varphi^{-1}(2.3) &= \{7, 9, 14, 18\}, \\ \varphi^{-1}(2.5) &= \{11, 22\}, \\ \varphi^{-1}(2.7) &= \emptyset, \\ \varphi^{-1}(2.11) &= \{23, 2.23\}, \\ \varphi^{-1}(2.13) &= \emptyset. \end{aligned}$$

Remark 4.2 If there exists a prime $\bar{p}$ such that $2p + 1 \equiv 0[\bar{p}]$, $p \equiv (\frac{\bar{p}-1}{2})[\bar{p}]$. In this case, $\varphi^{-1}(2p) = \emptyset$.

4.1.2 $a_1 \equiv 0[2]$

Then $a_1 = 2a_1'$ for some $a_1 \in \mathbb{N}$. Hence $2p^{2a_1'} + 1 \equiv 0[3]$. On the other hand, $m_0 = p^\beta$ shows that $\beta = 3$ so that $\varphi^{-1}\left(2p^{2a_1'}\right) = \{2.3^{2a_1'+1}, 3^{2a_1'+1}\}$ provided $p = 3$ and if $p > 3$ then $\varphi^{-1}\left(2p^{2a_1'}\right) = \emptyset$. Together we thus have established the next theorem.

Theorem 4.1 *If $a_1' \in \mathbb{N}$ and $p > 3$ is odd prime then $\varphi^{-1}\left(2p^{2a_1'}\right) = \emptyset$.*

As a consequence, we now have an alternative primality test.

Deduction 4.3 Let $p > 3$. If $\varphi^{-1}\left(2p^{2a_1'}\right)$ is non-empty for some $a_1' \in \mathbb{N}$, then p is composite.

Theorem 4.2 *If $a_1' \in \mathbb{N}$ and $p > 3$ be odd prime such that $p \equiv 1[3]$, the set $\varphi^{-1}\left(2p^{2a_1'-1}\right) = \emptyset$.*

Proof Since $p > 3$ and $p \equiv 1[3]$ therefore $2p^{2a_1'-1} + 1 \equiv 2.1 + 1 \equiv 0[3]$. Thus 3 is a divisor of $2p^{2a_1'-1}$ for all $a_1' \in \mathbb{N}$. Hence $\varphi^{-1}\left(2p^{2a_1'-1}\right) = \emptyset$. ■

Remark 4.3 The only case that remains is the set $\varphi^{-1}(2p^{a_1})$ provided a_1 is odd positive integer and $p \equiv -1[3]$. Interested reader can work on it for further improvement.

4.2 *When $a = 2$*

In the set $\varphi^{-1}\left(2^2 p^{a_1}\right)$, any element x will be of the form $2^2 m_0, 2m_1, m_1$ where $m_0 \in O(2p^{a_1})$, $m_1 \in O(2^2 p^{a_1})$. Proceeding in a similar manner as we did before, we obtain $2^2 m_0 \in E(2^2 p^{a_1})$ where

$$m_0 = \begin{cases} 3^{a_1+1}, \text{ if } p = 3 \\ (2p^{a_1} + 1)_*. \end{cases}$$

On the other hand, $2m_1 \in E(2^2 p^{a_1})$ where

$$m_1 = \begin{cases} 5^{a_1+1} \text{ if } p = 5, \\ (2^2 p^{a_1} + 1)_*, \\ 3_1^{\beta}(2.3^{a_1+1-\beta_1} + 1)_* \text{ if } \beta_1 \in \mathbb{N}_{a_1}, \\ (2p^{\gamma_1} + 1)_*(2p^{a_1-\gamma_1} + 1)_* \text{ if } \gamma \in \mathbb{W}_{a_1} \end{cases}$$

For example, let us consider $\varphi^{-1}\left(2^2 3\right)$. Here $m_0 \in \{3^3, (2.3 + 1)_*\}$, i.e., $m_0 \in \{3^2, 7\}$ and

$$m_1 = \begin{cases} 13, \\ 3^{\beta_1}(2.3^{2-\beta_1}+1)_*, \beta_1 \in \mathbb{N}_1, \\ (2.3^{\gamma_1}+1)_*(2.3^{1-\gamma_1}+1)_*, \gamma_1 \in \mathbb{W}_1. \end{cases}$$

Now for $\beta_1 = 1$, we get $2.3 + 1 = 7$ which is admissible. But for $\beta_1 = 2$, we get $2.3^0 + 1$, i.e., 3 which is unacceptable because it is not distinct from 3^{β_1}.

On the other hand, if $q_1 = (2.3^{\gamma_1}+1)_*, q_2 = (2.3^{1-\gamma_1}+1)_*$, keeping in mind that $q_1 < q_2$ are distinct odd primes other than 3, we get

$$\gamma_1 = 0 \Rightarrow (q_1, q_2) = (3, 7)$$
$$\gamma_1 = 1 \Rightarrow (q_1, q_2) = (7, 3)$$

and hence both results are rejected. As a consequence, now we get $m_1 \in \{13, 3.7\}$ and therefore

$$\begin{aligned} \varphi^{-1}\left(2^2 3\right) &= \{2^2 3^2, 2^2 7, 2.13, 2.3.7, 13, 3.7\} \\ &= \{13, 21, 26, 28, 36, 42\} \end{aligned}$$

4.3 When $a = 3$

Following the steps as we did earlier,

$$\varphi^{-1}\left(2^3 p^{a_1}\right) \subseteq \{x \in \mathbb{N} : x = 2^e m_{3-e} : e \in \mathbb{W}_3\}.$$

- For $e = 3$, we have

$$x = \begin{cases} 2^3 3^{a_1+1} \text{ if } p = 3, \\ 2^3(2p^{a_1}+1)_* \end{cases}.$$

- For $e = 2$, we have

$$x = \begin{cases} 2^2 5^{a_1+1} \text{ if } p = 5, \\ 2^2(2^2 p^{a_1}+1)_*, \\ 2^2 3^{\beta_1}(2.3^{a_1+1-\beta_1}+1)_*, \beta_1 \in \mathbb{N}_{a_1}, \text{ if } p = 3, \\ (2p^{\gamma_1} 1)_*(2p^{a_1-\gamma_1}+1)_*, \gamma_1 \in \mathbb{W}_{a_1} \end{cases}.$$

- For $e = 0$ we have $x = 2m_0, m_0$ where

$$m_0 = \begin{cases} (2^3 p^{a_1} + 1)_*, \\ 3^{\beta_1}(2^2 3^{a_1+1-\beta_1} + 1)_*, \beta_1 \in \mathbb{N}_{a_1} \text{ if } p = 3, \\ 5^{\beta_1}(25^{a_1+1-\beta_1} + 1)_*, \beta_1 \in \mathbb{N}_{a_1} \text{ if } p = 5, \\ (2p^{\gamma_1} + 1)_*(2^2 p^{a_1-\gamma_1} + 1)_*, \gamma_1 \in \mathbb{W}_{a_1}, \\ (2^2 p^{\gamma_1} + 1)_*(2p^{a_1-\gamma_1} + 1)_*, \gamma_1 \in \mathbb{W}_{a_1}, \\ 3^{\beta_1}(2.3^{\gamma_1} + 1)_*(2.3^{a_1+1-\beta_1-\gamma_1} + 1)_*, \beta_1 \in \mathbb{N}_{a_1}, \gamma_1 \in \mathbb{W}_{a_1-\beta_1}, \\ \prod\limits_{\substack{i=1 \\ \gamma_1+\gamma_2+\gamma_3=a_1}}^{3} (2p^{\gamma_1} + 1)_* \end{cases}$$

Proceeding in similar manner one can derive the other results corresponding to higher values of $a \in {}_4\mathbb{N}$.

5 Conclusion

Thus, we have answered the problem of construction of the set $\varphi^{-1}\left(2^a p_1^{a_1}\right)$ for different values of $a \in \mathbb{N}$. Techniques as well as different types of combination of prime factors of the elements of the set $\varphi^{-1}\left(2^a p_1^{a_1}\right)$ have been discussed. The interested reader can work on special cases for further improvement. Moreover, special types of primes in place of p can be considered for deriving interesting results.

References

1. Coleman, R., *On the image of Euler's Totient Function.* http://arxiv.org/pdf/0910.2223v1.pdf
2. H. Gupta, Euler's totient function and its inverse. Indian J. Pure Appl. Math. **12**(1), 22–30 (1981)
3. FermatSearch.org, http://www.fermatsearch.org
4. Keller, W., *Prime Factors of Fermat Numbers.* http://www.prothsearch.net/fermat.html
5. Weisstein, E.W., Fermat number. http://mathworld.wolfram.com/FermatNumber.html, *Math-World*
6. Tsang, C., Fermat numbers. M414 Number Theory
7. http://www.prothsearch.net/fermat.html/#Summary

An EOQ Model with Carbon Constraints Without Loss of Generality with Uncertain Cost and Uncertain Carbon Emission Associated with Some Fuzzy Parameters

Anuradha Sahoo and Arati Nath

Abstract Economic Order Quantity (EOQ) models without loss of generality for single items and multi-items are presented to choose an order quantity that minimizes its cost per unit time subject to the constraint on the amount of carbon emitted. Here, the proposed model is discussed in an uncertain environment. So, the EOQ model with a fixed cost, holding cost, and purchased cost (or produced cost) has been considered in a fuzzy environment. Also, fixed emitted carbon, holding emitted carbon, and purchased (or produced) emitted carbon are taken in a fuzzy environment with the limitation of total carbon emission per unit production time. Here we considered a fixed cost, holding cost, and purchased cost (or produced cost) as trapezoidal fuzzy numbers. The computational procedure for the defined EOQ model is carried out by using the signed-distance method and expected value technique. Numerical examples are also given to exemplify the proposed model.

Keywords Carbon-constrained EOQ · Trapezoidal fuzzy number · Signed-distance method · Expected value technique

1 Introduction

In real life, the costs of the items depend upon many factors like item's quality, stock level, selling price, the period of storage, etc. In reality, the cost can't be fixed. Nowadays almost every single real-world problem comprises the cost in an uncertain environment. Quite a few investigators have studied the inventory models by taking different parameters in an uncertain environment in different situations.

By using the GP approach, Kotb et al. [1] proposed a multi-item EOQ model by considering holding cost to be a continuous function of the order quantity under varying holding cost. In this paper, a classical EOQ model and an EOQ model by taking holding costs as constant are derived without any constraint.

A. Sahoo (✉) · A. Nath
Department of Mathematics, ITER, Siksha 'O' Anusandhan University, Bhubaneswar, Odisha 751030, India
e-mail: anuradha25anu@gmail.com

S. R. Mishra et al. (eds.), *Recent Trends in Applied Mathematics*, Lecture Notes in Mechanical Engineering, https://doi.org/10.1007/978-981-15-9817-3_6

To determine the optimal solution, a hybrid discrete particle swarm algorithm is developed and tested by Weianyg et al. [2] based on real data collected from the pharmacological initiative. Their methods focus on local optimization for carbon emission reduction in enterprises and may require additional time, money, and effort for satisfactory implementation.

To maximize the total profit and to determine the optimal order size Hung-Chi Chang [3], proposed a model with a fuzzy defective rate. After that, he presented his proposed model with defective rate and annual demand in a fuzzy environment. So, they use a signed-distance method to estimate the total profit in fuzzy environment per unit production time.

An inventory model without shortage is considered by taking ordering cost and holding cost in the fuzzy environment by Sayed and Aziz [4]. Triangular fuzzy numbers are used to obtain the optimum order quantity. Also signed-distance method used for the defuzzification of the discussed fuzzy model.

Sadegheih [5] proposed an optimization model to minimize the total costs and provides the best solutions under the carbon emission trading program. So, he considered different types of costs like; the capital investment cost in discrete form, the cost of transmission losses, the power generation costs, and carbon emission costs as the cost function in his model.

Dutta and Kumar [6] proposed an inventory model in a fuzzy environment where shortages are not allowed and determine the optimal total cost and the optimal order quantity. They use trapezoidal fuzzy numbers. The computation of economic order quantity is carried out through the defuzzification process by using the signed-distance method.

A multi-objective optimization technique based on the bacterial colony chemotaxis (MOBCC) algorithm proposed by Lu and Pingli [7] to study the constrained emission or economic dispatch problem involving competing objectives in electric power systems with carbon capture system (CCS) technology.

Based on the investigation of existing tools and sustainability demands in building, a new computer calculation system has been established by Fu et al. [8] to compute the carbon emission for optimizing sustainability during the construction. They also described the system structure and detailed functions. Lastly, a case study is analyzed to establish the designed LCA framework and system functions.

Zhang et al. [9] projected a model to minimize the scheduling cost and greenhouse gas emission cost which includes both thermal generators and wind farms.

Sahoo and Dash [10] consider purchasing cost as a fuzzy number and demand as a random variable in a fuzzy environment to formulate a single-period inventory model for multi-item newsboy problem where there is the occurrence of randomness and fuzziness. Buckley's minimization concept is used to obtain the expected profit and optimum order quantity.

To design a multi-product closed-loop green supply chain network, Talaei et al. [11] have introduced a mixed-integer linear programming model which is taking into account and incapable of minimizing total costs. So, they developed a model to consider such environmental concurs to reduce the amount of CO_2 emission in the environment throughout the network in question.

Rajalakshmi and Michael Rosario [12] examine an inventory model in a fuzzy environment with allowable shortage which is fully backlogged. To estimate the fuzzy total cost, they fuzzify the ordering cost, holding cost, and backorder cost by using triangular, trapezoidal, pentagonal fuzzy numbers. After that, they use a signed-distance method for the defuzzification of their proposed fuzzy model.

A multi-item two-warehouse deterministic inventory model described by Garai et al. [13] for deteriorating items with stock-dependent to define possibility, necessity, and credibility measures of an exponential fuzzy number, and its expected value. So, they considered different cost values and other parameters in a fuzzy environment. Solution methodology by using the expected value technique of their proposed model also has been discussed.

Sutherland et al. [14] present a methodology to optimize the tool path for high efficiency, low energy consumption, and carbon footprint in the milling process. First, they introduced the description and influencing factors of the tool path. After that, they proposed a multi-objective tool path optimization model with maximum machining efficiency, minimum energy consumption, and carbon emission.

Taking into consideration the effect of the price of carbon and ecological awareness of consumers, Li et al. [15] explore constructor and seller's decision under the centralized and decentralized conclusion. So, they analyze the condition of whether a constructer transfers the emission reduction task to the seller. Also, they use computational researches to analyze compassion by changing the cost function.

In this paper, we consider the carbon-constrained EOQ model in an uncertain environment under the limitation of total carbon emission per unit production time. We use the signed-distance method and expected value technique to obtain the optimal solutions of our proposed models. In both methods, first, we defuzzify the fuzzy inventory model to get the crisp inventory model and then obtain the optimal solution of that crisp model using LINGO software. So, we considered a fixed cost, holding cost, and purchased (or produced) cost as trapezoidal fuzzy numbers in objective function as well as in the carbon constraint to achieve our goal.

2 Mathematical Model

In this paper, the carbon-constrained economic order quantity (EOQ) model is developed in an uncertain environment with uncertain costs under the limitation of the total carbon emissions per unit production time. Here, the proposed model is discussed in two cases by describing the model in an uncertain environment. In case-1, an EOQ model with fixed cost, holding cost, purchased cost (or produced cost), fixed carbon emission, holding carbon emission, and purchased (or produced) carbon emission for the proposed model are taken as trapezoidal fuzzy numbers. The computational procedure for the defined EOQ model is carried out by using the signed-distance method. In case 2, an EOQ model with fixed cost, holding cost, purchased cost (or produced cost), fixed carbon emission, holding carbon emission purchased (or produced) carbon emission, and the limitation of total carbon emission in a unit production time for the proposed model are taken as trapezoidal fuzzy numbers. Here, we use the expected value technique for the defuzzification of the proposed model. So, we have considered different assumptions and different notations to construct the carbon-constrained EOQ model in an uncertain environment to minimize the total annual cost under the limitation of total carbon emission per unit production time.

2.1 General Economic Order Quantity Models

(i) **Single-item economic order quantity (SEOQ) model**

Assumption:

- Without loss of generality.
- With zero lead time.
- Positive lead time can be included and does not affect the solution to the problem.
- The firm must satisfy all the demands.
- The analysis can be easily extended to the settings with backorders.

Notation:

$d =$ Demand per unit time,
$q =$ Number of order quantity per unit time,
$a_1 =$ Fixed cost per order,
$h_1 =$ Holding cost per unit item per unit time,
$p_1 =$ Purchased cost or produced cost per unit time,
$a_2 =$ Fixed amount of emitted carbon associated per order,
$h_2 =$ Holding emitted carbon per unit item per unit time,
$p_2 =$ Purchased emitted carbon or produced emitted carbon per unit time,
$l =$ Limitation of the total carbon emission per unit time,
$c =$ Total cost per unit time.

Our main objective is to minimize the total annual cost. Here, the sum of fixed cost, holding cost (or carrying cost), and purchase cost (or produced cost) is recognized as the annual cost.

i.e. Total annual cost $= \frac{a_1 d}{q} + \frac{h_1 q}{2} + p_1 d$.

The limitation of the total carbon emission for the proposed inventory model is given as follows.

Carbon emission constraint: $\frac{a_2 d}{q} + \frac{h_2 q}{2} + p_2 d \leq l$.

Hence, an inventory model with a fixed cost, holding costs, and purchased cost (or produced cost) corresponding to the carbon emission constraints are given as follows.

$$\text{SEOQ}: \quad \min c = \frac{a_1 d}{q} + \frac{h_1 q}{2} + p_1 d$$

$$\text{S.t}\,\frac{a_2 d}{q}+\frac{h_2 q}{2}+p_2 d\ \le l$$
$$q\ge 0$$

(ii) **Multi-item economic order quantity (MEOQ) model**

Assumption:

- Without loss of generality.
- With zero lead time.
- Positive lead time can be included and does not affect the solution to the problem.
- The firm must satisfy all the demands.
- The analysis can be easily extended to the settings with backorders.

Notation:

d_i = Demand per unit time
q_i = Number of order quantity per unit time
a_{1i} = Fixed cost per order
h_{1i} = Holding cost per unit item per unit time
p_{1i} = Purchased cost or produced cost per unit time
a_{2i} = Fixed amount of emitted carbon associated per order
h_{2i} = Holding emitted carbon per unit item per unit time
p_{2i} = Purchased emitted carbon or produced emitted carbon per unit time
l = Limitation of the total carbon emission per unit time
c = Total cost per unit time

Our main objective is to minimize the total annual cost. Here, the sum of fixed cost, holding cost (or carrying cost), and purchase cost (or produced cost) is recognized as the annual cost.

i.e. Total annual cost $=\sum_{i=1}^{n}\frac{a_{1i}d}{q_i}+\frac{h_{1i}q}{2}+p_{1i}d_i$

The limitation of the total carbon emission for the proposed inventory model is given as follows.

Carbon emission constraint: $\sum_{i=1}^{n}\frac{a_{2i}d_i}{q_i}+\frac{h_{2i}q_i}{2}+p_{2i}d_i\le l$

Hence, an inventory model with fixed cost, holding costs, and purchased cost (or produced cost) corresponding to the carbon emission constraints is given as follows.

$$\text{MEOQ}:\ \ \min c=\sum_{i=1}^{n}\frac{a_{1i}d}{q_i}+\frac{h_{1i}q}{2}+p_{1i}d_i$$
$$\text{S.t}\sum_{i=1}^{n}\frac{a_{2i}d_i}{q_i}+\frac{h_{2i}q_i}{2}+p_{2i}d_i\le l$$
$$q_i\ge 0;\ i=1,\ 2,\ 3,\ \ldots,\ n$$

In real-life situations, uncertainty arises. So, we considered parameters as the trapezoidal fuzzy number to describe an inventory EOQ model for single items and multi items. In this paper, we described the above model in two cases as follows.

2.2 *Fuzzy Economic Order Quantity Models*

(i) **Fuzzy single-item economic order quantity (FSEOQ) model**

Notations:

$\tilde{a}_1$ = Fuzzy fixed cost per order
$\tilde{h}_1$ = Fuzzy holding cost per unit item per unit time
$\tilde{p}_1$ = Fuzzy purchased cost or produced cost per unit time
$\tilde{a}_2$ = Fuzzy fixed amount of emitted carbon associated per order
$\tilde{h}_2$ = Fuzzy holding emitted carbon per unit item per unit time
$\tilde{p}_2$ = Fuzzy Purchased emitted carbon or produced emitted carbon per unit time
$\tilde{c}$ = Fuzzy total annual cost
$\tilde{l}$ = Limitations of total carbon emission in a unit production time

When uncertainty arises in real-life situations, the previously discussed SEOQ model becomes FSEOQ. So, the FSEOQ model in an uncertain environment by considering the fixed cost, holding cost, production cost (or purchased cost), fixed emitted carbon, holding emitted carbon, and purchased (or produced) emitted as a trapezoidal fuzzy number can be described as follows.

$$\text{FSEOQ 1}: \quad \min \tilde{c} = \frac{\tilde{a}_1 d}{q} + \frac{\tilde{h}_1 q}{2} + \tilde{p}_1 d$$
$$\text{S.t} \frac{\tilde{a}_2 d}{q} + \frac{\tilde{h}_2 q}{2} + \tilde{p}_2 d \leq l$$
$$q \geq 0$$

Now considering the total carbon emission in a unit production time in a fuzzy environment, the above FSEOQ 1 becomes FSEOQ 2 as follows.

$$\text{FSEOQ 2}: \quad \min \tilde{c} = \frac{\tilde{a}_1 d}{q} + \frac{\tilde{h}_1 q}{2} + \tilde{p}_1 d$$
$$\text{S.t} \frac{\tilde{a}_2 d}{q} + \frac{\tilde{h}_2 q}{2} + \tilde{p}_2 d \leq \tilde{l}$$
$$q \geq 0$$

(ii) **Fuzzy multi-item economic order quantity (FMEOQ) model**

Notations:

$\tilde{a}_{1i}$ = Fuzzy fixed cost per order,
$\tilde{h}_{1i}$ = Fuzzy holding cost per unit item per unit time,
$\tilde{p}_{1i}$ = Fuzzy purchased cost or produced cost per unit time,
$\tilde{a}_{2i}$ = Fuzzy fixed amount of emitted carbon associated per order,
$\tilde{h}_{2i}$ = Fuzzy holding emitted carbon per unit item per unit time,
$\tilde{p}_{2i}$ = Fuzzy purchased emitted carbon or produced emitted carbon per unit time,
$\tilde{c}$ = Fuzzy total annual cost,
$\tilde{l}$ = Limitations of total carbon emission in a unit production time.

When uncertainty arises in real-life situations, the previously discussed MEOQ model becomes FMEOQ. So, the FMEOQ model in an uncertain environment by considering the fixed cost, holding cost, production cost (or purchased cost), fixed emitted carbon, holding emitted carbon, and purchased (or produced) emitted carbon as a trapezoidal fuzzy number can be described as follows.

$$\text{FMEOQ 1}: \quad \min \tilde{c} = \sum_{i=1}^{n} \frac{\tilde{a}_{1i}d}{q_i} + \frac{\tilde{h}_{1i}q}{2} + \tilde{p}_{1i}d_i$$

$$\text{S.t} \sum_{i=1}^{n} \frac{\tilde{a}_{2i}d_i}{q_i} + \frac{\tilde{h}_{2i}q_i}{2} + \tilde{p}_{2i}d_i \le l$$

$$q_i \ge 0; \; i = 1, 2, 3, \ldots, n$$

Now considering the total carbon emission in a unit production time in a fuzzy environment, the above FMEOQ 1 becomes FMEOQ 2 as follows.

$$\text{FMEOQ 2}: \quad \min \tilde{c} = \sum_{i=1}^{n} \frac{\tilde{a}_{1i}d}{q_i} + \frac{\tilde{h}_{1i}q}{2} + \tilde{p}_{1i}d_i$$

$$\text{S.t} \sum_{i=1}^{n} \frac{\tilde{a}_{2i}d_i}{q_i} + \frac{\tilde{h}_{2i}q_i}{2} + \tilde{p}_{2i}d_i \le \tilde{l}$$

$$q_i \ge 0; \; i = 1, 2, 3, \ldots, n$$

3 Methodology

3.1 Trapezoidal Fuzzy Number

The trapezoidal fuzzy number represented by four points as follows:

$$\tilde{B} = (b_1, b_2, b_3, b_4$$

Membership function of trapezoidal fuzzy number is defined by

$$\mu_{\tilde{A}}(x) = \begin{cases} 0 & ; x < b_1 \\ \dfrac{x - b_1}{b_2 - b_1} & ; b_1 \le x \le b_2 \\ 1 & ; b_2 \le x \le b_3 \\ \dfrac{b_4 - x}{b_4 - b_3} & ; b_3 \le x \le b_4 \\ 0 & ; x > b_4 \end{cases}$$

Now, α-cut of this trapezoidal fuzzy number is as follows (Fig. 1):

$$\tilde{B}_\alpha = [b_1 + (b_2 - b_1)\alpha, b_4 - (b_4 - b_3)\alpha]$$

Method 1:

Signed-distance Method:

Let, $\tilde{B} = (b_1, b_2, b_3, b_4)$ be a trapezoidal fuzzy number. Then the signed distance of $\tilde{B}$ is defined as follows.

$$d(\tilde{B}, 0) = \frac{1}{2}\int_0^1 [B_L(\alpha) + B_R(\alpha)]d\alpha$$

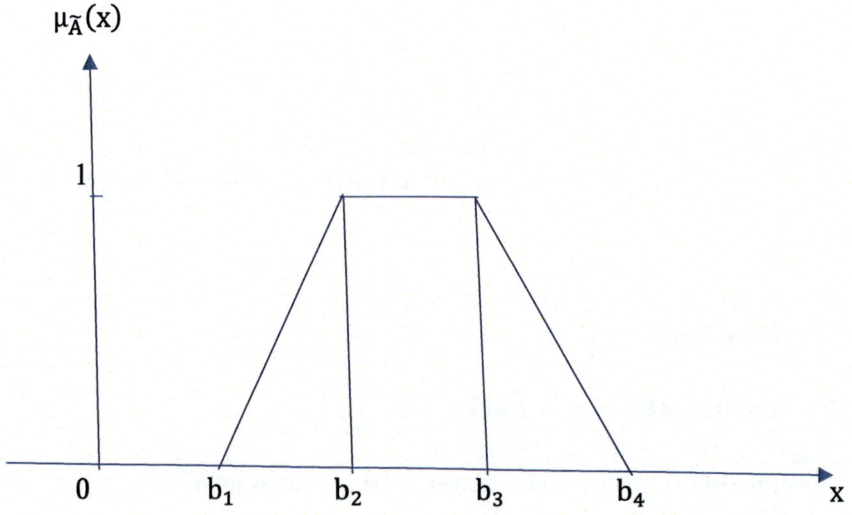

Fig. 1 Trapezoidal Fuzzy Number

$$\text{where, } B_L(\alpha) = n_1 + (n_2 - n_1)\alpha$$
$$B_R(\alpha) = n_4 - (n_4 - n_3)\alpha, \alpha \in [0, 1]$$

To defuzzify the fuzzy models FSEOQ 1 and FMEOQ 1, here, we have used the signed-distance method. Here, we considered fixed cost, holding cost, purchased cost (or produced cost), fixed emitted carbon, holding emitted carbon, and purchased (or produced) emitted carbon as trapezoidal fuzzy numbers to minimize the total annual cost under the limitation of total carbon emission per unit production time. By using the signed-distance method, our fuzzy models FSEOQ 1 and FMEOQ 1 become crisp models. Then the crisp models can be solved by using an optimization technique. But here, we have used LINGO software to obtain the optimal solution.

The model FSEOQ 1 can also be written as

$$\begin{aligned} \text{FSEOQ}' \ 1: \quad & \min \tilde{c} = (m_1, m_2, m_3, m_4) \\ & \text{S.t.} \ (n_1, n_2, n_3, n_4) \le l \\ & q \ge 0 \end{aligned}$$

The model FMEOQ 1 can also be written as

$$\begin{aligned} \text{FMEOQ}' \ 1: \quad & \min \tilde{c} = (m_1, m_2, m_3, m_4) \\ & \text{S.t} \ (n_1, n_2, n_3, n_4) \le l \\ & q_i \ge 0; \ i = 1, 2, 3, \ldots, n \end{aligned}$$

Let, $\tilde{c} = (m_1, m_2, m_3, m_4)$ be a trapezoidal fuzzy number. Then the signed distance of c̃ is defined as follows:

$$d(\tilde{c}(\tilde{m}), 0) = \frac{1}{2} \int_0^1 [B_L(\alpha) + B_R(\alpha)] d\alpha$$

$$\text{where, } B_L(\alpha) = m_1 + (m_2 - m_1)\alpha$$
$$B_R(\alpha) = m_4 - (m_4 - m_3)\alpha, \alpha \in [0, 1]$$

Let, $\tilde{n} = (n_1, n_2, n_3, n_4)$ be a trapezoidal fuzzy number. Then the signed distance of $\tilde{n}$ is defined as follows:

$$d(\tilde{n}, 0) = \frac{1}{2} \int_0^1 [B_L(\alpha) + B_R(\alpha)] \, d\alpha$$

$$\text{where, } B_L(\alpha) = n_1 + (n_2 - n_1)\alpha$$
$$B_R(\alpha) = n_4 - (n_4 - n_3)\alpha, \alpha \in [0, 1]$$

Method 2:

Expected value techniques:

Let, $\tilde{X} = (x_1, x_2, x_3, x_4)$ be a trapezoidal fuzzy number, then the expected value $\tilde{X}$ is defined as follows:

$$E(\tilde{X}) = \frac{(x_1 + x_2 + x_2 + x_3)}{4}$$

To defuzzify the fuzzy models FSEOQ 2 and FMEOQ 2, here, we have used the expected value technique. Here, we considered fixed cost, holding cost, purchased cost (or produced cost), fixed emitted carbon, holding emitted carbon, purchased (or produced) emitted carbon, and total carbon emission in a unit production time as trapezoidal fuzzy numbers to minimize the total annual cost under the limitation of total carbon emission per unit production time. By using the expected value technique, our fuzzy models FSEOQ 2 and FMEOQ 2 become crisp models as CSEOQ 2 and CMSEOQ 2, respectively. Then the crisp models can be solved by using an optimization technique. But here we have used LINGO software to obtain the optimal solution.

The fuzzy model FSEOQ 2 converted to crisp model CSEOQ 2 as follows.

$$\text{CSEOQ 2}: \quad \min \tilde{c} = E\left(\frac{\tilde{a}_1 d}{q} + \frac{\tilde{h}_1 q}{2} + \tilde{p}_1 d\right)$$

$$\text{S.t}\, E\left(\frac{\tilde{a}_2 d}{q} + \frac{\tilde{h}_2 q}{2} + \tilde{p}_2 d\right) \le E\left(\tilde{l}\right)$$

$$q \ge 0$$

The fuzzy model FMEOQ 2 converted to crisp model CMEOQ 2 as follows.

$$\text{CMEOQ 2}: \quad \min \tilde{c} = \sum_{i=1}^{n} E\left(\frac{\tilde{a}_{1i} d}{q_i} + \frac{\tilde{h}_{1i} q}{2} + \tilde{p}_{1i} d_i\right)$$

$$\text{S.t} \sum_{i=1}^{n} E\left(\frac{\tilde{a}_{2i} d_i}{q_i} + \frac{\tilde{h}_{2i} q_i}{2} + \tilde{p}_{2i} d_i\right) \le E\left(\tilde{l}\right)$$

$$q_i \ge 0; \; i = 1, 2, 3, \ldots, n$$

4 Numerical Example

A company uses 600 units of raw materials per unit production time. Holding each order costs $ 2. Fixed costs and purchased costs are $ 120 and $ 5 per item per year of the inventory, respectively. Holding emitted carbon, fixed emitted carbon, and purchased emitted carbon of the inventory are 3, 2, and 1, respectively. The limitation of total carbon emission in a unit production time is 50,000. Find the optimum total annual cost and order quantity for a single item and a multi-item inventory models.

Solution:

Given, $d = 600, a_1 = 0.120, h_1 = 2, p_1 = 5, a_2 = 2, h_2 = 3, p_2 = 1$ and $l = 50{,}000$
Taking, $\alpha = 0$

4.1 *General Economic Order Quantity Models*

(i) **Single-item economic order quantity (SEOQ) model**

By putting the given data, in SEOQ model under carbon constraint, we have,

$$\text{SEOQ}: \quad \min c = \frac{72,000}{q} + \frac{2q}{2} + 3000$$
$$\text{S.t.} \ \frac{1200}{q} + \frac{3q}{2} + 600 \leq 50,000$$
$$q \geq 0$$

By using LINGO software, the optimal solution of the model is as follows:
$c = 3536.656, q = 268.3282$

(ii) **Multi-item economic order quantity (MEOQ) model**

Consider,

$d_1 =$ Rs. 600, $d_2 =$ Rs. 700, $d_3 =$ Rs. 800,
$a_{11} =$ Rs. 120, $a_{12} =$ Rs. 130, $a_{13} =$ Rs. 140,
$h_{11} =$ Rs. 2, $h_{12} =$ Rs. 3, $h_{13} =$ Rs. 4,
$p_{11} =$ Rs. 5, $p_{12} =$ Rs. 6, $p_{13} =$ Rs. 7,
$a_{21} = 2, a_{22} = 3, a_{23} = 4,$
$h_{21} = 3, h_{22} = 4, h_{23} = 5,$
$p_{21} = 1, p_{22} = 2, p_{23} = 3$

By putting the given data, in MEOQ model under carbon constraint, we have,

$$\begin{aligned}\text{MEOQ}: \quad \min c = & \frac{72,000}{q_1} + \frac{2q_1}{2} + 3000 + \frac{91,000}{q_2} + \frac{3q_2}{2} \\ & + 4200 + \frac{112,000}{q_3} + \frac{4q_3}{2} + 5600\end{aligned}$$
$$\text{S.t.} \quad \frac{1200}{q_1} + \frac{3q_1}{2} + 600 + \frac{2100}{q_2} + \frac{4q_2}{2}$$

$$+ 1400 + \frac{3200}{q_3} + \frac{5q_3}{2} + 2400 \leq 50,000$$
$$q_i \geq 0;\ i = 1, 2, 3$$

By using LINGO software, the optimal solution of the model is as follows: $c = 29082.15$, $q_1 = 268.3282$, $q_2 = 246.2060$, $q_3 = 236.6432$.

4.2 Fuzzy Economic Order Quantity Models

Method 1 (Signed-distance Method)

(i) **Fuzzy single-item economic order quantity (FSEOQ) model**

Consider,

$$\tilde{a}_1 = (120, 120.2, 120.4, 120.6),\ \tilde{h}_1 = (2, 2.2, 2, 4, 2.6),\ \tilde{p}_1 = (5, 5.2, 5.4, 5.6),$$
$$\tilde{a}_2 = (2, 2.2, 2.4, 2.6),\ \tilde{h}_2 = (3, 3.2, 3.4, 3.6),\ \tilde{p}_2 = (1, 1.2, 1.4, 1.6)$$

By putting the given data, in FSEOQ 1 model under carbon constraint, we have,

$$\begin{aligned} \text{FSEOQ 1}: \quad \min \tilde{c} = & \frac{(120, 120.2, 120.4, 120.6)d}{q} \\ & + \frac{(2, 2.2, 2, 4, 2.6)q}{2} + (5, 5.2, 5.4, 5.6)d \\ \text{S.t.} & \frac{(2, 2.2, 2.4, 2.6)d}{q} + \frac{(3, 3.2, 3.4, 3.6)q}{2} \\ & + (1, 1.2, 1.4, 1.6)d \leq 50,000 \\ & q \geq 0 \end{aligned}$$

The above model by using a signed-distance method can also be written as follows:

$$\begin{aligned} \text{FSEOQ}'\ 1: \quad \min c = & (72,000/q + q + 3000, 72,120/q + 1.1q + 3120, \\ & 72,240/q + 1.2q + 3240,\ 72,360/q + 1.4q + 3360) \\ \text{S.t.} \quad & (2/q + 1.5q + 600, 2.2/q + 1.6q + 720, \\ & 2.4/q + 1.7q_1 + 840, 2.6/q + 1.8q + 960) \leq 50,000 \\ & q \geq 0 \end{aligned}$$

Now, by using the defuzzification technique, the above model is converted to the model CSEOQ 1 as follows:

$$\text{CSEOQ 1}: \quad \min c = \frac{144,360}{q} + 2.4q + 6360$$
$$\text{S.t.} \frac{4.6}{q} + 3.3q + 1560 \le 50,000$$
$$q \ge 0$$

By using LINGO software, the optimal solution of the model is as follows: $c = 7537.224$, $q = 245.2550$.

(ii) **Fuzzy multi-item economic order quantity (FMEOQ) model**

Consider,

$$\begin{aligned}
\tilde{a}_{11} &= (120, 120.2, 120.4, 120.6), \tilde{a}_{12} = (130, 130.2, 130.4, 130.6),\\
\tilde{a}_{13} &= (140, 140.2, 140.4, 140.6),\\
\tilde{h}_{11} &= (2, 2.2, 2.4, 2.6), \tilde{h}_{12} = (3, 3.2, 3.4, 3.6), \tilde{h}_{13} = (4, 4.2, 4.4, 4.6),\\
\tilde{p}_{11} &= (5, 5.2, 5.4, 5.6), \tilde{p}_{12} = (6, 6.2, 6.4, 6.6), \tilde{p}_{13} = (7, 7.2, 7.4, 7.6),\\
\tilde{a}_{21} &= (2, 2.2, 2.4, 2.6), \tilde{a}_{22} = (3, 3.2, 3.4, 3.6), \tilde{a}_{23} = (4, 4.2, 4.4, 4.6),\\
\tilde{h}_{21} &= (3, 3.2, 3.4, 3.6), \tilde{h}_{22} = (4, 4.2, 4.4, 4.6), \tilde{h}_{23} = (5, 5.2, 5.4, 5.6),\\
\tilde{p}_{21} &= (1, 1.2, 1.4, 1.6), \tilde{p}_{22} = (2, 2.2, 2.4, 2.6), \tilde{p}_{23} = (3, 3.2, 3.4, 3.6)
\end{aligned}$$

By putting the given data, in FMEOQ 1 model under carbon constraint, we have,

$$\begin{aligned}
\text{FMEOQ 1}: \quad \min \tilde{c} = &\frac{(120, 120.2, 120.4, 120.6)d_1}{q_1} + \frac{(2, 2.2, 2.4, 2.6)q_1}{2}\\
&+ (5, 5.2, 5.4, 5.6)d_1 + \frac{(130, 130.2, 130.4, 130.6)d_2}{q_2}\\
&+ \frac{(3, 3.2, 3.4, 3.6)q_2}{2} + (6, 6.2, 6.4, 6.6)d_2\\
&+ \frac{(140, 140.2, 140.4, 140.6)d_3}{q_3} + \frac{(4, 4.2, 4.4, 4.6)q_3}{2}\\
&+ (7, 7.2, 7.4, 7.6)d_3\\
\text{S.t.} \quad &\frac{(2, 2.2, 2.4, 2.6)d_1}{q_1} + \frac{(3, 3.2, 3.4, 3.6)q_1}{2}\\
&+ (1, 1.2, 1.4, 1.6)d_2 + \frac{(3, 3.2, 3.4, 3.6)d_2}{q_2}\\
&+ \frac{(4, 4.2, 4.4, 4.6)q_2}{2} + (2, 2.2, 2.4, 2.6)d_2\\
&+ \frac{(4, 4.2, 4.4, 4.6)d_3}{q_3} + \frac{(5, 5.2, 5.4, 5.6)q_3}{2}\\
&+ (3, 3.2, 3.4, 3.6)d_3 \le 50,000
\end{aligned}$$

$$q_i \geq 0;\ i = 1, 2, 3, \ldots, n$$

The above model by using a signed-distance method can also be written as follows:

$$\begin{aligned}
\text{FMEOQ}'\ 1:\ \min c =& (72,000/q_1 + q_1 + 91,000/q_2 + 1.5q_2 \\
& + 11,200/q_3 + 2q_3 + 12,800, \\
& 72,120/q_1 + 1.1q_1 + 91,140/q_2 \\
& + 1.6q_2 + 11,216/q_3 + 2.1q_3 13,220, \\
& 72,240/q_1 + 1.2q_1 + 91,280/q_2 + 1.7q_2 \\
& + 11,232/q_3 + 2.2q_3 + 13,640, \\
& 72,360/q_1 + 1.4q_1 + 91420/q_2 + 1.8q_2 \\
& + 11,248/q_3 + 2.3q_3 + 14,060) \\
\text{S.t.}\quad & (2/q_1 + 1.5q_1 + 3/q_2 + 2q_2 + 4/q_3 \\
& + 2.5q_3 + 4400, 2.2/q_1 + 1.6q_1 + 3.2/q_2 + 2.1q_2 \\
& + 4.2/q_3 + 2.6q_3 + 4820, 2.4/q_1 + 1.7q_1 + 3.4/q_2 \\
& + 2.2q_2 + 4.4/q_3 + 2.7q_3 + 5240, 2.6/q_1 + 1.8q_1 \\
& + 3.6/q_2 + 2.3q_2 + 4.6/q_3 + 2.8q_3 + 5660) \leq 50,000 \\
& q_i \geq 0; i = 1, 2, 3
\end{aligned}$$

Now, by using the defuzzification technique, the above model converted to the model CMEOQ 1 as follows:

$$\begin{aligned}
\text{CMEOQ 1}:\ \min c =& \frac{144,360}{q_1} + 2.4q_1 + \frac{182,420}{q_2} + 3.3q_2 \\
& + \frac{224,480}{q_3} + 4.3q_3 + 26,860 \\
\text{S.t.}\quad & \frac{4.6}{q_1} + 3.3q_1 + \frac{6.6}{q_2} + 4.3q_2 + \frac{8.6}{q_3} + 5.3q_3 \\
& + 10,060 \leq 50,000 \\
& q_i \geq 0;\ i = 1, 2, 3
\end{aligned}$$

By using LINGO software, the optimal solution of the model is given as follows: $c = 31553.94$, $q_1 = 245.2550$, $q_2 = 235.1144$, $q_3 = 228.4834$

Method 2 (Expected value ETFN technique):

(i) **Fuzzy single-item economic order quantity (FSEOQ) model**

Consider, $\tilde{l} = (50000, 50020,\ 50040,\ 50060)$

By putting the given data, in FSEOQ 2 model under carbon constraint, we have,

$$\text{FSEOQ 2}: \quad \min \tilde{c} = \frac{(120, 120.2, 120.4, 120.6)d}{q} + \frac{(2, 2.2, 2, 4, 2.6)q}{2} + (5, 5.2, 5.4, 5.6)d$$

$$\text{S.t.} \quad \frac{(2, 2.2, 2.4, 2.6)d}{q} + \frac{(3, 3.2, 3.4, 3.6)q}{2} + (1, 1.2, 1.4, 1.6)d \le (50,000, 50,020, 50,040, 50,060)$$

$$q \ge 0$$

Now, by using the defuzzification technique expected value technique, the above model FSEOQ 2 can be converted to the model CSEOQ 2 as follows:

$$\text{CSEOQ 2}: \quad \text{Min c} = \frac{72180}{\text{q}} + 1.15\text{q} + 3180$$

$$\text{S.t.} \frac{1380}{\text{q}} + 1.65\text{q} + 780 \le 50030$$

$$\text{q} \ge 0$$

By using LINGO software, the optimal solution of the model is as follows:
$c = 3756.219$, $q = 250.5299$.

(ii) **Fuzzy multi-item economic order quantity (FMEOQ) model**

Consider, $\tilde{l} = (50,000, 50,020, 50,040, 50,060)$

By putting the given data, in FMEOQ 2 model under carbon constraint, we have,

$$\text{FMEOQ 2}: \quad \min \tilde{c} = \frac{(120, 120.2, 120.4, 120.6)d_1}{q_1} + \frac{(2, 2.2, 2.4, 2.6)q_1}{2} + (5, 5.2, 5.4, 5.6)d_1 + \frac{(130, 130.2, 130.4, 130.6)d_2}{q_2} + \frac{(3, 3.2, 3.4, 3.6)q_2}{2} + (6, 6.2, 6.4, 6.6)d_2 + \frac{(140, 140.2, 140.4, 140.6)d_3}{q_3} + \frac{(4, 4.2, 4.4, 4.6)q_3}{2} + (7, 7.2, 7.4, 7.6)d_3$$

$$\text{S.t.} \quad \frac{(2, 2.2, 2.4, 2.6)d_1}{q_1} + \frac{(3, 3.2, 3.4, 3.6)q_1}{2} + (1, 1.2, 1.4, 1.6)d_2 + \frac{(3, 3.2, 3.4, 3.6)d_2}{q_2} + \frac{(4, 4.2, 4.4, 4.6)q_2}{2} + (2, 2.2, 2.4, 2.6)d_2 + \frac{(4, 4.2, 4.4, 4.6)d_3}{q_3} + \frac{(5, 5.2, 5.4, 5.6)q_3}{2}$$

$$+ (3, 3.2, 3.4, 3.6)d_3 \leq (50,000, 50,020, 50,040, 50,060)$$
$$q_i \geq 0;\ i = 1, 2, 3, \ldots, n$$

Now, by using the defuzzification technique expected value technique, the above model FMEOQ 2 can be converted to the model CMEOQ 2 as follows:

$$\text{CMEOQ 2}: \quad \min c = \frac{72,180}{q_1} + 1.15q_1 + \frac{91,210}{q_2} + 1.65q_2 + \frac{112,240}{q_3} + 2.15q_3 + 13,430$$
$$\text{S.t.} \quad \frac{1380}{q_1} + 1.65q_1 + \frac{2310}{q_2} + 2.15q_2 + \frac{3440}{q_3} + 2.65q_3 + 5030 \leq 50,030$$
$$q_i \geq 0;\ i = 1, 2, 3$$

By using LINGO software, the optimal solution of the model is given as follows: $c = 15764.57$, $q_1 = 250.5299$, $q_2 = 235.1144$, $q_3 = 228.4834$.

5 Results and Discussion

No. of items	Deterministic model		Fuzzy model			
			Signed-distance method		Expected value ETFN technique	
	Cost	Order quantity	Cost	Order quantity	Cost	Order quantity
Single item (for one = item)	$c =$ 3536.656	$q =$ 268.3282	$c =$ 7537.224	$q =$ 245.2550	$c =$ 3756.219	$q =$ 250.5299
Multi items (for three = items)	$c =$ 29082.15	$q_1 =$ 268.3282 $q_2 =$ 246.2060 $q_3 =$ 236.6432	$c =$ 31553.94	$q_1 =$ 245.2550 $q_2 =$ 235.1144 $q_3 =$ 228.4834	$c =$ 15764.57	$q_1 =$ 250.5299 $q_2 =$ 235.1144 $q_3 =$ 228.4834

6 Conclusion

In this paper, we discussed the EOQ models for single item and multi-items in an uncertain environment. By using the concept of the signed-distance method and expected value technique, we can obtain the corresponding crisp models to our proposed fuzzy models. The corresponding deterministic model can be solved by using an optimization technique. But here, we have used the LINGO software to obtain the optimal solutions. The defuzzification technique can also be applied to other fuzzy models.

References

1. M.O. Aboul-El-Ata, K.A.M. Kotb, Multi-item EOQ inventory model with varying holding cost under two restrictions: a geometric approach. Prod. Plann. Control **8**(6), 608–611 (1997)
2. Q. Weianyg et al., Optimization of carbon emission considering production planning at the enterprise level. J. Clean. Prod. **162**, 635–645 (2002)
3. H.-C. Chang, An application of fuzzy sets theory to the EOQ model with imperfect quality Items. Comput. Oper. Res. **31**(12), 2079–2092 (2004)
4. J.K. Sayed, L.A. Aziz, Fuzzy Inventory Model without shortage using signed distance method. Appl. Math. Inform. Sci. **1**(2) (2007)
5. A. Sadegheih, Optimal design methodologies under the carbon emission trading program using MIP, GA, SA, and TS. Renew. Sustain. Energy Levels **15**(1), 504–513 (2011)
6. D. Dutta, P. Kumar, Fuzzy inventory model without shortage using trapezoidal fuzzy number with sensitivity analysis. IOSR J. Math. **4**(3), 32–37 (2012). ISSN: 2278-5728
7. Z. Lu, T.X. Pingli, Low-carbon emission/economic power dispatch using the multi-objective bacterial colony chemotaxis optimization algorithm considering carbon capture power plant. Int. J. Electr. Power Energy Syst. **53**, 106–112 (2013)
8. F. Fu, H. Luo, et al., Development of a carbon emission calculations system for optimizing building plan based on the LCA framework. Optim. Indus. Syst. (2014)
9. Y. Zhang, H. Trinh et al., Wind–thermal systems operation optimization considering emission problem. Int. J. Electr. Power Energy Syst. **65**, 235–245 (2015)
10. A. Sahoo, J.K. Dash, optimal solution for a single period inventory model with fuzzy cost and demand as a fuzzy random variable. J. Intell. Fuzzy Syst. **28**, 1195–1203 (2015)
11. M. Talaei, B.F. Moghaddam et al., A robust fuzzy optimization model for carbon-efficient closed-loop supply chain network design problem: a numerical illustration in electronics industry. J. Clean. Prod. **113**, 662–673 (2016)
12. R.M. Rajalakshmi, G. Michael Rosario, A fuzzy inventory model with allowable shortage using different fuzzy number. Int. J. Comput. Appl. Math. **12**, 1 (2017). ISSN 1819-4966
13. T. Garai, D. Chakraborty, et al., Expected value of exponential fuzzy number and its application to multi-item deterministic inventory model for deteriorating items. J. Uncertainty Anal. Appl. **5**(8) (2017)
14. J.W. Sutherland et al., Multi-objective optimization of tool path considering efficiency, energy-saving and carbon-emission for free-form surface milling. J. Clean. Prod. **172**, 3311–3322 (2018)
15. G. Li, Y. Xiong et al., Carbon emissions reduction and transfer in supply chains under A cap-and-trade system with emissions-sensitive demand. Syst. Sci. Control Eng. **6**(2), 37–44 (2018)

Ion Acoustic Solitary Wave Propagation in Collisional Magnetized Nonthermal Plasma

B. Boro, A. N. Dev, B. K. Saikia, and N. C. Adhikary

Abstract The ion acoustic (IA) solitary wave (SW) propagation in collisional plasma is presented, in presence of magnetic field. The considered plasma is consisting with mobile positive and negative ions and nonthermal electrons. Using the reductive perturbation technique, the basic set of fluid equations are reduced to a three-dimensional damped Zakharov–Kuznetsov (DZK) nonlinear wave equation. The dissipation generated by ion-neutral collision is taken into the consideration. It is observed that the solitary wave amplitude diminishes with time as the ion-neutral collision frequency increases. Also, the characteristic features of rarefactive solitary wave amplitudes are observed for the parameters like negative ion concentration ratio (μ_n) and nonthermal electrons (α_e). This analysis is suitable for understanding the astrophysical plasma environments.

1 Introduction

The nonlinear ion acoustic wave (IAW) propagation in plasma has received much attention of researchers due to its existence in space and laboratory plasmas [1]. Also, the investigation on this area has reached a greater height. The existence of IAW mode is observed in plasma which is comprised of mainly cold (mobile) ions and hot electrons. In such plasma, the wave inertia is provided by ions, and restoring force is generated by hot electrons. The phase velocity of IAW is usually much larger than ion thermal speed but much smaller than electron thermal speed. Meanwhile,

B. Boro · B. K. Saikia
Centre of Plasma Physics-Institute for Plasma Research (CPP-IPR), Nazirakhat, Tepesia, Sonapur, Guwahati 782402, Assam, India

A. N. Dev (✉)
Center for Applied Mathematics and Computing, Siksha 'O' Anusandhan (Deemed to be University), Khandagiri, Bhubneswar 751030, Odisha, India
e-mail: apulnarayan@gmail.com

N. C. Adhikary
Physical Sciences Division, Institute of Advanced Study in Science and Technology, Vigyan Path, Paschim Boragaon, Garchuk, Guwahati 751030, Assam, India

S. R. Mishra et al. (eds.), *Recent Trends in Applied Mathematics*, Lecture Notes in Mechanical Engineering, https://doi.org/10.1007/978-981-15-9817-3_7

depending on different regions of plasma environment, the IAW excitation in plasma leads to the generations of many nonlinear wave structures such as soliton, shock, double layers, rouge wave, turbulence, and wave modulations, etc. [2–5], which mainly depend on the initial perturbations of the wave. These distinct wave modes propagation in plasma can be distinguished from their frequency range [6]. Considering the case of collisionless plasma, initially, Washimi and Tanuiti [7] studied the IAW mode propagation, by solving the one-dimensional Korteweg–de Vries (KdV) nonlinear wave equation. They revealed that the suitable condition for the generation of solitary wave is when there is a balance between the wave steepening due to weak nonlinearity and dispersion effect of the wave. In the year 1973, Ikezi [8] came up with the experimental poof of the existing IAW mode. Later, in the year 1992, Shukla and Silin [9] studied the dust ion acoustic wave (DIAW) mode propagation in a dusty plasma medium. They reported that, in such plasma condition, the number density of electrons is much smaller than that of ions by following the charge neutrality condition. Where in usual IAW, the number density of electron is almost equal to that of ions present in plasma. Meanwhile, in collisionless plasma, many research works have been carried out to study the IAW propagation in different plasma environments [10–12]. However, in real-life situation, the collisional effect generated by colliding between various plasma constituents such as ion-neutral, ion–dust, electron-neutral, dust-neutral is very obvious. In such condition, while propagating through a collisional plasma medium, the wave amplitude, width, and speed of the solitary wave changes due to the dissipation effect generated by collisions and the wave amplitude can diminish with time [13]. Therefore, the study of dissipative IA solitary wave propagation in collisional plasma has received great attention from the researchers due to its real physical situation [13–16].

Moreover, from the study of space and astrophysical plasma, the observed evidences reveal that particles are usually in high energy state and get accelerated easily. In such environments, the nonequilibrium state of the system can be achieved and the energetic particle distribution can be represented by kappa, q-non-extensive, and Cairns distribution function, which describes the nonthermal state [16]. Therefore, in the present work, our emphasis is to study the nonlinear IAW propagation in collisional magnetized plasma by considering pair ion of hydrogen $\left(H^{+}, H^{-}\right)$. In formulating the mathematical model of the plasma system. Reductive perturbation technique is adopted to solve the damped Zakharov-Kuznetsov (DZK) nonlinear wave equation. ZK equation is useful in studying the IAW propagation in magnetized plasma.

2 Fluid Model

The collisional magnetized plasma consisting of mobile positive and negative ions, nonthermal electrons are considered to study the nonlinear IAW propagation in plasma. It is assumed that the static magnetic field is along z-direction, i.e., $B = B_0\hat{z}$. The three-dimensional DZK nonlinear wave equation is derived by solving the basic fluid equations. The normalized set of equations representing the dynamics of mobile positive and negative ions are given by

Continuity equation for pair ions (i.e., N_+ and N_-)

$$\frac{\partial N_\pm}{\partial T} + \frac{\partial}{\partial X}(N_\pm V_{x\pm}) + \frac{\partial}{\partial Y}(N_\pm V_{y\pm}) + \frac{\partial}{\partial Z}(N_\pm V_{z\pm}) = 0 \tag{1}$$

X-, Y-, Z-directions of momentum Eq. for pair ion:

$$\frac{\partial(V_{x\pm})}{\partial T} + V_{x\pm}\frac{\partial}{\partial X} + V_{y\pm}\frac{\partial}{\partial Y} + V_{z\pm}\frac{\partial}{\partial Z}(V_{x\pm}) = \mp\frac{\partial\varphi}{\partial X} + \left(\overrightarrow{V_{x\pm}} \times \Omega_{B+}\right) \tag{2}$$

$$\frac{\partial(V_{y\pm})}{\partial T} + V_{x\pm}\frac{\partial}{\partial X} + V_{y\pm}\frac{\partial}{\partial Y} + V_{z\pm}\frac{\partial}{\partial Z}(V_{y\pm}) = \mp\frac{\partial\varphi}{\partial Y} + \left(\overrightarrow{V_{y\pm}} \times \Omega_{B+}\right) \tag{3}$$

$$\frac{\partial(V_{z\pm})}{\partial T} + V_{x\pm}\frac{\partial}{\partial X} + V_{y\pm}\frac{\partial}{\partial Y} + V_{z\pm}\frac{\partial}{\partial Z}(V_{z\pm}) = \mp\frac{\partial\varphi}{\partial Z} + \left(\overrightarrow{V_{z\pm}} \times \Omega_{B+}\right) - \nu_\pm V_{z\pm} \tag{4}$$

The number density of nonthermal electron N_e is,

$$N_e = \left\{1 - \alpha_e\varphi + \alpha_e\varphi^2\right\}\exp(\varphi) \tag{5}$$

where $\alpha_e = \frac{4\beta_e}{(1+3\beta_e)}$

Poisson's equation:

$$\nabla^2\varphi = \left[\mu_n N_- - N_+ + \mu_e\{1 - \alpha_e\varphi + \alpha_e\varphi\}\exp(\varphi)\right] \tag{6}$$

where, $N_{\pm,e}$ denotes the number density of positive ion, negative ion, and electrons normalized by the equilibrium value $n_{(\pm,e)}$, $V_\pm$ is the fluid velocity of positive and negative ions having x, y, z components and is normalized by $c_s\left(=\frac{\kappa_b T_e}{m_+}\right)^{\frac{1}{2}}$, time-variable T is normalized by plasma frequency $\omega_{pi}\left(=\frac{4\pi n_{+0}e^2}{m_+}\right)^{1/2}$, the space variable X is normalized by $\frac{c_s}{\omega_{pi}}$, where $\omega_{pi} = \frac{c_s}{\lambda_d}$ giving Debye length $\lambda_D = \sqrt{\frac{\kappa_b T_e}{4\pi n_{+0}e^2}}$, φ is the wave potential normalized by $\frac{\kappa_b T_e \psi}{e}$, $\Omega_{B+} = \frac{Z_+ e B_0}{m_+\omega_{pi}}$ is the normalized ion cyclotron frequency, $\nu_\pm$ is the collisional frequency of positive and negative ions normalized by $\upsilon^\pm = \frac{\upsilon_0^\pm}{\omega_{pi}}$, $\mu_n = z_- n_{0-}/z_+ n_{0+}$, $\mu_e = n_{e0}/z_+ n_{0+}$ are the concentration ratios of negative ions, electrons, and κ_b is the Boltzmann constant. The overall charge neutrality condition at equilibrium is given by: $(1 - \mu_n - \mu_e) = 0$.

For studying the nonlinear IA wave propagation in a dissipative magnetized plasma, we adopted reductive perturbation technique (RPT) to obtain the damped ZK nonlinear wave equation. The two independent, space and time variables are stretched as

$$\xi = \varepsilon^{\frac{1}{2}} X, \eta = \varepsilon^{\frac{1}{2}} Y, \varsigma = \varepsilon^{\frac{1}{2}} (z - M\tau), T = \varepsilon^{\frac{3}{2}} \tau \tag{7}$$

where ε is a small parameter which measures the strength of nonlinearity. It is assumed that ion plasma frequency is greater than that of ion-neutral collision frequency and can be expanded as

$$\nu_{\pm} = \frac{\upsilon_0^{\pm}}{\omega_{pi}} = \nu_0 \varepsilon^{\frac{1}{2}} \tag{8}$$

The expansion of the perturb quantities about the equilibrium point in power of ε is given by

$$N_{\pm} = 1 + \varepsilon N_{\pm 1} + \varepsilon^2 N_{\pm 2} + \cdots \tag{9}$$

$$V_{\pm x,y} = \varepsilon^{\frac{3}{2}} V_{\pm x,y1} + \varepsilon^2 V_{\pm x,y2} + \varepsilon^{\frac{5}{2}} V_{\pm x,y3} + \cdots \tag{10}$$

$$V_{\pm z} = \varepsilon V_{\pm z1} + \varepsilon^2 V_{\pm z2} + \cdots \tag{11}$$

$$\varphi = \varepsilon \varphi^1 + \varepsilon^2 \varphi^2 + \cdots \tag{12}$$

Now, using the stretched co-ordinates and perturbed quantities in the normalized set of Eqs. (1)–(6), we equate the lowest power of ε from continuity and X-, Y-, Z-directions of momentum equations, which are obtained as

$$\left\{ N_+^{(1)} = \frac{\varphi^1}{M^2}, N_-^{(1)} = -\frac{\mu_+ Z_\beta \varphi^1}{M^2}, V_{x+}^{(1)} = \frac{1}{\Omega_B}\frac{\partial \varphi^1}{\partial \eta}, V_{x-}^{(1)} = \frac{1}{\Omega_B}\frac{\partial \varphi^1}{\partial \eta}, V_{y+}^{(1)} = \frac{1}{\Omega_B}\frac{\partial \varphi^1}{\partial \xi}, V_{y-}^{(1)} \right.$$
$$\left. = \frac{1}{\Omega_B}\frac{\partial \varphi^1}{\partial \xi}, V_{z+}^{(1)} = \frac{\varphi^1}{M}, V_{z+}^{(1)} = -\frac{\mu_+ Z_\beta \varphi^1}{M} \right\} \tag{13}$$

From Poissons Eq., we get

$$0 = \left[\mu_n N_-^{(1)} - N_+^{(1)} + \mu_e (1 - \alpha_e) \varphi^1 \right] \tag{14}$$

Using Eq. (13) in Eq. (14), we obtained the phase velocity M of IA waves in the form of

$$M = \pm \left[\frac{(\mu_n \mu_+ Z_\beta + 1)}{\mu_e (1 - \alpha_e)} \varphi^1 \right] \tag{15}$$

where $\mu_+ = \frac{m_+}{m_-}$ is the mass ratio of positive to negative ion, $z_\beta = \frac{z_-}{z_+}$ is the charge state ratio of negative to positive ions.

Again, equating the next higher order of ε from the normalized set of Eqs. (1)–(6), we get

$$\frac{\partial N_{+}^{(2)}}{\partial \varsigma} = \frac{1}{M}\frac{\partial N_{+}^{(1)}}{\partial \tau} + \frac{1}{M}\frac{\partial V_{x+}^{(2)}}{\partial \xi} + \frac{1}{M}\frac{\partial V_{y+}^{(2)}}{\partial \eta} + \frac{1}{M}\frac{\partial V_{z+}^{(2)}}{\partial \varsigma} + \frac{1}{M}\frac{\partial}{\partial \varsigma}\left(N_{+}^{(1)} V_{z+}^{(1)}\right) \tag{16}$$

$$V_{y+}^{(2)} = -\frac{M}{\Omega_B}\frac{\partial V_{x+}^{(1)}}{\partial \varsigma} \tag{17}$$

$$V_{x+}^{(2)} = -\frac{M}{\Omega_B}\frac{\partial V_{y+}^{(1)}}{\partial \varsigma} \tag{18}$$

$$\frac{\partial V_{z+}^{(2)}}{\partial \varsigma} = \frac{1}{M}\frac{\partial V_{z+}^{(1)}}{\partial \tau} + \frac{1}{M} V_{z+}^{(1)}\frac{\partial V_{z+}^{(1)}}{\partial \varsigma} + \frac{1}{M}\frac{\partial \varphi^{2}}{\partial \varsigma} + \frac{1}{M}\nu_0 V_{z+}^{(1)} \tag{19}$$

$$\frac{\partial N_{-}^{(2)}}{\partial \varsigma} = \frac{1}{M}\frac{\partial N_{-}^{(1)}}{\partial \tau} + \frac{1}{M}\frac{\partial V_{x-}^{(2)}}{\partial \xi} + \frac{1}{M}\frac{\partial V_{y-}^{(2)}}{\partial \eta} + \frac{1}{M}\frac{\partial V_{z-}^{(2)}}{\partial \varsigma} + \frac{1}{M}\frac{\partial}{\partial \varsigma}\left(N_{-}^{(1)} V_{z-}^{(1)}\right) \tag{20}$$

$$V_{y-}^{(2)} = -\frac{M}{\mu_{+} Z_{\beta}\Omega_B}\frac{\partial V_{x-}^{(1)}}{\partial \varsigma} \tag{21}$$

$$V_{x-}^{(2)} = \frac{M}{\mu_{+} Z_{\beta}\Omega_B}\frac{\partial V_{y-}^{(1)}}{\partial \varsigma} \tag{22}$$

$$\frac{\partial V_{z-}^{(2)}}{\partial \varsigma} = \frac{1}{M}\frac{\partial V_{z-}^{(1)}}{\partial \tau} + \frac{1}{M} V_{z-}^{(1)}\frac{\partial V_{z-}^{(1)}}{\partial \varsigma} + \frac{1}{M}\frac{\partial \varphi^{2}}{\partial \varsigma} + \frac{1}{M}\nu_0 V_{z-}^{(1)} \tag{23}$$

$$\left(\frac{\partial^2}{\partial \xi^2} + \frac{\partial^2}{\partial \eta^2} + \frac{\partial^2}{\partial \varsigma^2}\right)\varphi^1 = \left[\mu_n N_{-}^{(2)} - N_{+}^{(2)} + \mu_e(1-\alpha_e)\varphi^{(2)} + \frac{\mu_e}{2}\varphi^{(1)^2}\right] \tag{24}$$

Again, equating the coefficients of next higher order of ε from the Poisson's Eq. (6) and differentiating this equation with respect to ς, we get

$$\frac{\partial^3\varphi^1}{\partial \varsigma^3} + \frac{\partial}{\partial \varsigma}\left(\frac{\partial^2\varphi^1}{\partial \xi^2} + \frac{\partial^2\varphi^1}{\partial \eta^2} +\right) = \left[\mu_n\frac{\partial N_{-}^{(2)}}{\partial \varsigma} - \frac{\partial N_{+}^{(2)}}{\partial \varsigma} + \mu_e(1-\alpha_e)\frac{\partial \varphi^{(2)}}{\partial \varsigma} + \frac{\mu_e}{2}\frac{\partial \varphi^{(1)^2}}{\partial \varsigma}\right] \tag{25}$$

Substituting Eqs. (16)–(24) in Eq. (25), we get the DZK equation, which is in the form of

$$\frac{\partial \varphi^1}{\partial \tau} + A\varphi^1\frac{\partial \varphi^1}{\partial \varsigma} + B\frac{\partial^3\varphi^1}{\partial \varsigma^3} + C\frac{\partial}{\partial \varsigma}\left(\frac{\partial^2\varphi^1}{\partial \xi^2} + \frac{\partial^2\varphi^1}{\partial \eta^2}\right) + D\varphi^1 = 0 \tag{26}$$

where

$$A = \frac{3\left(1 + \mu_+^2 Z_\beta^2 \mu_n\right) - \mu_e M^4}{2M(1 + \mu_+ Z_\beta \mu_n)}$$

$$B = \frac{M^3}{2(1 + \mu_+ Z_\beta \mu_n)}$$

$$C = \frac{M^3\left(Z_\beta \mu_+ \Omega_B^2 - (\mu_n - \mu_+ Z_\beta)\right)}{2\mu_+ Z_\beta \Omega_B^2 (1 + \mu_+ Z_\beta \mu_n)}$$

$$D = \frac{\nu_0}{2}$$

$$\nabla^2 = \left(\frac{\partial^2 \varphi^1}{\partial \xi^2} + \frac{\partial^2 \varphi^1}{\partial \eta^2} + \frac{\partial^2 \varphi^1}{\partial \varsigma^2}\right)$$

In Eq. (26), A denotes the nonlinear coefficient, B is the dispersion co-efficient. C denote the transverse co-efficient, and D denotes the dissipative coefficient due to ion-neutral collision. If $D = 0$, Eq. (26) simply represents the Zakharov–Kuznetsov (ZK) nonlinear wave equation.

3 Solution of Damped Zakharov–Kuznetsov (DZK) Equation

To obtain the solution of the DZK Eq. (26), we used the transformation $\chi(\xi, \zeta, \eta, \tau) = \varphi(l\xi + m\zeta + n\eta - U\tau)$ moving with velocity U and l, m, n are the direction cosines along which the wave propagates and $\left(l^2 + m^2 + n^2\right) = 1$. Substituting $\varphi(\chi) = \varphi(\xi, \zeta, \eta, \tau)$ gives $\frac{\partial}{\partial \xi} = l\frac{d}{d\chi}$, $\frac{\partial}{\partial \zeta} = m\frac{d}{d\chi}$, $\frac{\partial}{\partial \eta} = n\frac{d}{d\chi}$, $\frac{\partial}{\partial \tau} = -U\frac{d}{d\chi}$. Now, Eq. (26) can be rewritten as

$$-U\frac{d\varphi}{d\chi} + An\varphi\frac{d\varphi}{d\chi} + F\frac{d^3\varphi}{d\chi^3} - D\varphi = 0 \tag{27}$$

where $F = \left[Bn^3 + Cn\left(l^2 + m^2\right)\right]$

Integrating Eq. (27) w.r.t χ and putting the integration constant as 0, Eq. (27) can be transformed as

$$F\frac{d^2\varphi}{d\chi^2} + \frac{An}{2}(\varphi)^2 - U\varphi - D\varphi\chi = 0 \tag{28}$$

Now, using the boundary condition as $\varphi \to 0$ at $\chi \to \infty$, the time evolution equation is in the form of [17]

$$\varphi^1 = \varphi_0 Sech^2 \frac{\chi}{\Delta}$$

where $\varphi_0 = \frac{3U}{lA} e^{(-D\tau)}$, $\Delta = \sqrt{\frac{4lF}{U} e^{D\tau}}$ are amplitude and width of the solitary wave, respectively.

4 Results and Discussions

The variation of nonlinear and dispersion coefficients A and B with respect to the ion concentration ratio μ_n and nonthermal electron parameter α_e are shown in Fig. 1a and Fig. 1b. From Fig. 1a, it is observed that the nonlinear term A is negative and is decreases gradually with increasing μ_n value. On the other hand, Fig. 1b represents that with varying nonthermal electron parameter α_e, there is a sharp decrease in the nonlinear term A from positive to negative value. However, in both cases (Fig. 1a, b), the dispersion coefficient B is positive and increases with increasing α_e and μ_n value. Which implies that for the chosen values of plasma parameters, the IA wave excitations in plasma lead to the evolution of rarefactive solitary wave.

In Fig. 2a, b, the potential profile φ of rarefactive solitary wave excitation has shown with varying time τ (i.e., $\tau = 0, 1, 2$) values and at different values of collisional frequency ν_0 (i.e., $\nu_0 = 0, 0.08, 0.16$). In Fig. 2a, keeping the fixed value of $\nu_0 = 0.04$, it is observed that when $\tau = 2$, the decrease in the solitary wave amplitude is more than that of the case when $\tau = 0$. This implies that in a collisional plasma, the damped in solitary wave amplitude is observed as time passes. While in Fig. 2b, the solitary wave potential φ is plotted at a fixed time value of $\tau = 1$, for different ν_0 values (i.e., $\nu_0 = 0, 0.08, 0.16$). It is observed that under the presence of a strong collisional effect, i.e., at $\nu_0 = 0.16$, the prominent decrease in solitary wave amplitude is observed than that of the case when there is any collision, i.e., at $\nu_0 = 0$. Hence, from Fig. 2a, b, it is observed that in a collisional plasma, the amplitude of

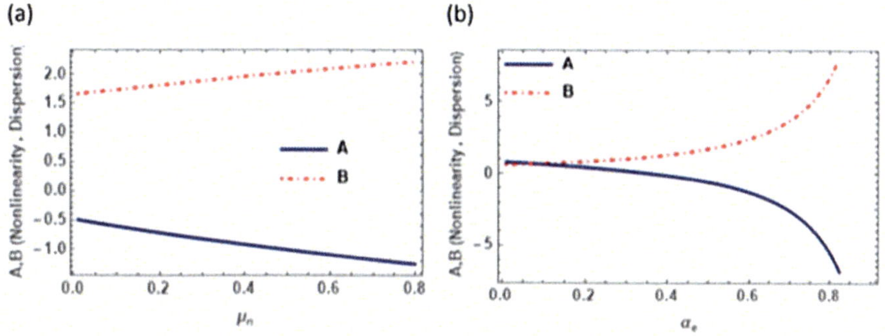

Fig. 1 Variation of nonlinearity and dispersion coefficients (A & B) **a** with α_e **b** with μ_n. The plasma parameters are $\mu_e = 0.9$, $\mu_+ = 1$, $\alpha_e = 0.5$ $\Omega_B = 0.3$

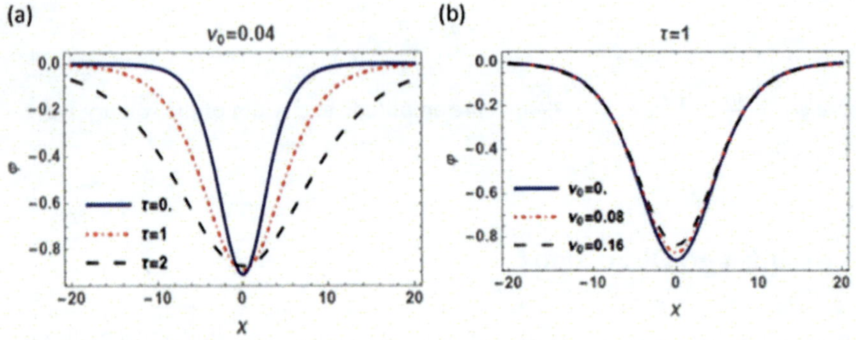

Fig. 2 **a** Solitary wave potential φ at different time $\tau(0, 1, 2)$ values. **b** Solitary wave potential φ at different values of ion-neutral collision frequency $\nu_0(0, 0.08, 0.16)$. The plasma parameters are $\mu_n = 0.1$, $\mu_e = 0.9$, $\mu_+ = 1$, $U = 0.1$, $l = 0.2$, $m = 0.55$, $n = 0.25$, $\Omega_B = 0.3$

solitary wave profile decreases with time and also due to collisional effect generated by ions. In three-dimensional portraits of Fig. 3a, b, the influence of ν_0 is studied to observe the solitary wave propagation in collisionless plasma medium considering $\nu_0 = 0$ and in case of collisional medium considering $\nu_0 = 0.05$. It is observed

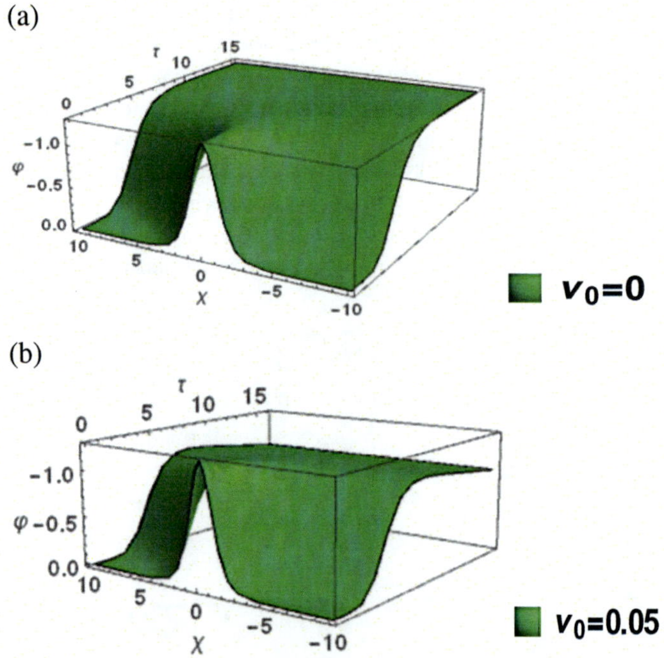

Fig. 3 3D plot of solitary wave potential φ at **a** $\nu_0 = 0$ **b** $\nu_0 = 0.05$. The other plasma parameters are same as Fig. 2

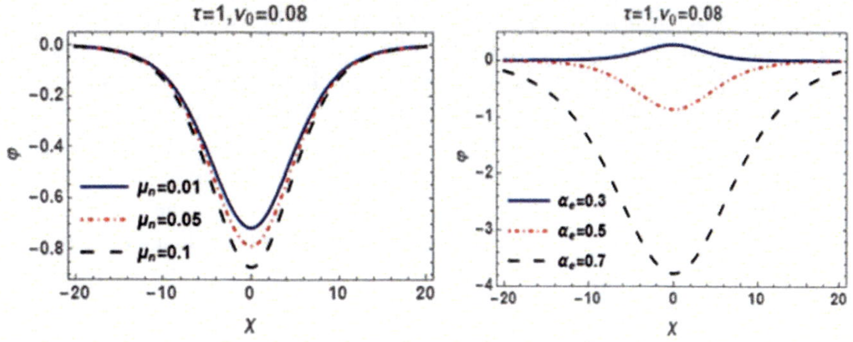

Fig. 4 Solitary wave potential φ at $\nu_0 = 0.08$, $\tau = 1$ **a** for different $\mu_n(0.01, 0.05, 0.1)$ **b** for different $\alpha_e(0.3, 0.5, 0.7)$. The other plasma parameters are same as Fig. 2

that for the case of $\nu_0 = 0$ (Fig. 3a), the solitary wave amplitude remains same as time passes, however, the change in the width of the solitary wave is observed with varying time values. On the other hand, when it is considered in a collisional medium considering the value of ν_0 as $\nu_0 = 0.05$, it is noticed that the solitary wave amplitude decreases with time and the width of the solitary wave also changes as shown in Fig. 3b. In Fig. 4a, b, the effect of ion concentration ratio μ_n and nonthermal electron parameter α_e have shown for the changes in the characteristics features of solitary wave potential profile φ. It is observed that at a fixed value of $\nu_0 = 0.08$ and $\tau = 1$, the increasing value of ion concentration ratio μ_n increases the wave amplitude of rarefactive solitary wave. It is because, the nonlinearity in plasma system decreases with increasing negative ion concentration, which is shown in Fig. 1a. While Fig. 3b represents that at a fixed value of $\nu_0 = 0.08$ and $\tau = 1$, the increase in α_e value also increases the wave amplitude of rarefactive solitary wave. However, at a lower value of α_e, i.e., at $\alpha_e \leq 0.3$, the rarefactive solitary wave changes to compressive one. This represents that the chosen α_e value has strong influences in the generation of rarefactive or compressive solitary wave in plasma.

5 Conclusion

We have presented the theoretical study on nonlinear IA solitary wave propagation in collisional magnetized plasma. It is considered that plasma is consisting of mobile pair ion of hydrogen $\left(H^+, H^-\right)$, and Crain's nonthermal electron distribution. In this study, the dissipation generated by ion-neutral collision frequency is considered. By adopting the reductive perturbation technique, three-dimensional DZK nonlinear wave equation is derived and the time-dependent solution is obtained by using co-ordinate transformation. It is observed that in collisional plasma, the solitary wave amplitude diminishes by collisions between different plasma constituents. And with passing time the decrease in solitary wave amplitude is also observed. Also, it is observed that when collision frequency is zero, i.e., $\nu_0 = 0$, the amplitude of solitary wave remains same as time progresses, however, the width of the wave changes. On the other hand, when $\nu_0 = 0.05$, the solitary wave amplitude as well as width changes with respect to time. And finally, we have shown the effect of ion concentration ratio μ_n and nonthermal electron parameter α_e, on solitary wave potential profile φ. It is observed that amplitude of the rarefactive solitary wave increases with increase in negative ion concentrations and with increasing value of nonthermal electron. However, for lower value of α_e, i.e., at $\alpha_e \leq 0.3$, change in the solitary wave profile from rarefactive to compressive wave is observed. This investigation is helpful in studying space plasma environments.

References

1. P.K. Shukla, A.A. Mamun, *Introduction to Dusty Plasma Physics* (IOP, Bristol, 2002)
2. Y. Nakamura, A. Sarma, Phys. Plasmas **8**, 9 (2001)
3. A.P. Misra, N.C. Adhikary, P.K. Shukla, Phys. Rev. E **86**, 056406 (2012)
4. M. Bacha, L.A. Gougam, M. Tribeche, Phys. A 17509 (2016)
5. N.S. Saini, I. Kourakis, Phys. Plasmas **15**, 123701 (2008)
6. F.F. Chen, *Introduction to Plasma Physics* (Plenum, New York, 1974)
7. H. Washimi, T. Taniuti, Phys. Rev. Lett. **17**, 19 (1966)
8. H. Ikezi, Phys. Phys. Fluids **16**, 10 (1973)
9. P.K. Shukla, V.P. Silin, Phys. Scr. **45**, 508 (1992)
10. N.J. Zabusky, M.D. Kruskal, Phys. Rev. Lett. **15**, 6 (1965)
11. S.G. Tagare, R.V. Reddy, J. Plasma Phys. **35**, 219 (1986)
12. S. Ghosh, T.K. Chaudhuri, S. Sarkar, M. Khan, M.R. Gupta, Phys. Rev. E **65**, 037401 (2001)
13. M.R. Hassan, T.I. Rajib, S. Sultana, arXiv:1912.04756v4 [physics.plasm-ph] (2019)
14. S.A. Krapak, G. Morfill, Phys. Plasmas **8**, 2629 (2001)
15. S.V. Vladimirov, K.N. Ostrikov, M.Y. Yu, Phys. Rev. E **60**, 3257 (1999)
16. M. Farooq, A. Mushtaq, J. Qasim, Contrib. Plasma Phys. **59**, 122 (2019)
17. W.M. Moslem, Phys. Plasmas **10**, 8 (2003)

Particle–Antiparticle Trapping in a Magnetically Quantized Plasma and Its Effect on the Evolution of Solitary Wave

Manoj Kr. Deka and Apul N. Dev

Abstract In an ion beam plasma system, the effect of the magnetically quantized degenerate trapped electron (Particle) and positron (Antiparticle) on small-amplitude ion acoustic solitary waves are studied with the help of the Korteweg–de-Vries (K–dV) equation. Here, the magnetized positive ions and beam ions are considered as non-degenerate. The effect of magnetic quantization, degenerate temperature, normalized positron concentration, normalized ion beam concentration along with other relevant physical plasma parameters of the origin of astrophysical plasma environment, on solitary wave propagation is studied, especially in the non-linear regime. On the other hand, in the linear regime, the dependency of frequency and wave number on the above-mentioned plasma parameters are discussed. It is found that both compressive, as well as rarefactive type solitary waves can exist in such a plasma environment. Three different modes of propagation, the fast beam and slow beam mode and the inherent ion acoustic mode co-exist in such plasma system and the role of different degenerate plasma parameters on these wave modes are also discussed in detail. The normalized positron concentration and, velocity, as well as normalized ion beam concentration along with different degenerate plasma parameters, has an astounding control on the polarity of the solitary wave as well as the amount of compression and rarefaction, a typical plasma wave mode would undergo. Within the chosen degenerate and non-degenerate plasma parameters, for a compressive (rarefactive) fast beam mode, the slow mode appears as rarefactive (compressive). For some typical combinations of degenerate parameters, the fast mode can propagate with a hypersonic phase velocity which shows distinctive characteristics with positron density.

Keywords Trapped electron–positron · Ion beam · Hypersonic soliton

---

M. Kr. Deka
Vill-Borka, PO- Borka, Kamrup, Guwahati 781101, Assam, India
e-mail: manojd143@gmail.com

A. N. Dev (✉)
Centre for Applied Mathematics and Computing, Siksha 'O' Anusandhan (Deemed to be University), Bhubaneswar 751030, Odisha, India
e-mail: apulnarayan@gmail.com

S. R. Mishra et al. (eds.), *Recent Trends in Applied Mathematics*, Lecture Notes in Mechanical Engineering, https://doi.org/10.1007/978-981-15-9817-3_8

1 Introduction

Due to symmetry and anti-symmetry in the mass and charge of electron and positron, the characteristics of solitary wave in electron–positron–ion (e–p–i) plasma has been always dealt with substantial importance than electron–ion (e–i) plasmas. Moreover, due to the abundance of e–p–i plasma in different space plasma environment such as the early universe [1, 2], centre of our galaxy [3], active galactic nuclei [4], neutron stars [5] as well as in different laboratory plasma environment [6–10], the investigations on e–p–i plasmas has been getting boost throughout the world. On the other hand, from numerous investigations and in situ measurements on space and laboratory plasma environment, the rampant presence of non-Maxwellian populations of plasma species has been well established due to the long-range interaction [11, 12] and the presence of high energetic particles in such plasma environment [13]. Apart from this, the atoms or molecules undergo a significant change in the presence of a strong magnetic field. In quantum plasmas, in the presence of strong magnetic field, the cyclotron orbits of electrons are quantized [14], and as a result, electrons occupy discrete Landau levels, where, in each level, the number of electrons is directly proportional to the strength of the magnetic field. To produce the extreme state of matter which exists in space environment like neutron star, white dwarfs, etc., the intense ion beams are regarded as the indispensable tool [15–23]. Also, these ion beams are quite prevalent in earth magnetosphere, magnetosphere of different planets [24] in supernova-driven plasma flows [25] in pulsars, blazars, etc. [26].

Because of its abundance and occurrence in a different laboratory and space plasma, solitary waves in e–p–i and e–i plasma in degenerate/non-degenerate as well as in relativistic/ultrarelativistic plasmas has been widely investigated as mentioned in the above paragraph [27–33]. Shah et al. investigated the properties of solitary wave in degenerate plasmas in the presence of trapped electrons in relativistic, ultra-relativistic and non-relativistic regimes and concluded that, in the non-relativistic regime, the solitary wave potential has the highest value, whereas, for ultra-relativistic and relativistic regimes, the wave potential is in the intermediate and in the lowest level, respectively [34]. Abdikian and Mahmood investigated the properties of acoustic solitons in a relativistic degenerate e–p–i plasma in the presence of magnetic field and concluded that the width and height of the electron–positron acoustic solitary wave decrease with increasing relativistic effect and they also found that the potential of the solitary wave increases with increasing positron density and the width reduces with increasing strength of the magnetic field [35]. Rasheed et al. studied the propagation of solitary waves in a degenerate electron–positron plasma and concluded that only compressive solitary structure is obtained in such plasmas whose amplitudes are drastically reduced by the presence of positrons [36].

Recently, a lot of investigation has been carried out with trapped plasma species in degenerate plasmas including e–p–i in the presence of Landau quantized magnetic fields. For example, Tsintsadze et al. studied the properties of ion acoustic waves in a quantized degenerate plasma with trapped electron and magnetic field and concluded that singularities occur separating the regions of weak and strong quantization due to the ambient magnetic field and thereby affecting the solitary wave potential [37]. Irfan et al. considered ion pressure anisotropy and electron trapping in degenerate plasma to study the features of solitary wave and reported that the thermal correction for electron and the pressure anisotropy from ions strongly modify the wave amplitudes and width in such plasmas [38]. Arshad et al. studied the properties of solitary wave in degenerate plasma in the presence of electron trapping in degenerate plasma considering a modified Hasegawa Mima equation [39]. Masood et al. using the drift approximation in a magnetically quantized plasma with electron trapping showed the occurrence of one- and two-dimensional drift ion solitary structures for such plasma [40]. On the other hand, Iqbal et al. considered the electron and positron trapping with Landau quantized magnetic field to investigate the properties of solitary waves in linear and non-linear regimes and concluded that the wave characteristics are significantly modified by the presence of trapped positrons in such plasmas [41]. Shaukat et al., considering Landau quantized magnetic field and electron temperature effect, studied the properties of drift solitary wave in degenerate plasmas and concluded that both compressive and rarefactive solitary structure exists in such plasmas [42]. Shaukat observed drift solitary structures in the presence of Landau quantization in degenerate e–p–i plasma and observed only compressive solitary structures in such plasmas [43]. Shan et al. studied the properties of solitary wave in the presence of cold positron beam and q-distributed trapped electron and concluded that the cold positron beams can significantly modify the wave propagation characteristics in such plasmas [44]. Last but not the least, very recently, Iqbal et al. investigated the nature of ion acoustic waves in degenerate relativistic and ultra-relativistic plasmas in the presence of Landau diamagnetism and electron trapping and concluded that in both the regime, only compressive solitary structures are formed [45]. Here in this report, in the presence of ion and ion beams and magnetically quantized degenerate electrons as well as positrons, the salient features of solitary wave propagation governed by a K–dV equation is described in detail. We would be happy if our companion researcher investigates the striking features of such plasma numerically as well as experimentally though it is beyond the scope of the present Manuscript.

1.1 Theoretical Formulations

The number density of electron and positrons for Fermi–Dirac distribution within the energy range ε and $\varepsilon + d\varepsilon$, for all the Landau levels, can be written as [14, 46, 47]

$$n_j = \frac{p_{F_j}^2 \eta_j}{2\pi^2\hbar^3}\sqrt{\frac{m_j}{2}}\sum_{l=0}^{\infty}\int_0^{\infty}\frac{\varepsilon^{-1/2}}{1+\exp(\varepsilon - U/T)}d\varepsilon \tag{1}$$

with $U = \mu - l\hbar\omega_{cj} - q_j\varphi$, where μ is the chemical potential, $\eta = \hbar\omega_{ce}/\varepsilon_{Fe}$, $\varepsilon_{Fe} = (\hbar^2/2m_e)(3\pi^2 n_{e0})^{2/3}$, $n_{e0} = (p_{Fe}^3/3\pi^2\hbar^3)$, $\varphi = e\phi/\varepsilon_{Fe}$ is the equilibrium number density, p_{Fe} is the momentum on the Fermi surface, $T = \left(\pi T/2\sqrt{2}\varepsilon_{Fe}\right)$, quantizing magnetic field is defined by appears through η.

The Fermi integrals are evaluated following the general treatment for such type of integral and, finally, we derive the expression for degenerate electron density as [14, 46, 47]

$$\begin{aligned} N_e =& \frac{\eta}{2}(3 - T^2) + (1-\eta)^{\frac{3}{2}} + T^2(1-\eta)^{-\frac{1}{2}} \\ &+ \frac{3\phi}{2}\left\{\frac{\eta}{2}(1+T^2) + (1-\eta)^{\frac{1}{2}} - \frac{T^2}{3}(1-\eta)^{-\frac{3}{2}}\right\} \\ &+ \frac{3\phi^2}{8}\left\{-\frac{\eta}{2}(1+5T^2) + (1-\eta)^{-\frac{1}{2}} + T^2(1-\eta)^{-\frac{5}{2}}\right\} \end{aligned} \tag{2}$$

Similarly, to obtain positron density, it is clear that $U = \mu - q_p\varphi - l\hbar\omega_{ce} = U = \left(1 - \phi\mu_p^{-\frac{2}{3}} - l\eta\mu_p^{-\frac{2}{3}}\right)\varepsilon_{Fp}$, $q_p = e$, So we obtain $l_{\max} = \frac{1-\phi\mu_p^{-\frac{2}{3}}}{\eta\mu_p^{-\frac{2}{3}}}$ so that the integral of Eq. (1) becomes a real quantity and then following the same procedures as mentioned in the above references, we obtain the positron density as

$$\begin{aligned} n_p = n_{p0}\Bigg[& \frac{3\eta}{2}\mu_p^{-\frac{2}{3}}\left(1-\phi\mu_p^{-\frac{2}{3}}\right)^{1/2} - \frac{\eta T^2}{2}\mu_p^{-2}\left(1-\phi\mu_p^{-\frac{2}{3}}\right)^{-3/2} \\ & + \left(1-\phi\mu_p^{-\frac{2}{3}} - \eta\mu_p^{-\frac{2}{3}}\right)^{3/2} + \mu_p^{-\frac{4}{3}}T^2\left(1-\phi\mu_p^{-\frac{2}{3}} - \eta\mu_p^{-\frac{2}{3}}\right)^{-1/2}\Bigg] \end{aligned} \tag{3}$$

where $n_{p0} = p_{Fp}^3/3\pi^2\hbar^3$, $\eta_p = \eta\mu_p^{-\frac{2}{3}}$. On expanding, Eq. (1) becomes

$$\begin{aligned} \frac{n_p}{n_{p0}} =N_p =& \frac{\eta\mu_p^{-\frac{2}{3}}}{2}\left(3-\mu_p^{-\frac{4}{3}}T^2\right) + \left(1-\mu_p^{-\frac{2}{3}}\eta\right)^{\frac{3}{2}} + \mu_p^{-\frac{4}{3}}T^2\left(1-\mu_p^{-\frac{2}{3}}\eta\right)^{-\frac{1}{2}} \\ &+ \frac{3\phi}{2}\left\{-\frac{\eta\mu_p^{-\frac{4}{3}}}{2}\left(1+\mu_p^{-\frac{4}{3}}T^2\right) - \mu_p^{-\frac{2}{3}}\left(1-\mu_p^{-\frac{2}{3}}\eta\right)^{\frac{1}{2}} + \frac{\mu_p^{-2}T^2}{3}\left(1-\mu_p^{-\frac{2}{3}}\eta\right)^{-\frac{3}{2}}\right\} \\ &+ \frac{3\phi^2}{8}\left\{-\frac{\eta\mu_p^{-2}}{2}\left(1+5\mu_p^{-\frac{4}{3}}T^2\right) + \mu_p^{-\frac{4}{3}}\left(1-\mu_p^{-\frac{2}{3}}\eta\right)^{-\frac{1}{2}} + \mu_p^{-\frac{8}{3}}T^2\left(1-\mu_p^{-\frac{2}{3}}\eta\right)^{-\frac{5}{2}}\right\} \end{aligned} \tag{4}$$

The basic set of equations can be written as follows:

$$\frac{\partial n_i}{\partial t} + \vec{\nabla}.(n_i \vec{v}_i) = 0 \quad (5a)$$

$$\frac{\partial n_b}{\partial t} + \vec{\nabla}.(n_b \vec{v}_b) = 0 \quad (5b)$$

$$\frac{\partial \vec{v}_i}{\partial t} + \left(\vec{v}_i.\vec{\nabla}\right)\vec{v}_i = \frac{q_i}{m_i}\left(-\nabla\phi + \vec{v}_i \times \vec{B}_0\right) \quad (5c)$$

$$\frac{\partial \vec{v}_b}{\partial t} + \left(\vec{v}_b.\vec{\nabla}\right)\vec{v}_b = \frac{q_b}{m_b}\left(-\nabla\phi + \vec{v}_b \times \vec{B}_0\right) \quad (5d)$$

$$\nabla^2\varphi = 4\pi e\left(n_e - n_i - n_b - n_p + Z_h n_{ho}\right) \quad (5e)$$

Assume that $n_{eo} = n_{io} + n_{po} + n_{bo}$ where $n_{i0},\ n_{b0},\ n_{p0}, n_{eo}$ are the unperturbed number density for ions, ions beams, positrons electrons, respectively.

1.2 Linear Analysis

Adopting the standard procedures for linearization techniques, the above equations, i.e. (5a)–(5e) reduce to

$$\frac{\partial n_{i1}}{\partial t} + n_{i0}\frac{\partial}{\partial x}v_{i1} = 0,\ \ n_i = n_{i1} + n_{i0},\ v_i = v_{i1} \quad (6a)$$

$$m_i\frac{\partial v_{i1}}{\partial t} = -e\frac{\partial \varphi}{\partial x} + v_{i1} \times B \quad (6b)$$

$$\frac{\partial n_{b1}}{\partial t} + n_{b0}\frac{\partial}{\partial x}v_{b1} = 0,\ \ n_b = n_{b1} + n_{b0},\ v_b = v_{b1} \quad (6c)$$

$$m_b\frac{\partial v_{b1}}{\partial t} = -e\frac{\partial \varphi}{\partial x} + v_{b1} \times B \quad (6d)$$

$$\frac{\partial^2}{\partial x^2}\varphi = 4\pi e\left(n_e - n_{i1} - n_{i0} - n_{b1} - n_{b0} - n_p\right) \quad (6e)$$

where the subscript “0” stands for the unperturbed part whereas subscript “1” stands for the perturbed part. Assuming a sinusoidal solution proportional to $e^{i(-wt+kx)}$ for oscillating quantities, we have $\frac{\partial}{\partial x} = ik,\ \frac{\partial}{\partial t} = -iw$, and then above set of basic Eqs. (6a)–(6e) becomes

$$wn_{i1} = n_{i0}kv_{i1}, m_s wv_{i1} = ek\varphi, wn_{b1} = n_{b0}kv_{b1}, m_b wv_{b1} = ek\varphi \quad (7a)$$

$$-k^2\varphi = 4\pi e\left(n_e - n_{i1} - n_{i0} - n_{b1} - n_{b0} - n_p\right) \tag{7b}$$

Evaluating Eqs. (7a) and (7b), we get the dispersion relation as

$$\frac{w}{k} = C_s\left\{\frac{(\alpha_0 - \mu_p \Upsilon_0)}{k^2\lambda_{Fe}^2 + (\alpha_1 - \mu_p \Upsilon_1)}\right\}^{\frac{1}{2}} \tag{8}$$

1.3 Evolution Equation of Non-linear Wave

The set of Eqs. (5a)–(5e) are normalized using the following normalized parameter $\varphi e = \varepsilon_{Fe}\phi$, $t = \tau\omega_j^{-1}$, $x = X\lambda_{Fe}$, $N_j = n_j/n_{j0}$, $V_j = v_j/C_s$. Thus $\lambda_{Fe}\omega_j = C_s$, $\omega_j = \left(4\pi n_{j0}e^2/m_j\right)^{1/2}$, $\lambda_{Fe} = \left(\varepsilon_{Fe}/4\pi n_{i0}e^2\right)^{1/2}$, $C_s = \sqrt{\varepsilon_{Fe}/m_i}$.

After normalization, the normalized set of equations are obtained as follows:

$$\frac{\partial N_i}{\partial T} + \nabla(N_i V_i) = 0 \tag{9a}$$

$$\frac{\partial N_b}{\partial T} + \nabla(N_b V_b) = 0 \tag{9b}$$

$$\frac{\partial V_i}{\partial T} + (V_i\nabla)V_i = -\nabla\phi + V_i\Omega_{Bi} \tag{9c}$$

$$\frac{\partial V_b}{\partial T} + (V_b\nabla)V_b = -\nabla\phi + V_b\Omega_{Bb} \tag{9d}$$

$$\nabla^2\phi = \left(N_e - N_b\mu_b - N_p\mu_p - \mu_i N_i + \mu_{hn}\right) \tag{9e}$$

where $\Omega_{Bk} = eB_0/\omega_k m_k, k = (i = \text{ion}, b = \text{beam}), \mu_i = \frac{n_{i0}}{n_{e0}}, \ \mu_b = \frac{n_{b0}}{n_{e0}}, \ \mu_p = \frac{n_{p0}}{n_{e0}}$.

Adopting the standard reductive perturbation technique and using the following stretched co-ordinates, we derive the evolution equation of solitary wave in terms of a K–dV equation.

$$\xi = \varepsilon^{1/2}\left(I_x x + I_y y + I_z z - \lambda T\right) \text{ and } \tau = \varepsilon^{3/2}T \tag{10}$$

where λ is the Mach Number and I_x, I_y, I_z are the direction cosines for x, y, z axes, respectively. The physical variables in equations are expanded in a power series in terms of the expansion parameter ε as

$$\left.\begin{aligned} N_k &= 1 + \varepsilon N_k^{(1)} + \varepsilon^2 N_k^{(2)} + \varepsilon^3 N_k^{(3)} + \ldots \\ V_{bx} &= V_{b0} + \varepsilon V_{bx}^{(1)} + \varepsilon^2 V_{bx}^{(2)} + \varepsilon^3 V_{bx}^{(3)} + \ldots \\ V_{kx} &= \varepsilon V_{kx}^{(1)} + \varepsilon^2 V_{kx}^{(2)} + \varepsilon^3 V_{kx}^{(3)} + \ldots \\ V_{ky,z} &= \varepsilon^{\frac{3}{2}} V_{ky,z}^{(1)} + \varepsilon^2 V_{ky,z}^{(2)} + \varepsilon^{\frac{5}{2}} V_{ky,z}^{(3)} + \ldots \\ \phi &= \varepsilon \phi^{(1)} + \varepsilon^2 \phi^{(2)} + \varepsilon^3 \phi^{(3)} + \ldots \end{aligned}\right\} \tag{11}$$

Substituting expansions from Eq. (11) and stretching co-ordinates through Eq. (9a)–(9e) and equating the power of lowest power of ε, we get $V_{ix}^{(1)} = \phi^{(1)}/\lambda$, $V_{iy}^{(1)} = \frac{I_x}{\Omega_B}\frac{\partial \phi^{(1)}}{\partial \xi}$, $V_{iz}^{(1)} = \frac{I_x}{\Omega_B}\frac{\partial \phi^{(1)}}{\partial \xi}$ $N_i^{(1)} = \phi^{(1)}/\lambda^2$, $V_{bx}^{(1)} = \phi^{(1)}/(\lambda - V_{b0})$, $N_b^{(1)} = \phi^{(1)}/(\lambda - V_{b0})^2$, $N_e^{(1)} = \alpha_1 \phi^{(1)}$, $N_p^{(1)} = \Upsilon_1 \phi^{(1)}$, where $\alpha_1 = \frac{3}{2}\left\{\frac{\eta}{2}(1+T^2) + (1-\eta)^{\frac{1}{2}} - \frac{T^2}{3}(1-\eta)^{-\frac{3}{2}}\right\}$, $\Upsilon_1 = \frac{3}{2}\left\{-\frac{\eta \mu_p^{-\frac{4}{3}}}{2}\left(1+\mu_p^{-\frac{4}{3}}T^2\right) - \mu_p^{-\frac{2}{3}}\left(1-\mu_p^{-\frac{2}{3}}\eta\right)^{\frac{1}{2}} + \frac{\mu_p^{-2}T^2}{3}\left(1-\mu_p^{-\frac{2}{3}}\eta\right)^{-\frac{3}{2}}\right\}$ and the expression for Mach number or the phase velocity(normalized) of the non-linear wave is given by

$$\lambda^2 = \left[\left\{\mu_b/\left(\alpha_1 - \mu_p \Upsilon_1\right)(1-\delta)^2\right\} + \left(\mu_i/\alpha_1 - \mu_p \Upsilon_1\right)\right], \textit{with } \delta = V_{b0}/\lambda \tag{12}$$

Similarly, equating the coefficient of higher power of ε from the Eqs. (9a) to (9e), we get

$$\frac{\partial N_i^{(1)}}{\partial \tau} - M\frac{\partial N_i^{(2)}}{\partial \xi} + \sum_{l=x,y,z} I_l \frac{\partial}{\partial \xi} V_{il}^{(2)} = 0 \tag{13}$$

$$\frac{\partial V_{ix}^{(2)}}{\partial \xi} = \frac{1}{\lambda}\frac{\partial V_{ix}^{(1)}}{\partial \tau} + \frac{I_x}{\lambda} V_{ix}^{(1)} \frac{\partial V_{ix}^{(1)}}{\partial \xi} + \frac{I_x}{\lambda}\frac{\partial \phi^{(2)}}{\partial \xi} \tag{14}$$

$$\frac{\partial V_{iy}^{(2)}}{\partial \xi} = \frac{\lambda I_x}{\Omega_B^2}\frac{\partial^3 \phi^{(1)}}{\partial \xi^3} \text{ and } \frac{\partial V_{iz}^{(2)}}{\partial \xi} = \frac{M I_x}{\Omega_B^2}\frac{\partial^3 \phi^{(1)}}{\partial \xi^3} \tag{15}$$

$$\frac{\partial N_b^{(1)}}{\partial \tau} - (\lambda - I_x V_{b0})\frac{\partial N_b^{(2)}}{\partial \xi} + \sum_{l=x,y,z} I_l \frac{\partial V_{bl}^{(2)}}{\partial \xi} = 0 \tag{16}$$

$$\frac{\partial V_{bx}^{(1)}}{\partial \tau} - (\lambda - I_x V_{b0})\frac{\partial V_{bx}^{(2)}}{\partial \xi} + I_x V_{bx}^{(1)}\frac{\partial V_{bx}^{(1)}}{\partial \xi} + I_x \frac{\partial \phi^{(2)}}{\partial \xi} = 0 \tag{17}$$

$$\frac{\partial V_{by}^{(2)}}{\partial \xi} = \frac{I_x(\lambda - I_x V_{b0})}{\Omega_B^2}\frac{\partial^3 \phi^{(1)}}{\partial \xi^3} \text{ and } \frac{\partial V_{bz}^{(2)}}{\partial \xi} = \frac{I_x(\lambda - I_x V_{b0})}{\Omega_B^2}\frac{\partial^3 \phi^{(1)}}{\partial \xi^3} \tag{18}$$

$$I_x^2 \frac{\partial^3 \phi^{(1)}}{\partial \xi^3} = \frac{\partial}{\partial \xi} N_e^{(2)} - \mu_b \frac{\partial}{\partial \xi} N_b^{(2)} - \mu_p \frac{\partial}{\partial \xi} N_p^{(2)} - \mu_i \frac{\partial}{\partial \xi} N_i^{(2)} \tag{19}$$

Now solving Eqs. (13)–(19) along with first-order equations, we obtain the following K–dV equation

$$p\frac{\partial\phi^{(1)}}{\partial\tau}+q\phi^{(1)}\frac{\partial\phi^{(1)}}{\partial\xi}+r\frac{\partial^3\phi^{(1)}}{\partial\xi^3}=0 \quad (20)$$

where $p=\frac{2\mu_b}{(\lambda-V_{b0})^3}+\frac{2\mu_i}{\lambda^3}$

$$q=\frac{\mu_b I_x^2}{(\lambda-I_x V_{b0})^4}+\frac{\mu_i I_x^2}{\lambda^4}-\frac{(\alpha_2-\mu_p\Upsilon_2)}{I_x^2}$$

$$r=1+2\mu_b/\Omega_{Bb}^2+2\mu_i/\Omega_{Bi}^2$$

$$\alpha_2=\frac{3}{8}\left\{-\frac{\eta}{2}(1+5T^2)+(1-\eta)^{-\frac{1}{2}}+T^2(1-\eta)^{-\frac{5}{2}}\right\}$$

$$\Upsilon_2=\frac{3}{8}\left\{-\frac{\eta\mu_p^{-2}}{2}\left(1+5\mu_p^{-\frac{4}{3}}T^2\right)+\mu_p^{-\frac{4}{3}}\left(1-\mu_p^{-\frac{2}{3}}\eta\right)^{-\frac{1}{2}}+\mu_p^{-\frac{8}{3}}T^2\left(1-\mu_p^{-\frac{2}{3}}\eta\right)^{-\frac{5}{2}}\right\}$$

By using the tanh method [48], the stationary wave solution of Eq. (20) is derived as

$$\phi^{(1)}=\phi_m \sec h^2\left(\frac{\chi}{\omega}\right) \quad (21)$$

where quantities $\phi_m=\frac{3u}{A}$ and $\omega=2\left(\frac{B}{u}\right)^{1/2}$ and $A=\frac{q}{p}$ & $B=\frac{r}{p}$ are the amplitude and width of the solitary waves, respectively.

2 Results and Discussion

2.1 *Including the Effect of Positron*

All the results and discussions are basically based on the analysis of Eq. (21) which is analytically studied to unlock the salient features of the solitary wave propagation in such plasma. Here, plasma density is considered as 10 cm^{-3} [26–28], magnetic field is in the range of 10^9~10^{11} G and the Fermi temperature is calculated to be around 3.6277×10^7K [49, 50]. For obvious reasons of its abundance, helium positive ions are considered for the plasma environment of our interest and also considered ion beams with some initial beam velocity.

In Fig. 1, we discuss the dependency of non-normalized wave frequency on the wave number for different degenerate and non-degenerate plasma parameters, and

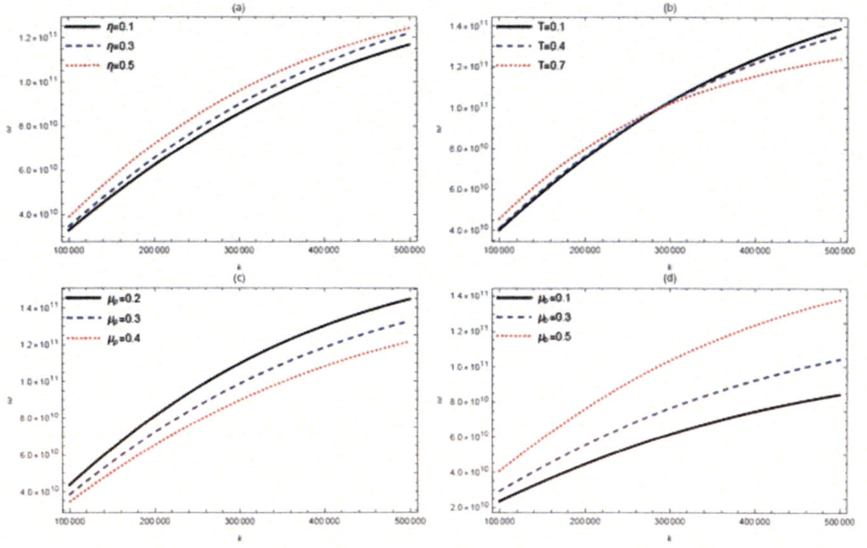

Fig. 1 Dispersion relation of ion beam plasma with **a** magnetic quantization, η, **b** degenerate Temperature, T, **c** positron density and **d** ion beam density

as seen from the figure, with increasing magnetic quantization (Fig. 1a), the wave frequency increases. This is because an increase in the magnetic field will always increase the amount of confinement of the plasma species and with a greater number of confined plasma species along a particular direction, the number of complete wave cycles will tend to increase. On the other hand, with increasing degenerate temperature, initially, the wave frequency increases with increasing wave number but after a certain wave number, the frequency decreases.

This is because, as the wave number increases, the wavelength will decrease which means frequency will increase which is also the case initially in Fig. 1b, but as the wavelength continues to decrease with the increase of wave number, the increase of temperature which is nothing but the thermal velocity, can result in a rapid interaction or collision among the charged particle which eventually leads to the decrease in frequency. On the other hand, as shown in Fig. 1c, with increasing positron density, the frequency goes down, which is primarily due to the decrease in the restoring force due to the addition of positron, whereas, as shown in Fig. 1d, the frequency increases with increasing beam ion density as the addition of ion beams compensate the loss of positive ions due to addition of positron.

In Fig. 2, we describe the variation in phase velocity with ion beam velocity at different concentrations of the positron, and as expected the phase velocity goes down with increasing positron density. Here, we can see the propagation of namely, the fast beam mode, the slow beam mode and the inherent ion acoustic mode. The characteristic of these modes under different physical plasma conditions is described

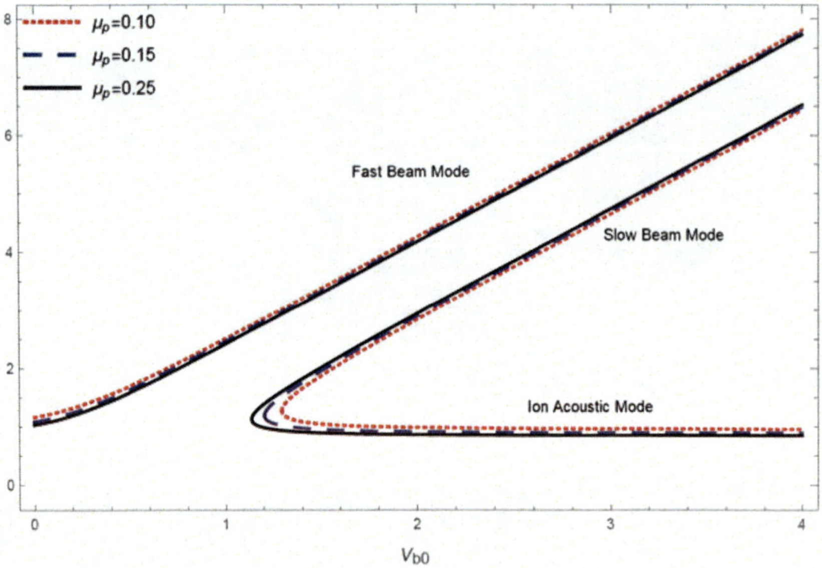

Fig. 2 Effect of positron density on three different modes of ion beam plasma. Here, the various plasma parameters considered are $\eta = 0.2, T = 0.2, \mu_b = 0.2$

in subsequent figures later on. As already described in different literature, the dominant mode of wave propagation in basically the fast beam mode with the background inherent ion acoustic plasma mode, whereas the slow beam mode will also be there having some peculiar characteristics which we believe is quite interesting (to be discussed later on).

Here, in this figure, we can see that the fast beam mode and the background plasma ion acoustic mode almost bears the similar characteristic with increasing positron density, whereas the slow beam mode shows an increase in the phase velocity with increasing positron density, this is primarily because as the restoring force decreases, the slow beam mode is free to propagate with more ease than compared to the case of absence or decrease of positron density. Moreover, with increasing positron density, the reduction of ions and hence ion beams, will always make the slow beam mode freer to move since this mode appears only at the expense of ion beams. This is also evident from Fig. 1d, where, we can see that with increasing ion beam density, the frequency increases, and also it is very much simple that, as the wave number increases (i.e. the wavelength decreases), the frequency will increase.

Figure 3a describes the typical conditions under which phase velocity can go up to the hypersonic range. The typical values of the magnetic quantization parameter, degenerate temperature, normalized positron, ion and ion beam density under which the plasma system supports fast beam mode with phase velocity only in the hypersonic range, are obtained numerically solving equations of phase velocity for a certain range of beam velocity. The phase velocity can also reach up to the hypersonic range with

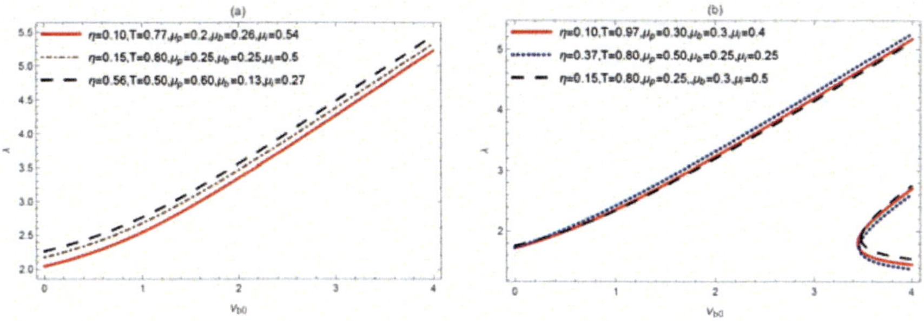

Fig. 3 **a** Phase velocity sketched in the hypersonic range for only fast mode solitary wave for different relevant combination degenerate plasma parameters, positron density, beam density, etc., **b** Phase velocity approaching hypersonic speed for all the plasma modes for a different combination of degenerate and non-degenerate plasma parameters

other degenerate plasma parameters, as shown in Fig. 3b, but in that case, the other two modes also appear for the same beam velocity.

It is clear from Fig. 3a that the phase velocity increases and reaches the highest value when the quantization, temperature are almost comparable with the highest value of positron density among the possible parameters which support hypersonic phase velocity obtained numerically. On the other hand, if the positron density decreases with increasing either of the degenerate plasma parameters, the phase velocity decreases and it is the least when the rate of decrease of positron density is the highest, irrespective of the rate of increase of the degenerate plasma parameter.

Figure 4 shows the variation of the non-linear hypersonic solitary waves for the range of combinations of plasma parameters considered for the hypersonic wave propagation. In the non-linear regime, the solitary wave potential reaches the lowest value for the comparable range of degenerate plasma parameters along with the highest chosen normalized positron density, whereas the potential of the solitary wave is the least when the value of degenerate plasma parameter along with positron density is the lowest among the chosen values. This is primarily due to the change in the non-linearity of the plasma system under the chosen set of plasma parameters. Now, if we consider the individual plots of Fig. 4, we see that with decreasing the restoring force with increasing positron density, the plasma particle can undergo frequent interaction and this interaction is more effective in the presence of magnetic quantization which can do this in a more channelized way due to the presence of strong magnetic field by confining the plasma particles along the path of the magnetic field.

Due to this frequent and prolonged interaction among the plasma charged particles, the plasma system can become more and more non-linear, and thereby the amplitude of the solitary wave may decrease as it is also evident from the evolution equation that the maximum amplitude is controlled by the non-linear coefficient. Now from the variation of the other plots in the same graph can be understood from the fact that as the degenerate plasma parameters, as well as the positron density decreases, the wave become free to move and due to this the rate of interaction

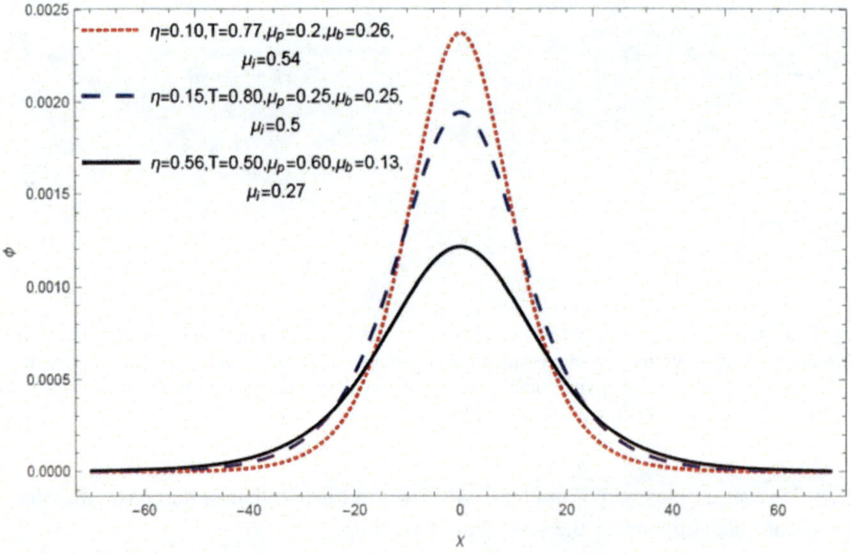

Fig. 4 Characteristic propagation of solitary wave in the hypersonic range of phase speed for the plasma parameter as shown in Fig. 2a

among the plasma particle decreases which makes the plasma system less non-linear which in turn helps the solitary wave to propagate with a higher potential.

Figure 5 describes the variation of solitary wave potential with the magnetic quantization parameter η at three different ion beam to Fermi ion sound velocity ratio V_{b0}, where $V_{b0} = V_b / C_s$. Here, we have chosen three different regimes of beam velocity, i.e. when $V_b / C_s < 1$, $V_b / C_s = 1$, and $V_b / C_s > 1$. Here, as seen from the three plots, the solitary wave potential is the highest when the beam velocity is equal to the Fermi ion sound velocity and apart from that, it is seen that solitary wave potential is the lowest when $V_{b0} = V_b = 2C_s$. This may be because when the ion beam velocity is nearly equal to the Fermi ion sound speed, the beam ions can couple to the background ions easily and thus can energize the wave which may allow the wave to propagate with higher amplitude. On the other hand, when the beam velocity is much greater than Fermi ion sound velocity (=$2C_s$), the ion beams can't couple to the background ions and, with increasing velocity of the beam ions, the system can behave more non-linearly and hence the amplitude of the solitary wave decreases. Similar is the reason for Fig. 5a, i.e. the case when $V_b / C_s < 1$.

On the other hand, in all the three cases discussed, we can see that with increasing quantization parameter, the amplitude of the solitary wave goes down. This may be because, with increasing magnetic quantization, the interaction among the charged plasma particle can be more and also, frequent due to the presence of a strong magnetic field which tends to increase the non-linear behaviour and thus eventually results in the reduction of the solitary wave potential.

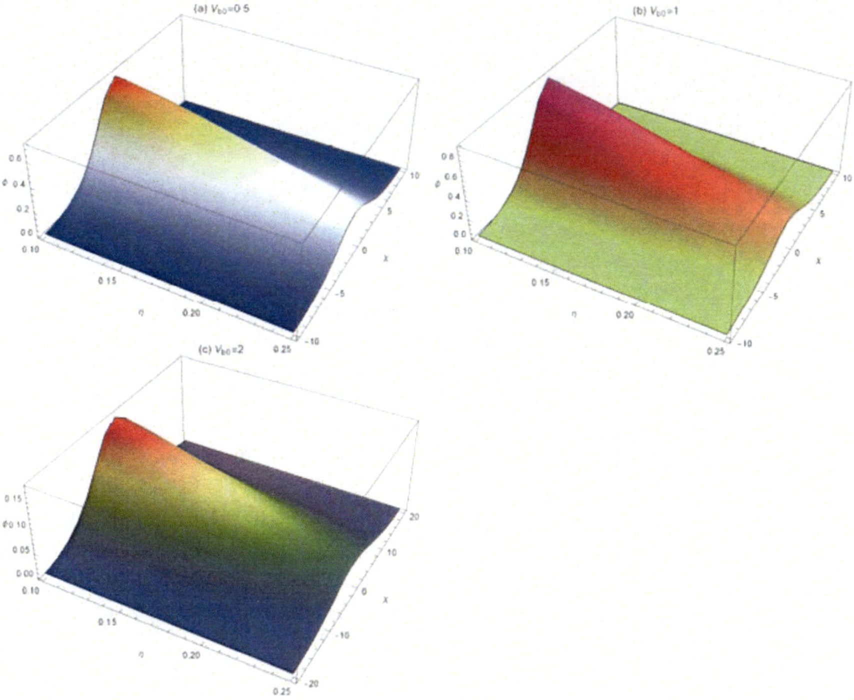

Fig. 5 Fast mode solitary wave propagation with quantized magnetic field η when **a** $V_{b0} = V_b/C_s < 1$, **b** $V_{b0} = V_b/C_s = 1$ and **c** $V_{b0} = V_b/C_s > 1$ along with other plasma parameters considered are $\eta = 0.2$, $T = 0.2$, $\mu_b = 0.2$, $\mu_p = 0.3$

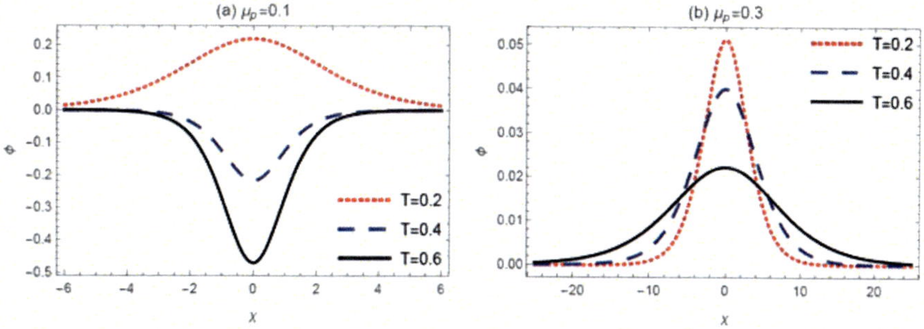

Fig. 6 Fast mode solitary wave propagation with degenerate temperature T, at $V_{b0} = V_b/C_s = 1$ when **a** $\mu_p = 0.1$ and **b** $\mu_p = 0.3$ and **c** $V_{b0} = V_b/C_s > 1$ along with other plasma parameters considered are $\eta = 0.2$, $\mu_b = 0.2$

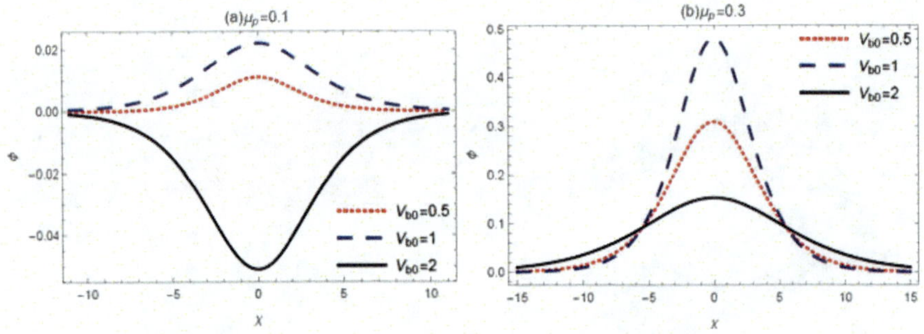

Fig. 7 Fast mode solitary wave propagation with ion beam velocity, when **a** $\mu_p = 0.1$ and **b** $\mu_p = 0.3$ along with other plasma parameters considered are $\eta = 0.2$, $\mu_b = 0.2$, $T = 0.2$

Figure 6 shows the variation of solitary wave potential with degenerate temperature. Figure 6a shows the variation of solitary wave potential with degenerate temperature for a lower normalized positron concentration and Fig. 6b shows the variation of solitary wave potential with degenerate temperature for a higher normalized positron concentration. As can be seen from Fig. 6a, both compressive as well as a rarefactive solitary wave can co-exist at a lower concentration of positron, whereas, at a higher concentration of positrons, only compressive solitary wave exists. It is also seen that the solitary wave amplitude decreases with increasing degenerate temperature and positron density. This may be because the positron density increases, the restoring force decreases which in turn decreases the frequency and if, in this condition, the degenerate temperature increases, there may be a rapid interaction among the charged particles of the plasma and thereby may cause an enhancement in the non-linearity of the plasma system and thus, we can expect to notice a decrease of the solitary wave potential with increasing degenerate temperature.

Figure 7 describes the characteristics of solitary wave propagation with the streaming velocity of ion beams at different positron density.

Figure 7a describes the solitary wave propagation with increasing streaming velocity of beam ions when the positron density is lower i.e. $\mu_p = 0.1$ and it is seen that in the compressive regime the amplitude is the highest when $V_{b0} = V_b \big/ C_s = 1$ and with increasing ion beam streaming speed, the amount of rarefaction increases. This may be because, in the situation $V_{b0} = 1$, the beam ions can undergo an effective coupling with the background plasma ions, energizing the wave further, whereas, as the ion beam speed increases, becomes twice of the Fermi ion sound speed, the ion beams can add to the amount rarefaction suffered by the wave in terms of additional streaming speed and thus the width as well as the amount of rarefaction increases as the ion beams can drive the wave further overcoming positron contribution. This can be better understood in Fig. 7b. As seen from the figure, when the positron density increases, the solitary wave has the highest amplitude when $V_{b0} = 1$ for obvious reason as discussed above and further when the beam velocity becomes twice of the Fermi ion sound speed, the wave amplitude (or amount of compression) decreases but continues to be compressive may be due to the higher amount of positron contribution.

On the other hand, in Figs. 8, 9 and 10, we try to summarize the variation of solitary wave potential with normalized positron density for three different regimes of beam velocity, i.e. when $V_b/C_s < 1$, $V_b/C_s = 1$, and $V_b/C_s > 1$ at two different concentrations of ion beams. For instance, the variation of solitary wave potential when $V_b/C_s < 1$ at two different ion beam concentration is plotted in Fig. 8a, b, respectively. As seen from Fig. 8a, b, with an increase in the positron density, the solitary wave potential increases with lower and higher concentrations of beam ions.

This increase in the solitary wave potential can be attributed to the fact that as the positron density increases, a lesser restoring force helps in a better coupling of the ion beams to the plasma ions thereby energizing the wave potential. However, keeping the same physical situation intact as shown in Fig. 8a, if we increase the ion beam density, a slight reduction in the solitary wave potential is seen as depicted in Fig. 8b. This is primarily because as the beam ions density increases, there may be an increase in the charged particle collision which eventually decreases the solitary wave potential.

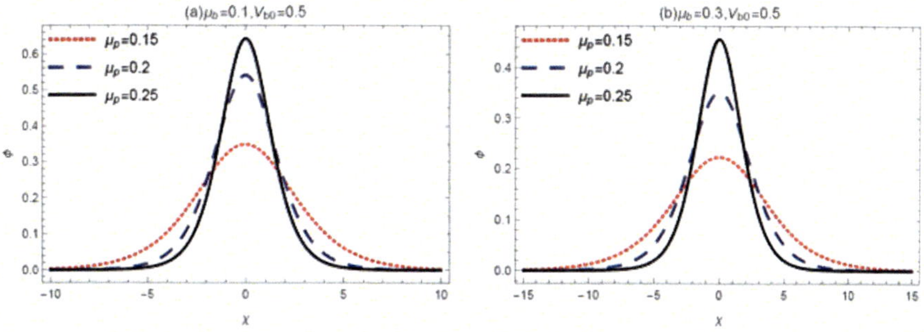

Fig. 8 Fast mode solitary wave propagation with positron density, when **a** $\mu_b = 0.1$ and $V_{b0} = V_b/C_s < 1$, **b** $\mu_b = 0.3$ and $V_{b0} = V_b/C_s < 1$ along with other plasma parameters considered are $\eta = 0.2$, $T = 0.2$

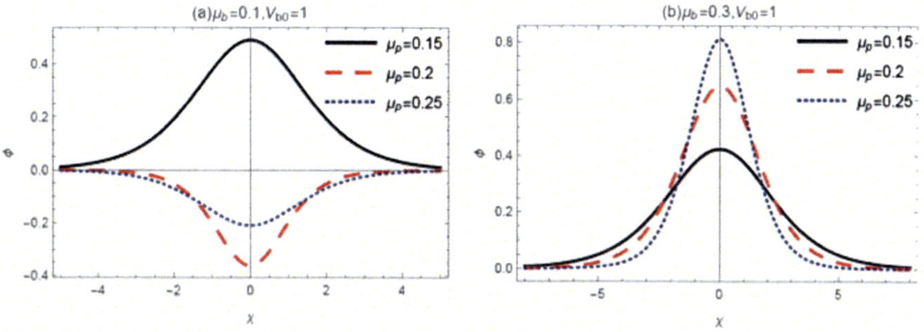

Fig. 9 Fast mode solitary wave propagation with positron density, when **a** $\mu_b = 0.1$ and $V_{b0} = V_b/C_s = 1$, **b** $\mu_b = 0.3$ and $V_{b0} = V_b/C_s = 1$ along with other plasma parameters considered are $\eta = 0.2$, $T = 0.2$

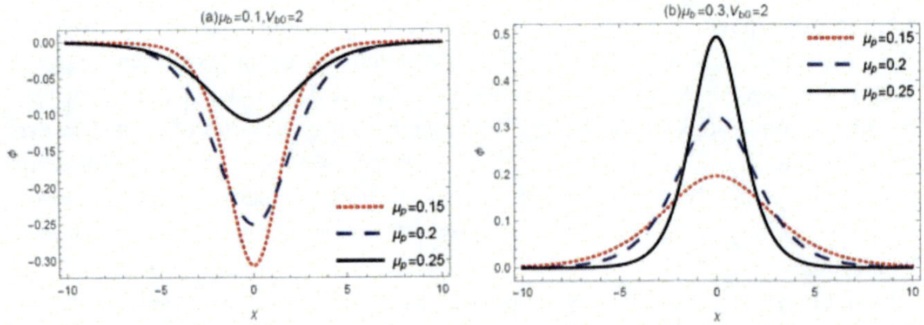

Fig. 10 Fast mode solitary wave propagation with positron density, when **a** $\mu_b = 0.1$ and $V_{b0} = V_b/C_s > 1$, **b** $\mu_b = 0.3$ and $V_{b0} = V_b/C_s > 1$ along with other plasma parameters considered are $\eta = 0.2$, $T = 0.2$

In Fig. 9a and b, we describe the solitary wave propagation with positron density when $V_b/C_s = 1$. As it is already described that when the ion beam velocity and ion Fermi sound velocity becomes equal, the ion beams can easily couple to the background plasma ions and primarily due to this reason, we see the highest possible solitary wave potential with increased positron density for a higher ion beam concentration among all the chosen physical plasma situation. With increased ion beams, this coupling may be the most effective which is the case portrayed in Fig. 9b.

Also, as seen from Fig. 9a, at lower ion beam concentration, with increasing positron density, both compressive and rarefactive solitary waves can co-exist. Now as the positron concentration increases the restoring force increases and as this increase in restoring force is effective due to better coupling of background ions and beam ions, due to significant rate of frequency decrease, the amount of rarefaction increases than compression. Again, on the other hand, the rate of decrease of ions due to positrons can be expected to be compensated by the addition of more beam ions which in turn helps in the reduction of rarefaction and as a result, the solitary wave can become compressive again at higher ion beam concentration as shown in Fig. 9b.

On the other hand, in Fig. 10a and b, the similar plasma parameters are considered as Figs. 8 and 9 except the case $V_b = 2C_s$. Now, since in this case, the beam ions are much more energetic than background plasma ions, the rate of loss of ions with increasing positron density will take place more quickly and hence the frequency, and thus the amount of rarefaction will tend to increase even in the lower positron concentration unlike the case of Fig. 9a. However, as shown in Fig. 10b with the addition of energetic ion beam density at a faster rate (i.e. with a higher velocity), the rate of loss of ion due to the addition of positrons will be compensated and again we can expect a higher amount of compression which is exactly the situation described in the Figure. In Fig. 11a and b, we discuss the characteristics of ion acoustic mode which shows more or less similar characteristics of dominant fast mode. Here, we sketch the nature of propagation of ion acoustic mode with ion beam velocity at lower and higher positron concentration in Fig. 11a and b, respectively. The wave

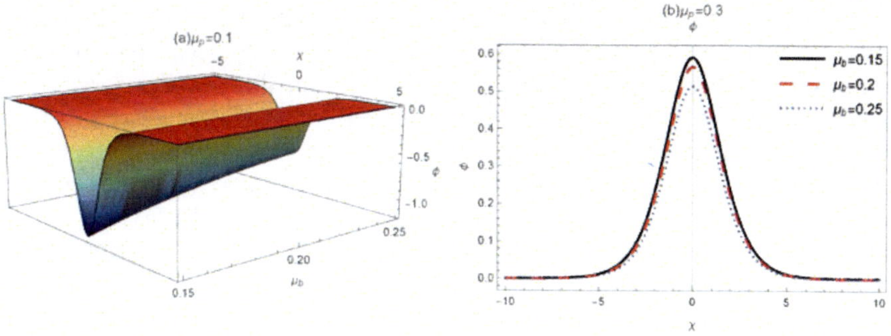

Fig. 11 Solitary wave of ion acoustic mode with ion beam density at **a** $\mu_p = 0.1$ and $V_{b0} = V_b / C_s > 1$, **b** $\mu_p = 0.3$ and $V_{b0} = V_b / C_s > 1$ along with other plasma parameters such as $\eta = 0.2$, $T = 0.2$

potential continues to be rarefactive in nature with low positron concentration and on careful observation, we notice that with increasing ion beam concentration, the amount of rarefaction decreases which is for obvious reason as discussed above. On the other hand, for a higher positron concentration, when the ion beam density increases, which results in compensation of the rate of decrease of ion density, the amount of compression increases and as a result, the wave continues to propagate in a compressive mode.

Figure 12a and b demonstrate the features of solitary waves in the slow beam mode at different ion beam density with increasing positron density. To understand the nature of the slow mode, we have to correlate Fig. 12 with Fig. 10.

In both cases, the same physical parameter is used to understand the nature of the two different beam modes. It has been already established that for compressive type fast mode solitary wave in ion beam plasma, the slow beam mode will always appear bearing the opposite characteristics, i.e. rarefactive type [51]. But, till now, it has never been demonstrated what would happen if the fast beam mode is originally a

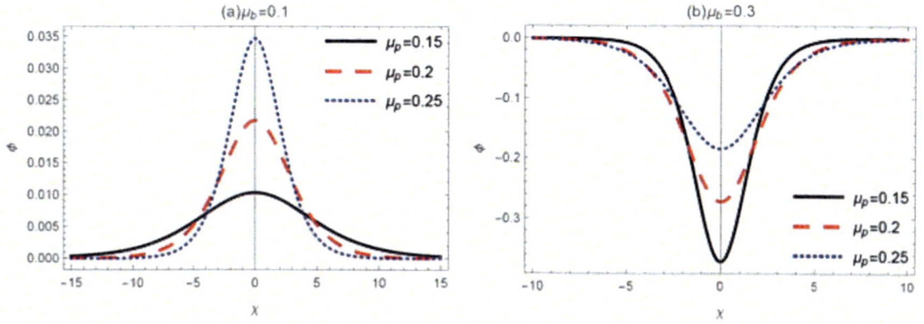

Fig. 12 Solitary wave of slow mode with positron density when **a** $\mu_b = 0.1$ and $V_{b0} = V_b / C_s > 1$, **b** $\mu_b = 0.3$ and $V_{b0} = V_b / C_s > 1$ along with other plasma parameters considered are $\eta = 0.2$, $T = 0.2$

rarefactive type. Here, we have demonstrated that the slow beam mode always carries a distinctive characteristic from that of a compressive type. As seen from both the figures, for a rarefactive type fast beam mode (Fig. 10a), the slow beam mode appears compressive type (Fig. 12a), whereas, for a compressive type fast mode (Fig. 10b), the slow beam mode appears rarefactive type (Fig. 12b). Interestingly, in both the figures we see that in the case of a rarefactive type compressive fast mode solitary wave, the amount of rarefaction decreases with increasing positron density (Fig. 10a), whereas, in the compressive type slow beam mode solitary wave, the amount of compression increases with increasing positron density (Fig. 12a). On the other hand, for a compressive type fast beam mode, the amount of compression increases with increasing positron density (Fig. 10b), while, for a rarefactive type slow beam mode, the amount of rarefaction decreases with increasing positron density (Fig. 12b). Thus, we see that the fast beam mode and slow beam mode will be always of differing characteristics, irrespective of any physical plasma situation in ion beam plasma. We believe that this is the first report where this typical swing between compression and rarefaction in ion beam-driven ion acoustic plasma is discussed in detail though we still believe that proper numerical scheme will reveal the exact nature and type of different modes in such plasma system which is beyond the scope for the present manuscript.

2.2 *Excluding the Effect of Positron*

Here in this section, we discuss the evolution of Hypersonic soliton without the effect of the positron, and it is seen that the requirement of the value of different parameters such as magnetic quantization, degenerate temperature, ion beam concentration, etc., are quite different in the case of without positron, for the propagation of the solitary wave with hypersonic velocity.

If we compare Figs. 3 and 13, we see that with positron, we can have solitary waves whose Mach number may vary from supersonic to hypersonic whereas, without

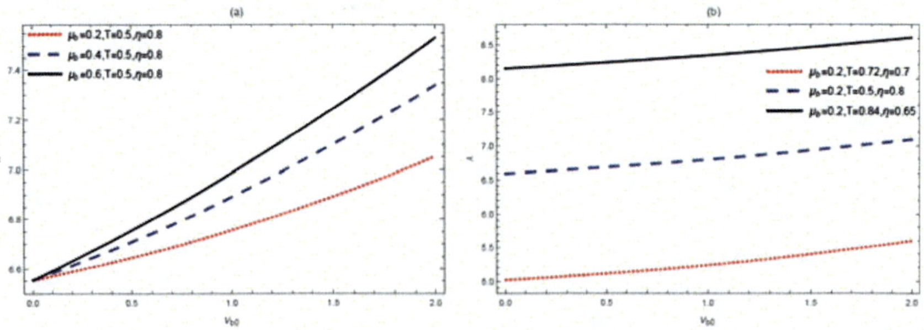

Fig. 13 **a** Variation of Mach number with ion beam concentration for the different chosen combinations of the magnetic quantization and temperature. **b** Variation of Mach number only in the Hypersonic range

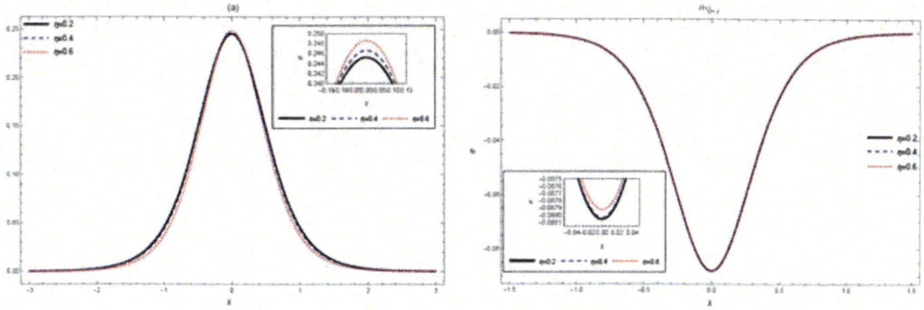

Fig. 14 **a** Spatial variation of ion acoustic mode and **b** slow mode with magnetic quantization

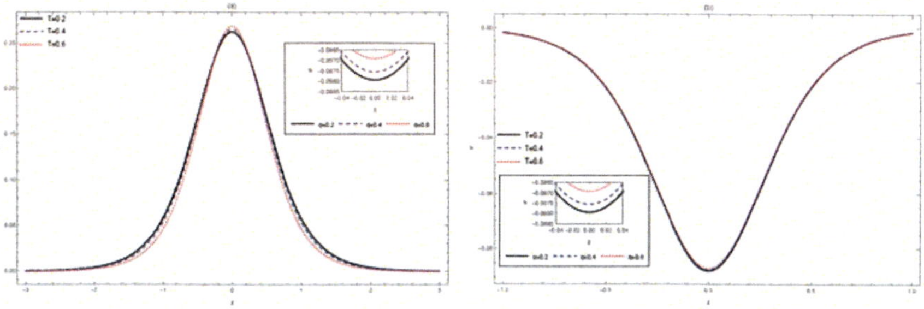

Fig. 15 **a** Variation of ion acoustic mode and **b** slow mode with degenerate temperature with $\eta = 0.2$, $\mu_b = 0.2$

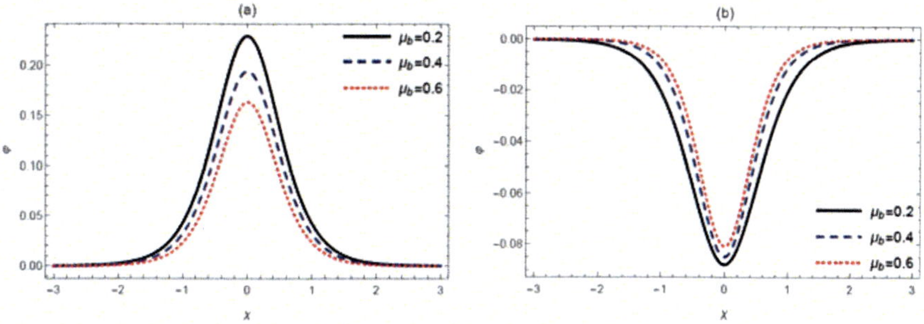

Fig. 16 **a** Variation of ion acoustic mode and **b** slow mode with ion beam density with $T = 0.2$, $\eta = 0.2$

positron, we can have Mach number only in the Hypersonic range. Moreover, in the presence of positrons (Fig. 3), we need higher values of beam velocity to obtain solitary waves in the hypersonic range than compared to the case of without positrons (Fig. 13).

On the other hand, the ion acoustic mode (Figs. 14a, 15a, 16a) is always compressive in the absence of positron whereas, we can have compressive, as well as the rarefactive solitary wave of ion acoustic mode with positrons (Fig. 11). Apart from these, the slow mode (Figs. 14b, 15b, 16b) is always rarefactive in the absence of positrons, whereas the slow mode can be compressive as well as rarefactive in the presence of positrons (Fig. 12).

3 Conclusion

Here, we have discussed the outstanding characteristics of wave propagation in an ion beam plasma with and without positron in the presence of magnetic quantization and degenerate temperature. Depending on different chosen physical plasma parameter of magnetic quantization, degenerate temperature, ion beam density and positron density, we can encounter only hypersonic fast beam mode propagating in such plasma, whereas, depending on some other physical parameter, we can have hypersonic fast beam mode along with slow beam mode and ion acoustic mode. All of these three modes can be either compressive or rarefactive type depending on ion beam density, velocity, positron density, etc. On the other hand, the ion acoustic mode, slow mode bears complete opposite characteristics with and without positron. In the presence of degenerate plasma parameter, the non-degenerate plasma parameter such as the beam ion's concentration as well positron density has a substantial role in the amount of compression and rarefaction suffered by either fast beam mode or slow beam mode. The slow beam mode becomes compressive for a rarefactive type fast beam mode, whereas, for a compressive type fast beam mode, the slow beam mode appears rarefactive type. However, the ion acoustic mode which is the integral background plasma mode continues to follow the same trend of variation with the dominant plasma mode, i.e. fast mode. The results presented here can be of significant importance for different astrophysical plasma environments where such plasma ingredient has an abundance as discussed in the introduction part of the manuscript.

Acknowledgements Manoj Kumar Deka would like to thank the Department of Science and Technology, Govt of India, for providing financial assistance to carry out the research work under a sanction by Science and Engineering Research Board via Grant No. YSS/2015/001896.

References

1. W.K. Misner, S. Thorne, J.A. Wheeler, (Gravitation San Francisco: Freeman 1973)
2. M.J. Rees, G.W. Gibbons, S.W. Hawking, S. Siklaseds, *The Early Universe* (Cambridge University Press, Cambridge, 1983)
3. M.I. Barns, *Positron Electron Pairs in Astrophysics* (American Institute of Physics, New York, 1983)
4. H.R. Miller, P.J. Witta, *Active Galactic Nuclei* (Springer, Berlin, 1978)
5. F.C. Michel, *Theory of Neutron star Magnetosphere* (Chicago University Press, Chicago, 1991)
6. N. Ghosh, A. Saha, N. Pal, P. Chatterjee, J. Theor. Appl. Phys. **9**, 321 (2015)
7. J.K, Mishra, Phys. Plasmas. **17**, 102315 (2010)

8. C.M. Surko, M. Leventhal, A. Passner, Phys. Rev. Lett. **62**, 901 (1989)
9. H. Boehmer, M. Adams, N. Rynn, Phys. Plasmas **2**, 4369 (1995)
10. E.P. Liang, S.C. Wilks, M. Tabak, Phys. Rev. Lett. **81**, 4887 (1998)
11. S.A. El-Tantawy, M. Tribeche, W.M. Moslem, Phys. Plasmas **19**, 032104 (2012)
12. C. Tsallis, J. Stat. Phys. **52**, 479 (1988)
13. R. Lundin, A. Zakharov, R. Pellinen, H. Borg, B. Hultqvist, N. Pissarenko, E.M. Dubinin, S.W. Barabash, I. Liede, H. Koskinen, Nature **341**, 609 (1968)
14. M.K. Deka, A.N. Dev, Ann. Phys. **395**, 45 (2018)
15. P.K. Shukla, B. Eliasson, Rev. Mod. Phys. **83**, 885 (2011)
16. R. Redmer, G. R€opke, Contrib. Plasma Phys. **50**, 970 (2010)
17. V.E. Fortov, D.H.H. Hoffmann, B.Y. Sharkov, Phys. Usp. **51**, 109 (2008)
18. D.H.H. Hoffmann, A. Blazevic, P. Ni, O. Rosmez, M. Roth, N.A. Tahir, A. Tauschwitz, S. Udrea, D. Varentsov, K. Weyrich, Y. Maron, Laser Part. Beams **23**, 47 (2005)
19. A. Rahman, S.A. Khan, A. Qamar, Plasma Sci. Technol. **17**, 1000 (2015)
20. M.T. Gabdullin, S.K. Kodanova, T.S. Ramazanov, M.K. Issanova, T.N. Ismagambetova, Nukleonika **61**, 125 (2016)
21. V.E. Fortov, *Extreme States of Matter on Earth and in the Cosmos* (Springer, Berlin/Heidelberg, 2011)
22. B.Y. Sharkov, D.H.H. Hoffmann, A.A. Golubev, Y. Zhao, Matter Radiat. Extremes **1**, 28 (2016)
23. V.E. Fortov, *Extreme States of Matter High Energy Density Physics* (Springer Series in Materials Science, 2016)
24. K. Sauer, E. Dubinin, Phys. Scr. T. **107**, 167 (2004)
25. M.E. Dieckmann, A. Meli, P.K. Shukla, L.O.C. Drury, A. Mastichiadis, Plasma Phys. Controlled Fusion **50**, 065020 (2008)
26. M.E. Dieckmann, B. Eliasson, P.K. Shukla, Phys. Rev. E **70**, 036401 (2004)
27. M.K. Deka, A.N. Dev, A.P. Misra, N.C. Adhikary, Phys. Plasmas **25**, 012102 (2018)
28. E.F. El-Shamy, R.C. Al-Chouikh, A. El-Depsy, S. Al-Wadie, Phys. Plasmas **23**, 122122 (2016)
29. A. El-Depsy, M.M. Selim, IEEE Trans. Plasma Sci. **44**, 2901 (2016)
30. A.U. Rahman, S.A. Khan, A. Qamar, Plasma Sci. Tech. **17**, 12 (2015)
31. S.K. El-Labany, W.F. El-Taibany, A.E. El-Samahy, A.M. Hafez, A. Atteya, IEEE Trans. Plasma Sci. **44**, 842 (2016)
32. K. Aoutou, M. Tribeche, T.H. Zerguini, Astrophys Space Sci. **340**, 355 (2012)
33. S. Sadiq, S. Mahmood, Q. Haque, M.Z. Ali, Astrophy. J. **793**, 27 (2014)
34. H.A. Shah, W. Masood, M.N.S. Qureshi, N.L. Tsintsadze, Phys. Plasmas **18**, 102306 (2011)
35. A. Abdikian, S. Mahmood, Phys. Plasmas **23**, 122303 (2016)
36. A. Rasheed, N.L. Tsintsadze, G. Murtaza, Phys. Plasmas **18**, 112701 (2011)
37. N.L. Tsintsadze, H.A. Shah, M.N.S. Qureshi, M.N. Tagviashvili, Phys. Plasmas **22**, 022303 (2015)
38. M. Irfan, S. Ali, A.M. Mirza, Phy. Plasmas **24**, 052108 (2017)
39. S. Arshad, H. Shah, M.N.S. Qureshi, Phys. Scr. **89**, 075602 (2014)
40. W. Masood, M.I. Shaukat, H.A. Shah, A.M. Mirza, Phys. Plasmas **22**, 032305 (2015)
41. M.J. Iqbal, W. Masood, H.A. Shah, N.L. Tsintsadze, Phys. Plasmas **24**, 014503 (2017)
42. M.I. Shaukat, W. Masood, H.A. Shah, M.J. Iqbal, A.M. Mirza, Phys. Plasmas **23**, 102310 (2016)
43. M.I. Shaukat, Phys. Plasmas **24**, 062305 (2017)
44. S. Ali Shan, A.U. Rehman, A. Mushtaq, Phys. Plasmas **24**, 032104 (2017)
45. M.J. Iqbal, H.A. Shah, W. Masood, N.L. Tsintsadze, Eur. Phys. J. D **72**, 192 (2018)
46. H.A. Shah, M.J. Iqbal, N. Tsintsadze, W. Masood, M.N.S. Qureshi, Phys. Plasmas **119**, 092304 (2012)
47. L.D. Landau, E.M. Lifshitz, *Statistical Physics, Part 1* (Butterworth-Heinemann, Oxford, 1980)

48. W. Malfliet, Am. J. Phys. **60**, 650 (1992)
49. D. Koester, G. Chanmugam, Rep. Prog. Phys. **53**, 837 (1990)
50. S.S. Ghosh, G.S. Lakhina, Nonlinear Processes Geophys. **11**, 219 (2004)
51. M.K. Deka, N.C. Adhikary, A.P. Misra, H. Bailung, Y. Nakamura, Phys. Plasmas **19**, 103704 (2012)

Study on Analytical Solutions of K-dV Equation, Burgers Equation, and Schamel K-dV Equation with Different Methods

Sanjaya Kumar Mohanty and Apul N. Dev

Abstract The study on analytical solutions of differential equations is quite useful in Modeling in fluid dynamics, physics, etc. In this review work we studied the analytical solutions of Korteweg–de Vries equation (K-dV), Burgers equation, Schamel equation, and Schamel–Korteweg–de Vries equations by using different analytical methods such as tanh method, sech method, sine-Gordon equation method, $\left(\frac{G'}{G}\right)$ expansion method, and tanh–coth methods. The $\left(\frac{G'}{G}\right)$ method has different types that are used to solved Schamel equation and Schamel–K-dV equation.

Keywords K-dV equation · Burgers equation · Schamel equation · Schamel–K-dV equation · tanh method · coth method · sech method · $\left(\frac{G'}{G}\right)$ methods · Sine-Gordon method · tanh-coth method

1 Introduction

Nonlinear evolution equations are used to describe the physical existence or physical models. The study of applications study on analytical solutions of the nonlinear partial differential equations is in fluid dynamics, plasma physics, nonlinear optics, engineering, mathematical physics and modeling, and so on. The important and applied nonlinear evolution equations are K-dV equation, Burgers equation, Schamel equation. Schamel–K-dV equation, and so on. To find out the analytical solutions of the nonlinear equations, many authors provide many methods and out of those

S. K. Mohanty
Department of Mathematics, Siksha 'O' Anusandhan (Deemed to be University), Bhubaneswar 751030, Odisha, India

A. N. Dev (✉)
Center for Applied Mathematics and Computing, Siksha 'O' Anusandhan (Deemed to be University), Bhubaneswar 751030, Odisha, India
e-mail: apulnarayandev@soa.ac.in

S. R. Mishra et al. (eds.), *Recent Trends in Applied Mathematics*, Lecture Notes in Mechanical Engineering, https://doi.org/10.1007/978-981-15-9817-3_9

methods tanh method, sech method, tanh–coth method, sine–cosine method, $\left(\frac{G'}{G}\right)$ method, and sine-Gordon method are quite famous.

The tanh method is used to solve different nonlinear evolution equations in recent years. In 1996, Willy Malfliet solved K-dV Burgers, Dissipative-Dispersion, combined K-dV-MKdV, and extended MKdV–Bureges equations [1]. In 2000, Fan explores the solution of K-dV–Burgers–Kuramoto, 2-dimensional K-dV–Burgers and generalized Burgers–Fisher equations [2]. In 2004, Wazwaz solved generalized K-dV, generalized Fishers equations [3]. In 2005, Evans and Raslan studied the improved K-dV equation, equal width wave equation (EWE), Regularized long Wave and Coupled Burgers equations [4]. In 2007, Wazwaz studied the analytical solutions of the fifth-order K-dV equation, Lax equation, Sawada–Kotera (SK) equations, etc. [5]. In 2009, Sarma solved K-dV equation and MKdV equations [6, 7]. In 2010, Jawad et al. solved Burgers, K-dV–Burgers, Coupled Burgers, generalized time-delayed Burgers, Perturbed Burgers equations [8]. In 2013, Karimi solved a modified K-dV equation [9]. In 2016, Zhang and Yin solved Burgers equation [10]. Adem solved Coupled KP equation [11], Tariq and Akram solved Cahn-Allen and Phi-4 equations [12] and Ralson et al. solved time-fractional EW and MEW equations [13]. In 2019, Ali et al. solved variable coefficients of Burgers equation [14] and so on.

The tanh–coth method is used to solve analytical solutions of nonlinear evolutions equations is quite useful. In 2007 and In 2008, Wazwaz solved Fisher, Newell–Whitehead, Allen–Cahn and Fitz–Hugh–Nagumo, Burgers–Fisher, Burgers, Kodomtsev–Petviashvili, Pochhammer–Chree equations [15–17]. In 2009, Wazzan solved K-dV and K-dV Burgers equations [18]. In 2010, Parkes solved the MKdV equation, Salas and Gomezs solved K-dV equation of sixth order and MKdV equation of sixth order [19, 20]. In 2013, Jawad solved one-dimensional Burgers equation, K-dV–Burgers equation, Coupled Burgers equation, and generalized time-delayed Burgers equation [21]. In 2017 Chukkol et al. solved K-dV–Burgers equation with forcing term [22]. Asokan and Vinodh solved Sawad–Kotera equation [23]. In 2018, Asokan and Vinodh solved generalized K-dV–BBM and potential K-dV–BBM equations [24] and so on.

The sine-Gordon method is used to solved nonlinear evolution equations. In 2003, Yan solved K-dV equation, MKdV equation, and Complex NLS positive equation [25]. In 2016, Alquran and Krishnan solved generalized Phi-4 equation, generalized regularized long-wave equation and equal width equation and regularized long-wave equations [26]. In 2020, Korkmaz et al. solved conformable time-fractional RLW equation [27]. Guirao et al. solved $(3+1)$-dimensional B-type Kadomtsev-Petviashvili–Boussinesq equation [28].

The sine–cosine method is used to get analytical solutions of nonlinear equations. In 2004 and 2005, Wazwaz solved K-dV equation, generalized K-dV equation, Boussinesq equation, RLW equation, and Phi-4 equation, complex modified K-dV equation and complex generalized K-dV equation [29, 30]. In 2014, Bibi and Mohyudi-Din solved a modified K-dV equation [31]. In 2015, Yang and Tang solved the Schamel–K-dV equation [32]. In 2017, Raslan solved coupled general equal width wave equation [33]and so on.

The sech method is used to solve nonlinear evolution equations. In 2004, Baldwin et al. solved Hirota Satsuma System of Coupled K-dV equation [34]. In 2007, Wazwaz solved Jaulent–Miodek equation [35]. In 2008, Ganji and Abdollahzadeh solved Lax seventh-order K-dV equation [36]. In 2011, Wei and Tang solved coupled ZK equation [37]. In 2016, Jawad solved modified ZK equations, Dubrovsky equations [38], and so on.

$\left(\frac{G'}{G}\right)$ method is used to solve many nonlinear evolution equations out of those equations such equations are K-dV equation [19, 39, 40], modified-K-dV equation [41], fifth-order K-dV equation [42], seventh-order K-dV equation [43], ninth-order K-dV equation [44], K-dV–Burgers equation [45], the 2D-K-dV equation [46], Burgers equation [47, 48], equal width Burgers equation [49], K-dV-MKdV equation [50], coupled MKdV equation [51], Schrodinger–K-dV equation [52], coupled MKdV equation [53], Schamel–K-dV equation [54], and so on.

2 Solutions of K-dV Equation by *tanh* Method

Sarma in 2009 [6] evaluated the solutions of the K-dV equations of 3rd order using *tanh* method. K-dV equation of third order is of the form

$$\frac{\partial u}{\partial t} + Au\frac{\partial u}{\partial x} + B\frac{\partial^3 u}{\partial x^3} = 0 \tag{1}$$

where A and B are nonzero constants and $u = u(x, t)$. Now we are using wave transformation $X = a(x - kt)$ to convert the partial differential equation into ordinary differential equation, where a and k are nonzero constants.

$$Ba^2\frac{d^2u}{dX^2} + A\frac{u^2}{2} - ku = 0, \tag{2}$$

where $u = u(X)$. Now introducing the independent variable $Z = tanh X$ and $u(X) = \omega(Z)$. Now substituting these values in (2),

$$Ba^2(1 - Z^2)(1 - Z^2)\frac{d^2\omega}{dZ^2} - 2Ba^2Z(1 - Z^2)\frac{d\omega}{dZ} + A\frac{\omega^2}{2} - k\omega = 0 \tag{3}$$

The above equation has power series solution as it has a singular point +1 and −1.

$$\omega(Z) = \sum_{r=0}^{\infty} a_r Z^{\rho+r}$$

Then Eq. (3) becomes

$$\begin{aligned}&Ba^2(1-Z^2)(1-Z^2)\sum_{r=0}^{\infty}a_r(\rho+r)(\rho+r-1)Z^{\rho+r-2}\\&-2Ba^2Z(1-Z^2)\sum_{r=0}^{\infty}a_r(\rho+r)Z^{\rho+r-1}\\&+\tfrac{A}{2}(\sum_{r=0}^{\infty}a_rZ^{\rho+r})^2-k\sum_{r=0}^{\infty}a_rZ^{\rho+r}=0\end{aligned}\tag{4}$$

Now equating the highest order derivative and highest power on nonlinear term in (4) and taking $\rho=0$.

$$\begin{aligned}&Z^{4+\rho+r-2}=Z^{(\rho+r)2}\\&\Rightarrow 4+r-2=2r\\&\Rightarrow r=2.\end{aligned}$$

For $r=2$, we have

$$\omega(Z)=a_0+a_1Z+a_2Z^2,\ a_2\neq 0.\tag{5}$$

Now (3) becomes

$$\begin{aligned}&2a_2Ba^2Z^4+4Ba^2a_2Z^4+\frac{A}{2}a_2^2Z^4+2Ba^2a_1Z^3+Aa_1a_2Z^3\\&-4a_2Ba^2Z^2-4Ba^2a_2Z^2+\frac{A}{2}a_1^2Z^2+Aa_0a_2Z^2-ka_2Z^2\\&-2Ba^2a_1Z+Aa_0a_1Z-ka_1Z+\frac{A}{2}a_0^2-ka_0+2a_2Ba^2=0\end{aligned}\tag{6}$$

Now equating the coefficient of Z^4, Z^3, Z^2, Z and constant terms in (6) to get the value of a_0, a_1, a_2 and k. Here $a_0=\frac{1}{A}(k+8Ba^2)$, $a_1=0$, $a_2=-\frac{1}{A}12Ba^2$ and $k=\pm 4Ba^2$. Now substituting the values of a_0, a_1 and a_2 in (5).

Case 1 $K=4Ba^2$ and $Z=tanhX$.

$$\begin{aligned}&\Rightarrow u(X)=\frac{12}{A}Ba^2(1-tanh^2X)\\&\Rightarrow u(X)=\frac{12}{A}Ba^2sech^2X\\&\Rightarrow u(x,t)=\frac{12}{A}Ba^2sech^2[a(x-kt)]\\&\Rightarrow u(x,t)=\frac{12}{A}Ba^2sech^2[a(x-4Ba^2t)]\end{aligned}\tag{7}$$

Case 2 $K = -4Ba^2, Z = tanhX$

$$u(x,t) = \frac{4}{A}Ba^2(-2 - 3sech^2[a(x-kt)]) \tag{8}$$

Equations (7) and (8) are the required solutions of the K-dV equation (1).

3 Solutions of K-dV Equation by *sech* Method

Sarma in 2009 [7] evaluated the numerical solution of the K-dV equations of third order y using *tanh* method. The well-known K-dV equation in the simplest form is

$$\frac{\partial u}{\partial t} + Au\frac{\partial u}{\partial x} + B\frac{\partial^3 u}{\partial x^3} = 0 \tag{9}$$

Now using transformation $X = a(x - kt)$.

Equation (9) is of the form

$$Ba^2\frac{d^2u}{dX^2} + A\frac{u^2}{2} - ku = 0, \tag{10}$$

where $u = u(X)$. Now introducing the independent variable $Z = sechX$ and $u(X) = \omega(Z)$.

Then Eq. (10) can be written as

$$Ba^2Z^2(1-Z^2)\frac{d^2\omega}{dZ^2} + Ba^2(Z - 2Z^3)\frac{d\omega}{dZ} + A\frac{\omega^2}{2} - k\omega = 0 \tag{11}$$

Let us assume the power series solution of the Eq. (11) as follows:

$$\omega(Z) = \sum_{r=0}^{\infty} a_r Z^{\rho+r}.$$

Now Eq. (11) becomes

$$\begin{aligned}
&Ba^2Z^2(1-Z^2)\sum_{r=0}^{\infty} a_r(\rho+r)(\rho+r-1)Z^{\rho+r-2} \\
&+Ba^2(Z-2Z^3)\sum_{r=0}^{\infty} a_r(\rho+r)Z^{\rho+r-1} \\
&+\frac{A}{2}\left(\sum_{r=0}^{\infty} a_r Z^{\rho+r}\right)^2 - k\sum_{r=0}^{\infty} a_r Z^{\rho+r} = 0
\end{aligned} \tag{12}$$

Now equating the highest order derivative and highest power on nonlinear term in (12) and taking $\rho = 0$.

$$Z^{4+\rho+r-2} = Z^{(\rho+r)2}$$
$$\Rightarrow 4 + r - 2 = 2r$$
$$\Rightarrow r = 2.$$

For $r = 2$, then

$$\omega(Z) = a_0 + a_1 Z + a_2 Z^2,\ a_2 \neq 0 \tag{13}$$

Now Eq. (11) becomes

$$\begin{aligned}&-2Ba^2a_2Z^4 - 4Ba^2a_2Z^4 + \frac{A}{2}a_2^2Z^4 - 2Ba^2a_1Z^3 + Aa_1a_2Z^3\\&+4Ba^2a_2Z^2 + \frac{A}{2}a_1^2Z^2 + Aa_0a_2Z^2 - ka_2Z^2 + Ba^2a_1Z\\&+Aa_0a_1Z - ka_1Z + \frac{A}{2}a_0^2 - ka_0 = 0\end{aligned} \tag{14}$$

Now equating the coefficient of Z^4, Z^3, Z^2, Z and the constant terms in (14) to get the value of a_0, a_1, a_2, and k. We have

$$a_0 = -\frac{8}{A}Ba^2,\quad a_1 = 0,\quad a_2 = \frac{12}{A}Ba^2 \text{ and } k = \pm 4Ba^2 \tag{15}$$

Now substituting these values in Eq. (13)

Case 1 $K = 4Ba^2$

$$u(x,t) = \frac{12}{A}Ba^2 sech^2[a(x - 4Ba^2t)] \tag{16}$$

Case 2 $K = -4Ba^2$

$$u(x,t) = \frac{12}{A}Ba^2 sech^2[a(x + 4Ba^2t)] \tag{17}$$

Equations (16) and (17) are the required solutions of the K-dV equation (9).

4 Solutions of K-dV Equation by *sine–cosine* Method

Wazwaz in 2009 [29] solved the solutions of the kdv equation by using *sine–cosine* method. The well-known K-dV equation in the simplest form is

$$\frac{\partial u}{\partial t} + Au\frac{\partial u}{\partial x} + B\frac{\partial^3 u}{\partial x^3} = 0 \tag{18}$$

Now using transformation $X = a(x - kt)$.

Equation (18) is of the form

$$Ba^2\frac{d^2u}{dX^2} + A\frac{u^2}{2} - ku = 0, \tag{19}$$

where $u = u(X)$.

Case 1

Let us assume the solutions of the Eq. (19) as follows:

$$u(X) = \lambda \cos^{\beta}(\mu X), \text{ where } |X| \leqslant \frac{\pi}{\mu}, \tag{20}$$

λ, μ are nonzero constants. Now Eq. (19) becomes

$$\begin{aligned}&-Ba^2\lambda\mu^2\beta\cos^{\beta}(\mu X) - Ba^2\lambda\mu^2\beta(\beta-1)\cos^{\beta}(\mu X) - ka\lambda\cos^{\beta}(\mu X)\\&+Ba^2\lambda\mu^2\beta(\beta-1)\cos^{\beta-2}(\mu X) + \frac{A}{2}a\lambda^2\cos^{2\beta}(\mu X) = 0\end{aligned} \tag{21}$$

Now equating the coefficient of Eq. (21), $\beta = -2$, $\lambda = \frac{3k}{2A}$ and $\mu = \sqrt{\frac{-k}{8Ba}}$, where $k < 0$.

Now substituting these values in Eq. (20), it becomes

$$\begin{aligned}u(X) &= \lambda\cos^{\beta}(\mu X)\\&= \frac{3k}{2A}\cos^{-2}(\mu X)\\\Rightarrow u(x,t) &= \frac{3k}{2A}\cos^{-2}\{\mu[a(x-kt)]\}\end{aligned} \tag{22}$$

Case 2

Let us assume the solutions of the Eq. (19) as follows:

$$u(X) = \lambda \sin^{\beta}(\mu X), \text{ where } |X| \leqslant \frac{\pi}{\mu}, \tag{23}$$

λ, μ are nonzero constants. Now Eq. (19) becomes

$$\begin{aligned}&-Ba^2\lambda\mu^2\beta\sin^{\beta}(\mu X) - Ba^2\lambda\mu^2\beta(\beta-1)\sin^{\beta}(\mu X) - ka\lambda\sin^{\beta}(\mu X)\\&+Ba^2\lambda\mu^2\beta(\beta-1)\sin^{\beta-2}(\mu X) + \frac{A}{2}a\lambda^2\sin^{2\beta}(\mu X) = 0\end{aligned} \tag{24}$$

Now equating the coefficient of Eq. (24), $\beta = -2$, $\lambda = \frac{3k}{A}$ and $\mu = \sqrt{\frac{-k}{4Ba^2}}$, where $k < 0$.

Now substituting these values in Eq. (23), it becomes

$$\begin{aligned} u(X) &= \lambda \sin^{\beta}(\mu X) \\ &= \frac{3k}{A} \sin^{-2} \mu(X) \\ \Rightarrow u(x,t) &= \frac{3k}{A} \sin^{-2}\{\mu[a(x-kt)]\} \end{aligned} \tag{25}$$

Equations (24) and (25) are the required solutions of the K-dV equation (18).

5 Solutions of K-dV Equation by *Sine–Gordon* Method

Hepson, Korkmaz, Hosseini, Rezazadeh and Eslami together solved the K-dV equation bu using *sine–Gordon* method in 2017. The well-known K-dV equation in the simplest form is

$$\frac{\partial u}{\partial t} + Au\frac{\partial u}{\partial x} + B\frac{\partial^3 u}{\partial x^3} = 0 \tag{26}$$

Now using transformation $X = a(x - kt)$.

Equation (26) is of the form

$$Ba^2\frac{d^2u}{dX^2} + A\frac{u^2}{2} - ku = 0, \tag{27}$$

where $u = u(X)$.

The sine-Gordon equation is

$$\frac{\partial^2 u}{\partial x^2} - \frac{\partial^2 u}{\partial t^2} = m^2 sinu, \tag{28}$$

where $u = u(x,t)$ and using the transformation $X = a(x - ct)$. Equation (29) becomes

$$\frac{d^2u}{dX^2} = \frac{m^2 sinu}{a^2(1-k^2)} \tag{29}$$

Again introducing the new variable $\omega(X)$=$u(X)$, $\frac{d\omega}{dX} = sin\omega$, $\sin\omega(X) = sech(X)$, $\cos\omega(X) = tanh(X)$ and using some integral calculation, the predicated solution of (29) is

$$u(\omega) = A_0 + \sum_{j=1}^{n} \cos^{j-1}(A_j \cos\omega + B_j sin\omega) \tag{30}$$

Let us assume the solution of Eq. (27) of type (30). Now taking the derivatives of (30) and substituting the derivative values in (27). Then in order to balance the two-term equation it has j = 2.

Then Eq. (30) becomes

$$u(\omega) = A_0 + A_1 \cos\omega + A_2 \cos^2\omega + B_1 \sin\omega + B_2 \sin\omega \cos\omega \tag{31}$$

Now taking the derivatives of (31) and substituting in (27) then Eq. (27) becomes

$$\begin{aligned}
&\sin\omega(a^2BB_2 + AA_2B_2 + AA_0B_2 + AA_1B_1 - kB_2)\\
&+\sin^2\omega(-4a^2BB_1 + 2AB_1B_2 - 3AA_1A_2 - AA_0A_1 - kA_1)\\
&+\sin^3\omega(-20a^3BB_2 - 5AA_2B_2 - 2AA_0B_2 - AA_1B_1 + 2kB_2)\\
&+\sin^4\omega(4a^2BB_1 - 3AB_1B_2 + 3AA_1A_2 + 2a^2BA_1)\\
&+\sin^5\omega(24a^2BB_2 + 4AA_2B_2)\\
&+\cos\omega(16a^2BA_2 - AB_2^2 - 2AA_0A_2 + AB_1^2 - A^2A_1 + 2KA_2)\\
&+\cos^3\omega(-40a^2BA_2 - 2AA_2^2 + 3AB_2^2 + 2AA_0A_2 + AB_1^2 - AA_1^2 + 2KA_2)\\
&+\cos^5\omega(24a^2BA_2 + 2AA_2^2 - 2AB_2^2)\\
&+\sin\omega\cos\omega(AA_0B_1 - kB_1 - 5a^2BB_1 - 2AA_2B_1 - 2AA_1B_2)\\
&+\sin\omega\cos^3\omega(6a^2BB_1 + 3AA_1B_2 + 3AA_2B_1) = 0
\end{aligned} \tag{32}$$

Now equating the coefficients of (32) to get the value of A_0, A_1, A_2, B_1, B_2 and k. Here

$$A_1 = 0,\ A_2 = \frac{-6a^2B}{A},\ B_1 = 0,\ B_2 = 0,\ k = -14a^2B + AA_0$$

Then

$$\begin{aligned}
u(\omega) &= A_0 - \frac{6a^2B}{A}\cos^2\omega\\
u(X) &= A_0 - \frac{6a^2B}{A} tanh^2(X)\\
u(x,t) &= A_0 - \frac{6a^2B}{A} tanh^2[a(x-kt)]\\
u(x,t) &= A_0 - \frac{6a^2B}{A} tanh^2[a(x-(-14a^2B + AA_0)t)].
\end{aligned} \tag{33}$$

Equation (33) is the solution of Eq. (26).

6 Solutions of K-dV Equation by $\frac{G'}{G}$ Method

Mehdipoor and Neirameh [39] in 2015 studied the analytic solution of K-dV equations by using $\frac{G'}{G}$ method. The well-known K-dV equation in the simplest form is

$$\frac{\partial u}{\partial t} + Au\frac{\partial u}{\partial x} + B\frac{\partial^3 u}{\partial x^3} = 0 \tag{34}$$

Now using transformation $X = (x - kt)$.

Equation (34) is of the form

$$B\frac{d^2u}{dX^2} + A\frac{u^2}{2} - ku = 0, \tag{35}$$

where $u = u(X)$.

Now introducing the independent variable $Z = \frac{G'(X)}{G(X)}$ and $u(x) = \omega(X)$, where $G(X)$ satisfies the second-order differential equation

$$G'' + \lambda G' + \mu G = 0, \tag{36}$$

where λ and μ are constants. Then Eq. (35) is converted into

$$B(-\mu - \lambda Z - Z^2)^2\frac{d^2\omega}{dZ^2} + B(-\lambda - 2Z)(-\lambda Z - \mu - Z^2)\frac{d\omega}{dZ} \\ + \frac{A}{2}\omega^2 - k\omega + C_1 = 0 \tag{37}$$

Let us assume the power series solution of the Eq. (37) as follows:

$$u(Z) = \alpha_{m1}\left(\frac{G'}{G}\right)^{m1} + \cdots, \tag{38}$$

where α_i's are constant and m1 is the positive integer, which can be determined by considering the highest order derivatives and nonlinear terms. Now Eq. (37) becomes

$$B(-\mu - \lambda Z - Z^2)^2[\alpha_{m1}m_1(m_1 - 1)Z^{m_1-2} + \cdots] \\ + B(-\lambda - 2Z)(-\lambda Z - \mu - Z^2)[\alpha_{m1}m1Z^{m1-1} + \cdots] \\ + \frac{A}{2}\omega^2 - k\omega + C_1 = 0. \tag{39}$$

Considering the homogeneous balance between u^2 and $\frac{d^2u}{dx^2}$ in Eq. (39), $m1 = 2$ then Eq. (38) becomes and substitute $\frac{G'}{G} = Z$.

$$u(Z) = \alpha_0 + \alpha_1 Z + \alpha_2 Z^2, \quad \alpha_2 \neq 0. \tag{40}$$

Now by using (40) in (37), it becomes

$$\begin{aligned}&6B\alpha_2 Z^4 + \frac{A}{2}\alpha_2 Z^4 + 2B\alpha_1 Z^3 + 10B\lambda\alpha_2 Z^3 + A\alpha_1\alpha_2 Z^3\\&+3B\alpha_1\lambda Z^2 + 4B\alpha_2\lambda^2 Z^2 + 8B\alpha_2\mu Z^2 + \frac{A}{2}\alpha_1^2 Z^2\\&+A\alpha_2\alpha_0 Z^2 - k\alpha_2 Z^2 + B\alpha_1\lambda^2 Z + 2B\mu\alpha_1 Z + 6B\lambda\mu\alpha_2 Z + B\lambda\alpha_1\mu\\&+2B\alpha_2\mu^2 + \frac{A}{2}\alpha_0^2 - k\alpha_0 + C_1 = 0\end{aligned} \tag{41}$$

Now equating the coefficient of Z^4, Z^3, Z^2, Z and the constant terms in (41) to get the value of $\alpha_2 = \frac{-12B}{A}$, $\alpha_1 = \frac{-12\lambda B}{A}$, $k = A\alpha_0 + B\lambda^2 + 8\mu B$.

Then Eq. (40) becomes

$$u(Z) = \alpha_0 - \frac{12\lambda B}{A}Z - \frac{12B}{A}Z^2, \quad \alpha_2 \neq 0, \tag{42}$$

where $X = x - (B\lambda^2 + A\alpha_0 + 8B\mu)$.

Now considering the general solution of Eq. (36).

Case 1

Hyperbolic function traveling wave solutions when $\lambda^2 - 4\mu > 0$.

$$G(X) = C_2 e^{(\frac{-\lambda}{2} + \frac{\sqrt{\lambda^2-4\mu}}{2})X} + C_3 e^{(\frac{-\lambda}{2} - \frac{\sqrt{\lambda^2-4\mu}}{2})X} \tag{43}$$

Now by using Eq. (43) in Eq. (42)

$$u(X) = -\frac{3B}{A}(\lambda^2 - 4\mu)\frac{d_1 \sinh\frac{1}{2}\sqrt{\lambda^2-4\mu}X + d_2\cosh\frac{1}{2}\sqrt{\lambda^2-4\mu}X}{d_2 \sinh\frac{1}{2}\sqrt{\lambda^2-4\mu}X + d_1\cosh\frac{1}{2}\sqrt{\lambda^2-4\mu}X} + \frac{3B\lambda^2}{A} + \alpha_0 = 0. \tag{44}$$

In particular, if $d_1 \neq 0$ and $d_2 = 0$ then Eq. (44) becomes

$$u(X) = \frac{3B}{A}(\lambda^2 - 4\mu)sech^2\left(\frac{1}{2}\sqrt{\lambda^2 - 4\mu}X\right) + \frac{B}{12A}\mu + \alpha_0 \tag{45}$$

Case 2

Trigonometric function traveling wave solutions when $\lambda^2 - 4\mu < 0$.

$$G(X) = e^{(\frac{-\lambda}{2})X}\{C_2\cos\frac{\sqrt{\lambda^2-4\mu}}{2}X + C_3\sin\frac{\sqrt{\lambda^2-4\mu}}{2}X\} \tag{46}$$

Now by using Eq. (46) in Eq. (42)

$$u(X) = -\frac{3B}{A}(4\mu - \lambda^2)\left\{\frac{-C_2\sinh\frac{1}{2}\sqrt{\lambda^2-4\mu}X + C_3\cosh\frac{1}{2}\sqrt{\lambda^2-4\mu}X}{C_3\sinh\frac{1}{2}\sqrt{\lambda^2-4\mu}X + C_2\cosh\frac{1}{2}\sqrt{\lambda^2-4\mu}X}\right\}^2 + \frac{3B\lambda^2}{A} + \alpha_0. \tag{47}$$

In particular, if $C_2 \neq 0$ and $C_3 = 0$ in Eq. (47), it is

$$u(X) = -\frac{3B}{A}(4\mu - \lambda^2)tanh^2\frac{1}{2}\sqrt{\lambda^2 - 4\mu X} + \frac{3B\lambda^2}{A} + \alpha_0 \tag{48}$$

Case 3
Trigonometric function traveling wave solutions when $\lambda^2 - 4\mu = 0$.

$$G(X) = (C_2 + C_3 X)e^{\frac{-\lambda}{2}X} \tag{49}$$

Now by using Eq. (49) in Eq. (42)

$$u(X) = -\frac{12B}{A}(\frac{C_3}{C_2 + C_3 X})^2 + \frac{3B}{A}\lambda^2 + \alpha_0. \tag{50}$$

Here (44), (47), (50) are the types of solutions of the K-dV equations by using $\frac{G'}{G}$ method.

7 Solutions of K-dV Equation by *tanh−coth* Method

K-dV equation is

$$\frac{\partial u}{\partial t} + Au\frac{\partial u}{\partial x} + B\frac{\partial^3 u}{\partial x^3} = 0 \tag{51}$$

Now using transformation $X = (x - kt)$.

Equation (51) is of the form

$$B\frac{d^2u}{dX^2} + A\frac{u^2}{2} - ku = 0, \tag{52}$$

where $u = u(X)$.

Let the power series solution of Eq. (52) of the form

$$u(X) = a_0 + \sum_{j=1}^{n}[a_j Y^j(X) + b_j Y^{-j}(X)], \tag{53}$$

$Y(X)$ is the solution of the Riccati equation

$$Y' = A_1 + B_1 Y + C_1 Y^2, \tag{54}$$

where A_1, B_1, C_1 are constants.

Now using Eq. (53) and substituting the values in Eq. (52).

$$
\begin{aligned}
& Ba^2\left[\sum_{j=1}^{n}\left[a_j j(j-1)Y^{j-2}(X)+b_j(j)(j+1)Y^{-j-2}(X)\right]\right] \\
& +\frac{A}{2}\left[\sum_{j=1}^{n}\left[a_jY^j(X)+b_jY^{-j}(X)\right]\right]^2 \\
& -k\left[a_0+\sum_{j=1}^{n}\left[a_jY^j(X)+b_jY^{-j}(X)\right]\right]=0
\end{aligned}
\tag{55}
$$

The parameter n is the positive constant that can be determined by balancing the linear highest term of highest order with the nonlinear term, here $n = 2$. Then Eq. (53) becomes

$$u(X) = a_0 + a_1Y + a_2Y^2 + b_1Y^{-1} + b_2Y^{-2}. \tag{56}$$

Now using Eqs. (54) and (58) then Eq. (52) is

$$
\begin{aligned}
& Y^4\left(6Ba^2a_2C_1^2+\frac{A}{2}a_2^2\right) \\
& +Y^3\left(2Ba^2a_1C_1^2+10Ba^2a_2B_1C_1+Aa_1a_0\right) \\
& +Y^2\left(3Ba^2a_1B_1C_1+4Ba^2a_2B_1^2+8Ba^2a_2B_1C_1-ka_2+\frac{A}{2}a_1^2+Aa_0a_2\right) \\
& +Y\left(Ba^2a_1B_1^2+2Ba^2a_1A_1C_1+6Ba^2a_2A_1B_1-ka_1+Aa_0a_1+Aa_0a_2\right) \\
& +Y^{-1}\left(Ba^2b_1B_1^2+2Ba^2b_1A_1C_1+6Ba^2b_2B_1C_1-2Ba^2b_2B_1^2-kb_1+Aa_0b_1+Aa_1b_2\right) \\
& +Y^{-2}\left(3Ba^2b_1A_1B_1+6Ba^2b_2B_1^2+8Ba^2b_2A_1C_1-kb_2+\frac{A}{2}b_1^2Aa-0b_2\right) \\
& +Y^{-3}\left(2Ba^2b_1+10Ba^2b_2A_1B_1+Ab_1b_2\right) \\
& +Y^{-4}\left(6Ba^2b_2A_1^2+\frac{A}{2}b_2^2\right) \\
& +Ba^2a_1A_1B_1+2Ba^2a_2A_1^2+Ba^2b_1B_1C_1+2Ba^2b_2C_1^2-ka_0+\frac{A}{2}a_0^2+Aa_1b_1+Aa_2b_2=0
\end{aligned}
\tag{57}
$$

Now equating the coefficients of $Z^4, Z^3, Z^2, Z, Z^{-1}, Z^{-2}, Z^{-3}, Z^{-4}$ and the constant terms in (57) to get the values of a_0, a_1, a_2, b_1, b_2.

Then $a_1 = 60\frac{B}{A}a^2B_1C_1, a_2 = -12\frac{B}{A}a^2C_1^2, b_1 = \frac{60B^2a^4A_1^3B_1}{ABa^2-6ABa^2A_1^2}, b_2 = -12\frac{B}{A}a^2A_1^2$.
Now substituting the values of the coefficients in

$$u(X) = a_0 + a_1Y + a_2Y^2 + b_1Y^{-1} + b_2Y^{-2}, \tag{58}$$

where Y is the solution of the Riccati equation.

8 Solutions of Burgers Equation by *tanh* Method

Consider the Burgers equation of the form.

$$\frac{\partial u}{\partial t} + Au\frac{\partial u}{\partial x} - B\frac{\partial^2 u}{\partial x^2} = 0 \tag{59}$$

where A, B are constants and $u = u(x,t)$. Now using wave transformation $X = a(x - kt)$ to convert the partial differential equation into ordinary differential equation, where a and k are nonzero constants.

Here,

$$X = a(x - kt) \Rightarrow \frac{dX}{dx} = a \quad and \quad \frac{dX}{dt} = -ka$$

$$\frac{\partial}{\partial t} = \frac{d}{dX}\frac{dX}{dt} = \frac{d}{dX}(-ka) = -ka\frac{d}{dX}$$
$$\frac{\partial}{\partial x} = \frac{d}{dX}\frac{dX}{dx} = \frac{d}{dX}(a) = a\frac{d}{dX}$$
$$\frac{\partial^2}{\partial x^2} = \frac{\partial}{\partial x}\frac{\partial}{\partial x} = \frac{\partial}{\partial x}\left(a\frac{d}{dX}\right) = a\frac{d}{dX}\frac{dX}{dx}\left(\frac{d}{dX}\right) = a^2\frac{d}{dX}\left(\frac{d}{dX}\right) = a^2\frac{d^2}{dX^2}$$

Now substituting these values in (59), we have

$$-ka\frac{du}{dX} + Aua\frac{du}{dX} - Ba^2\frac{d^2u}{dX^2} = 0$$
$$\Rightarrow -k\frac{du}{dX} + Au\frac{du}{dX} - Ba\frac{d^2u}{dX^2} = 0$$
$$\Rightarrow -ku + A\frac{u^2}{2} - Ba\frac{du}{dX} = 0, \tag{60}$$

where $u = u(X)$. Now introducing the independent variable $Z = tanhX$ and $u(X) = \omega(Z)$

$$\Rightarrow \frac{dZ}{dX} = sech^2X \quad and \quad \frac{d\omega}{dX} = \frac{d\omega}{dZ}\frac{dZ}{dX} = \frac{d\omega}{dZ}sech^2X$$

Now substituting these values in (60), then

$$-k\omega + A\frac{\omega^2}{2} - Ba(1 - Z^2)\frac{d\omega}{dZ} = 0 \tag{61}$$

Let us assume the power series solution of the Eq. (61) as follows:

$$\omega(Z) = \sum_{r=0}^{\infty} a_r Z^{\rho+r}$$

$$\Rightarrow \frac{d\omega}{dZ} = \sum_{r=0}^{\infty} a_r(\rho+r)Z^{\rho+r-1}$$

Now substituting these values in Eq. (61)

$$-k\sum_{r=0}^{\infty} a_r Z^{\rho+r} + \frac{A}{2}\left(\sum_{r=0}^{\infty} a_r Z^{\rho+r}\right)^2 - Ba(1-Z^2)\sum_{r=0}^{\infty} a_r(\rho+r)Z^{\rho+r-1} = 0 \tag{62}$$

Now equating the highest order derivative and highest power on nonlinear term in (62) and taking $\rho = 0$.

$$2\rho + 2r = \rho + r + 1$$
$$\Rightarrow r = 1.$$

For $r = 1$, we have

$$\omega(Z) = a_0 + a_1 Z, \; a_1 \neq 0$$
$$\Rightarrow \frac{d\omega}{dZ} = a_1 \tag{63}$$

Now (61) becomes

$$\frac{A}{2}a_0^2 - ka_0 - Baa_1 + Aa_0a_1Z - ka_1Z + \frac{A}{2}a_1^2Z^2 + Baa_1Z^2 = 0 \tag{64}$$

Now equating the coefficient of Z^2, Z and the constant terms in (64) to get the value of a_0, a_1, and k.

We have $a_0 = \frac{k}{A}$, $a_1 = -\frac{2Ba}{A}$, $k = 2Ba$ and $k = -2Ba$.

Case 1 $K = 2Ba$

$$u(x,t) = \frac{2Ba}{A}(1 - \sqrt{1 - sech^2[a(x - 2Bat)]}) \tag{65}$$

Case 2 $K = -2Ba$

$$u(x,t) = \frac{-2Ba}{A}(1 + \sqrt{1 - sech^2[a(x + 2Bat)]}) \tag{66}$$

Equations (65) and (66) are the required solutions of the Burgers equation (59).

9 Solutions of Burgers Equation by *sech* Method

Consider the Burgers equation of the form.

$$\frac{\partial u}{\partial t} + Au\frac{\partial u}{\partial x} - B\frac{\partial^2 u}{\partial x^2} = 0 \tag{67}$$

where A, B are constants and $u = u(x, t)$. Now using wave transformation $X = a(x - kt)$ to convert the partial differential equation in to ordinary differential equation, where a and k are nonzero constants.

Then Eq. (67) is of the form

$$-ku + A\frac{u^2}{2} - Ba\frac{du}{dX} = 0, \tag{68}$$

where $u = u(X)$. Now introducing the independent variable $Z = sechX$ and $u(X) = \omega(Z)$

$$\Rightarrow \frac{dZ}{dX} = -sechXtanhX \quad and \quad \frac{d\omega}{dX} = \frac{d\omega}{dZ}\frac{dZ}{dX} = -ZtanhX\frac{d\omega}{dZ}$$

Now substituting these values in (68), then

$$-k\omega + A\frac{\omega^2}{2} + Baz\sqrt{1 - Z^2}\frac{d\omega}{dZ} = 0 \tag{69}$$

Let us assume the power series solution of the Eq. (69) as follows:

$$\omega(Z) = \sum_{r=0}^{\infty} a_r Z^{\rho+r}$$

$$\Rightarrow \frac{d\omega}{dZ} = \sum_{r=0}^{\infty} a_r(\rho + r)Z^{\rho+r-1}$$

Now substituting these values in Eq. (69)

$$-k\sum_{r=0}^{\infty} a_r Z^{\rho+r} + \frac{A}{2}\left(\sum_{r=0}^{\infty} a_r Z^{\rho+r}\right)^2 - Ba(1 - Z^2)\sum_{r=0}^{\infty} a_r(\rho + r)Z^{\rho+r-1} = 0 \tag{70}$$

Now equating the highest order derivative and highest power on nonlinear term in (62) and taking $\rho = 0$.

$$2\rho + 2r = \rho + r - 1 + 2$$

$$\Rightarrow r = 1.$$

For $r = 1$, we have

$$\omega(Z) = a_0 + a_1 Z,\ a_1 \neq 0$$
$$\Rightarrow \frac{d\omega}{dZ} = a_1 \tag{71}$$

Now (69) becomes

$$-ka_0 + \frac{A}{2}a_0^2 - ka_1 Z + Aa_0 a_1 Z + Baa_1 Z + \frac{A}{2}a_1^2 Z^2 - \frac{1}{2}Baa_1 Z^3 = 0 \tag{72}$$

Now equating the coefficient of Z^2, Z and the constant terms in (72) to get the value of a_0, a_1, and k.

We have $a_0 = \frac{2k}{A}$, $a_1 = -\frac{Ba}{A}$, $k = 2Ba$ and $k = -Ba$.

$$u(x,t) = \frac{-Ba}{A}(1 - secha(x + Bat)) \tag{73}$$

Equation (73) is the required solutions of the Burgers equation (67).

10 Solutions of Burgers Equation by $\left(\frac{G'}{G}\right)$ Method

The well-known Burgers equation in the simplest form is

$$\frac{\partial u}{\partial t} + Au\frac{\partial u}{\partial x} - B\frac{\partial^2 u}{\partial x^2} = 0 \tag{74}$$

Now using transformation $X = a(x - kt)$. Equation (74) is of the form

$$-ku + A\frac{u^2}{2} - Ba\frac{du}{dX} = 0, \tag{75}$$

where $u = u(X)$.

Now introducing the independent variable $Z = \frac{G'(X)}{G(X)}$ and $u(x) = \omega(X)$, where $G(X)$ satisfies the second-order differential equation

$$G'' + \lambda G' + \mu G = 0, \tag{76}$$

where λ and μ are constants. Then Eq. (75) is converted into

$$-k\omega + \frac{A}{2}\omega^2 - Ba(-\mu - \lambda Z - Z^2)\frac{d\omega}{dZ} = 0 \tag{77}$$

Let us assume the power series solution of the Eq. (77) as follows:

$$\omega(Z) = \alpha_{m1}(\frac{G'}{G})^{m1} + \cdots, \tag{78}$$

where α_i's are constant and m1 is the positive integer, which can be determined by considering the highest order derivatives and nonlinear terms. Now Eq. (77) becomes

$$-k(\alpha_{m1}Z^{m1} + \cdots) + \frac{A}{2}(\alpha_{m1}Z^{m1} + \cdots)^2 - Ba(-\mu - \lambda Z^2)(\alpha_{m1}m_1 Z^{m1-1} + \cdots) = 0 \tag{79}$$

Considering the homogeneous balance between u^2 and $\frac{du}{dx}$ in Eq. (79), $m1 = 1$ then Eq. (78) becomes and substitute $\frac{G'}{G} = Z$.

$$\omega(Z) = \alpha_0 + \alpha_1 Z, \quad \alpha_1 \neq 0.$$
$$\Rightarrow \frac{d\omega}{dZ} = \alpha_1 \tag{80}$$

Now by using (80) in (77), it becomes

$$\frac{A}{2}\alpha_1^2 Z^2 + Ba\alpha_1 Z^2 - k\alpha_1 Z + A\alpha_0\alpha_1 Z + Ba\alpha_1\lambda Z - k\alpha_0 + \frac{A}{2}\alpha_0^2 + Ba\mu\alpha_1 = 0. \tag{81}$$

Now equating the coefficient of Z^2, Z and constant terms in (81) to get the value of $\alpha_1 = \frac{-2Ba}{A}, \alpha_0 = \frac{K}{A} - \frac{Ba}{A}\lambda$.

Then Eq. (80) becomes

$$\omega(Z) = \frac{K}{A} - \frac{aB}{A}\lambda - \frac{2B}{A}aZ, \tag{82}$$

Now considering the general solution of Eq. (76).

Case 1
Hyperbolic function traveling wave solutions when $\lambda^2 - 4\mu > 0$.

$$G(X) = C_2 e^{\left(\frac{-\lambda}{2} + \frac{\sqrt{\lambda^2 - 4\mu}}{2}\right)X} + C_3 e^{\left(\frac{-\lambda}{2} - \frac{\sqrt{\lambda^2 - 4\mu}}{2}\right)X} \tag{83}$$

Now by using Eq. (83) in Eq. (82)

$$u(X) = \frac{K}{A} - \frac{B}{A}a\lambda - \frac{B}{A}a\sqrt{\lambda^2 - 4\mu}\frac{d_1 \sinh\frac{1}{2}\sqrt{\lambda^2 - 4\mu}X + d_2 \cosh\frac{1}{2}\sqrt{\lambda^2 - 4\mu}X}{d_2 \sinh\frac{1}{2}\sqrt{\lambda^2 - 4\mu}X + d_1 \cosh\frac{1}{2}\sqrt{\lambda^2 - 4\mu}X} = 0. \tag{84}$$

Case 2
Trigonometric function traveling wave solutions when $\lambda^2 - 4\mu < 0$.

$$G(X) = e^{(\frac{-\lambda}{2})X}\left\{C_2 \cos\frac{\sqrt{\lambda^2 - 4\mu}}{2}X + C_3 \sin\frac{\sqrt{\lambda^2 - 4\mu}}{2}X\right\} \tag{85}$$

Now by using Eq. (85) in Eq. (82)

$$u(X) = \frac{k}{A} - \frac{Ba}{A}\sqrt{4\mu - \lambda^2}\frac{-C_2 \sinh \frac{1}{2}\sqrt{\lambda^2 - 4\mu}X + C_3 \cosh \frac{1}{2}\sqrt{\lambda^2 - 4\mu}X}{C_3 \sinh \frac{1}{2}\sqrt{\lambda^2 - 4\mu}X + C_2 \cosh \frac{1}{2}\sqrt{\lambda^2 - 4\mu}X}. \quad (86)$$

Case 3
Trigonometric function traveling wave solutions when $\lambda^2 - 4\mu = 0$.

$$G(X) = (C_2 + C_3 X)e^{\frac{-\lambda}{2}X} \quad (87)$$

Now by using Eq. (87) in Eq. (82)

$$u(X) = \frac{k}{A} - \frac{2Ba}{A}\left(\frac{C_3}{C_2 + C_3 X}\right). \quad (88)$$

Here (84), (86), (88) are the types of solutions of the Burgers equations by using $\frac{G'}{G}$ method.

11 Solutions of Burgers Equation by *Sine–Gordon* Method

$$\frac{\partial u}{\partial t} + Au\frac{\partial u}{\partial x} - B\frac{\partial^2 u}{\partial x^2} = 0 \quad (89)$$

Now using transformation $X = a(x - kt)$.

Equation (89) is of the form

$$-Ba\frac{du}{dX} + A\frac{u^2}{2} - ku = 0, \quad (90)$$

where $u = u(X)$.

The sine–Gordon equation is

$$\frac{\partial^2 u}{\partial x^2} - \frac{\partial^2 u}{\partial t^2} = m^2 sinu, \quad (91)$$

where $u = u(x, t)$ and using the transformation $X = a(x - ct)$. Equation (92) becomes

$$\frac{d^2 u}{dX^2} = \frac{m^2 sinu}{a^2(1 - k^2)} \quad (92)$$

Again introducing the new variable $\omega(X) = u(X)$, $\frac{d\omega}{dX} = sin\omega$, $\sin\omega(X) = sech(X)$, $\cos\omega(X) = tanh(X)$ and using some integral calculation, the predicated solution of (92) is

$$u(\omega) = A_0 + \sum_{j=1}^{n} \cos^{j-1}(A_j \cos\omega + B_j sin\omega) \tag{93}$$

Let as assuming the solution of Eq. (90) of type (93). Now taking the derivatives of (93) and substituting the derivative values in (90). Then in order to balance the two-term equation it has j = 1.

Then Eq. (93) becomes

$$u(\omega) = A_0 + A_1 \cos\omega + B_1 \sin\omega \tag{94}$$

Now taking the derivatives of (94) and substituting in (90) then Eq. (90) becomes

$$\begin{aligned} &-kA_0 + \frac{A}{2}A_0^2 + \frac{A}{2}A_1^2 - kA_1\cos\omega + AA_0A_1\cos\omega - kB_1\sin\omega \\ &+AA_0B_1\sin\omega - \frac{A}{2}A_1^2\sin^2\omega + \frac{A}{2}B_1^2\sin^2\omega + BA_1a\sin^2\omega \\ &+AA_1B_1\cos\omega\sin\omega - BB_1a\cos\omega\sin\omega = 0 \end{aligned} \tag{95}$$

Now equating the coefficients of (95) to get the value of A_0, A_1, B_1 and k. Here

$$A_1 = \frac{Ba}{A},\ B_1 = \pm i\frac{Ba}{A},$$

Then

$$u(x,t) = A_0 + \frac{aB}{A}\sqrt{1 - sech^2 X} \pm i\frac{Ba}{A}sech. \tag{96}$$

Equation (96) is the solution of Eq. (89).

12 Solutions of Schamel–K-dV Equation by $\frac{G'}{G}$ Method

Consider the simplest form of Schamel–K-dV equation as

$$\frac{\partial u}{\partial t} + Au^{\frac{1}{2}}\frac{\partial u}{\partial x} + B\frac{\partial^3 u}{\partial x^3} = 0, \tag{97}$$

where A and B are arbitrary coefficients and $u = u(x,t)$.

Now using the wave transformation $X = x - kt$, where k is constant.

Then Eq. (97), becomes

$$-k\frac{dU}{dX}+AU^{1/2}+B\frac{d^3U}{dX^3}=0, \tag{98}$$

where $U = U(X)$. Now integrating Eq. (98), then we have

$$-kU+\frac{2}{3}AU^{3/2}+B\frac{d^2U}{dX^2}+c=0, \tag{99}$$

where c is the integration constant.

Let us consider $U^{1/2}(X) = V(X)$ Eq. (99) becomes

$$\Rightarrow -kV^2+\frac{2}{3}AV^3+2B\left(\frac{dV}{dX}\right)^2+2BV\frac{d^2V}{dX^2}+c=0 \tag{100}$$

Let us assume the solution of (100) of the form

$$V(X)=\sum_{i=0}^{n}a_i\left(\frac{G'}{G}\right)^i \tag{101}$$

where $G = G(X)$ satisfies the the second-order differential equation

$$G''+\lambda G'+\mu G=0 \tag{102}$$

where λ and μ are constants.

Now substituting the value of $\frac{d^2V}{dX^2}$, $\frac{dV}{dX}$, $V(X)$ in Eq. (100) balancing the highest order nonlinear term with highest order derivative we get $n = 2$. Then Eq. (101) becomes

$$V(X)=a_0+a_1\left(\frac{G'}{G}\right)+a_2\left(\frac{G'}{G}\right)^2 \tag{103}$$

Now substituting the values of $V(X)$, $\frac{d^2V}{dX^2}$, $\frac{dV}{dX}$ in (100) and equating the coefficients of $\left(\frac{G'}{G}\right)^i$ to get the values of a_0, a_1, a_2, k we have

$$a_0=a_0, a_1=-30\frac{B}{A}\lambda, a_2=-30\frac{B}{A} \tag{104}$$

Substituting these values in the assuming solution and applying the transformation $U = V^2$ we have the following different types of solution of the form:

$$U(X)=\left[a_0-30\frac{B}{A}\lambda\left(\frac{G'}{G}\right)-30\frac{B}{A}\left(\frac{G'}{G}\right)^2\right]^2 \tag{105}$$

where

Case 1 When $\lambda^2 - 4\mu > 0$

$$\left(\frac{G'}{G}\right) = \frac{\sqrt{\lambda^2 - 4\mu}}{2}\left(\frac{d_2 \cosh \frac{\sqrt{\lambda^2-4\mu}}{2}X + d_1 \sinh \frac{\sqrt{\lambda^2-4\mu}}{2}X}{d_1 \cosh \frac{\sqrt{\lambda^2-4\mu}}{2}X + d_2 \sinh \frac{\sqrt{\lambda^2-4\mu}}{2}X}\right) - \frac{\lambda}{2}$$

Case 2 When $\lambda^2 - 4\mu < 0$

$$\left(\frac{G'}{G}\right) = \frac{\sqrt{4m - \lambda^2}}{2}\frac{\left(-d_1 \sin \frac{\sqrt{4m-\lambda^2}}{2}X + d_2 \cos \frac{\sqrt{4m-\lambda^2}}{2}X\right)}{\left(d_1 \cos \frac{\sqrt{4m-\lambda^2}}{2}X + d_2 \sin \frac{\sqrt{4m-\lambda^2}}{2}X\right)} - \frac{\lambda}{2}$$

Case 3 When $\lambda^2 - 4\mu = 0$

$$\left(\frac{G'}{G}\right) = -\frac{\lambda}{2} + \frac{c_2 X}{c_1 + c_2 X}$$

13 Solutions of Schamel–K-dV Equation by Different Form of $\frac{G'}{G}$ Method

Consider the simplest form of Schamel–K-dV equation as

$$\frac{\partial u}{\partial t} + Au^{\frac{1}{2}}\frac{\partial u}{\partial x} + B\frac{\partial^3 u}{\partial x^3} = 0, \tag{106}$$

where A and B are arbitrary coefficients and $u = u(x, t)$.

Now using the wave transformation $X = x - kt$, where k is constant.

Then Eq. (106), becomes

$$-k\frac{dU}{dX} + AU^{1/2} + B\frac{d^3U}{dX^3} = 0, \tag{107}$$

where $U = U(X)$. Now integrating Eq. (107), then we have

$$-kU + \frac{2}{3}AU^{3/2} + B\frac{d^2U}{dX^2} + c = 0, \tag{108}$$

where c is the integration constant.

Let us consider $U^{1/2}(X) = V(X)$ Eq. (108) which becomes

$$\Rightarrow -kV^2 + \frac{2}{3}AV^3 + 2B\left(\frac{dV}{dX}\right)^2 + 2BV\frac{d^2V}{dX^2} + c = 0 \tag{109}$$

Let us assume the solution of (109) of the form

$$V(X) = \sum_{i=0}^{n} a_i\left(\frac{G'}{G}\right)^i + \sum_{i=1}^{n} b_i\left(\frac{G'}{G}\right)^{-i} \tag{110}$$

where $G = G(X)$ satisfies the the second-order differential equation

$$G'' + \mu G = 0 \tag{111}$$

μ is constants.

Now substituting the value of $\frac{d^2V}{dX^2}$, $\frac{dV}{dX}$, $V(X)$ in Eq. (109) and balancing the highest order nonlinear term with highest order derivative we get $n = 1$. Then Eq. (110) becomes

$$V(X) = a_0 + a_1\left(\frac{G'}{G}\right) + b_1\left(\frac{G'}{G}\right)^{-1} \tag{112}$$

Now using Eq. (112) and its derivatives into Eq. (109) and equating the coefficients of $\left(\frac{G'}{G}\right)^i$, where $i = 0, \pm1, \pm2, \pm3, \pm4, \pm5, \pm6$ to find out the values of a_0, a_1, b_1, μ, b, c. By using Mathematica we got the values as follows

$$a_0 = \frac{3k}{2A},\quad a_1 = \frac{-6B}{A}, a_2 = 0, b_1 = 0, b_2 = 0, \mu = \frac{-k}{4B}, c = 0.$$

Substituting these values in the assuming solution and applying the transformation $U = V^2$, we have the following different types of solution of the form:

$$U(X) = \left[\frac{3k}{2A} - \frac{6B}{A}\left(\frac{G'}{G}\right)\right]^2 \tag{113}$$

where

Case 1 When $-\mu > 0$

$$\left(\frac{G'}{G}\right) = \sqrt{\mu}\left(\frac{d_2\cosh\sqrt{-\mu}X + d_1\sinh\sqrt{-\mu}X}{d_1\cosh\sqrt{-\mu}X + d_2\sinh\sqrt{-\mu}X}\right)$$

Case 2 When $-\mu < 0$

$$\left(\frac{G'}{G}\right) = \sqrt{\mu}\frac{(-d_1\sin\sqrt{\mu}X + d_2\cos\sqrt{\mu}X)}{(d_1\cos\sqrt{\mu}X + d_2\sin\sqrt{\mu}X)}$$

14 Solutions of Coupled Schamel–K-dV Equation by $\left(\frac{G'}{G}\right)$ Method

Consider the coupled Schamel–K-dV equation of the form

$$\frac{\partial u}{\partial t} + au^{\frac{1}{2}}\frac{\partial u}{\partial x} + bu\frac{du}{dx} + p\frac{\partial^3 u}{\partial x^3} = 0 \tag{114}$$

where a, b, and p are arbitrary coefficients. Now using the wave transformation $X = x - kt$, where k is constant. Then Eq. (114) becomes

$$p\frac{dU^3}{dX} - k\frac{dU}{dX} + aU^{\frac{1}{2}}\frac{dU}{dX} + bU\frac{dU}{dX} = 0 \tag{115}$$

Integrating Eq. (115), it becomes

$$p\frac{dU^2}{dX^2} - kU + \frac{2}{3}aU^{\frac{3}{2}} + \frac{1}{2}bU^2 + c = 0 \tag{116}$$

where c is the integration constant.

Let $U^{\frac{1}{2}} = V$ then Eq. (116) becomes

$$V\frac{d^2V}{dX^2} + \left(\frac{dV}{dX}\right)^2 - \frac{k}{2p}V^2 + \frac{a}{3p}V^3 + \frac{b}{4p}V^4 + \frac{c}{2p} = 0 \tag{117}$$

Let us assume the solution of (117) of the form

$$V(X) = \sum_{i=0}^{n} a_i \left(\frac{G'}{G}\right)^i \tag{118}$$

where $G = G(X)$ satisfies the the second-order differential equation

$$G'' + \lambda G' + \mu G = 0 \tag{119}$$

where λ and μ are constants.

Balancing the highest order nonlinear term with highest order derivative of $V\frac{d^2V}{dX^2}$ and V^4, then it comes out that n = 1. Then

$$V(X) = a_0 + a_1\left(\frac{G'}{G}\right) \tag{120}$$

Now using Eq. (120) and its derivatives into Eq. (117) and equating the coefficients of $\left(\frac{G'}{G}\right)^i$, where $i = 0, 1, 2, 3, 4$ to find out the values of $a_0, a_1, \lambda, \mu, b, c$. By using

Mathematica, we got the values as follows:

$$a_0 = \frac{15}{8a}(k \pm 2\lambda\sqrt{pk}), \;\; a_1 = \pm\frac{15\sqrt{pk}}{2a},$$
$$\lambda^2 - 4\mu = \frac{k}{4p}, b = \frac{-16a^2}{75k}, c = 0. \tag{121}$$

Substituting these values in the assuming solution and applying the transformation $U = V^2$ we have the following different types of solution of the form:

$$U(X) = \left[\frac{15}{8a}(k \pm 2\lambda\sqrt{pk}) \pm \frac{15\sqrt{pk}}{2a}\left(\frac{G'}{G}\right)\right]^2 \tag{122}$$

where

Case 1 When $\lambda^2 - 4\mu > 0$

$$\left(\frac{G'}{G}\right) = \frac{\sqrt{\lambda^2 - 4\mu}}{2}\left(\frac{d_2 \cosh\frac{\sqrt{\lambda^2-4\mu}}{2}X + d_1 \sinh\frac{\sqrt{\lambda^2-4\mu}}{2}X}{d_1 \cosh\frac{\sqrt{\lambda^2-4\mu}}{2}X + d_2 \sinh\frac{\sqrt{\lambda^2-4\mu}}{2}X}\right) - \frac{\lambda}{2}$$

Case 2 When $\lambda^2 - 4\mu < 0$

$$\left(\frac{G'}{G}\right) = \frac{\sqrt{4m - \lambda^2}}{2}\frac{\left(-d_1 \sin\frac{\sqrt{4m-\lambda^2}}{2}X + d_2 \cos\frac{\sqrt{4m-\lambda^2}}{2}X\right)}{\left(d_1 \cos\frac{\sqrt{4m-\lambda^2}}{2}X + d_2 \sin\frac{\sqrt{4m-\lambda^2}}{2}X\right)} - \frac{\lambda}{2}$$

Case 3 When $\lambda^2 - 4\mu = 0$

$$\left(\frac{G'}{G}\right) = -\frac{\lambda}{2} + \frac{c_2 X}{c_1 + c_2 X}$$

15 Solutions of Coupled Schamel–K-dV Equation by Different Form of $\left(\frac{G'}{G}\right)$ Method

Consider the Coupled Schamel–K-dV equation of the form

$$\frac{\partial u}{\partial t} + au^{\frac{1}{2}}\frac{\partial u}{\partial x} + bu\frac{du}{dx} + p\frac{\partial^3 u}{\partial x^3} = 0 \tag{123}$$

where a, b and p are arbitrary coefficients. Now using the wave transformation $X = x - kt$, where k is constant. Then Eq. (123) becomes

$$p\frac{dU^3}{dX} - k\frac{dU}{dX} + aU^{\frac{1}{2}}\frac{dU}{dX} + bU\frac{dU}{dX} = 0 \tag{124}$$

Integrating Eq. (124), it becomes

$$p\frac{dU^2}{dX^2} - kU + \frac{2}{3}aU^{\frac{3}{2}} + \frac{1}{2}bU^2 + c = 0 \tag{125}$$

where c is the integration constant.

Let $U^{\frac{1}{2}} = V$ then Eq. (125) becomes

$$V\frac{d^2V}{dX^2} + \left(\frac{dV}{dX}\right)^2 - \frac{k}{2p}V^2 + \frac{a}{3p}V^3 + \frac{b}{4p}V^4 + \frac{c}{2p} = 0 \tag{126}$$

Let us assume the solution of (126) of the form

$$V(X) = \sum_{i=0}^{n} a_i\left(\frac{G'}{G}\right)^i + \sum_{i=1}^{n} b_i\left(\frac{G'}{G}\right)^{-i} \tag{127}$$

where $G = G(X)$ satisfies the the second-order differential equation

$$G'' + \mu G = 0 \tag{128}$$

where μ is constants.

Balancing the highest order nonlinear term with highest order derivative of $V\frac{d^2V}{dX^2}$ and V^4, then it comes out that n = 1. Then

$$V(X) = a_0 + a_1\left(\frac{G'}{G}\right) + b_1\left(\frac{G'}{G}\right)^{-1} \tag{129}$$

Now using Eq. (129) and its derivatives into Eq. (126) and equating the coefficients of $\left(\frac{G'}{G}\right)^i$, where $i = 0, \pm1, \pm2, \pm3, \pm4$ to find out the values of a_0, a_1, b_1, μ, b, c. By using Mathematica, we got the values as follows:

$$\begin{aligned} a_0 &= \frac{15k}{8a}, \quad a_1 = \pm\frac{15\sqrt{pk}}{2a}, b_1 = \pm\frac{15k\sqrt{pk}}{64ap}, \\ \mu &= \frac{k}{32p}, b = \frac{-16a^2}{75k}, c = \frac{225k^3}{512a^2} \end{aligned} \tag{130}$$

Substituting these values in the assuming solution and applying the transformation $U = V^2$, we have the following different types of solution of the form:

$$U(X) = \left[\frac{15k}{8a} \pm \frac{15\sqrt{pk}}{2a}\left(\frac{G'}{G}\right) \pm \frac{15k\sqrt{pk}}{64ap}\left(\frac{G'}{G}\right)\right]^2 \tag{131}$$

where

Case 1 When $-\mu > 0$

$$\left(\frac{G'}{G}\right) = \sqrt{\mu}\left(\frac{d_2 \cosh\sqrt{-\mu}X + d_1 \sinh\sqrt{-\mu}X}{d_1 \cosh\sqrt{-\mu}X + d_2 \sinh\sqrt{-\mu}X}\right)$$

Case 2 When $-\mu < 0$

$$\left(\frac{G'}{G}\right) = \sqrt{\mu}\frac{\left(-d_1 \sin\sqrt{\mu}X + d_2 \cos\sqrt{\mu}X\right)}{\left(d_1 \cos\sqrt{\mu}X + d_2 \sin\sqrt{\mu}X\right)}$$

References

1. W. Malfliet, Phys. Scr. **54**, 563 (1996)
2. E. Fan, Phys. Lett. A **277**, 212 (2000)
3. A.M. Wazwaz, Appl. Math. Comput. **154**, 713 (2004)
4. D.J. Evans, R.K. Raslan, Int. J. Comput. Math. **82**, 897 (2005)
5. A.M. Wazwaz, Appl. Math. Comput. **184**, 1002 (2007)
6. J. Sarma, Chaos. Solitons Fractals **39**, 277 (2009)
7. J. Sarma, Chaos. Solitons Fractals **42**, 1599 (2009)
8. A.J.M. Jawad, M.D. Petkovic, A. Biswas, Appl. Math. Comput. **216**, 3370 (2010)
9. M. Karimi, Math. Sci. J. **9**, 47 (2013)
10. J. Zhang, L. Yin, Gen. Math. Notes **36**, 19 (2016)
11. A.R. Adem, Comput. Math. Appl. **74**, 1897 (2017)
12. H. Tariq, G. Akram, Phys. A **473**, 352 (2017)
13. K.R. Ralsan, K.K. Ali, Chaos. Solitons Fractals **103**, 404 (2017)
14. M.E. Ali, F. Bilkis, G.C. Paul, H. Naher, N. Taghizadeh, Int. J. Sci. Eng. Res. **10**, 1140 (2019)
15. A.M. Wazwaz, Appl. Math. Comput. **188**, 1467 (2007)
16. A.M. Wazwaz, Appl. Math. Comput. **190**, 633 (2007)
17. A.M. Wazwaz, Appl. Math. Comput. **195**, 24 (2008)
18. L. Wazzan, Commun. Nonlinear Sci. Numer. Simul. **14**, 443 (2009)
19. E.J. Parkes, Appl. Math. Comput. **217**, 1749 (2010)
20. A.H. Salas, C.A. Gomezs, Appl. Appl. Math. **5**, 1504 (2010)
21. A.J.M. Jawad, Y.S. Ali, Int. J. Comput. Eng. Res. **3**, 30 (2013)
22. Y.B. Chukkol, M.N. Mohamad, M.I. Munior, AIP Conf. Proc. **1870**, 040024 (2017)
23. R. Asokan, D. Vinodh, Int. J. Pure Appl. Math. **117**, 19 (2017)
24. R. Asokan, D. Vinodh, Int. J. Appl. Comput. Math. **4**, 100 (2018)
25. Z. Yan, Chaos. Solitons Fractals **16**, 291 (2003)
26. M. Alquran, E.V. Krishnan, Non Linear Stud. **23**, 639 (2016)
27. A. Korkmaz, O.E. Hepson, K. Hosseini, H. Rezazadeh, M. Eslami, J. King Saud Univ. Sci. **32**, 567 (2020)
28. J.L.G. Guirao, H.M. Baskomus, A. Kumar, M.S. Rawat, G. Yel, Symmetry **12**, 17 (2020)
29. A.M. Wazwaz, Math. Comput. Model. **40**, 499 (2004)
30. A.M. Wazwaz, Comput. Math. Appl. **49**, 1101 (2005)
31. S. Bibi, J. Assoc. Arab Univ. Basic Appl. Sci. **15**, 90 (2014)
32. J. Yang, S. Tang, J. Math. Sci. Adv. Appl. **31**, 25 (2015)
33. K.R. Ralsan, T.S. El-Danal, K.K. Ali, J. Math. Soc. **25**, 350 (2017)

34. D. Baldwin, U. Goktas, W. Hereman, L. Hong, R.S. Martino, J.C. Miller, J. Symb. Comput. **37**, 669 (2004)
35. A.M. Wazwaz, Phys. Lett. A **366**, 85 (2007)
36. D.D. Ganji, M. Abdollahzadeh, Appl. Math. Comput. **206**, 438 (2008)
37. M. Wei, S. Tang, J. Appl. Anal. Comput. **1**, 267 (2011)
38. A.J.M. Jawad, J. Innov. Res. Sci. Eng. **4**, 162 (2016)
39. M. Mehdipoor, A. Neirameh, Astor Phys. Space Sci. **337**, 269 (2012)
40. M. Mehdipoor, Astro Phys. Space Sci. **338**, 73 (2012)
41. M. Wang, X. Li, J. Zhang, Phys. Lett. A **372**, 417 (2008)
42. H. Gao, R.X. Zhao, Appl. Math. Comput. **215**, 2781 (2009)
43. B. Agheli, R. Darzi, A. Dabbaghian, Opt. Quant. Electron. **49**, 387 (2017)
44. I. Aslan, T. Ozis, Appl. Math. Comput. **211**, 531 (2009)
45. I. Aslan, Appl. Math. Comput. **215**, 857 (2009)
46. I. Aslan, T. Ozis, Appl. Math. Comput. **209**, 425 (2009)
47. H. Kheiri, M.R. Moghaddam, V. Vafaei, Pramana J. Phys. **217**, 1376 (2010)
48. J. Manafian, M. Lakestain, Pranama J. Phys. **85**, 31 (2015)
49. S. Javadi, E. Moradi, M. Fardi, S. Abbasian, J. Math. Comput. Sci. **11**, 246 (2014)
50. X. Liu, L. Tian, Y. Wu, Appl. Math. Comput. **217**, 1376 (2010)
51. S. Kutluay, A. Esen, O. Tasbozan, Appl. Math. Comput. **217**, 384 (2010)
52. B.S. Bahrami, H. Abdollahzadeh, I.M. Berijani, D.D. Ganji, M. Abdollahzadeh, Pramana J. Phys. **77**, 263 (2011)
53. Z.M. Zuo, Y.M. Zhang, Appl. Math. Comput. **217**, 5936 (2011)
54. O. Donmea, D. Daghan, Sulemar Demirmir Univ. J. Nat. Appl. Sci. **21**, 208 (2017)

Controllability Study on the Symplectic Lie Group $Sp(2, \mathbb{R})$

Archana Tiwari

Abstract Lie groups and control theory are closely related as various engineering problems have been modelled as a control problem on Lie groups. Lie groups play a vital role in studying various concepts of control theory. Keeping in view the contribution of Lie groups in the study of controllability and optimal control, a control system is designed on the Symplectic Lie group $SP(2, \mathbb{R})$. A brief study of optimal control by minimizing the cost function is done. The stability of the system dynamics is thoroughly analysed. And finally, two unconventional numerical integrators, Kahan and Lie–Trotter, have been applied on the system dynamics to study some related properties.

Keywords Lie group · Lie algebra · Controllability · Stability

1 Introduction

Lie groups were first introduced into motion control problems by Brockett [2]. The controllability and observability aspects of Lie groups were focused in his theories. Jurdjevic and Sussmann [8, 9] further discussed controllability aspects for various Lie groups. Study of control theory and optimal control problems on Lie groups from geometric point of view is discussed in detail in [1]. There are various dynamical systems whose configuration space is Lie groups, few examples are spacecraft attitude control [4], problem of parallel car parking, autonomous underwater vehicle [12] and switching control problems. One of the significant classes is the control problem on matrix Lie group, which can be studied from a geometric point of view. Study of control system on $SO(3)$ was done by Remsing [15]. Pop [3, 14] examined control problems on $SE(3)$ and $H(3)$ matrix Lie groups. Most of the recent works in non-linear control theory have emphasized a driftless control system with fewer control than state variables.

A. Tiwari (✉)
National Institute of Technology, Rourkela, Odisha, India
e-mail: archanatiwari2010@gmail.com

S. R. Mishra et al. (eds.), *Recent Trends in Applied Mathematics*, Lecture Notes in Mechanical Engineering, https://doi.org/10.1007/978-981-15-9817-3_10

Table 1 Lie bracket commutation for $Sp(2, \mathbb{R})$

[., .]	m_1	m_2	m_3
m_1	0	$4m_3$	$2m_1$
m_2	$-4m_3$	0	$2m_2$
m_3	$-2m_1$	$2m_2$	0

The objective of this work is to study an optimal control problem associated with the Symplectic group $Sp(2, \mathbb{R})$. A left-invariant driftless controllable system is considered. In particular, stability of the system is studied in detail. Finally, two unconventional numerical integrators are applied to the system dynamics and few of the dynamical and geometrical properties are pointed out.

2 Symplectic Group $Sp(2, \mathbb{R})$

The Symplectic group $Sp(2, \mathbb{R})$ is the group of 2×2 matrices which preserve a nondegenerate antisymmetric bilinear form. It consists of the 2×2 matrices P satisfying $P^T AP = A$, where A is a fixed invertible skew-symmetric matrix.

The Symplectic Lie algebra $sp(2, \mathbb{R})$ consists of 2×2 matrices of the form

$$\{X \in gl_2 | X^T A + AX = 0\}. \tag{1}$$

The generators of the Lie algebra $sp(2, \mathbb{R})$ are.

$$m_1 = \begin{pmatrix} 0 & 0 \\ -2 & 0 \end{pmatrix}, m_2 = \begin{pmatrix} 0 & 2 \\ 0 & 0 \end{pmatrix}, \text{ and } m_3 = \begin{pmatrix} 1 & 0 \\ 0 & -1 \end{pmatrix}.$$

The Lie bracket commutation of the generators $\{m_1, m_2, m_3\}$ is presented in Table 1.

3 Driftless Control System on $Sp(2, \mathbb{R})$

A control system which is drift-free and left-invariant on the dynamics of $Sp(2, \mathbb{R})$ is defined on a three-dimensional manifold, which, in this case, is the Lie group $Sp(2, \mathbb{R})$ as [13]

$$\dot{X} = X\left(\sum_{i=1}^{3} m_i u_i\right), \tag{2}$$

where $X \in Sp(2, \mathbb{R})$, for $i = 1, 2, 3$, m_i are the basis elements of Lie algebra $sp(2, \mathbb{R})$ and u_i are the controls.

A control system which is drift-free, on the dynamics of $Sp(2, \mathbb{R})$ with less controls, is given in the following form:

$$\dot{X} = X(m_1 u_1 + m_2 u_2). \tag{3}$$

3.1 Controllability

Controllability portrays the ability of a control $u(t)$ which steers the system from an initial state to a desired final state in finite time. To study the system controllability, defined on a Lie group, a well-know theorem is discussed.

Chow–Rashevsky Theorem. 'If M is a connected manifold and the control distribution $\Delta = span\{f_1, f_2, ..., f_n\}$ is bracket generating, then the drift-free system

$$\dot{X} = X \sum_{i=1}^{n} x_i f_i(x), \forall x \in M$$

is controllable [7]'.

Proposition 1 *The control system* (3) *is controllable.*

Proof The algebra generated by $\{m_1, m_2\}$ under Lie bracket commutation coincides with $sp(2, \mathbb{R})$. Hence, by Chow's theorem, the system (3) which has less control terms is controllable.

3.2 The Optimal Control Problem

When a system is controllable, it ensures that a steering control exists that drives the system from an initial state X_0 to a desired final state X_f. But controllability analysis fails to address the uniqueness of the input options. There may exist many control options which can drive the system. Choosing control inputs in an optimizing manner ensures in improved performance. Minimum effort problem is considered in this section and the input cost function is designed in a manner that it minimizes the cost function.

The input cost function considered is

$$F = \frac{1}{2} \int_{0}^{x_f} (a_1 u_1^2 + a_2 u_2^2) dt, a_1, a_2 > 0. \tag{4}$$

To minimize F, the controlled Hamiltonian is defined as

$$\bar{\mathcal{H}} = x_1 u_1 + x_2 u_2 - \frac{1}{2}\left(a_1 u_1^2 + a_2 u_2^2\right). \tag{5}$$

According to Krishnaprasad's theorem [11], to obtain the optimized inputs, the controlled Hamiltonian $\bar{\mathcal{H}}$ is partially differentiated with respect to each control input and equated to zero as given below:

$$\frac{\partial \bar{\mathcal{H}}}{\partial u_1} = \frac{\partial \bar{\mathcal{H}}}{\partial u_2} = 0.$$

Hence, the optimized control inputs obtained are

$$u_1 = \frac{x_1}{a_1}, u_2 = \frac{x_2}{a_2}. \tag{6}$$

Substituting the optimized control inputs obtained from (6) to Eq. (5), the optimal Hamiltonian is found to be

$$\mathcal{H}(x_1, x_2, x_3) = \frac{1}{2}\left(\frac{x_1^2}{a_1} + \frac{x_2^2}{a_2}\right).$$

The restricted dynamics followed by the system with the optimized input can be found by Krishnaprasad's theorem [11], which is

$$[\dot{x}_1, \dot{x}_2, \dot{x}_3]^t = \Theta_- \cdot \nabla \mathcal{H}. \tag{7}$$

Here, Θ_- is the minus Lie–Poisson matrix and is defined by

$$\begin{bmatrix} 0 & -4x_3 & -2x_1 \\ 4x_3 & 0 & 2x_2 \\ 2x_1 & -2x_2 & 0 \end{bmatrix}.$$

Explicitly, (7) can be written as

$$\begin{cases} \dot{x}_1 = -\frac{4}{a_2} x_2 x_3 - \frac{2}{a_3} x_1 x_3 \\ \dot{x}_2 = \frac{4}{a_1} x_1 x_3 + \frac{2}{a_3} x_2 x_3 \\ \dot{x}_3 = \frac{2}{a_1} x_1^2 - \frac{2}{a_2} x_2^2 \end{cases}. \tag{8}$$

4 Casimir

Casimir function [5, 16] on the Poisson manifold commutes with every linear functional defined on the manifold. Casimir can be determined by solving a system of partial differential equation (PDE). The system of PDE for the system under consideration is

$$\Theta_- \cdot (\nabla C((x_1, x_2, x_3)))^t = 0. \tag{9}$$

Simplifying Eq. (9), the following system of PDE is obtained:

$$\begin{aligned} -4x_3 \frac{\partial C}{\partial x_2} - 2x_1 \frac{\partial C}{\partial x_3} &= 0 \\ 4x_3 \frac{\partial C}{\partial x_1} + 2x_2 \frac{\partial C}{\partial x_3} &= 0 \\ 2x_1 \frac{\partial C}{\partial x_1} - 2x_2 \frac{\partial C}{\partial x_2} &= 0. \end{aligned} \tag{10}$$

The Casimir is obtained by solving Eq. (10) analytically, which is

$$C = x_1 x_2 - 2x_3. \tag{11}$$

5 Stability

The stationary states, where the motion of the body freezes while following the dynamics, are called equilibrium states. For simplicity, all a_i's are replaced with a. The equilibrium states obtained here are not unique. The set equilibrium states $\{e_1^n, e_2^n, e_3^n\}$ are a subset of all equilibrium states of the system.

The system dynamics (8) has the equilibrium states

$$\begin{aligned} e_1^n &= (n, n, 0), n \in \mathbb{R}; \\ e_2^n &= (0, 0, n), n \in \mathbb{R}; \\ e_3^n &= (n, -n, 0), n \in \mathbb{R}. \end{aligned}$$

The linearized jacobian matrix of the system (8) is

$$J = \begin{bmatrix} -\frac{2}{a}x_3 & -\frac{4}{a}x_3 & -\frac{2}{a}x_1 - \frac{4}{a}x_2 \\ \frac{4}{a}x_3 & \frac{2}{a}x_3 & \frac{4}{a}x_1 + \frac{2}{a}x_2 \\ \frac{4}{a}x_1 & -\frac{4}{a}x_2 & 0 \end{bmatrix}.$$

The eigenvalues of the linearized matrix around

i. e_1^n are

$$\mu_1 = 0, \mu_2 = i4\sqrt{3}\frac{n}{a}, \mu_3 = -i4\sqrt{3}\frac{n}{a};$$

ii. e_2^n are

$$\mu_1 = 0, \mu_2 = \sqrt{20}\frac{n}{a}, \mu_3 = -\sqrt{20}\frac{n}{a};$$

iii. e_3^n are

$$\mu_1 = 0, \mu_2 = i4\frac{n}{a}, \mu_3 = -i4\frac{n}{a}.$$

For $n, a \neq 0$, the following proposition holds.

Proposition 2 *The equilibrium states, e_1^n, e_3^n, are spectrally stable whereas e_2^n are unstable.*

Proof Since none of the eigenvalues around e_1^n and e_3^n have no positive real part, the system is spectrally stable. But one of the eigenvalues around e_2^n has positive real part, so the system is unstable.

For $n = 0$, the stability of the system is inconclusive.

6 Numerical Integration of Dynamics

Equation (8) involves simultaneous nonlinear ordinary differential equations (ODEs), which makes it difficult to compute the analytic solution. Hence, some unconventional numerical techniques are proposed to solve the nonlinear ODEs and subsequent results are analysed.

Poisson, Casimir and Hamiltonian (Energy) preservation

Let $X^n = \left[x_1^n, ..., x_m^n\right] = y(t)$, for step length$h$, $y(t+h) = X^{n+1}$.

An integrator $\varphi : \mathbb{R}^n \to \mathbb{R}^n$ is Poisson preserving when it satisfies the following condition [6]:

$$\varphi_y(y)P(y)\varphi_y^t(y) = P(\varphi(y)), \tag{12}$$

where P is a Poisson tensor and $\varphi_y(y)$ denotes the Fréchet derivative.

Hence,

$\varphi_y(y) = \frac{\partial(y(t+h))}{\partial(y(t))} = \frac{\partial X^{n+1}}{\partial X^n} = P'$,

where P' is the Jacobian.
Thus, (12) reduces to $P' \cdot P(X^n) \cdot (P')^T = P(X^{n+1})$.
Further, the Casimir function C and the Hamiltonian $\mathcal{H}$ are preserved when

$$C(X^{n+1}) = C(X^n)$$

and

$$\mathcal{H}(X^{n+1}) = \mathcal{H}(X^n).$$

6.1 Kahan's Integrator

This numerical integrator inherits certain integrability properties from Runge–Kutta method, such as it preserves all affine symmetric integrals.

Kahan's integrator [10] when applied to the Symplectic Poisson system takes the following form:

$$\begin{cases} x_1^{n+1} - x_1^n = -\frac{2h}{a_2}\left(x_2^{n+1}x_3^n + x_3^{n+1}x_2^n\right) - \frac{2h}{a_3}\left(x_1^{n+1}x_3^n + x_3^{n+1}x_1^n\right) \\ x_2^{n+1} - x_2^n = \frac{2h}{a_1}\left(x_1^{n+1}x_3^n + x_3^{n+1}x_1^n\right) + \frac{h}{a_3}\left(x_2^{n+1}x_3^n + x_3^{n+1}x_2^n\right) \\ x_3^{n+1} - x_3^n = \frac{h}{a_1}\left(x_1^{n+1}x_1^n\right) - \frac{h}{a_2}\left(x_2^{n+1}x_2^n\right). \end{cases} \quad (13)$$

Proposition 3 *Kahan's integrator has the following properties:*

1. *The Poisson structure is not preserved.*
2. *The Casimir C of the system is not preserved.*
3. *The Hamiltonian $\mathcal{H}$ of the system is not preserved.*

Proof Simultaneous equations in (13) are solved, and it has been explicitly computed and shown that.

$P' \cdot P(X^n) \cdot (P')^T \neq P(X^{n+1})$.
Hence, Kahan's integrator is not Poisson preserving.
Also, it is noted that

$$C(X^{n+1}) \neq C(X^n)$$

and

$$\mathcal{H}(X^{n+1}) \neq \mathcal{H}(X^n),$$

which concludes that the integrator is neither Casimir preserving nor Hamiltonian preserving.

6.2 Lie–Trotter Integrator

In this section, another numerical integration has been discussed which is given by Trotter [17]. As per this integrator, Hamiltonian $\mathcal{H}$ of the dynamical system has to be split into $\mathcal{H}_1, \mathcal{H}_2, \ldots, \mathcal{H}_n$ in a way that the dynamics generated by $\mathcal{H}_1, \mathcal{H}_2, \ldots, \mathcal{H}_n$ can be explicitly found.

The Hamiltonian vector field $X_{\mathcal{H}}$ splits as

$$X_{\mathcal{H}} = X_{\mathcal{H}_1} + X_{\mathcal{H}_2},$$

where

$$\mathcal{H}_1 = \frac{x_1^2}{2a_1}, \mathcal{H}_2 = \frac{x_2^2}{2a_2}.$$

The corresponding integral curves are given by

$$\begin{bmatrix} x_1(t) \\ x_2(t) \\ x_3(t) \end{bmatrix} = \Upsilon_i \begin{bmatrix} x_1(0) \\ x_2(0) \\ x_3(0) \end{bmatrix}, i = 1, 2,$$

where

$$\begin{cases} \Upsilon_1 = \begin{bmatrix} 1 & 0 & 0 \\ 4\alpha^2 t^2 & 1 & 4at \\ 2at & 0 & 1 \end{bmatrix}; \alpha = \frac{x_1(0)}{a_1} \\ \Upsilon_2 = \begin{bmatrix} 1 & 4\beta^2 t^2 & -4\beta t \\ 0 & 1 & 0 \\ 0 & -2\beta t & 1 \end{bmatrix}; \beta = \frac{x_2(0)}{a_2} \end{cases} . \tag{14}$$

The Lie–Trotter integrator is presented as

$$\begin{bmatrix} x_1^{n+1} \\ x_2^{n+1} \\ x_3^{n+1} \end{bmatrix} = \Upsilon_1 \Upsilon_2 \begin{bmatrix} x_1^n \\ x_2^n \\ x_3^n \end{bmatrix},$$

i.e.

$$\begin{cases} x_1^{n+1} = x_1^n + 4\beta^2 t^2 x_2^n - 4\beta t x_3^n \\ x_2^{n+1} = 4\alpha^2 t^2 x_1^n + \left(16\alpha^2\beta^2 t^4 - 8\alpha\beta t^2 + 1\right)x_2^n + \left(-16\alpha^2\beta t^3 + 4\alpha t\right)x_3^n \\ x_3^{n+1} = 2\alpha t x_1^n + \left(4\alpha\beta^2 t^3 - 2\beta t\right)x_2^n + \left(1 - 8\alpha\beta t^2\right)x_3^n \end{cases} . \tag{15}$$

Proposition 4 *The Lie–Trotter integrator has the following properties:*

1. *The Poisson structure is preserved.*
2. *The Casimir C of the Poisson configuration is preserved.*
3. *The Hamiltonian $\mathcal{H}$ of the system is not preserved.*

Proof The Lie–Trotter integrator preserves the Poisson structure and Casimir because the flows of the Hamiltonian vector fields $X_{\mathcal{H}_1}$ and $X_{\mathcal{H}_2}$ are Poisson maps.

It does not preserve the Hamiltonian of the system because of the fact that

$$\{\mathcal{H}_1, \mathcal{H}_2\}_- \neq 0.$$

7 Trajectory of the Dynamical System

In this section, trajectory of the dynamics has been found using the above-mentioned integrators. For comparison, fourth-step Runge–Kutta integrator has also been implemented. The intial values for the numerical integrators is given in Table 2.

Figure 1 shows the trajectory for the system described in Eq. (13).

Figure 2 shows the trajectory for the system described in Eq. (15).

Table 2 Initial values for the integrators

Integrator	x_1	x_2	x_3	a_1	a_2	h
Kahan's integrator	1	1	1	1	1	1
Lie–Trotter integrator	1	1	1	1	1	–
Runge–Kutta integrator	1	1	1	1	1	1

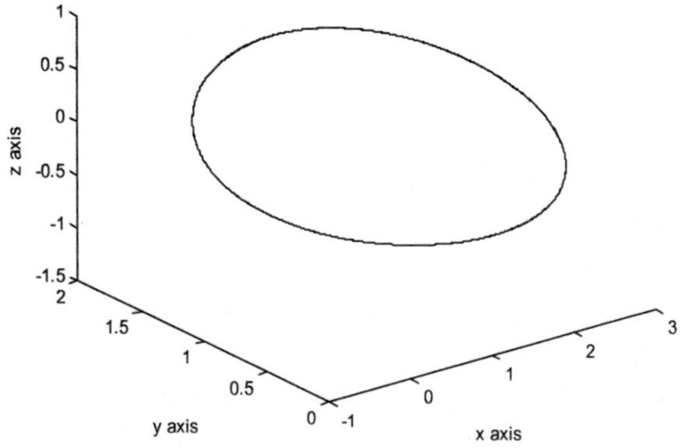

Fig. 1 Trajectory of Kahan's integrator

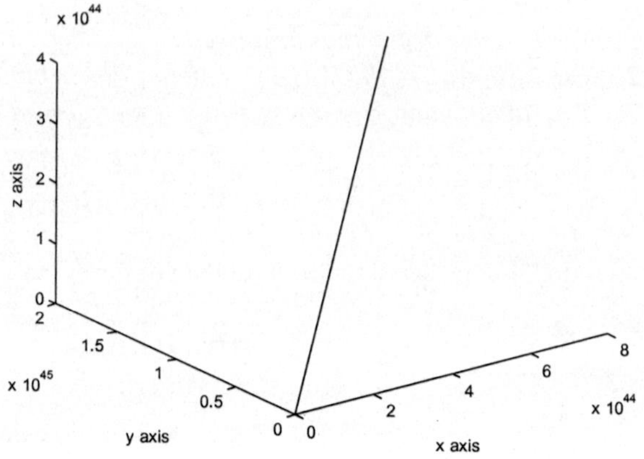

Fig. 2 Trajectory of Lie–Trotter integrator

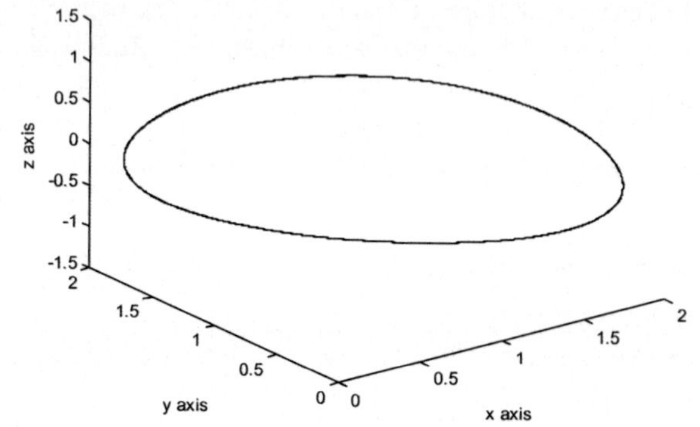

Fig. 3 Trajectory of RK 4 integrator

Figure 3 shows the trajectory for the system described in Eq. (8) for Runge–Kutta integrator.

From the figures, it can deduced that Kahan's integrator provided a favourable approximation of the dynamics, when compared with the Runge–Kutta (fourth step) method. However, Kahan's integrator has the benefit of easier implementation.

8 Conclusion

In this paper, a controllable driftless system on the Symplectic Lie group $Sp(2, \mathbb{R})$ is considered. An optimal control problem has been defined. The Casimir function and stability of the system are discussed in detail. Finally, a comparison is presented between the unconventional integrators and Runge–Kutta method.

References

1. A.A. Agrachev, Y.L. Sachkov, *Control Theory from the Geometric Viewpoint* (Springer, Berlin)
2. R.W. Brockett, System theory on group manifold and coset spaces. SIAM J. Control. **10**, 265–284 (1972)
3. M. Craivoveanu, C. Pop, A. Aron, C. Petrisor, An optimal control problem on the special Euclidean group SE(3;R), in *The International Conference of Differential Geometry and Dynamical Systems* (2009), pp. 68–78
4. R. Gupta, U.V. Kalabic, S.D. Cairano, A.M. Bloch, I.V. Kolmanovsky, Constrained spacecraft attitude control on SO(3) using fast nonlinear model predictive control, in *American Control Conference (ACC)* (Chicago., IL, 2015), pp. 2980–2986
5. B. Hernandnez Bermejo, V. Fairen (1998) Simple evaluation of Casimir invariants in finite dimensional Poisson systems. Phys. Lett. A. **3**, 148–154
6. L.O. Jay, Preserving Poisson structure and orthogonality in numerical integration of differential equations. Comput. Math. Appl. **48**, 237–255 (2004)
7. F. Jean, *Control of Nonholonomic Systems and Sub-Riemannian Geometry*. Lecture Notes (CIMPA School, Beirut, 2012)
8. V. Jurdjevic, *Geometric Control Theory* (Cambridge University Press, UK, 1997).
9. V. Jurdjevic, H.J. Sussmann, Control system on Lie groups. J. Differ. Eqn. **12**, 313–329 (1972)
10. W. Kahan, *Unconventional Numerical Methods for Trajectory Calculation*. Lecturer notes (1993)
11. P.S. Krishnaprasad, Optimal control and poisson reduction. Technical report, T.R. 93–87 (1993)
12. N.E. Leonard, Averaging and motion control systems on Lie groups. Ph.D thesis. University of Maryland, College Park, MD, United States (1994)
13. N.E. Leonard, P.S. Krishnaprasad, Motion control of drift-free, left-invariant systems on Lie groups. IEEE Trans. Automat. Control **40**(9), 1539–1554 (1995)
14. C. Pop, An optimal control problem on the Heisenberg Lie group H(3). Gen. Math. **5**, 323–330 (1997)
15. C. Remsing, Control and integrability on SO(3), in *Proceedings of the World Congress on Engineering* (London, UK, 2010)
16. J.L. Thiffeault, P.J. Morrison, Classification and Casimir Invariants of Lie-Poisson Brackets. Physica D **136**(3–4), 205–244 (2000)
17. H.F. Trotter, On the product of semi-groups of operators. Proc. Am. Math. Soc. **10**, 545–551 (1959)

Pattern Formation from Reaction–Diffusion Equation Using Discretization Method

Atanu Maji

Abstract We know that mathematics has wide applications in the areas of environmental science, medical science, ecology, biology, etc. One very useful term in the problem of prey–predator relation is the Lotka–Volterra predator–prey equation. Especially the differential equation plays a very important role in all areas of science. But it is also true that maximum biological and chemical problems are defined in form of some unknown functions. Here, in this paper, an environmental case involving two related populations of prey and predator species is discussed. As the classic Lotka–Volterra assumptions are imaginary, it is assumed that there is logistic behavior for both the existing species. We see that the number of two populations are too much dependent on each other.

Keywords Discretization · Spatiotemporal model · Reaction–diffusion equation · Turing pattern

1 Finite Differences

1.1 Introduction

The derivative of a function $v(x)$ or the rate of change of v at x is defined by

$$v'(x) = \lim_{\delta_x \to 0} \frac{v(x + \delta_x) - v(x)}{\delta_x}$$

There are various methods to solve a differential equation and finite difference method is one of the simplest and useful methods by approximating them. Sometimes, we have a differentiable function $v(x)$ for which there is no method to compute $v'(x)$.

A. Maji (✉)
Center for Applied Mathematics, Computing Institute of Technical Research and Education, Siksha O Anusandhan Deemed to Be University, Bhubaneswar, India
e-mail: atanumaji@soa.ac.in

S. R. Mishra et al. (eds.), *Recent Trends in Applied Mathematics*, Lecture Notes in Mechanical Engineering, https://doi.org/10.1007/978-981-15-9817-3_11

Or perhaps the formula for $v'(x)$ is simply unwieldy or difficult to compute. In this case one can approximate $v'(x)$ by using finite differences, i.e., we need to accept discretization of the continuous case [1]. In every discretization step, we will take a finite set of points and function is defined for every point of the set. Also, function will have an output corresponding to every point of the discrete set of points. Now from this discrete grid of the function, we can measure the approximations to the derivative [5]. Suppose $v_i = v(x_i)$ is the value of the function $v(x)$ at the ith grid of the computational table. Now consider all the equal intervals which will divide the x-axis and the length of the interval is $\delta_{x_i} = x_{i+1} - x_i$.

1.2 Taylor's Series

Let $v^{(n)}(x)$ is continuous on $\{x : p < x < q\}$ then for $p < a, a + h < q$

$$v(a+h) = \sum_{k=0}^{n-1} \frac{h^k}{k!} v^{(k)}(a) + R_n,\ where\ R_n = \frac{h^n(1-\theta)^{n-p}}{(n-1)!p} v^{(n)}(a+\theta h)$$

Here p is a given positive integer and $0 < \theta < 1$, also we can see $R_n = O(h^n)$. That means, if we have the value of a function $v(x)$ and its first $(n-1)$ derivatives at an initial point a, then it is very simple to find the function value at any point $a + h$ by using above equation. But there is some unknown value $O(h^n)$, we can treat as an error of the approximation to $v(a + h)$.

1.3 Constructing Finite Difference Approximation

Suppose there is a function v of two variables x and t. We approximate partial derivative of v with respect to x so we have to keep t fixed and the function $v(x, t)$ can be converted into a single variable function like $f(x)$ [6]. So from the Taylor's theorem with the step size δ_x we can write the expansion:
$v(a + \delta_x, t) = v(a, t) + \delta_x v_x(a, t) + \frac{(\delta_x)^2}{2!} v_{xx}(a, t) + \cdots +$
$\frac{(\delta_x)^{(n-1)}}{(n-1)!} v^{(n-1)}(a, t) + O((\delta_x)^n)$
Taking up to $O({\delta_x}^2)$ we get

$$v(a + \delta_x, t) = v(a, t) + \delta_x v_x(a, t) + O((\delta_x)^2)$$

$$\Rightarrow v_x(a, t) = \frac{v(a + \delta_x, t) - v(a, t)}{\delta_x} - \frac{O((\delta_x)^2)}{O(\delta_x)}$$

$$\Rightarrow v_x(a, t) = \frac{v(a + \delta_x, t) - v(a, t)}{\delta_x} - O(\delta_x)$$

Now when we approximate the solution of a partial differential equation by the methods of numerical analysis we must consider a grid of discrete x values, namely, $x_1, x_2, \ldots, x_n$ and also values of t as $t_0 = 0, t_1, t_2, \ldots$. We assume that the grid spacing is constant and it is δ_x in x such that $x_{i+1} = x_i + \delta_x$. We use a notation as $V_i^n = v(x_i, t_n)$. So we will now use

$$v_x(x_i, t_n) = \frac{V_{i+1}^n - V_i^n}{\delta_x} - O(\delta_x).$$

Similarly, we can use

$$v_x(x_i, t_n) = \frac{V_i^n - V_{i-1}^n}{\delta_x} - O(\delta_x)$$

So, from the above two equations we get

$$v_x(x_i, t_n) \approx \frac{V_{i+1}^n - V_{i-1}^n}{2\delta_x}. \text{ (neglecting the error term)}$$

For second-order unmixed partial derivatives,
$v(a + \delta_x, t) = v(a, t) + \delta_x v_x(a, t) + \frac{(\delta_x)^2}{2!} v_{xx}(a, t) + \frac{(\delta_x)^3}{3!} v_{xxx}(a, t) + O((\delta_x)^4)$
and also,

$$v(a - \delta_x, t) = v(a, t) - \delta_x v_x(a, t) + \frac{(\delta_x)^2}{2!} v_{xx}(a, t) - \frac{(\delta_x)^3}{3!} v_{xxx}(a, t) + O((\delta_x)^4)$$

From the addition of above two equations, we get,

$$v(a + \delta_x, t) + (a - \delta_x, t) = 2v(a, t) + {\delta_x}^2 v_{xx}(a, t) + O((\delta_x)^4)$$

If we use the discrete notation and calculate at (x_i, t_n) we get,

$$V_{i+1}^n + V_{i-1}^n = 2V_i^n + {\delta_x}^2 v_{xx}(x_i, t_n) + O((\delta_x)^4)$$

neglecting the error term and by rearranging we get

$$v_{xx}(x_i, t_n) \approx \frac{V_{i+1}^n - 2V_i^n + V_{i-1}^n}{(\delta_x)^2}$$

Again if we fix x at $x = x_i$ and derive the approximation of v_t and v_{tt} we get

$$v_t \approx \frac{V_i^{n+1} - V_i^n}{\delta_t} \text{ and } v_{tt} \approx \frac{V_i^{n+1} - 2V_i^n + V_i^{n-1}}{(\delta_t)^2}$$

1.4 Notation for Function of Several Variables

Let us consider a function v of two independent variables x and t. Also consider the sets of exactly same rectangles with the sides $\delta_x = h,\ \delta_t = k$ which can be generated by the equidistant grid lines parallel to y-axis, specified as $x_i = ih,\ i = 0, \pm1, \pm2, \ldots$ and the equidistant grid lines parallel to x-axis, specified as $y_j = jk,\ ,j = 0, 1, 2, \ldots$. Now if we use the notation $v_p = v(ih, jk) = v_{i,j}$ for the value of $v(x, t)$ at the mess point $p(ih, jk)$, so we can express [4],

$$(v_{xx})_p = (v_{xx})_{i,j} \approx \frac{v((i+1)h, jk) - 2v(ih, jk) + v((i-1)h, jk)}{h^2}$$

i.e.,

$$v_{xx} = \frac{v_{i+1,j} - 2v_{i,j} + v_{i-1,j}}{(\delta_x)^2}$$

with an approximation error of $O((\delta_x)^2)$. Similarly,

$$v_{tt} = \frac{v_{i,j+1} - 2v_{i,j} + v_{i,j-1}}{(\delta_t)^2}$$

with an approximation error of $O((\delta_t)^2)$.

2 Spatiotemporal Model: Pattern Formation

Predator–prey model is one of the backbones of the bio and ecosystem as bio-masses are grown out of their resource masses. The species fight against themselves, evolve, and disperse for the purpose of collecting resources for their existence. The population density of prey and predator are not uniform all over the world and sometimes for excessive population densities they shift from one place to another. Their portability mainly depends on the food chain they are connected [3]. And it is obvious that predators must move to the locations where the density of prey is maximum. Now if we consider a particular area and discuss that kind of potency of prey as well as the predator then a system of equations, namely, reaction–diffusion equations arises as

$$Q_T = gQ(1 - \frac{Q}{k}) - \frac{mQR}{Q + aR} + D_1\nabla^2 Q$$
$$R_T = sR(1 - h\frac{R}{Q}) + D_2\nabla^2 R$$

with the initial conditions $Q(0)$ and $R(0)$ both are positive, where $Q,\ R$ are the functions of T, which represents the population density of prey and predator, respectively at any immediate time 'T' and at any point $(X, Y) \in \Omega$ (Ω is $2D$ bounded rectangular domain with boundary $\partial\Omega$). Here g is the inherent growth rate of prey, k is carrying capacity, m is the capturing rate, q is half-saturation constant, s is the inherent growth rate of predator, and h is the conversion rate of prey into predator biomass. Also $D_1 > 0$ and $D_2 > 0$ denotes the constant diffusion coefficients of the prey N and predator P.

Now we can construct a model belonging to both space and time by using the transformation of some variables

$$Q = kq,\ R = \frac{gk}{m}r,\ X = \lambda x,\ T = \frac{t}{g},\ Y = \lambda y \text{ and } \lambda = \sqrt{\frac{D_1}{g}}$$

putting these in reaction–diffusion equations [2] we get

$$q_t = q(1-q) - \frac{qr}{q+\alpha r} + \nabla^2 q$$
$$r_t = \delta r(\beta - \frac{r}{q}) + d\nabla^2 r$$

where q_t, r_t represents the partial derivative of q, r, respectively, with respect to t, $\alpha = \frac{ga}{m}$ (a stands for handling time), $\beta = \frac{m}{hg}$, $\delta = \frac{sh}{m}$ are the dimensionless parameters, $d = \frac{D_2}{D_1}$ and ∇^2 is the Laplace operator.

Now we will draw various type of pattern made of strips and spots formed by the ratio-dependent Holling–Tanner model

$$q_t = q(1-q) - \frac{qr}{q+\alpha r} + \nabla^2 q$$
$$r_t = \delta r(\beta - \frac{r}{q}) + d\nabla^2 r$$

for different values of δ, where $\delta = \frac{sh}{m}$ is a dimensionless system parameter.

If we discretize the above two equations with $\beta = 1$ then we get $\frac{q_{i,j,k+1}-q_{i,j,k}}{\delta_t} = q_{i,j,k}(1-q_{i,j,k}) - \frac{q_{i,j,k}r_{i,j,k}}{q_{i,j,k}+\alpha r_{i,j,k}} + \frac{q_{i+1,j,k}-2q_{i,j,k}+q_{i-1,j,k}}{(\delta_x)^2} + \frac{q_{i,j+1,k}-2q_{i,j,k}+q_{i,j-1,k}}{(\delta_y)^2}$
$\frac{r_{i,j,k+1}-r_{i,j,k}}{\delta_t} = \delta r_{i,j,k}(1-\frac{r_{i,j,k}}{q_{i,j,k}}) + d[\frac{r_{i+1,j,k}-2r_{i,j,k}+r_{i-1,j,k}}{(\delta_x)^2} + \frac{r_{i,j+1,k}-2r_{i,j,k}+r_{i,j-1,k}}{(\delta_y)^2}]$.

For numerical calculation, we take the initial population distribution as $q_{i,j,1} = q_* + \epsilon\xi_{i,j}$ and $r_{i,j,1} = r_* + \epsilon\eta_{i,j}$, where (q_*, r_*) are the interior equilibrium point, $\epsilon = 0.001$ and $\xi_{i,j}, \eta_{i,j}$ are random in space.

For the pattern formation we will use $d = 16,\ \alpha = 0.4,\ \delta_x = \delta_y = 1,\ \delta_t = 0.01$, length of each x and y axis is 200, length of t axis is 10000. We will vary the value of δ as 0.7, 0.9 and 1.1. Then we get the patterns as follows:

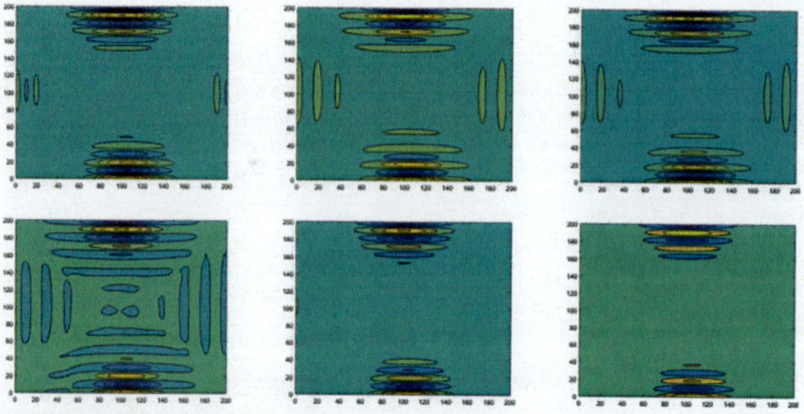

Population distribution of prey (upper row) and predator (lower row) over $2-D$ space for three different values of δ .The first column represents the pattern for δ=0.7 ,second column for $\delta=0.9$,third column for $\delta=1.1$ at 1000th time step.

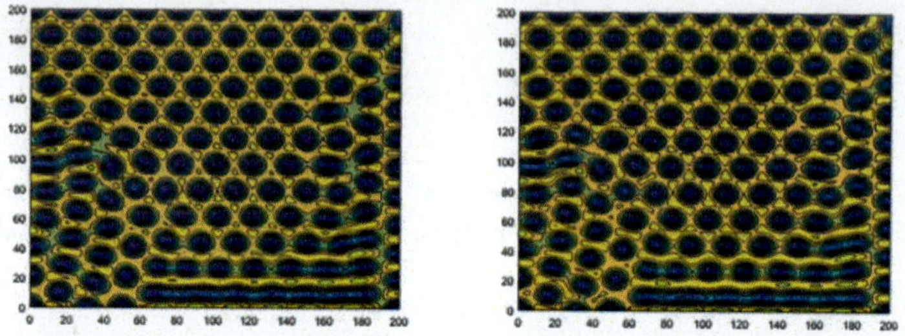

Population distribution of prey (left) and predator (right) tor $\delta=1.1$ at 45000th time step.

Spatiotemporal model is very popular in the recent days for its huge applications in the environment and health science and increasing computational power. This is mainly used for disease mapping, acid rain analysis, sulfate depositions, regional ozone monitoring, and analysis of satellite data. We can use it to develop a visual design of population density and population mobility of prey as well as predator. If we choose a particular region to verify their population density and state of living we can find out that it depends on various factors of nature. A pure prey–predator relation appears when a particular predator species is dependent on a particular species of prey and also the prey species has enough food supply to survive. But there are very few examples of such a prey–predator relation. Here I have only considered one factor that is δ. You can see from the visual representation that the pattern is changing slowly from the stripe pattern to spot pattern for an increasing value of δ. But it is more clearly visible for the increasing time steps. The differences of maximum and minimum cluster of both populations increase with the increase of δ, and the area filled up with higher population density increases gradually. Now if we consider more values of δ than 1.1 it will affect the homogeneous distribution of the population of two species at their stable state value over the space. I have used some algorithm to get a proper visual representation of population distribution. But it took too much time for higher time steps. There should be a more appropriate algorithm to form a pattern with the same model but with less amount of time to be taken.

References

1. L.J.S. Allen, *An Introduction to Mathematical Biology* (Pearson Prentice Hall, 2007)
2. M. Banerjee, Spatial pattern formation in ratio-dependent model: higher-order stability analysis (2010)
3. P. Bastani, Pattern formation in the Gray-Scott model (Department of Mathematics, Simon Fraser University)
4. R.L. Burden, J.D. Faires, *Numerical Analysis*, vol. 173–184, 9th edn (Brooks/Cole Cengag Learning, 2011), pp. 193–203
5. J.D. Murray, *Mathematical Biology I. An Introduction*, 3rd edn. (Springer, 2002), pp. 79–88
6. G.D. Smith, *Numerical Solution of Partial Differential Equations*, 3rd edn. (Clarendon Press, Oxford, 1985), pp. 17–23

Semi-analytical Approach to Solve the System of Nonlinear Differential Equations

B. Nayak and R. S. Tripathy

Abstract It is merely encountered that in several branches of sciences, including physics, chemistry, mathematics, and biology, the appearance of ordinary and partial differential equations plays an important role. The dynamical behaviors are based on the physical significance of the problem. Qualitatively, these are the several branches of physics. Such applications are laid down as population growth, potential field, electrical circuit, biological behavior of trees, etc. However, differential equations are obtained from the physical laws. The solution of linear differential equation is quite easy to handle but though some of the nonlinear differential equation admits analytical solution, it is difficult for many. Some classical methods are proposed to get approximate analytical solution. In the present situation, we adopt the Laplace transformation technique embodies with Adomian decomposition method (ADM). Validation of the present result is also obtained with earlier published work and conformity of the solution achieved.

Keywords Nonlinear differential equation · Laplace transform technique · Adomian decomposition method (ADM)

1 Introduction

In every area of engineering and physical science, the occurrence of linear and nonlinear ordinary or partial differential equations plays a vital role. The common areas are solid-state physics, bioengineering, plasma physics, analytical chemistry, fluid dynamic, mathematical biology, chemical kinetics, and so on. Based on the physical significance, the problem leads to a mathematical model of nonlinear differential equation. For the solution of these nonlinear differential equations, several authors have used analytical or semi-analytical methods. Such methods are homotopy analysis method (HAM), homotopy perturbation method (HPM), variational iteration method (VIM), etc.

B. Nayak (✉) · R. S. Tripathy
Department of Mathematics, Siksha 'O'Anusandhan Deemed to be University, Khandagiri, Bhubaneswar 751030, Odisha, India
e-mail: bimaleshnayak@gmail.com

S. R. Mishra et al. (eds.), *Recent Trends in Applied Mathematics*, Lecture Notes in Mechanical Engineering, https://doi.org/10.1007/978-981-15-9817-3_12

Numerical solution of various problems like functional differential equations (FDEs) is a major challenge in mathematics, in particular, the delayed differential equations of Cauchy problem. In the last two decades, various methods are taken care of for the solution of the nonlinear differential equations. These numerical methods are Runge–Kutta fourth-order method associated with shooting technique, finite difference technique, finite element method, finite volume, etc. Moreover, many researchers have also handled the similar type of complex nonlinear differential equations by using semi-analytical approach. These are methods such as variation parameter method, Adomian decomposition method, Laplace transformation technique, etc.; the basis of many such methods is Taylor series, i.e., polynomial approximation, Taylor collocation method, and differential transformation method (DTM); and Pade approximation method have been considered to get approximate analytical solutions for particular classes of differential equations. In general, every convergence function within their domain can be expressed as a power series form with the help of Taylor series and the idea of our Taylor series concept is to combine general method of steps which is suitable for Cauchy problems. The details of few methods are described later in the thesis. The said approach tells us to rewrite the terms containing delay with its initial guess function and its derivatives. However, the main objective of the methods used is to transform the partial differential equation to ordinary or to reduce the order of the differential equation. It is necessary because sometimes the number of initial conditions are also lacking for the solution of boundary value problems. However, delayed (or neutral) differential equation is reduced to particularly the type of Cauchy problem for ordinary differential equation. Moreover, in boundary value problems for the solution while using ADM, HAM, HPM, and VIM, it requires initial approximation guess from the prescribed initial conditions and at the time of computation the unknown initial conditions are to be obtained which satisfy the boundary conditions generally consisting of derivatives.

The Adomian decomposition method (ADM) [1–3] is useful to get an approximate analytical solution of the various differential equations those are nonlinear. It is more useful because of its solution procedure where it is not necessary to simplify the nonlinear differential equations for their solution and the series solutions for the physical model those are rapid convergence. Since the traditional perturbation or linearization technique we have to avoid. To get exact approximate analytical solution for nonlinear differential equation with adequate/inadequate initial conditions, solitary wave solutions, rational solutions, and few more exact solution described in [4–8], the method ADM has been applied. Many such techniques are available for micro- and nano-scale flows including finite element techniques [9] and shooting quadrature [10]. Another group of semi-numerical methods has in recent years also become popular. For the numerical computation of higher order approximations, we have used symbolic software, i.e., MAPLE, MATHEMATICA, and MATLAB. Many researchers have developed their code for various methods of solutions such as "*homotopy methods*" Bég, and his coworkers ([11, 12]) and the "*successive Taylor series linearization method (STSLM)*" use "Chebyshev interpolating polynomials" and "Gauss–Lobatto collocation," as investigated by Bhatti et al. [13]. An alternative approach for the polynomial functions to get much faster convergence than that of other is conducted by Adomian [14]. The method called "*Adomian decomposition method (ADM)*" has been developed recently in various problems of field of research, in particular, fluid dynamics problems. These studies are Kezzar and Sar [15] (nanofluids), Bég et al. [16] (magneto-rheological squeeze films), Ebaid et al. [17] (nanofluid boundary layers), and Aaboubi et al. [18] (electrochemical mass transport). There are many advantages of ADM for example, in the solution of nonlinear differential equation it presents approximate analytical solution without discretizing the problems.

The assumed initial guess can be obtained for the iterative solutions for the nonlinear differential equation where the higher order differential term becomes linear. Depending upon the complementary function and auxiliary equation is chosen and using the initial conditions initial guess functions is obtained. These guess solutions are used for the developed recursive equation or the iterative scheme for the rest of the nonlinear term appeared in the governing equations and the computation is going on for the particular number of steps where the error is optimized.

In the present paper, the governing differential equations or the system of differential equations are solved using Laplace transform technique and then Adomian decomposition is used for the rest of the nonlinear terms. Afterward, the inverse transform is obtained using Laplace inverse method. For few nonlinear problems, direct ADM is used and for the unknown values we used numerical technique so that the method is called semi-analytical approach.

To perform a solution by ADM, it is important to create a function in the form with the terms $u_k(x,t), k = 0, 1, 2, ...$ and further the series is used to get the exact solutions. The nth term approximation $u_{appr} = \sum_{k=0}^{n-1} u_k(x,t)$ can be referred to find numerical computation.

Let us consider

$$\frac{du_n}{dt} = (\alpha - u_n^2)(u_{n+1} - u_{n-1}) \tag{1}$$

the discretized nonlinear Schrodinger's equation:

$$i\frac{du_n}{dt} = (u_{n+1} + u_{n-1} - 2u_n) - |u_n|^2(u_{n+1} + u_{n-1}) \tag{2}$$

and the equations

$$\begin{aligned} \frac{du_n}{dt} &= u_n(v_n - v_{n-1}) \\ \frac{dv_n}{dt} &= v_n(u_{n+1} - u_n) \end{aligned} \tag{3}$$

where n in Eqs. (1)–(3) represents subscript.

2 Analysis of the Method

Our objective is to solve nonlinear differential equations by ADM. The considered nonlinear delay differential equation can be solved by the said method.

First of all, the general form of the differential equation can be written using particular operators as

$$Lu = R(u) + N(u) \tag{4}$$

Here, L indicates the linear operator which is reversible,

$$R(u) = \alpha(u(n+1) - u(n-1)) \tag{5}$$

is the remainder term and

$N(u) = -u(n)^2(u(n+1) - u(n-1))$ is the rest of the nonlinear terms.

The inverse operator is denoted as L^{-1} and it shows the integral w.r.t t' from 0 to t, such that $L^{-1} = \int_0^t ()dt'$.

Afterward, inverse operator is used in Eq. (1) and we get

$$u = f_0 + L^{-1}(R(u) + N(u)) \tag{6}$$

where the initial guess solution f_0 satisfies the condition $Lf_0 = 0$.

In ADM, the decomposition is used for the unknown function u and

$$u = \sum_{m=0}^{\infty} u_m(n, t) \tag{7}$$

In particular, u_0 satisfy the initial condition $u(n, 0)$.

The nonlinear term $N(u)$ can be decomposed into

$$N(u) = \sum_{m=0}^{\infty} A_m(u_0, u_1, ...u_m) \tag{8}$$

where the Adomian's polynomials are $A_m(u_0, u_1, ...u_m)$ and defined as

$$A_m = \frac{1}{m!}\frac{d^m}{d\lambda^m}\left(N\left(\sum_{m=0}^{\infty} \lambda^i u_i\right)\right) \tag{9}$$

Therefore, the terms of A_m are as follows:

$$A_0 = -u_0(n)^2(u_0(n+1) - u_0(n-1)) \tag{10}$$

$$A_1 = -2u_0(n)(u_0(n+1) - u_0(n-1))u_1(n) - u_0(n)^2(u_1(n+1) - u_1(n-1)) \tag{11}$$

and so on.

Hence, the first few terms of the decomposed series $u_m(n, t)$ can be solved as

$$u_0(n, t) = f_0$$

$$u_1(n, t) = L^{-1}(\alpha(u_0(n+1, t) - u_0(n-1, t)) + A_0) \tag{12}$$

$u_2(n, t) = L^{-1}(\alpha(u_1(n+1, t) - u_1(n-1, t)) + A_1)$ and so on.

To get the maximum number of term, we use the symbolic software MAPLE.

3 Application of ADM

Example 1 Let us assume.

$$f'''(\tau) - 2f'^2(\tau) + f(\tau)f''(\tau) - Mf'(\tau) = 0 \tag{13}$$

The corresponding boundary conditions are

$$f(0) = 0,\ f'(0) = 1,\ f'(\infty) = 0 \tag{14}$$

The given differential equation is of third-order ordinary differential equation and two initial conditions with one boundary condition. Therefore, we need one more initial condition such as

$$f''(0) = \alpha$$

where the unknown α is to be determined.

Using Laplace transformation of Eq. (13), we get

$$\begin{aligned} &L\{f'''(\tau)\} - L\{2f'^2(\tau) - f(\tau)f''(\tau)\} - ML\{f'(\tau)\} = 0 \\ &\Rightarrow s^3 f(s) - s^2 f(0) - sf'(0) - f''(0) - M(sf(s) - f(0)) \\ &-L\{2f'^2(\tau) - f(\tau)f''(\tau)\} = 0 \end{aligned} \tag{15}$$

Using initial conditions (14)

$$\Rightarrow (s^3 - Ms)f(s) - s - \alpha - L\{2f'^2(\tau) - f(\tau)f''(\tau)\} = 0$$
$$\Rightarrow f(s) = \frac{1}{(s^3 - Ms)}\left(s + \alpha + L\{2f'^2(\tau) - f(\tau)f''(\tau)\}\right) \tag{16}$$

Following the standard procedure of ADM,
let us assume

$$\begin{aligned} f(\tau) &= \sum_{m=0}^{\infty} f_m(\tau) \\ &= f_0 + f_1 + f_2 \ldots\ldots\ldots + f_m + \ldots\ldots \end{aligned} \tag{17}$$

The nonlinear terms present in Eq. (13) can be defined in the Adomian polynomial form as

$$\begin{aligned} f'^2(\tau) &= \sum_{m=0}^{\infty} A_m \\ f(\tau)f''(\tau) &= \sum_{m=0}^{\infty} B_m \\ f'(\tau) &= \sum_{m=0}^{\infty} C_m \end{aligned} \tag{18}$$

By using above relation the components of the Adomian polynomials can be expressed as

$$\begin{aligned} A_0(\tau) &= f_0'^2(\tau) \\ A_1(\tau) &= 2f_0'(\tau)f_1'(\tau) \\ A_2(\tau) &= f_1'^2(\tau) + 2f_0'(\tau)f_2'(\tau) \\ &\ldots\ldots\ldots\ldots \\ &\ldots\ldots\ldots\ldots \end{aligned} \tag{19}$$

$$\begin{aligned} B_0(\tau) &= f_0(\tau)f_0''(\tau) \\ B_1(\tau) &= f_0(\tau)f_1''(\tau) + f_1(\tau)f_0''(\tau) \\ B_2(\tau) &= f_0(\tau)f_2''(\tau) + f_1(\tau)f_1''(\tau) + f_2(\tau)f_0''(\tau) \\ &\ldots\ldots\ldots\ldots \\ &\ldots\ldots\ldots\ldots \end{aligned} \tag{20}$$

and

$$C_0(\tau) = f_0'(\tau)$$
$$C_1(\tau) = f_1'(\tau)$$
$$C_2(\tau) = f_2'(\tau) \tag{21}$$
$$\ldots\ldots\ldots\ldots$$
$$\ldots\ldots\ldots\ldots$$

The initial guess solution

$$\begin{aligned} f(\tau) &= f(0) + \tau f'(0) + \frac{\tau^2}{2!} f''(0) \\ &= 0 + \tau + \frac{\tau^2}{2!}\alpha \end{aligned} \tag{22}$$

$$f_0(\tau) = \tau + \frac{\tau^2}{2}\alpha$$

The recursive relation from Eq. (16) becomes

$$\Rightarrow f_{m+1}(\tau) = L^{-1}\left(\frac{1}{(s^3 - Ms)}(s + \alpha + L\{2A_m - B_m\})\right) \tag{23}$$

For $m = 0, 1, 2, \ldots.$

$$\Rightarrow f_1(\tau) = L^{-1}\left(\frac{1}{(s^3 - Ms)}(s + \alpha + L\{2A_0 - B_0\})\right)$$

$$\Rightarrow f_1(\tau) = L^{-1}\left(\frac{1}{(s^3 - Ms)}\left(s + \alpha + L\left\{2f_0'^2(\tau) - f_0(\tau)f_0''(\tau)\right\}\right)\right)$$

$$\begin{aligned} f_1(\tau) = \frac{(\alpha^2 + M^2 + M)}{M^{5/2}} \sinh(\sqrt{M}\tau) + \frac{1}{M^2}\alpha((M+1)\cosh(\sqrt{M}\tau) - \tau\alpha - 1) \\ - \frac{1}{6M}\left(\alpha^2\tau^3 + 3\alpha\tau^2 + 6\alpha + 6\tau\right) \end{aligned}$$

Similarly,

$$\Rightarrow f_2(\tau) = L^{-1}\left(\frac{1}{(s^3 - Ms)}(s + \alpha + L\{2A_1 - B_1\})\right)$$

$$\Rightarrow f_2(\tau) = L^{-1}\left(\frac{1}{(s^3 - Ms)}\left(s + \alpha + L\left\{4f_0'(\tau)f_1'(\tau) - f_0(\tau)f_1''(\tau) - f_1(\tau)f_0''(\tau)\right\}\right)\right)$$

$$f_2(\tau) = \left(\frac{1}{M^{1/2}}\sinh(\sqrt{M}\tau) + \frac{\alpha}{M}\left(\cosh(\sqrt{M}\tau) - 1\right)\right) + 4\frac{(\alpha^2 + M^2 + M)}{M^2}$$
$$\left(-\frac{1}{M^{3/2}}(3\alpha\tau + 2)\sinh(\sqrt{M}\tau) + \frac{-4\alpha + (\alpha M\tau^2 + 2M\tau + 4\alpha)\cosh(\sqrt{M}\tau)}{M^2}\right) +$$

....

Hence,

$$f(\tau) = f_0(\tau) + f_1(\tau) + f_2(\tau) + - - - - -$$

is the complete solution of the assumed boundary value problem.

Example 2 Let us consider the system of nonlinear equations,

$$f''(\eta) + f'(\eta)g'(\eta) + Kf'(\eta) + Mf(\eta) = 0$$
$$g''(\eta) + f'(\eta)g'(\eta) + g'^2(\eta) + Kg(\eta) = 0$$

With boundary conditions

$$f(0) = A, f(\infty) = 0$$
$$g(0) = B, g(\infty) = 0$$

Applying the standard procedure of Adomian decomposition method (ADM), next we introduce $L_1 = \frac{d^2}{d\eta^2}()$ and its inverse operators $L_1^{-1}() = \int_0^\eta \int_0^\eta ()d\eta d\eta$ the above system can be written as

$$f(\eta) = L_1^{-1}\left(-f'(\eta)g'(\eta) - Kf'(\eta) - Mf(\eta)\right)$$
$$g(\eta) = L_1^{-1}\left(-f'(\eta)g'(\eta) - g'^2(\eta) - Kg(\eta)\right)$$

The unknown functions $f(\eta), g(\eta)$ can be expressed as infinite series of the following form:

$$f(\eta) = \sum_{m=0}^{\infty} f_m, g(\eta) = \sum_{m=0}^{\infty} g_m,$$

The linear and nonlinear terms of ()-() can now be decomposed by an infinite series of polynomials as follows:

$$\left.\begin{array}{l}\sum_{m=0}^{\infty} A_m = f'g', \sum_{m=0}^{\infty} B_m = f', \sum_{m=0}^{\infty} C_m = f, \\ \sum_{m=0}^{\infty} D_m = g'^2, \sum_{m=0}^{\infty} E_m = g\end{array}\right\}$$

Invoking the boundary conditions:

$$\left.\begin{array}{l} f(0) = A,\ f'(0) = p\ , \\ g(0) = B, g'(0)\ = q \end{array}\right\}$$

The unknown values of p *and* q are to be determined.

The initial guess solutions and the successive order solutions are expressed as follows:

$$f_0(\eta) = A + \eta p$$

$$g_0(\eta) = B + \eta q$$

and

$$f_{m+1}(\eta) = L_1^{-1}(-A_m - KB_m - MC_m)$$
$$g_{m+1}(\eta) = L_1^{-1}(-A_m - D_m - KE_m)$$

Using $m = 0, 1, 2$

$$f_1 = T_1\eta^2 + T_2\eta^3$$

$$g_1 = T_3\eta^2 + T_4\eta^3$$

$$f_2 = T_5\eta^3 + T_6\eta^4 + T_7\eta^5$$
$$g_2 = T_8\eta^3 + T_9\eta^4 + T_{10}\eta^5$$
$$f_3 = T_{11}\eta^4 + T_{12}\eta^5 + T_{13}\eta^6 + T_{14}\eta^7$$
$$g_3 = T_{15}\eta^4 + T_{16}\eta^5 + T_{17}\eta^6 + T_{18}\eta^7$$

…………………

…………………

After getting all the unknowns, the required solution can be written as

$$f(\eta = \sum_{m=0}^{\infty} f_m(\eta)$$

$$g(\eta = \sum_{m=0}^{\infty} g_m(\eta)$$

All the $T_i's, i = 1 - 18$ are presented in Appendix 1.

Example 3 Let us consider the mKdV lattice equation.

$$\frac{du_n}{dt} = (1 - u_n^2)(u_{n+1} - u_{n-1})$$

With initial condition

$$u_0 = \tanh(k)\tanh(kn)$$

Using Eqs. (11) and (12), the standard procedure of decomposition method is

$$\begin{aligned} u_1 &= \tanh(k)[\tanh(k(n+1)) - \tanh(k(n+1))]t \\ &\quad - \tanh(k)^2 \tanh(kn)^2[\tanh(k)\tanh(k(n+1)) - \tanh(k)\tanh(k(n+1))]t \end{aligned}$$

$$\begin{aligned} u_2 = \frac{1}{2}\big[&\tanh(k)\tanh(k(n+2)) - 2\tanh(k)\tanh(kn) - \tanh(k)^2\tanh(k(n+1))^2(\tanh(k)\tanh(k(n+2)) \\ &- \tanh(k)\tanh(kn) + \tanh(k)\tanh(k(n-2)) + \tanh(k)^2\tanh(k(n-1))^2(\tanh(k)\tanh(kn) \\ &- \tanh(k)\tanh(k(n-2))) - 2\tanh(k)\tanh(kn)(\tanh(k)\tanh(k(n+1)) \\ &- \tanh(k)\tanh(k(n-1))) - \tanh(k)\tanh(k(n+1)) - \tanh(k)\tanh(k(n-1)) \\ &- \tanh(k)^2\tanh(kn)^2(\tanh(k)\tanh(k(n+1)) - \tanh(k)\tanh(k(n-1)))) \\ &- \tanh(k)^2\tanh(kn)^2(\tanh(k)\tanh(k(n+2)) - 2\tanh(k)\tanh(kn) \\ &- \tanh(k)^2\tanh(k(n+1))^2(\tanh(k)\tanh(k(n+2)) - \tanh(k)\tanh(kn)) \\ &- \tanh(k)\tanh(k(n-2)) + \tanh(k)^2\tanh(k(n-1))^2(\tanh(k)\tanh(kn) - \tanh(k)\tanh(k(n-2))))\big]t^2 \end{aligned}$$

Similarly, the other components are obtained using symbolic software MAPLE. Hence, the approximate solution is

$$u = \sum_{m=0}^{i} u_m$$

The exact analytical solution of the considered problem is

$$u = \tanh(k)\tanh(kn + 2\tanh(k)t)$$

However, for the validation of the present solution, the numerical results of the approximate solution are compared with the exact analytical solution and the error is calculated and presented in Table 1.

Table 1 The values of unknown α for various values of M

M	α
0	−1.28271
1	−1.62921
3	−2.15875
5	−2.58115

4 Results and Discussion

In the present analysis (Example 1), we have considered a nonlinear boundary value problem where the differential equation is third order and consists of two initial conditions. Due to lack of initial condition we have assumed one more initial for its possible solution. At present, instead of using numerical technique, semi-analytical method like Adomian decomposition method is imposed and the unknown initial value is obtained for the constant M and presented in Table 1. However, the effect of the parameter M on the profiles of $f'(\eta)$ and $f(\eta)$ is shown in Figs. 1 and 2, respectively. It is observed that with an increasing M, the profile of $f'(\eta)$ decreases smoothly to meet the boundary condition which confirms the choice of unknown initial condition. Also, the profile of $f(\eta)$ retards but the boundary layer thickness decreases. In the particular example (Example 2) system, nonlinear differential equations are considered with appropriate boundary conditions. Due to the lack of initial conditions we have assumed two more initials and then used ADM for their possible solution. The unknown initial conditions are obtained numerically and presented in Table 2. However, the behavior of various parameters M, K, A, and B on $f(\eta)$ and $g(\eta)$ is shown in Figs. 3, 4, 5 and 6.

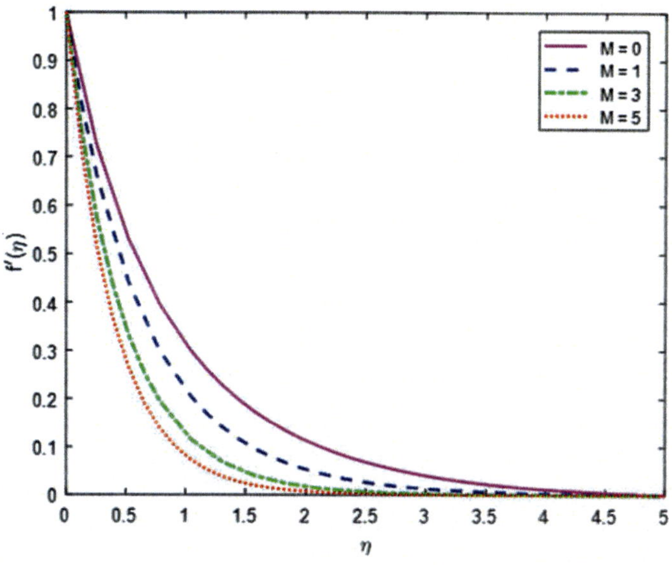

Fig. 1 Variation of parameter M on $f'(\eta)$

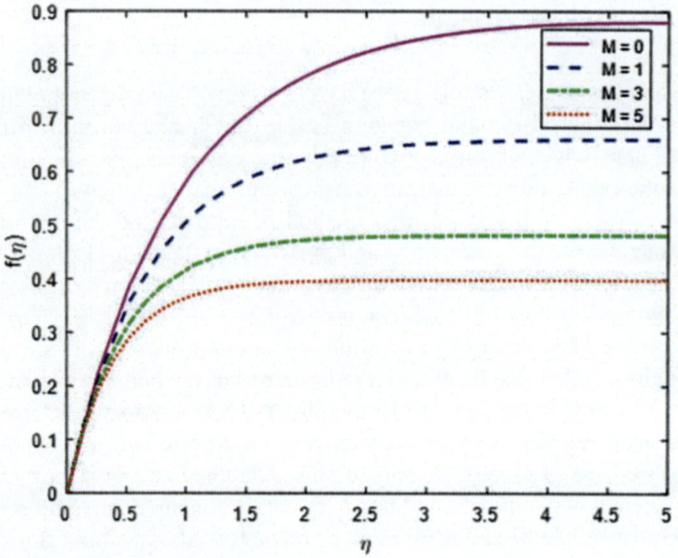

Fig. 2 Variation of parameter M on $f(\eta)$

Table 2 The values of unknown p and q

M	K	A	B	p	q
0.1	1	1	1	−1.6767	1.641
0.2				−1.4953	1.5045
0.3				−1.3967	1.5661
0.4				−1.6397	2.5303
1	0.1			0.6926	−0.182
	0.2			0.6902	−0.0696
	0.3			0.6812	0.0601
	0.4			0.6645	0.2136
	1	0.5		0.1212	0.5216
		0.6		0.0238	1.195
		0.7		−0.028	0.3987
		0.8		−0.0067	1.1996
		1	0.5	−0.1957	0.553
			0.6	−0.1607	0.8116
			0.7	−0.1172	1.4506
			0.8	−0.101	0.7494

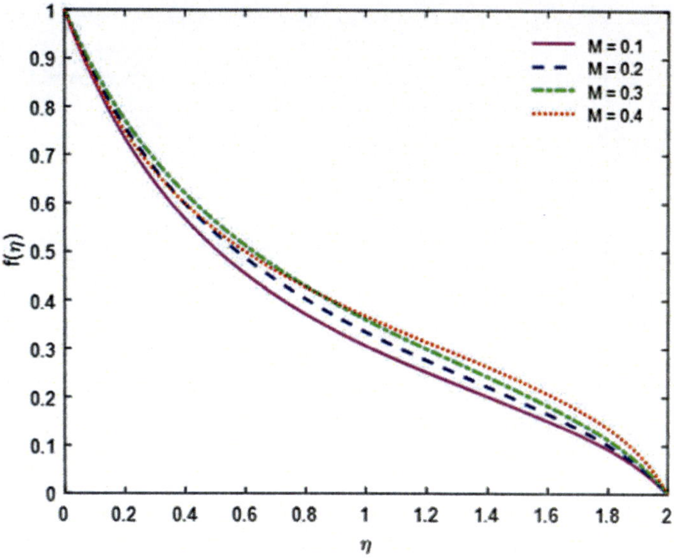

Fig. 3 Variation of parameter M on $f(\eta)$

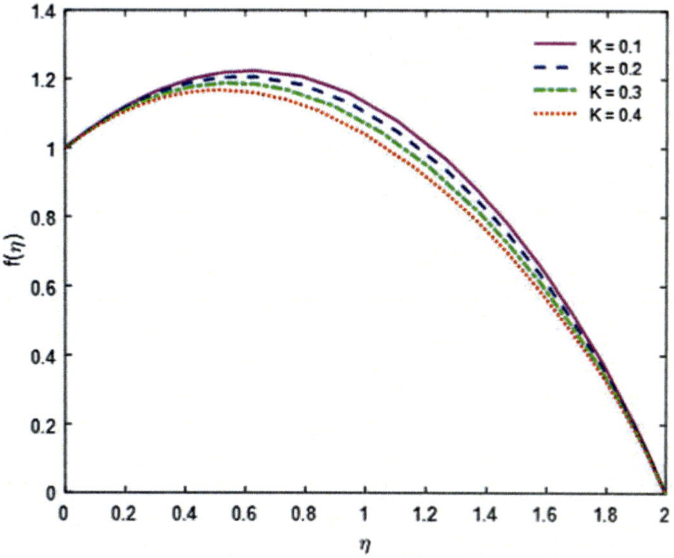

Fig. 4 Variation of parameter K on $f(\eta)$

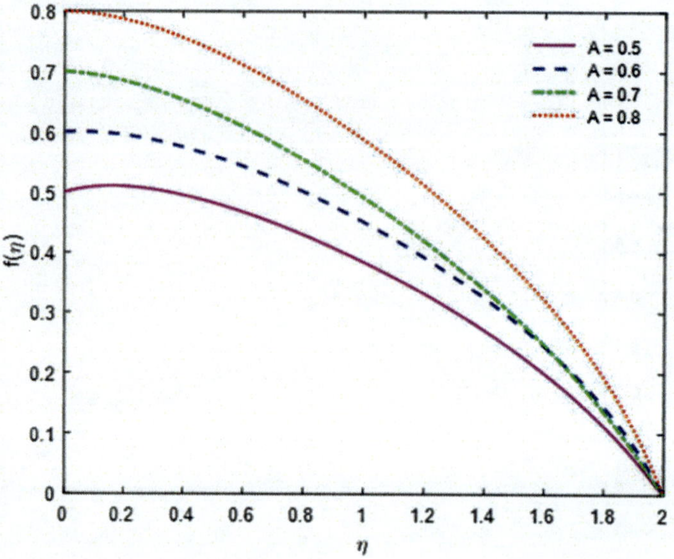

Fig. 5 Variation of parameter A on $f(\eta)$

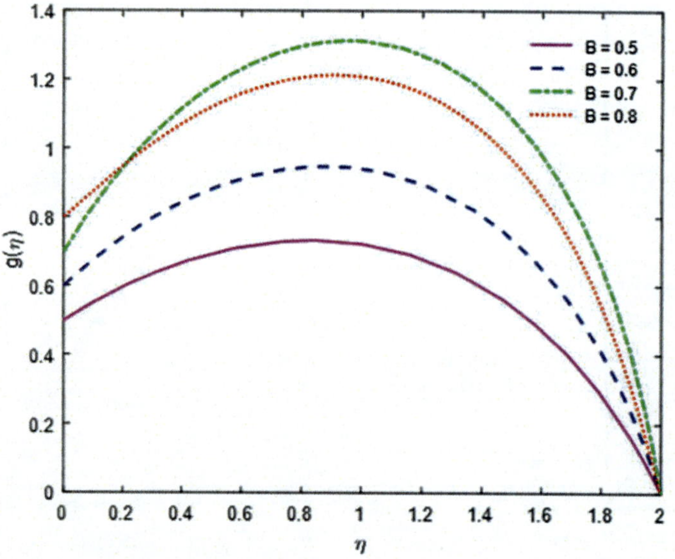

Fig. 6 Variation of parameter B on $g(\eta)$

Appendix 1

$$T_1 = -\frac{1}{2}(pq + AM + pK), T_2 = -\frac{1}{6}MpT_3 = -\frac{1}{2}\left(pq + q^2 + BK\right), T_4 = -\frac{1}{6}Kq,$$

$$T_5 = -\frac{1}{6}(2pT_3 + 2qT_1 + 2KT_1), T_6 = -\frac{1}{12}(3pT_4 + 3qT_2 + 3KT_2 + MT_1), T_7 = -\frac{1}{20}MT_2,$$

$$T_8 = -\frac{1}{6}(2pT_3 + 2qT_1 + 4qT_3), T_9 = -\frac{1}{12}(3pT_4 + 3qT_2 + 6qT_4 + KT_3), T_{10} = -\frac{1}{20}KT_4,$$

$$T_{11} = -\frac{1}{12}(3pT_8 + 4T_1T_3 + 3qT_5 + 3KT_5), T_{12} = -\frac{1}{20}(4pT_9 + 6T_1T_4 + 6T_2T_3 + 4qT_6 + 4KT_6 + MT_5),$$

$$T_{13} = -\frac{1}{30}(5pT_{10} + 9T_2T_4 + 5qT_7 + 5KT_7 + MT_6), T_{14} = -\frac{1}{42}MT_7,$$

$$T_{15} = -\frac{1}{12}(3pT_8 + 4T_1T_3 + 3qT_5 + 4T_3^2 + 6qT_8), T_{16} = -\frac{1}{20}(4pT_9 + 6T_1T_4 + 6T_2T_3 + 4qT_6 + 12T_3T_4 + 8qT_9 + KT_8).$$

$$T_{17} = -\frac{1}{30}\left(5pT_{10} + 9T_2T_4 + 5qT_7 + 9T_4^2 + 10qT_{10} + KT_9 + MT_6\right), T_{18} = -\frac{1}{42}KT_{10}$$

References

1. G. Adomian, *Stochastic System* (Academic Press, New York, 1983).
2. G. Adomian, *Solving Frontier Problem of Physics: The Decomposition Method* (Kluwer Academic Publishers, Boston, 1994).
3. G. Adomian, Solution of the Thomas-Fermi equation. Appl. Math. Lett. **11**, 131–133 (1998)
4. W. Abdul-Majid, A computational approach to soliton solutions of the Kadomtsev-Petviashvili equation. Appl. Math. Comput. **123**, 205–217 (2001)
5. W. Abdul-Majid, The decomposition method applied to systems of partial differential equations and to the reaction–diffusion Brusselator model. Appl. Math. Comput. **110**, 251–264 (2000)
6. D. Kaya, S.M. El-Sayed, Numerical soliton-like solutions of the potential Kadomtsev-Petviashvili equation by the decomposition method. Phys. Lett. A **320**, 192–199 (2003)
7. El-Danaf Talaat, S., Ramadan Mohamed, A., Abd Alaal Faysal, E.I.: The use of Adomian decomposition method for solving the regularized long-wave equation. Chaos Soliton Fract **26**, 747–757 (2005)
8. M.A. Abdou, J. Quant. Spectrosc. Radiat. Transf. **95**, 407 (2005)
9. M.J. Ablowitz, J.F. Ladic, On the solution of a class of nonlinear partial difference equations. Stud. Appl. Math. **57**, 1–12 (1977)
10. F.T. Zohra, M.J. Uddin, A.I. Ismail, O. Anwar Bég, A. Kadir, Boundary layer anisotropic slip magneto-bio convection flow from a rotating cone to a nanofluid with Stefan blowing effects. Chin. J. Phys. (2017). In press
11. T.A. Bég, O. Anwar Bég, M.M. Rashidi, M. Asadi, Homotopy semi-numerical modelling of nanofluid convection flow from an isothermal spherical body in a permeable regime. Int. J. Microscale Nanoscale Thermal Fluid Transp. Phenomena **3**(4), 67–96 (2012)
12. J. Srinivas, O. Anwar Bég, Homotopy study of entropy generation in magnetized micropolar flow in a vertical parallel plate channel with buoyancy effect. Heat Transf. Res. (2017). In press
13. M.M. Bhatti, A. Shahid, O. Anwar Bég, A. Kadir, Numerical study of radiative Maxwell viscoelastic magnetized flow from a stretching permeable sheet with the Cattaneo–Christov

heat flux model. Neural Comput. Appl. 12 pages (2017). https://doi.org/10.1007/s00521-017-2933-8
14. G. Adomian, *Solving Frontier Problems in Physics: The Decomposition Method* (Kluwer, Dordrecht, USA, 1994).
15. M. Kezzar, M.R. Sar, Series solution of nanofluid flow and heat transfer between stretchable/shrinkable inclined walls. Int. J. Appl. Comput. Math. **3**, 2231–2255 (2017)
16. O. Anwar Bég, D. Tripathi, T. Sochiand, P.K. Gupta, Adomian decomposition method (ADM) simulation of magneto-bio-tribological squeeze film with magnetic induction effects. Mech. Med. Biol. **15**, 1550072.1–1550072.23 (2015)
17. A. Ebaid, M.D. Aljoufi, A.M. Wazwaz, An advanced study on the solution of nanofluid flow problems via Adomian's method. Appl. Math. Lett. **46**, 117–122 (2015)
18. O. Aaboubi et al., Application of Adomian method for the magnetic field effects on mass transport at vertical cylindrical electrode. Electrochim. Acta **184**, 276–284 (2015)

Existence and Ulam Stability Criteria for Antiperiodic Boundary Value Problem of Fractional Difference Equation

A. George Maria Selvam and R. Dhineshbabu

Abstract This present work is concerned with the existence and Ulam stability criteria for a discrete antiperiodic boundary value problem (BVP) of fractional order $1 < \sigma \leq 2$ with Caputo fractional difference operator. Finally, some suitable examples are presented to demonstrate the main results.

Keywords Discrete fractional calculus · Existence · Ulam stability · Boundary value problem

Mathematics Subject Classification 34A08 · 34B15 · 34K10 · 34K20

1 Introduction

Fractional order differential equations (FODEs) introduce the notion of a non-integer order derivative which provides a new modeling approach for systems with extraordinary dynamical properties. Furthermore, since fractional calculus is a concept of conventional calculus, it has been found especially advantageous in automatic control and system theory, where FODEs are used to attain more accurate explanations of the dynamical systems, develop the characteristics of control loops, and enhance the novel control strategies. Then, during the past few decades, the FODEs found numerous applications in different types of complex systems in diverse disciplines, for example, diffusion, relaxation, turbulence, oscillation, and recently in statistical distribution theory [1, 2]. Moreover, they have many applications in various branches of sciences and engineering.

Recently, the exploration on stability theory of FODE has been productive and rapidly developed and it has drawn the attention of many analysts. A talk presented

A. G. M. Selvam (✉)
Department of Mathematics, Sacred Heart College (Autonomous), Tirupattur 635601, Tamil Nadu, India
e-mail: agmshc@gmail.com

R. Dhineshbabu
Department of Mathematics, Sri Venkateswara College of Engineering and Technology (Autonomous), Chittoor 517127, Andhra Pradesh, India

S. R. Mishra et al. (eds.), *Recent Trends in Applied Mathematics*, Lecture Notes in Mechanical Engineering, https://doi.org/10.1007/978-981-15-9817-3_13

by Ulam motivated the idea of carrying out the stability analysis of the functional equations [3] in 1940. Also, Ulam discussed some unsolved problems and one about the stability of homomorphism's was answered by Hyers [4] in 1941. Number of mathematicians was influenced by the concept of stability introduced by Th. M. Rassias [5]. Ulam's problem for functional equation is recently generalized with differential equations replacing functional equations. Obloza came up with idea to prove Hyers–Ulam (HU) and Hyers–Ulam–Rassias (HUR) stability of differential equations in [6]. Nowadays, there are many researchers who discussed that the study of HU stability of the integer order equation was generalized to the FODEs [7]. Further, Wang et al. [3] studied the pioneering work on the solutions of HU stability, existence and uniqueness results, and data dependence to a class of Cauchy BVPs.

The fractional antiperiodic boundary value problems (FABVPs) are used in various physical processes of stochastic transport with different boundary conditions (BCs). The study of 2, 3, 4, multipoint and nonlocal BVPs has been examined by many researchers using different methods which includes fixed point theorems in cones, Krasnoselskii theorem, degree theory, and Leray–Schauder type [8–10]. Integral BCs appear in semiconductors, thermal conduction, and hydrodynamic problems [11]. In recent years, the theory of discrete fractional calculus has been established by a very few researchers [12–15]. Some important results on fractional order sums and differences are given by Atici and Eloe [16, 17]. The comparison of the Caputo and Riemann–Liouville operators was carried out in [18]. However, there is few research papers published on existence and HU stability for discrete FABVPs with Caputo difference operator [19, 20]. Inspired by the above discussions, in this paper, existence and HU stability criteria for discrete FABVP have been investigated of the following form:

$$\begin{cases} {}_0^C\Delta_k^\sigma v(\kappa) = \phi(\kappa+\sigma-1, v(\kappa+\sigma-1)), \ \kappa \in [0, q+1]_{\mathbb{N}_0}, \ \sigma \in (1, 2] \\ v(\sigma-2) = -v(\sigma+q), \ \Delta v(\sigma-2) = -\Delta v(\sigma+q) \end{cases} \tag{1}$$

where $\phi : [\sigma-1, \sigma+q]_{\mathbb{N}_{\sigma-1}} \times \mathbb{R} \to \mathbb{R}$ is a continuous and ${}_0^C\Delta_\kappa^\sigma$ is the Caputo fractional difference operator (CFDO).

The paper is organized with basic concepts introduced in Sect. 2. Existence and uniqueness results for discrete FABVP of (1) are established in Sect. 3. In Sect. 4, we develop conditions for Ulam stability results and suitable examples are presented as the applications of our results in Sect. 5.

2 Preliminaries

We recollect some well-known useful definitions and lemmas needed in this study.

Definition 2.1 (see [19, 20]) The σth fractional sum for $\sigma > 0$ is defined by

$$\Delta^{-\sigma}\hbar(\kappa)=\frac{1}{\Gamma(\sigma)}\sum_{\ell=a}^{\kappa-\sigma}(\kappa-g(\ell))^{(\sigma-1)}\hbar(\ell),$$

for all $\kappa\in\mathbb{N}_{a+\sigma}$, $g(\ell)=\ell+1$, and $\kappa^{(\sigma)}:=\frac{\Gamma(\kappa+1)}{\Gamma(\kappa+1-\sigma)}$.

Definition 2.2 (see [19, 20]) Let $\sigma>0$, $n-1<\sigma\leq n$ and set $\alpha=n-\sigma$. The σth CFDO is defined as

$${}_{0}^{C}\Delta_{\kappa}^{\sigma}\Psi(\kappa)=\Delta^{-\alpha}\big(\Delta^{n}\Psi(\kappa)\big)=\frac{1}{\Gamma(\alpha)}\sum_{\ell=a}^{\kappa-\alpha}(\kappa-g(\ell))^{(\alpha-1)}\Delta^{n}\Psi(\ell),$$

for all $\kappa\in\mathbb{N}_{a+\sigma}$, where $n=\lceil\sigma\rceil$, $\lceil.\rceil$ ceiling of number.

Lemma 2.1 (see [19, 20]) *Suppose that $\sigma>0$ and Ψ is defined on $\mathbb{N}_a$. Then*

$${}_{0}^{C}\Delta_{\kappa}^{\sigma}\Delta^{-\sigma}\phi(\kappa)=\phi(\kappa)+A_0+A_1\kappa+\cdots+A_{n-1}\kappa^{(n-1)},$$

for some $A_i\in\mathbb{R}$, where $0\leq i\leq n-1$.

Lemma 2.2 [20] *One has*

1. $\displaystyle\sum_{\ell=0}^{\kappa-\sigma}(\kappa-g(\ell))^{(\sigma-1)}=\frac{\Gamma(\kappa+1)}{\sigma\,\Gamma(\kappa-\sigma+1)}.$
2. $\sum_{\ell=0}^{q}(\sigma+q-g(\ell))^{(\sigma-1)}=\frac{1}{\sigma}\frac{\Gamma(\sigma+q+1)}{\Gamma(q+1)}.$
3. $\sum_{\ell=0}^{q+1}(\sigma+q-g(\ell))^{(\sigma-2)}=\frac{1}{\sigma-1}\frac{\Gamma(\sigma+q+1)}{\Gamma(q+2)}.$

3 Existence and Uniqueness Solutions

In this section, existence and uniqueness solutions of a discrete FABVP (1) are established, if the solution exists.

Theorem 3.1 *Let $1<\sigma\leq 2$ and $\phi:[\sigma-1,\sigma+q]_{\mathbb{N}_{\sigma-1}}\times\mathbb{R}\to\mathbb{R}$ be given. Then the solution of a discrete FABVP*

$$\begin{cases}{}_{0}^{C}\Delta_{\kappa}^{\sigma}v(\kappa)=\phi(\kappa+\sigma-1),\ \ \kappa\in[0,q+1]_{\mathbb{N}_0},\\ v(\sigma-2)=-v(\sigma+q),\ \Delta v(\sigma-2)=-\Delta v(\sigma+q)\end{cases}\tag{2}$$

if $v(\kappa)$, for $\kappa\in[\sigma-2,\sigma+q]_{\mathbb{N}_{\sigma-2}}$, has the form

$$v(\kappa) = \frac{1}{\Gamma(\sigma)} \sum_{\ell=0}^{\kappa-\sigma} (\kappa - g(\ell))^{(\sigma-1)} \phi(\ell+\sigma-1) - \frac{1}{2\Gamma(\sigma)} \sum_{\ell=0}^{q} (\sigma+q-g(\ell))^{(\sigma-1)} \phi(\ell+\sigma-1)$$
$$+ \frac{[q+2(\sigma-1-\kappa)]}{4\Gamma(\sigma-1)} \sum_{\ell=0}^{q+1} (\sigma+q-g(\ell))^{(\sigma-2)} \phi(\ell+\sigma-1), \tag{3}$$

where $g(\ell) = \ell + 1$.

Proof Suppose that $v(\kappa)$ is a solution of (2). Applying Lemma 2.1, we obtain a general solution for (2) in the form

$$v(\kappa) = \frac{1}{\Gamma(\sigma)} \sum_{\ell=0}^{\kappa-\sigma} (\kappa - g(\ell))^{(\sigma-1)} \phi(\ell+\sigma-1) - A_0 - A_1\kappa \tag{4}$$

for some $A_0, A_1 \in \mathbb{R}$. Applying Δ operator to (4), we get

$$\Delta v(\kappa) = \frac{1}{\Gamma(\sigma-1)} \sum_{\ell=0}^{\kappa-\sigma+1} (\kappa - g(\ell))^{(\sigma-2)} \phi(\ell+\sigma-1) - A_1. \tag{5}$$

In view of the boundary conditions $v(\sigma-2) = -v(\sigma+q)$ and $v(\sigma-2) = -v(\sigma+q)$, we find the value of A_0 and A_1 as follows:

$$A_0 = \frac{1}{2\Gamma(\sigma)} \sum_{\ell=0}^{q} (\sigma+q-g(\ell))^{(\sigma-1)} \phi(\ell+\sigma-1) - \frac{(q+2\sigma-2)}{4\Gamma(\sigma-1)} \sum_{\ell=0}^{q+1} (\sigma+q-g(\ell))^{(\sigma-2)} \phi(\ell+\sigma-1) \text{ and}$$
$$A_1 = \frac{1}{2\Gamma(\sigma-1)} \sum_{\ell=0}^{q+1} (\sigma+q-g(\ell))^{(\sigma-2)} \phi(\ell+\sigma-1).$$

Substituting the value of A_0 and A_1 in $v(k)$, we obtain (3). Conversely, if (3) has a solution, it is clear that the solution satisfies the BVP (2). The proof is completed.

Define the operator

$$(Tv)(\kappa) = \left[\frac{1}{\Gamma(\sigma)} \sum_{\ell=0}^{\kappa-\sigma} (\kappa - g(\ell))^{(\sigma-1)} - \frac{1}{2\Gamma(\sigma)} \sum_{\ell=0}^{q} (\sigma+q-g(\ell))^{(\sigma-1)} \right] \phi(\ell+\sigma-1, v(\ell+\sigma-1))$$
$$+ \frac{[q+2(\sigma-1-\kappa)]}{4\Gamma(\sigma-1)} \sum_{\ell=0}^{q+1} (\sigma+q-g(\ell))^{(\sigma-2)} \phi(\ell+\sigma-1, v(\ell+\sigma-1)) \tag{6}$$

for $\kappa \in [\sigma-2, \sigma+q]_{\mathbb{N}_{\sigma-2}}$. It is obvious that $v(\kappa)$ is a solution of (1) if it is a fixed point of (6).

For computational convenience, let us consider the Banach space $\mathbb{E}$ with the norm $\|v\| = \max|v(\kappa)|$ for $\kappa \in [\sigma-2, \sigma+q]_{\mathbb{N}_{\sigma-2}}$.

We assume the following:

(H_1) Let us have constant $\eta > 0$, which satisfies $|\phi(\kappa, v) - \phi(\kappa, u)| \leq \eta|v-u|$, for all $u, v \in \mathbb{E}$ and $\kappa \in [\sigma-2, \sigma+q]_{\mathbb{N}_{\sigma-2}}$.

(H_2) There is a bounded function $Q : [\sigma-2, \sigma+q]_{\mathbb{N}_{\sigma-2}} \to \mathbb{R}$ such that $|\phi(\kappa, v)| \leq Q(\kappa)|v|$ for all $v \in \mathbb{E}$.
(H_3) For a non-decreasing function $\Psi \in [\sigma - 2, \sigma + q]_{\mathbb{N}_{\sigma-2}} \to \mathbb{R}^+$, there exists $\lambda > 0$ such that

$$\frac{\varepsilon}{\Gamma(\sigma)} \sum_{\ell=0}^{\kappa-\sigma} (\kappa - g(\ell))^{(\sigma-1)} \Psi(\ell + \sigma - 1) \leq \lambda \, \varepsilon \, \Psi(\kappa + \sigma - 1), \quad \kappa \in [0, q+1]_{\mathbb{N}_0}.$$

Theorem 3.2 *Assume that (H_1) holds. Then a discrete FABVP (1) has a unique solution on* $\mathbb{E}$ *provided that*

$$\Lambda = \eta \left[\frac{3\Gamma(\sigma+q+1)}{2\Gamma(\sigma+1)\Gamma(q+1)} + \frac{(q+2)\Gamma(\sigma+q+1)}{4\Gamma(\sigma)\Gamma(q+2)} \right] < 1. \tag{7}$$

Proof For each $\kappa \in [\sigma - 2, \sigma + q]_{\mathbb{N}_{\sigma-2}}$ and $u, v \in \mathbb{E}$, we have

$$\begin{aligned}
|(Tv)(\kappa) - (Tu)(\kappa)| \leq & \frac{1}{\Gamma(\sigma)} \sum_{\ell=0}^{\kappa-\sigma} (\kappa - g(\ell))^{(\sigma-1)} |\phi(\ell+\sigma-1, v(\ell+\sigma-1)) - \phi(\ell+\sigma-1, u(\ell+\sigma-1))| \\
& + \frac{1}{2\Gamma(\sigma)} \sum_{\ell=0}^{q} (\sigma + q - g(\ell))^{(\sigma-1)} |\phi(\ell+\sigma-1, v(\ell+\sigma-1)) - \phi(\ell+\sigma-1, u(\ell+\sigma-1))| \\
& + \frac{|q + 2\sigma - 2 - 2\kappa|}{4\Gamma(\sigma-1)} \sum_{\ell=0}^{q+1} (\sigma + q - g(\ell))^{(\sigma-2)} |\phi(\ell+\sigma-1, v(\ell+\sigma-1)) - \phi(\ell+\sigma-1, u(\ell+\sigma-1))| \\
\leq & \frac{\eta}{\Gamma(\sigma)} \sum_{\ell=0}^{\kappa-\sigma} (\kappa - g(\ell))^{(\sigma-1)} |v(\ell+\sigma-1) - u(\ell+\sigma-1)| \\
& + \frac{\eta}{2\Gamma(\sigma)} \sum_{\ell=0}^{q} (\sigma + q - g(\ell))^{(\sigma-1)} |v(\ell+\sigma-1)) - u(\ell+\sigma-1))| \\
& + \frac{\eta|q + 2\sigma - 2 - 2\kappa|}{4\Gamma(\sigma-1)} \sum_{\ell=0}^{q+1} (\sigma + q - g(\ell))^{(\sigma-2)} |v(\ell+\sigma-1) - u(\ell+\sigma-1)|
\end{aligned}$$

$$\begin{aligned}
\|Tv - Tu\| \leq & \frac{\eta}{\Gamma(\sigma)} \left[\sum_{\ell=0}^{\kappa-\sigma} (\kappa - g(\ell))^{(\sigma-1)} + \frac{1}{2} \sum_{\ell=0}^{q} (\sigma + q - g(\ell))^{(\sigma-1)} \right] \|v - u\| \\
& + \frac{\eta|q + 2\sigma - 2 - 2\kappa|}{4\Gamma(\sigma-1)} \left[\sum_{\ell=0}^{q+1} (\sigma + q - g(\ell))^{(\sigma-2)} \right] \|v - u\| \\
\|Tv - Tu\| \leq & \left[\frac{3\Gamma(\sigma+q+1)}{2\Gamma(\sigma+1)\Gamma(q+1)} + \frac{(q+2)\Gamma(\sigma+q+1)}{4\Gamma(\sigma)\Gamma(q+2)} \right] \eta \|v - u\|.
\end{aligned}$$

Clearly, T is a contraction mapping. It is clear from Lemma (7) in [19] that there is a fixed point T has a unique. The proof is complete.

Theorem 3.3 *Assume that the hypothesis (H_2) holds. Then, the FABVP (1) has at least a solution on* $\mathbb{E}$ *provided that*

$$\Gamma(\sigma+q+1)[6(q+1)+\sigma(q+2)] \le \tfrac{4\Gamma(\sigma+1)\Gamma(q+2)}{Q^*}, \tag{8}$$

where $Q^* = \max\big\{Q(\kappa) : \kappa \in [\sigma-2, \sigma+q]_{\mathbb{N}_{\sigma-2}}\big\}$.

Proof Let us define $S = \big\{v(\kappa)|[\sigma-2,\sigma+q]_{\mathbb{N}_{\sigma-2}} \to \mathbb{R},\ \|v\| \le M\big\}$, for $M > 0$. To prove this theorem, it is enough to show that $T : S \to S$. For $v(\kappa) \in S$, we have

$$|(Tv)(\kappa)| \le \left[\frac{1}{\Gamma(\sigma)}\sum_{\ell=0}^{\kappa-\sigma}(\kappa-g(\ell))^{(\sigma-1)} + \frac{1}{2\Gamma(\sigma)}\sum_{\ell=0}^{q}(\sigma+q-g(\ell))^{(\sigma-1)}\right]|\phi(\ell+\sigma-1, v(\ell+\sigma-1))|$$
$$+\frac{|q+2\sigma-2-2\kappa|}{4\Gamma(\sigma-1)}\sum_{\ell=0}^{q+1}(\sigma+q-g(\ell))^{(\sigma-2)}|\phi(\ell+\sigma-1, v(\ell+\sigma-1))|$$
$$\le \frac{Q(\kappa)}{\Gamma(\sigma)}\left[\sum_{\ell=0}^{\kappa-\sigma}(\kappa-g(\ell))^{(\sigma-1)} + \frac{1}{2}\sum_{\ell=0}^{q}(\sigma+q-g(\ell))^{(\sigma-1)}\right]|v(\ell+\sigma-1)|$$
$$+\frac{Q(\kappa)|q+2\sigma-2-2\kappa|}{4\Gamma(\sigma-1)}\left[\sum_{\ell=0}^{q+1}(\sigma+q-g(\ell))^{(\sigma-2)}\right]|v(\ell+\sigma-1)|$$
$$\|Tv\| \le \frac{Q(\kappa)}{\Gamma(\sigma)}\left[\sum_{\ell=0}^{\kappa-\sigma}(\kappa-g(\ell))^{(\sigma-1)} + \frac{1}{2}\sum_{\ell=0}^{q}(\sigma+q-g(\ell))^{(\sigma-1)}\right]\|v\|$$
$$+\frac{Q(\kappa)|q+2\sigma-2-2\kappa|}{4\Gamma(\sigma-1)}\left[\sum_{\ell=0}^{q+1}(\sigma+q-g(\ell))^{(\sigma-2)}\right]\|v\|$$
$$\|Tv\| \le \frac{Q(\kappa)}{\Gamma(\sigma)}\left[\frac{\Gamma(\kappa+1)}{\sigma\,\Gamma(\kappa-\sigma+1)} + \frac{1}{2}\cdot\frac{\Gamma(\sigma+q+1)}{\sigma\,\Gamma(q+1)}\right]\|v\|$$
$$+\frac{Q(\kappa)|q+2\sigma-2-2\kappa|}{4\Gamma(\sigma-1)}\left[\frac{\Gamma(\sigma+q+1)}{(\sigma-1)\Gamma(q+2)}\right]\|v\|$$
$$\|Tv\| \le \left[\frac{\Gamma(\sigma+q+1)[6(q+1)+\sigma(q+2)]}{4\Gamma(\sigma+1)\Gamma(q+2)}\right]Q^*M.$$

From (8), we get $\|Tv\| \le M$. Thus, it is clear that $T : S \to S$. According to Brouwer theorem in [19], Eq. (6) has at least one fixed point which is a solution of (1). This completes the proof.

4 The Ulam Stability

In this section, stability analysis is presented for the discrete FABVP (1). The following definitions for FDE are given on the basis of [19, 20].

Definition 4.1 If for every function $u(\kappa) \in \mathbb{E}$ of

$$\left|{}_0^C\Delta_\kappa^\sigma u(\kappa) - \phi(\kappa+\sigma-1, u(\kappa+\sigma-1))\right| \le \xi,\quad \kappa \in [0, q+1]_{\mathbb{N}_0}, \tag{9}$$

where $\xi > 0$, there is a solution$v \in \mathbb{E}$ of (1) and positive constant $K > 0$such that

$$|u(\kappa) - v(\kappa)| \leq K\xi, \quad \kappa \in [\sigma - 2, \sigma + q]_{\mathbb{N}_{\sigma-2}}. \tag{10}$$

Then, the FABVP (1) is HU stable.

Definition 4.2 If for every function $u(\kappa) \in \mathbb{E}$ of

$$\left|{}_0^C\Delta_\kappa^\sigma u(\kappa) - \phi(\kappa + \sigma - 1, u(\kappa + \sigma - 1))\right| \leq \xi\, \Psi(\kappa + \sigma - 1), \quad k \in [0, q + 1]_{\mathbb{N}_0}, \tag{11}$$

where $\xi > 0$, there is a solution$v \in \mathbb{E}$ of (1) and positive constant $K > 0$ such that

$$|u(\kappa) - v(\kappa)| \leq K\xi\, \Psi(\kappa), \quad \kappa \in [\sigma - 2, \sigma + q]_{\mathbb{N}_{\sigma-2}}. \tag{12}$$

Then, the FABVP (1) is HUR stable.

Theorem 4.1 *Suppose that the hypothesis (H_1) together with the inequality (9), then a discrete FABVP (1) is HU stable provided that*

$$\beta < \frac{4\Gamma(\sigma+1)\Gamma(q+2)}{\Gamma(\sigma+q+1)[6(q+1)+\sigma(q+2)]}. \tag{13}$$

Proof From inequality (9), for $\kappa \in [\sigma - 2, \sigma + q]_{\mathbb{N}_{\sigma-2}}$, we can find a function ${}_0^C\Delta_\kappa^\sigma u(\kappa) = \phi(\kappa + \sigma - 1, u(\kappa + \sigma - 1)) + f(\kappa + \sigma - 1)$ and $|f(\kappa + \sigma - 1)| \leq \xi$. It follows that

$$\begin{aligned}
&\left| u(\kappa) - \frac{1}{\Gamma(\sigma)}\sum_{\ell=0}^{\kappa-\sigma}(\kappa - g(\ell))^{(\sigma-1)}\phi(\ell + \sigma - 1, u(\ell + \sigma - 1)) \right.\\
&+ \frac{1}{2\Gamma(\sigma)}\sum_{\xi=0}^{q}(\sigma + q - g(\ell))^{(\sigma-1)}\phi(\ell + \sigma - 1, u(\ell + \sigma - 1))\\
&\left. -\frac{[q + 2\sigma - 2 - 2\kappa]}{4\Gamma(\sigma - 1)}\sum_{\ell=0}^{q+1}(\sigma + q - g(\ell))^{(\sigma-2)}\phi(\ell + \sigma - 1, u(\ell + \sigma - 1)) \right|\\
&\leq \frac{\xi}{\Gamma(\sigma)}\sum_{\ell=0}^{\kappa-\sigma}(\kappa - g(\ell))^{(\sigma-1)}\\
&\leq \frac{\xi\,\Gamma(\sigma + q + 1)}{\Gamma(\sigma + 1)\Gamma(q + 1)}.
\end{aligned} \tag{14}$$

With the help of solutions (3) and (14), for $\kappa \in [\sigma - 2, \sigma + q]_{\mathbb{N}_{\sigma-2}}$, we have

$$\begin{aligned}
|u(\kappa) - v(\kappa)| \leq &\left| u(\kappa) - \frac{1}{\Gamma(\sigma)}\sum_{\ell=0}^{\kappa-\sigma}(\kappa - g(\ell))^{(\sigma-1)}\phi(\ell + \sigma - 1, v(\ell + \sigma - 1)) \right.\\
&+ \frac{1}{2\Gamma(\sigma)}\sum_{\ell=0}^{q}(\sigma + q - g(\ell))^{(\sigma-1)}\phi(\ell + \sigma - 1, v(\ell + \sigma - 1))
\end{aligned}$$

$$
\left. - \frac{[q+2\sigma-2-2\kappa]}{4\Gamma(\sigma-1)} \sum_{\ell=0}^{q+1} (\sigma+q-g(\ell))^{(\sigma-2)} \phi(\ell+\sigma-1, v(\ell+\sigma-1)) \right|
$$

$$
\leq \left| u(\kappa) - \frac{1}{\Gamma(\sigma)} \sum_{\ell=0}^{\kappa-\sigma} (\kappa-g(\ell))^{(\sigma-1)} \phi(\ell+\sigma-1, u(\ell+\sigma-1)) \right.
$$

$$
+ \frac{1}{2\Gamma(\sigma)} \sum_{\ell=0}^{q} (\sigma+q-g(\ell))^{(\sigma-1)} \phi(\ell+\sigma-1, u(\ell+\sigma-1))
$$

$$
\left. - \frac{[q+2\sigma-2-2\kappa]}{4\Gamma(\sigma-1)} \sum_{\ell=0}^{q+1} (\sigma+q-g(\ell))^{(\sigma-2)} \phi(\ell+\sigma-1, u(\ell+\sigma-1)) \right|
$$

$$
+ \frac{1}{\Gamma(\sigma)} \sum_{\ell=0}^{\kappa-\sigma} (\kappa-g(\ell))^{(\sigma-1)} |\phi(\ell+\sigma-1, u(\ell+\sigma-1)) - \phi(\ell+\alpha-1, v(\ell+\alpha-1))|
$$

$$
+ \frac{1}{2\Gamma(\sigma)} \sum_{\ell=0}^{q} (\sigma+q-g(\ell))^{(\sigma-1)} |\phi(\ell+\sigma-1, u(\ell+\sigma-1)) - \phi(\ell+\sigma-1, v(\ell+\sigma-1))|
$$

$$
+ \frac{|q+2\sigma-2-2\kappa|}{4\Gamma(\sigma-1)} \sum_{\ell=0}^{q+1} (\sigma+q-g(\ell))^{(\sigma-2)}
$$

$$
\times |\phi(\ell+\sigma-1, u(\ell+\sigma-1)) - \phi(\ell+\sigma-1, v(\ell+\sigma-1))|
$$

$$
|u(\kappa)-v(\kappa)| \leq \frac{\xi\,\Gamma(\sigma+q+1)}{\Gamma(\sigma+1)\Gamma(q+1)} + \frac{\beta}{\Gamma(\sigma)} \sum_{\ell=0}^{\kappa-\sigma} (\kappa-g(\ell))^{(\sigma-1)} |u(\ell+\sigma-1) - v(\ell+\sigma-1)|
$$

$$
+ \frac{\beta}{2\Gamma(\sigma)} \sum_{\ell=0}^{q} (\sigma+q-g(\ell))^{(\sigma-1)} |u(\ell+\sigma-1) - v(\ell+\sigma-1)|
$$

$$
+ \frac{\beta|q+2\sigma-2-2\kappa|}{4\Gamma(\sigma-1)} \sum_{\ell=0}^{q+1} (\sigma+q-g(\ell))^{(\sigma-2)} |u(\ell+\sigma-1) - v(\ell+\sigma-1)|
$$

$$
\leq \frac{\xi\,\Gamma(\sigma+q+1)}{\Gamma(\sigma+1)\Gamma(q+1)} + \frac{\beta\|u-v\|}{\Gamma(\sigma)} \sum_{\ell=0}^{\kappa-\sigma} (\kappa-g(\ell))^{(\sigma-1)}
$$

$$
+ \frac{\beta\|u-v\|}{2\Gamma(\sigma)} \sum_{\ell=0}^{q} (\sigma+q-g(\ell))^{(\sigma-1)} + \frac{\beta\|u-v\||T+2\sigma-2-2\kappa|}{4\Gamma(\sigma-1)} \sum_{\ell=0}^{L+1} (\sigma+L-g(\ell))^{(\sigma-2)}
$$

$$
\|u-v\| \leq \frac{\xi\,\Gamma(\sigma+q+1)}{\Gamma(\sigma+1)\Gamma(q+1)} + \beta\left[\frac{3\Gamma(\sigma+q+1)}{2\Gamma(\sigma+1)(q+1)} + \frac{(q+2)\Gamma(\sigma+q+1)}{4\Gamma(\sigma)\Gamma(q+2)}\right]\|u-v\|.
$$

From the above inequality, we have

$$
\|u-v\| \leq \frac{4(q+1)\Gamma(\sigma+q+1)}{4\Gamma(\sigma+1)\Gamma(q+2) - \beta\,[6(q+1)+\sigma(q+2)]\Gamma(\sigma+q+1)}\xi,
$$

where $\frac{4(q+1)\Gamma(\sigma+q+1)}{4\Gamma(\sigma+1)\Gamma(q+2) - \beta\,[6(q+1)+\sigma(q+2)]\Gamma(\sigma+q+1)} > 0.$
Thus, FABVP (1) is HU stable.

Theorem 4.2 *If the hypotheses* (H_1) *and* (H_3) *and the inequality* (11) *are satisfied, then a discrete FABVP* (1) *is HUR stable provided that* (13) *holds.*

Proof With the argument used in the proof of Theorem 4.1, we obtain

$$\|u - v\| \leq \lambda \frac{4\Gamma(\sigma+1)\Gamma(q+2)}{4\Gamma(\sigma+1)\Gamma(q+2)-\beta[6(q+1)+\sigma(q+2)]\Gamma(\sigma+q+1)} \xi \, \Psi(\kappa+\sigma-1),$$

where $\lambda \frac{4\Gamma(\sigma+1)\Gamma(q+2)}{4\Gamma(\sigma+1)\Gamma(q+2)-\beta[6(q+1)+\sigma(q+2)]\Gamma(\sigma+q+1)} > 0$.
Thus, FABVP (1) is HUR stable.

5 Applications

As the applications of our results, we consider the following examples.

Example 5.1 Suppose that $\sigma = \frac{6}{5}$ and $q = 2$. Let $\phi(\kappa, v(\kappa)) = \frac{\sin v(\kappa)}{25+\kappa^2}$ and $\eta = \frac{1}{25}$. Then discrete FABVP (1) becomes

$$ {}_0^C\Delta_\kappa^{\frac{6}{5}} v(\kappa) = \frac{\sin\left(v\left(\kappa+\frac{1}{5}\right)\right)}{25+\left(\kappa+\frac{1}{5}\right)^2}, \quad \kappa \in [0, 3], \tag{15}$$

subject to the conditions

$$v\left(-\frac{4}{5}\right) = -v\left(\frac{16}{5}\right), \quad \Delta v\left(-\frac{4}{5}\right) = -\Delta v\left(\frac{16}{5}\right). \tag{16}$$

In this case, inequality (7) is

$$\Lambda = \eta\left[\frac{3\Gamma(\sigma+q+1)}{2\Gamma(\sigma+1)\Gamma(q+1)} + \frac{(q+2)\Gamma(\sigma+q+1)}{4\Gamma(\sigma)\Gamma(q+2)}\right] \approx 0.2675 < 1.$$

Therefore, from Theorem 3.2, we conclude that BVP (15), (16) has a unique solution.

When $\sigma \in (1, 2]$ with step size 0.2 and varying different Lipschitz constants η, conditions (7) of Theorem 3.2 are satisfied and BVP (15), (16) has different unique solutions, see Table 1 and Fig. 1.

Example 5.2 Suppose that $\sigma = \frac{17}{10}$, $q = 5$, and $M = 1000$ with $\phi(\kappa, v(\kappa)) = \frac{1}{27}\kappa e^{-\frac{\kappa}{100}|v(\kappa)|}$. Then discrete FABVP (1) takes the form

$$ {}_0^C\Delta_\kappa^{\frac{17}{10}} v(\kappa) = \frac{1}{27}\left(\kappa + \frac{7}{10}\right) e^{-\left(\kappa+\frac{7}{10}\right)\frac{1}{100}\left|v\left(\kappa+\frac{7}{10}\right)\right|}, \quad \kappa \in [0, 6], \tag{17}$$

subject to the conditions

$$v\left(-\frac{3}{10}\right) = -v\left(\frac{67}{10}\right), \quad \Delta v\left(-\frac{3}{10}\right) = -\Delta v\left(\frac{67}{10}\right). \tag{18}$$

Table 1 Illustration of $\sigma \in (1, 2]$ with step size 0.2 and varying η

σ	Λ			
	$\eta = \frac{1}{25}$	$\eta = \frac{1}{26}$	$\eta = \frac{1}{29}$	$\eta = \frac{1}{34}$
1.2	0.2675	0.2572	0.2306	0.1967
1.4	0.3210	0.3086	0.2767	0.2360
1.6	0.3806	0.3660	0.3281	0.2799
1.8	0.4469	0.4297	0.3852	0.3286
2	0.5200	0.5000	0.4483	0.3824

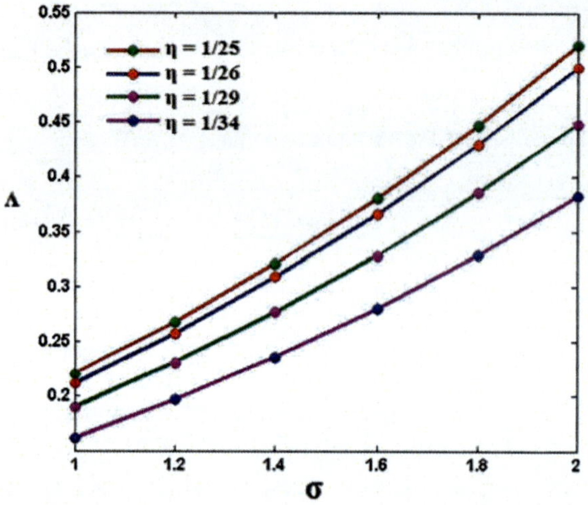

Fig. 1 σ versus Λ

The Banach space is $E := \left\{ \nu(\kappa) | \left[-\frac{3}{10}, \frac{67}{10} \right]_{\mathbb{N}_{\frac{-3}{10}}} \to \mathbb{R}, \|\nu\| \leq 1000 \right\}$. We note that

$$\frac{4\Gamma(\sigma+1)\Gamma(q+2)M}{\Gamma(\sigma+q+1)[6(q+1)+\sigma(q+2)]} = 1000(0.0335) \approx 33.5308.$$

It is clear that $|\phi(\kappa, \nu(\kappa))| \leq \frac{67}{270} < 33.5308$, whenever $\nu \in [-1000, 1000]$. So, ϕ satisfies the condition. Therefore, by Theorem 3.3, we conclude that the BVP (17), (18) has at least one solution.

Example 5.3 Assume that $\sigma = \frac{36}{20}$ and $q = 7$ with $\phi(\kappa, \nu(\kappa)) = \lambda \nu(\kappa)$. Then discrete FABVP (1) becomes

$$ {}_{0}^{C}\Delta_{\kappa}^{\frac{36}{20}} \nu(\kappa) = \lambda \nu\left(\kappa + \tfrac{16}{20}\right), \ \kappa \in [0, 8], \tag{19}$$

with the boundary conditions

$$v\left(-\frac{4}{20}\right)=-v\left(\frac{176}{20}\right),\ \Delta v\left(-\frac{4}{20}\right)=-\Delta v\left(\frac{176}{20}\right). \tag{20}$$

Since

$$\frac{4\Gamma(\sigma+1)\Gamma(q+2)}{[6(q+1)+\sigma(q+2)]\Gamma(\sigma+q+1)}\approx 0.0182.$$

If $\lambda < 0.0182$ and the inequality

$$\left|{}_0^C\Delta_\kappa^{\frac{36}{20}}u(\kappa)-\phi\left(\kappa+\frac{16}{20},u\left(\kappa+\frac{16}{20}\right)\right)\right|\leq\xi,\quad \kappa\in[0,8]_{\mathbb{N}_0},$$

holds, then the BVP (19), (20) is HU stable according to Theorem 4.1.

References

1. A.A. Kilbas, H.M. Srivastava, J.J. Trujillo, *Theory and Applications of Fractional Differential Equations*, vol. 204. North-Holland Mathematics Studies (Elsevier, Amsterdam, 2006)
2. I. Podlubny, *Fractional Differential Equations*. Mathematics in Science and Engineering (Academic Press, New York, 1999)
3. J. Wang, L. Lv, Y. Zhou, Ulam stability and data dependence for fractional differential equations with Caputo derivative. Electron. J. Qualitat. Theory Different. Equ. **63**, 1–10 (2011)
4. D.H. Hyers, On the stability of the linear functional equations. Proc. Nat. Acad. Sci. **27**, 222–224 (1941)
5. Th.M. Rassias, On the stability of the linear mappings in Banach spaces. Proc. Amer. Math. Soc. **72**, 297–300 (1978)
6. M. Obloza, Hyers stability of the linear differential equation. Rocznik Nauk-Dydakt. Prace Mat. **13**, 259–270 (1993)
7. S. Abbas, M. Benchohra, On the generalized Ulam—Hyers—Rassias stability for Darboux problem for partial fractional implicit differential equations. Appl. Math. E-Notes **14**, 20–28 (2014)
8. R.P. Agarwal, B. Ahmad, Existence theory for anti-periodic boundary value problems of fractional differential equations and inclusions. Comput. Math Appl. **62**, 1200–1214 (2011)
9. B. Ahmad, J.J. Nieto, Anti-periodic fractional boundary value problems. Comput. Math Appl. **62**, 1150–1156 (2011)
10. Y. Wang, Lyapunov-type inequalities for certain higher order differential equations with anti-periodic boundary conditions. Appl. Math. Lett. **25**, 2375–2380 (2012)
11. R.Y. Chegis, Numerical solution of the heat conduction problem with an integral condition. Litovsk. Mat. Sb. **24**, 209–215 (1984)
12. H. Chen, Z. Jin, S. Kang, Existence of positive solution for Caputo fractional difference equation. Adv. Diff. Equ. **2015**(44), 1–12 (2015)
13. A.G.M. Selvam, R. Dhineshbabu, Uniqueness of solutions of a discrete fractional order boundary value problem, in *AIP Conference Proceeding*, vol. 2095, No. 030008 (2019), pp. 1–7
14. C.S. Goodrich, Existence and uniqueness of solutions to a fractional difference equation with non local conditions. Comput. Math Appl. **61**, 191–202 (2011)
15. Y. Pan, Z. Han, S. Sun, C. Hou, The existence of solutions to a class of boundary value problems with fractional difference equations. Adv. Difference Equ. **2013**(275), 1–20 (2013)

16. F.M. Atici, P.W. Eloe, A transform method in discrete fractional calculus. Int. J. Differ. Equ. **2**(2), 165–176 (2007)
17. F.M. Atici, P.W. Eloe, Two-point boundary value problems for finite fractional difference equations. J. Differ. Equ. Appl. **17**, 445–456 (2011)
18. G.A. Anastassiou, Discrete fractional calculus and inequalities. http://arxiv.org/abs/0911.3370
19. F. Chen, Y. Zhou, Existence and Ulam stability of solutions for discrete fractional boundary value problem. Discrete Dyn. Nature Soc. 7 (2013)
20. A.G.M. Selvam, R. Dhineshbabu, Hyers-Ulam stability results for discrete antiperiodic boundary value problem with fractional order $2 < \delta \leq 3$. Int. J. Eng. Adv. Technol. **9**(1), 4997–5003 (2019)

Effect of Electrification on Boundary Layer Stagnation Point Flow of Nanofluid Over a Stretching Sheet

Kamala Kumar Pradhan, Ashok Misra, and Saroj Kumar Mishra

Abstract In this investigation, the stagnation point flow of silver water nanofluid over a linear stretching sheet using Buongiorno's two-component non-homogeneous nanofluid model is studied. The governing equations are reduced to ordinary differential equations by using similarity transformation and solved numerically by using bvp4c function of MATLAB package. The impact of electrification in presence of viscous dissipation on normalized velocity, temperature and nanoparticle concentration is analysed and examined through graphs. The physical parameters like skin friction coefficient, rate of heat transfer and nanoparticle concentration are derived and presented by tables. It is found that the higher electrification parameter reduces the normalized base fluid temperature and enhances the normalized velocity due to Lorentz force. So, it may be concluded that electrification of particles is an important and possible mechanism for enhancement of thermal conductivity of base fluid.

Keywords Nanofluids · BVP4C function · Electrification of particle · Viscous dissipation

1 Introduction

The flow over a stretching surface is an important problem in many engineering and industrial applications such as hot rolling, wire drawing, paper production, glass blowing, plastic films drawing, etc. The first investigation on two-dimensional axisymmetric boundary layer flow over a stretched surface was studied by Sakiadis

---

K. K. Pradhan (✉) · A. Misra · S. K. Mishra
School of Applied Science, Centurion University of Technology and Management, Paralakhemundi, Gajapati, Odisha, India
e-mail: mr.kml1411@gmail.com

A. Misra
e-mail: amisra1972@gmail.com

S. K. Mishra
e-mail: s1_mishra@yahoo.com

S. R. Mishra et al. (eds.), *Recent Trends in Applied Mathematics*, Lecture Notes in Mechanical Engineering, https://doi.org/10.1007/978-981-15-9817-3_14

[23]. The similarity solution in analytical form for steady two-dimensional incompressible boundary layer flow caused by the stretching sheet was studied by Crane [8]. Stagnation-point flow describes the motion of fluid near the stagnation region of a solid surface. The two-dimensional stagnation-point flow towards a semi-infinite wall was studied by Hiemenz [11] for first time. This problem was extended by Homann [12] to the case of axisymmetric stagnation-point flow. The study of stagnation-point flow over a stretching surface was first investigated by Mahapatra and Gupta [17]. After this pioneering work, the flow over a stagnation point towards a stretching/shrinking sheet has been investigated by several authors [2, 3, 13, 16].

The conventional heat transfer fluids like water, ethylene glycol and engine oil are having limited capacity in terms of thermal properties but the solids metals are having higher thermal conductivity as compared to them. So it can be expected that the fluids containing solid particles can increase the thermal conductivity of above-cited applications. In view of this, Maxwell [18] suggested that the thermal conductivity of fluids can be enhanced by dispersing tiny-sized solid particles in the conventional fluids. This leads to the creation of an innovative coolant through which solid particles are suspended in conventional fluids to change the thermophysical properties of base fluids. This suspension was coined by Choi [7] and named as nanofluid which is used to reduce or enhance the thermal conductivity as per the need. These fluids are also very stable and not having any additional problems like sedimentation, erosion and additional pressure drop [9]. These suspended nanoparticles can change the thermal properties of base fluid due to its tiny size and low volume fraction. Khan and Pop [14] are first to consider the problem of stretching sheet in nanofluids.

Viscous dissipation is a process of converting mechanical energy of downward flowing fluid into thermal and acoustical energy. The idea of viscous dissipation was first given by Brinkman [5] and used by Gebhart [10]. Later, many researchers have extended this work and took this effect in their experiments like Vajravelu and Hadjinicalaou [27] and Partha et al. [22].

There are several models for nanofluid available to study its flow and heat transfer characteristics, but among them Buongiorno's model [6] and Tiwari and Das model [26] are very popular. As the Tiwari and Das model [26] analyses the behaviour of nanofluids by solid volume fraction whereas Buongiorno's model [06] analyses the combined effects of Brownian motion and thermophoresis on heat transfer characteristics. Recently, Buongiorno's model [06] has been used by several authors like Nield and Kuznetsov [20], Kuznetsov and Neild [15], Khan and Pop [14], and Bachok et al. [1]. Wen [28] investigated that the non-homogeneity of nanofluid is well justified when the nanoparticle migration phenomena occur. So in our present investigation we have considered Buongiorno's two-component non-homogeneous nanofluid model.

Though the use of nanofluid is highly essential on thermal management, but still from the above literatures, it reveals that there is no investigation done to study the flow and heat transfer characteristic of nanofluids with the effect of electrification of nanoparticle. The static electrification of solid particles within the non-conducting fluid was investigated by Soo [24]. He suggested that due to the particle–particle collision and particle–wall interactions, an effective drag force is produced on the ions which have a pronounced effect on the boundary layer characteristics of a two-phase flow. Hence, the present investigation is the extension work of Baker et al. [4] with the influence of electrification of nanoparticles in presence of viscous dissipation effect on skin friction, Nusselt number and Sherwood's number using Buongiorno's two-component non-homogeneous model.

2 Mathematical Formulation

The present study considers two-dimensional boundary layer, steady state of stagnation point flow of viscous incompressible nanofluid in the region $y > 0$ driven by a stretching/shrinking surface located at $y = 0$ with a fixed stagnation point at x = 0 as shown in Fig. 1.

It is assumed that, the base fluid is pure water, electrically non-conducting copper nanoparticles are electrified and both base fluids and nanoparticles are in equilibrium. The sheet is stretched with velocity $U_w(x) = bx$, and the ambient velocity $U_\infty(x) = ax$ is assumed to vary linearly from the stagnation point 'O', where a and b are constants with $a > 0$. We note that $b > 0$ and $b < 0$ correspond to the stretching and shrinking sheets, respectively, and x is the coordinate measured along the stretching/shrinking sheet. It is also assumed that the temperature and concentration of the surface are $T_w and C_W$, respectively, while the temperature and concentration in the free-stream condition are $T_\infty and C_\infty$, respectively, where $T_\infty > T_w$. We assume that the nanoparticle concentration is dilute. The simplified

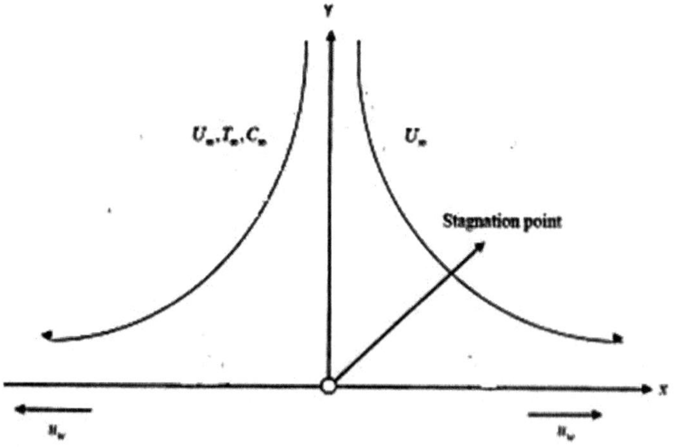

Fig. 1 Physical model and coordinate system

two-dimensional equations governing the flow in the boundary layer of a steady, laminar and incompressible nanofluid are derived as given below:

Continuity equation for fluid:

$$\frac{\partial u}{\partial x}+\frac{\partial v}{\partial y}=0 \tag{1}$$

Continuity equation for nanoparticle:

$$u\frac{\partial C}{\partial x}+v\frac{\partial C}{\partial y}=D_B\frac{\partial^2 C}{\partial y^2}+\frac{D_T}{T}\frac{\partial^2 T}{\partial y^2}+\left(\frac{q}{m}\right)\frac{1}{F}\left(\frac{\partial(CE_x)}{\partial x}+\frac{\partial(CE_y)}{\partial y}\right) \tag{2}$$

Momentum equation in the X-direction:

$$\rho_{nf}\left[u\frac{\partial u}{\partial x}+v\frac{\partial u}{\partial y}\right]=-\frac{\partial p}{\partial x}+\mu_{nf}\frac{\partial^2 u}{\partial y^2}+C\rho_s\left(\frac{q}{m}\right)E_x \tag{3}$$

Momentum equation in Y-direction:

$$\frac{\partial p}{\partial y}=O(\delta) \tag{4}$$

Energy equation

$$(\rho c)_{nf}\left[u\frac{\partial T}{\partial x}+v\frac{\partial T}{\partial y}\right]=k_{nf}\left[\frac{\partial^2 T}{\partial y^2}\right]+\rho_s c_s D_B\frac{\partial C}{\partial y}\frac{\partial T}{\partial y}+\frac{\rho_s c_s D_T}{T}\left(\frac{\partial T}{\partial y}\right)^2$$
$$+\left(\frac{q}{m}\right)\frac{Cc_s\rho_s}{F}\left(E_x\frac{\partial T}{\partial x}+E_y\frac{\partial T}{\partial y}\right)+\mu_{nf}\left(\frac{\partial u}{\partial y}\right)^2 \tag{5}$$

The E-field is given by the equation:

$$\frac{\partial E_x}{\partial x}+\frac{\partial E_y}{\partial y}=\frac{\rho_s}{\epsilon_0}\frac{q}{m}\ \text{where } \epsilon_0 \text{ is the permittivity.} \tag{6}$$

Following Soo [25] neglecting the change in electric field in the x-direction, the transverse electric field is given by

$$\frac{\partial E_y}{\partial y}=\frac{\rho_s}{\epsilon_0}\frac{q}{m} \tag{7}$$

And E_x in potential flow outside the boundary layer is assumed almost negligible with the boundary conditions

$$\left.\begin{array}{l} u = U_W = bx, v = 0, T = T_W, C = C_w \, at \; y = 0 \\ u = U_\infty = ax, v = 0, T \to T_\infty, C \to C_\infty \, as \; y \to \infty \end{array}\right\} \tag{8}$$

3 Similarity Transformation

The governing partial differential Eqs. (1–7) are transferred into corresponding ordinary differential equations by introducing the dimensionless variables

$$\eta = y\sqrt{\frac{a}{\nu_f}}, \theta = \frac{T - T_\infty}{T_w - T_\infty}, u = axf', v = -\sqrt{a\nu_f}\, f(\eta), \psi = \sqrt{a\nu_f}\, xf(\eta),$$
$$S = \frac{C - C_\infty}{C_w - C_\infty}$$

Since Eq. (1) is satisfied with the above variables, Eqs. (2), (3) and (5) are transformed into

$$S'' + ScfS' + \frac{N_t}{N_b}\theta'' + \frac{ScN_F}{N_{Re}}\left(S + N_C + \eta S'\right) = 0 \tag{9}$$

$$\left(f'\right)^2 - ff'' = \frac{1}{\varphi_1} f''' + 1 + \varphi_2 \frac{Sc}{N_F} MN_bS \tag{10}$$

$$-f\theta' = \frac{\varphi_4\varphi_3}{Pr}\theta'' + \varphi_3 Nb\theta' S' + \varphi_3 Nt\left(\theta'\right)^2 + \frac{N_F\varphi_3 Nb\eta\theta' Sc(S + Nc)}{N_{Re}} + Ec\varphi_3\varphi_5\left(f''\right)^2, \tag{11}$$

respectively, by considering $\left(\frac{T_w - T_\infty}{T_\infty}\right) \ll 1$

$$\text{where}M = \left(\frac{q}{m}\right)\frac{1}{FU}E_x, N_F = \frac{U}{Fx}, \frac{1}{N_{Re}} = \left(\left(\frac{q}{m}\right)^2 \frac{\rho_s x^2}{U^2\epsilon_0}\right),$$
$$N_b = \frac{\tau D_B(C_w - C_\infty)}{\nu_f}$$

$$Nt = \frac{\tau D_T(T_w - T_\infty)}{\nu_f T_\infty}, Pr = \frac{\nu_f}{\alpha_f}, Sc = \frac{\nu_f}{D_B}, N_c = \left(\frac{C_\infty}{(C_w - C_\infty)}\right), \tau = \frac{\rho_s c_s}{\rho_f c_f}$$

Using the Maxwell model [19] for thermal conductivity, we get the thermophysical constants $\varphi_1, \varphi_2, \varphi_3, \varphi_4, \varphi_5$ as

$$\varphi_1 = \frac{1}{(1 - C)^{2.5}\left[(1 - C) + C\frac{\rho_s}{\rho_f}\right]}, \varphi_2 = \frac{\rho_s}{\rho_f}\frac{1}{\left[(1 - C) + C\frac{\rho_s}{\rho_f}\right]},$$
$$\varphi_3 = \frac{1}{(1 - C) + C\tau},$$

$$\varphi_4 = \frac{k_s + 2k_f - 2C(k_f - k_s)}{k_s + 2k_f + C(k_f - k_s)}, \varphi_5 = \frac{1}{(1 - C)^{2.5}}$$

subjected to the boundary conditions

$$\left.\begin{array}{l} \eta = 0, f(0) = 0, f'(0) = \epsilon = \frac{b}{a}, \theta(0) = 1, s(0) = 1 \\ \eta = \infty, f'(\infty) = 1, \theta(\infty) = 0, s(\infty) = 0 \end{array}\right\} \quad (12)$$

4 Result and Discussion

The set of non-linear ordinary differential Eqs. (16–18), with the associated boundary conditions (19), are solved numerically by using boundary value problem default solver bvp4c of MATLAB package. The obtained results are presented by tables and analysed through graphs. In this investigation, we have considered silver (Ag) nanoparticles with pure water as a base fluid. The nanoparticle volume fraction (φ) is taken as 0.001 to interpret the result and $\varphi = 0$ represents the regular fluid with Prandtl number $Pr = 6.2$. The thermophysical properties of nanofluid (Ag) and base fluid (water) used in the numerical simulations are given in Table 1.

Extensive calculations have been performed to obtain the velocity, temperature, concentration profiles as well as skin friction, normalized Nusselt number and normalized Sherwood number for various values of physical parameters such asε,φ, Ec, Sc, Pr, N_F, N_b, N_t, N_{Re}, *Nc*, *M*. To determine the accuracy of our numerical results, the skin friction coefficient is compared with the published results of Bakar et al. [4] in Table 2. The calculated values show a favourable agreement for which we believe that the present results are correct and accurate and Table 3 describes the

Table 1 Thermophysical properties of base fluid and nanoparticles [21]

Physical properties	Base fluid	Nanoparticles
	Water	Ag
$C_p(J/kgK)$	4179	235
$\rho(kg/m^3)$	997.1	10,500
$k(W/mK)$	0.613	429

Table 2 Comparison of $f''(0)$ with different values of ε when $M = Ec = 0$, $Nb = NF = Nt = \varphi = 0.1$, $Sc = 3$, $Pr = 6.2$

ε	Bakar et al. [4]	Present
	$f''(0)$	$f''(0)$
0.0	1.510003	1.5100
0.5	0.83834	0.8383
1	0	0

Table 3 Numerical values of $f''(0)$, $-\theta'(0)$ and $-S'(0)$ for various flow parameters

ε	Ec	φ	Sc	Pr	N_F	N_b	N_t	N_{Re}	Nc	M	$f''(0)$	$-\theta'(0)$	$-S'(0)$
										0.1	0.14641	1.68587	0.36464
0.5	0.1	0.01	1.5	6.2	0.1	0.1	0.1	2.0	0.1	**0.3**	1.96994	1.77674	0.49710
										0.4	2.79832	1.81253	0.55000
						0.0					−0.16405	1.88228	−1.18014
1.5	0.1	0.01	1.5	6.2	0.1	**0.05**	0.1	2.0	0.1	0.1	0.19873	1.72669	0.15748
						0.1					0.48239	1.60555	0.54515
							0.05				1.86733	1.32572	0.57079
1.5	0.1	0.01	1.5	6.2	0.1	0.1	**0.1**	2.0	0.1	0.1	2.01041	1.24399	0.32295
							0.15				2.13799	1.16393	0.15257
	0.0										2.01041	1.24399	0.32295
1.5	**1.0**	0.01	2.0	6.2	0.1	0.1	0.1	2.0	0.1	0.1	1.60189	1.01256	1.30803
	2.0										1.31973	0.87405	1.91669
	3.0										1.11711	0.78381	2.32351
0.5											2.01041	1.24399	0.32295
1.5	0.1	0.01	1.5	6.2	0.1	0.1	0.1	2.0	0.1	0.1	0.19873	1.72669	0.15748
2.0											−0.91422	1.93622	0.09359

numerical results of different flow quantities on the friction factor coefficient ($f''(0)$), the rates of heat transfer ($-\theta'(0)$) and nanoparticle concentration coefficient ($-S'(0)$).

From Table 3, it is observed that the increase of electrification parameter (M), thermophoresis parameter (Nt) and Brownian motion parameter (Nb) increases the velocity of fluid flow and consecutively the rate of heat and mass transfer is decreasing. But the increase of Eckert number (Ec) and stretching parameter (ε) decreases the velocity of fluid flow.

4.1 Effect of Electrification (M)

Figures 2, 3 and 4 describe the impact of electrification parameter (M) on velocity, temperature and concentration distribution, respectively. It is observed that the fluid velocity is increasing as M is increasing whereas the temperature and concentration are decreasing. This is due to the existence of electrification of particles within the boundary layer which produces a drag force called Lorentz's force which acts as an accelerating force on the velocity field. As a result, the velocity increases and the temperature and concentration decrease. Hence, we concluded that the electrification may play an important role to control the velocity and heat transfer in various conducting fluids.

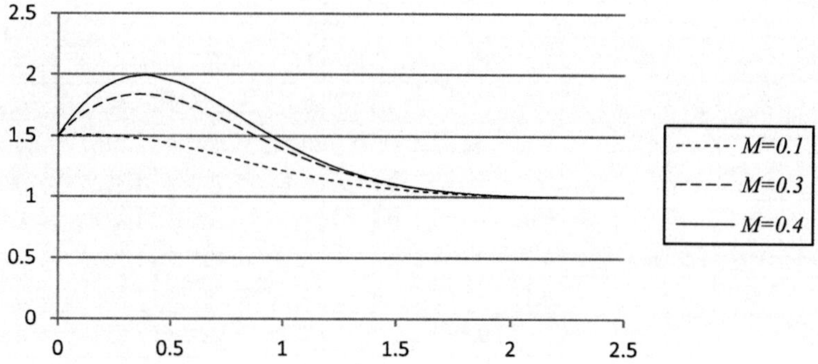

Fig. 2 Effect of electrification parameter on $f^{'}(\eta)$

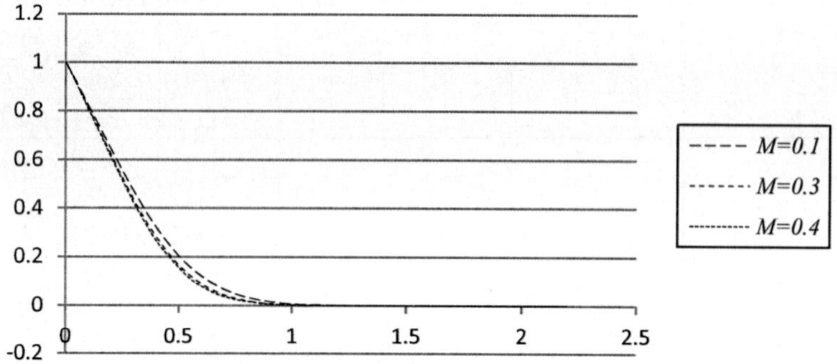

Fig. 3 Effect of electrification parameter on $\theta(\eta)$

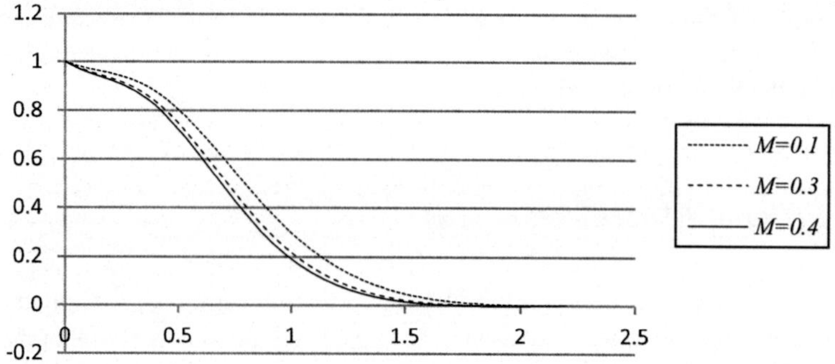

Fig. 4 Effect of electrification parameter on $S(\eta)$

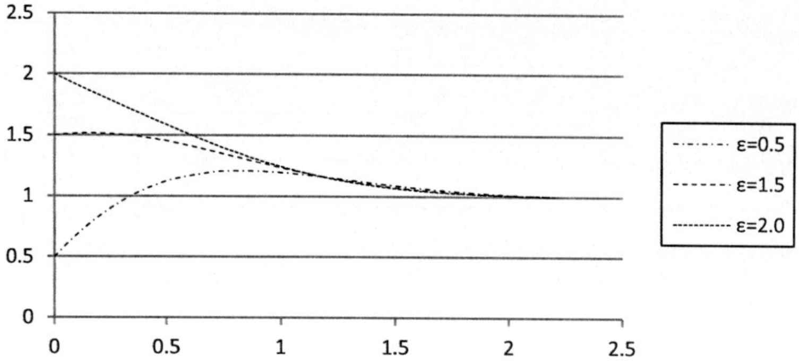

Fig. 5 Effect of stretching parameter on $f'(\eta)$

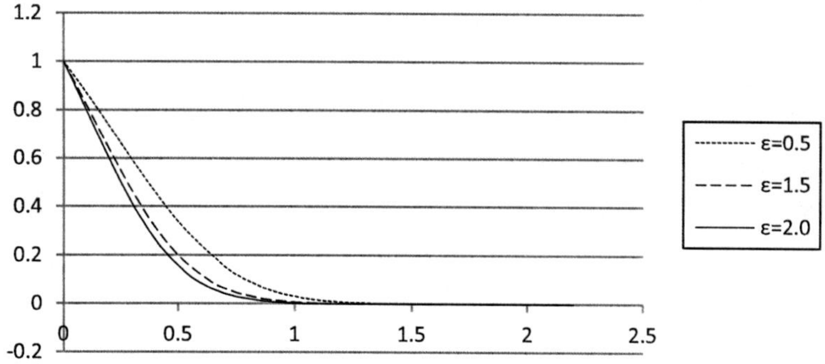

Fig. 6 Effect of stretching parameter on $\theta(\eta)$

4.2 Effect of Stretching Parameter (ε)

Here we noticed that $\epsilon > 1$ represents the free stream is directed towards negative x-direction and $0 < \epsilon < 1$ represents the movement of fluid and plate is in the same direction, while the plate moves towards the positive x- direction. The increase of ϵ shows the increase of velocity and decrease of temperature and concentration (Figs. 5, 6, 7).

4.3 Effect of Viscous Dissipation Parameter (Ec)

Figure 9 represents the impact of Eckert number (Ec) on temperature profiles. It is found that temperature decreases with the increase of Ec. It may conclude that the viscous dissipation has contributed to increase in velocity and thermal boundary layer thicknesses while the local Nusselt number decreases (Figs. 8, 10).

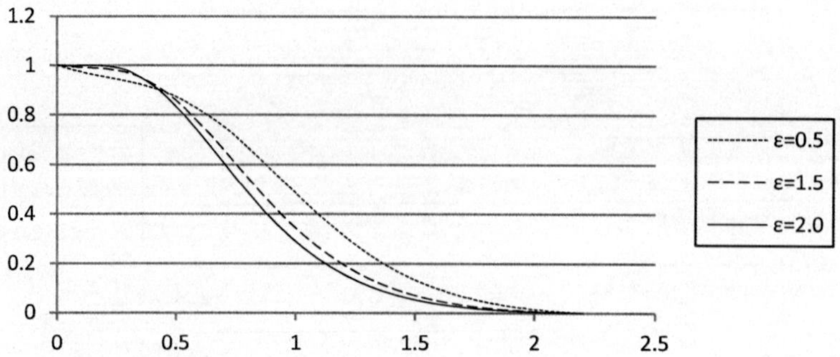

Fig. 7 Effect of stretching parameter on $S(\eta)$

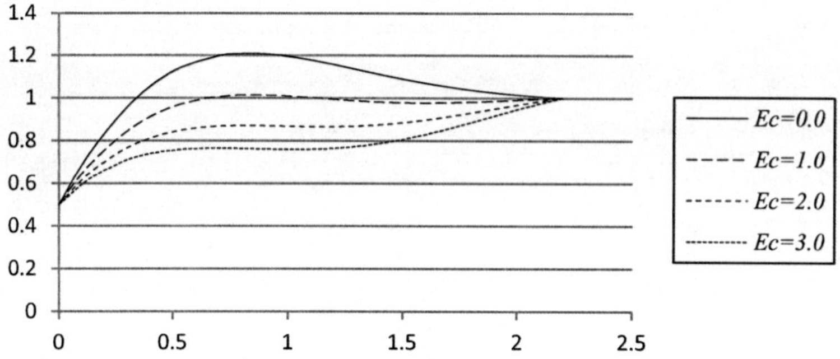

Fig. 8 Effect of viscous dissipation parameter on $f'(\eta)$

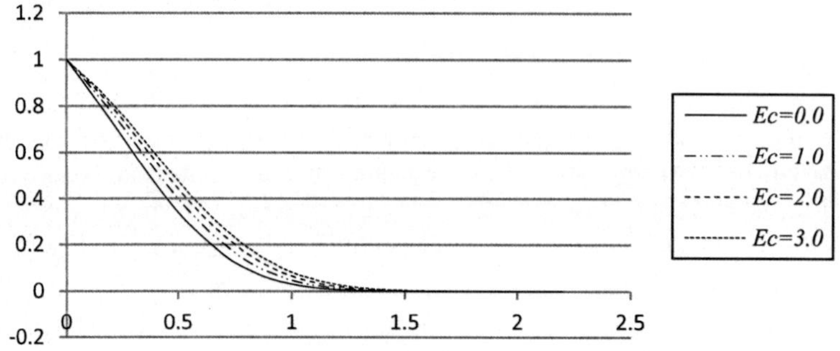

Fig. 9 Effect of viscous dissipation parameter on $\theta(\eta)$

*4.4 Effect of Thermophoresis Parameter (*Nt*)*

The impact of thermophoresis parameter (*Nt*) on temperature and nanoparticle concentration fields is shown in Figs. 11 and 12, respectively. It is observed that

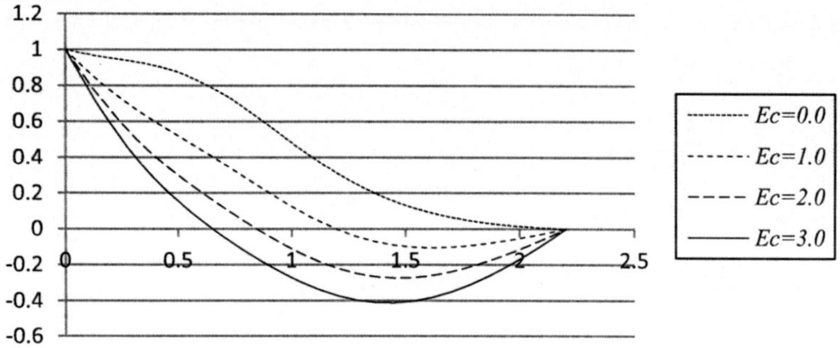

Fig. 10 Effect of viscous dissipation parameter on $S(\eta)$

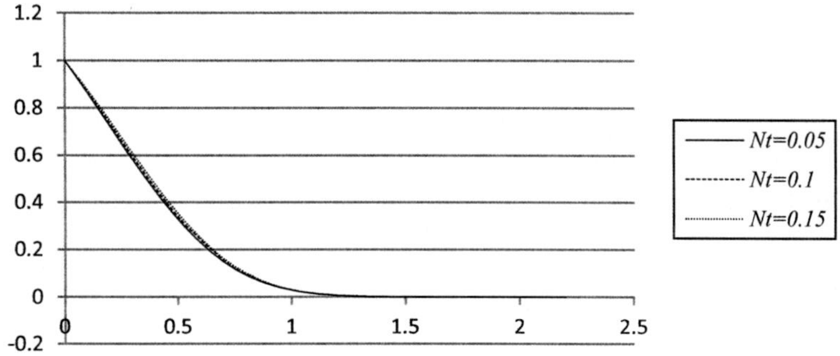

Fig. 11 Effect of thermophoresis parameter on $\theta(\eta)$

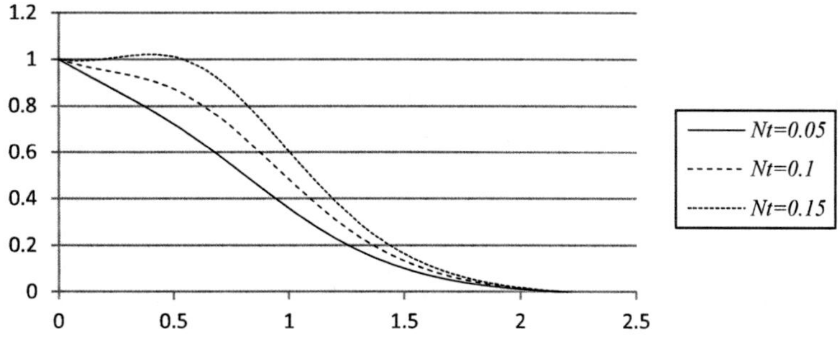

Fig. 12 Effect of thermophoresis parameter on $S(\eta)$

the fluid temperature and concentration are increasing due to the increase of thermophoresis. As we know that thermophoresis is a mechanism through which the tiny particles are dragging away from hot surface to cold surface. So the fluid temperature and concentration are increasing due to the increase of thermophoresis.

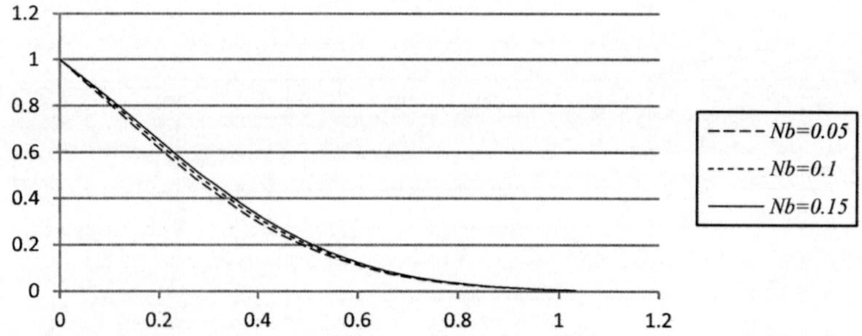

Fig. 13 Effect of Brownian diffusion parameter on $\theta(\eta)$

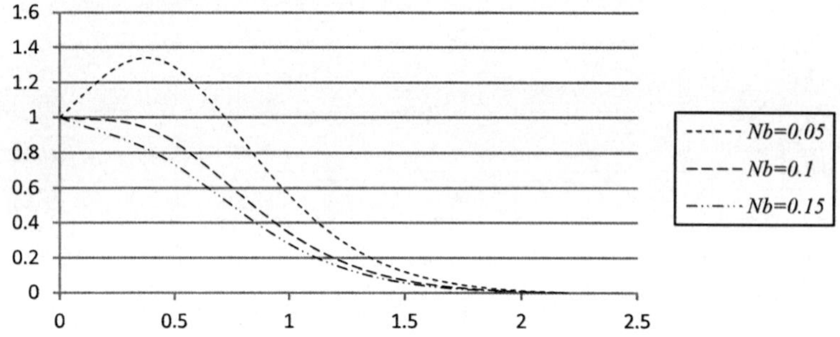

Fig. 14 Effect of Brownian diffusion parameter on $S(\eta)$

4.5 Effect of Brownian Diffusion Parameter (**Nb**)

Figures 13 and 14 display the effect of Brownian motion on temperature and nanoparticle concentration fields, respectively. It is found that due to the impact of Brownian motion, the temperature of fluid flow is increasing whereas the nanoparticle concentration is decreasing. Since the larger *Nb* corresponds to stronger random motion of nanoparticles within a fluid, the fluid temperature gets enhance.

5 Critical Analysis

5.1 Impact of Electrification in Absence of Viscous Dissipation

The impacts of electrification in absence of viscous dissipation are shown in Figs. 15, 16 and 17. From Table 4, it is observed that the increase of electrification in absence of viscous dissipation increases the fluid velocity and decreases the rates of heat transfer. It occurs due to the Lorentz force.

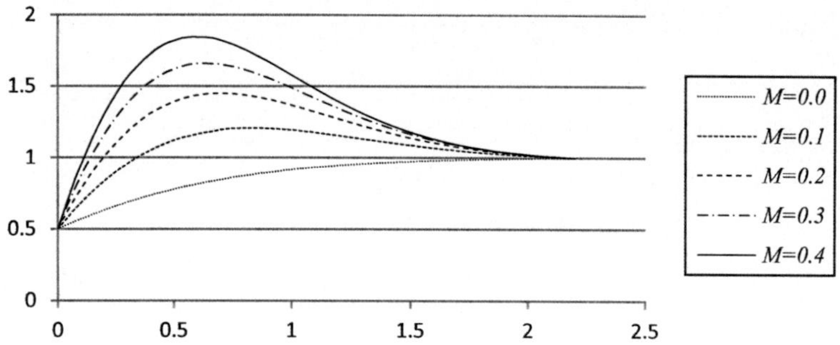

Fig. 15 Effect of electrification in absence of viscous dissipation on $f'(\eta)$

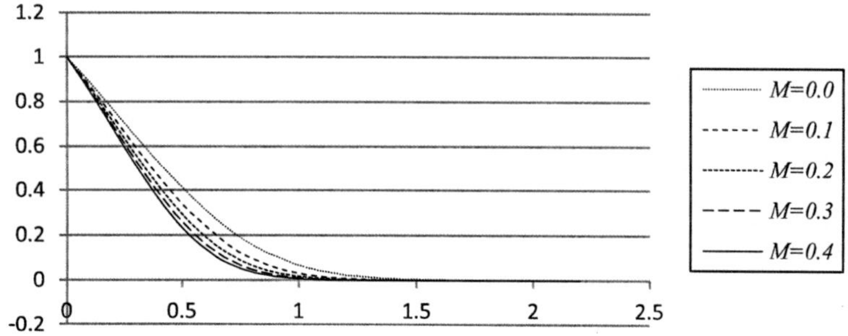

Fig. 16 Effect of electrification in absence of viscous dissipation on $\theta(\eta)$

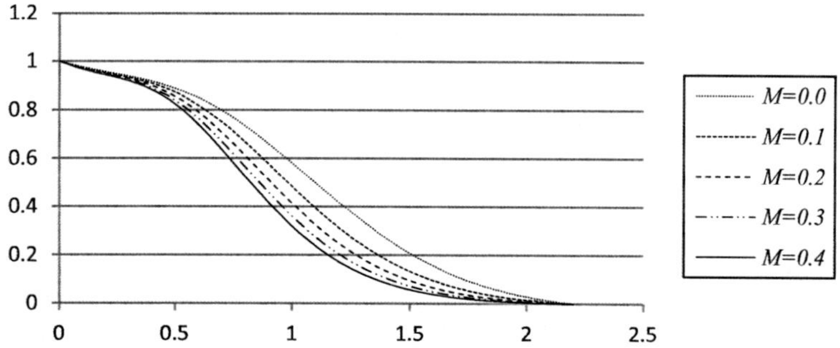

Fig. 17 Effect of electrification in absence of viscous dissipation on $S(\eta)$

Table 4 Values of $f''(0)$, $-\theta'(0)$ and $-S'(0)$ with different values of M

$Pr = 6.2, N_{Re}, = 2.0, Nt = 0.1, N_F = 0.1, Nb = 0.1, Sc = 1.5$			
M	$f''(0)$	$-\theta'(0)$	$-S'(0)$
0.0	0.73823	1.10181	0.28418
0.1	2.01041	1.24399	0.32295
0.2	3.12693	1.34042	0.35686
0.3	4.15119	1.41546	0.38542
0.4	5.11200	1.47784	0.40992

5.2 *Impact of Viscous Dissipation in Absence of Electrification*

The impacts of viscous dissipation in absence of electrification are shown in Figs. 18 and 19. From Table 5, it is observed that the increase of viscous dissipation effect in absence of electrification increases the rates of heat transfer.

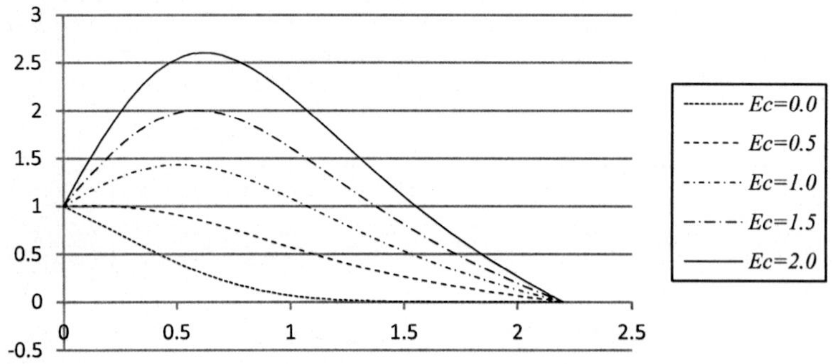

Fig. 18 Effect of viscous dissipation in absence of electrification on $\theta(\eta)$

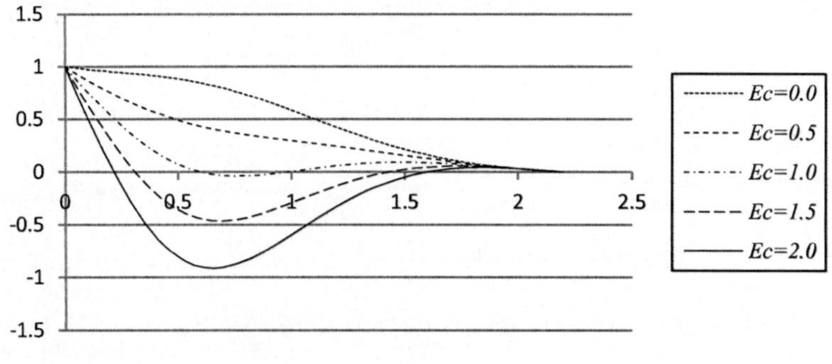

Fig. 19 Effect of viscous dissipation in absence of electrification on $S(\eta)$

Table 5 Values of $f''(0)$, $-\theta'(0)$ and $-S'(0)$ with different values of M

$Pr = 6.2$, $N_{Re'} = 2.0$, $Nt = 0.1$, $N_F = 0.1$, $Nb = 0.1$, $Sc = 1.5$			
Ec	$f''(0)$	$-\theta'(0)$	$-S'(0)$
0.0	0.73823	1.10240	0.31316
0.5	0.73823	−0.09896	1.33815
1.0	0.73823	−1.40065	2.45697
1.5	0.73823	−2.81303	3.67969
2.0	0.73823	−4.34760	5.01750

Table 6 shows the computed values of $f''(0)$, $-\theta'(0)$ and $-S'(0)$ in various situations for the different parametric values of the flow problem. It can be observed that with electrification and viscous dissipation the value of the Nusselt number with 1% volume fraction of nanoparticle is greater than that of the nanofluid with 10% volume fraction. Also, a larger Nusselt number corresponds to more active convection and the fluid motion enhances heat transfer by advection. Here, it is observed that smaller values of Schmidt number (increase in mass diffusivity will reduce Schmidt number) has an impact to increase the Nusselt number, i.e. enhances heat transfer by advection. Thus, electrification of particles in addition to Brownian motion and thermophoresis may be a potential mechanism for enhancement of thermal conductivity.

6 Conclusion

The flow over a stretching sheet plays an important role in the field of engineering and science due to its incredible applications. With this view, an attempt was made to investigate the effects of electrification on boundary layer stagnation-point flow towards a stretching/shrinking sheet in a nanofluid using Buongiorno's model. The impact of various flows of parameters was discussed through graphs and tables.

- The increment of electrification parameter (M) increases the flow of velocity and decreases the temperature of fluid flow. It is due to the effect of Lorentz's force.
- The increase in Brownian motion increases the temperature of fluid flow and decreases the nanoparticle concentration as the larger Nb corresponds to stronger random motion of nanoparticles within a fluid.
- The increase of thermophoresis increases the fluid temperature and the nanoparticle concentration.
- The large values of Ec cause the increase of nanofluid temperature as it has the contribution to increase in fluid velocity.
- The increment of velocity ratio parameter (ε) causes the reduction of surface temperature gradient. Hence, the fluid temperature is enhanced due to the effect of Lorentz's force.

Table 6 Values of $f''(0)$, $-\theta'(0)$ and $-S'(0)$ with for different parametric values

Present investigation $M = 0.1, Nb = 0.1, Nt = 0.1, Ec = 0.1, Pr = 6.2, Sc = 1.5, \varphi = 0.01, \varepsilon = 0.5$											
Without electrification and viscous dissipation			Without electrification and with viscous dissipation			With electrification and without viscous dissipation			With electrification and viscous dissipation		
$f''(0)$	$\theta'(0)$	$S'(0)$	$f''(0)$	$\theta'(0)$	$S'(0)$	$f''(0)$	$\theta'(0)$	$S'(0)$	$f''(0)$	$\theta'(0)$	$S'(0)$
0.73823	1.10507	0.31089	0.73823	1.10240	0.31316	0.73823	1.10181	0.28418	0.14641	**1.68587**	0.36464
Investigation by Baker et al. [4] $Nb = 0.1, Nt = 0.1, Pr = 6.2, Sc = 3.0, \varphi = 0.1, \varepsilon = 0.5$			Present investigation with $M = 0.1, Nb = 0.1, Nt = 0.1, Ec = 0.1, Pr = 6.2, Sc = 3.0, \varphi = 0.1, \varepsilon = 0.5$								
Without electrification and viscous dissipation			Without electrification and with viscous dissipation			With electrification and without viscous dissipation			With electrification and viscous dissipation		
0.8383	0.98695	0.81322	0.8383	0.73814	0.99430	2.32661	1.10215	0.92078	2.328089	**1.07684**	0.9013

References

1. N. Bachok, A. Ishak, I. Pop, Boundary layer flow of nanofluids over a moving surface in a flowing fluid. Int. J. Thermal. Sci. **49**, 1663–1668 (2010)
2. N. Bachok, A. Ishak, I. Pop, Melting heat transfer in boundary layer stagnation-point flow towards a stretching/shrinking sheet. Phys. Lett. A **374**, 4075–4079 (2010)
3. N. Bachok, A. Ishak, I. Pop, On the stagnation-point flow towards a stretching sheet with homogeneous-heterogeneous reactions effects. Comm. Nonlinear Sci. Numer. Simulat. **16**, 4296–4302 (2011)
4. N.A. Bakar, N. Bachok, N.M. Arfin, Nanofluid flow using buongiorno model over a stretching sheet and thermophsical properties of nanofluids. Indian J. Sci. Technol. **9**(31) (2016)
5. H.C. Brinkman, The viscosity of concentrated suspensions and solutions. J. Chem. Phys. **20**, 571–581 (1952)
6. J. Buongiorno, Convective transport in nanofluids. J. Heat Tran. **128**, 240–250 (2006)
7. S.U. Choi, J.A. Eastman, Enhancing thermal conductivity of fluids with nanoparticles, in *Proceedings of the ASME International Mechanical Engineering Congress and Exposition, 66*. Argonne National Lab., IL (United States) (1995)
8. L.J. Crane, Flow past a stretching plate. Zeitschrift Für Angewandte Mathematik Und Physik Zamp **21**(4), 645–647 (1970)
9. Das SK, Choi SUS, Yu W, Pradeep T: Nanofluids: Science and Technology NJ: Wiley; 2007.
10. B. Gebhart, Effects of viscous dissipation in natural convection. J. Fluid Mech. **14**(02), 225–232 (1962)
11. K. Hiemenz, Die Grenzschicht an einem in den gleichförmigen Flüssigkeitsstrom eingetauchten geraden Kreiszylinder. Dingler's Polytech. J. **326**, 321–324 (1911)
12. F. Homann, Der einfluss grosser zahigkeit bei der stromung um den zylinder und um die kugel. Z. Angew. Math. Mech. **16**, 153–164 (1936)
13. A. Ishak, Y.Y. Lok, I. Pop, Stagnation-point flow over a shrinking sheet in a micropolar fluid. Chem. Eng. Comm. **197**, 1417–1427 (2010)
14. A.V. Khan, I. Pop, Boundary-layer flow of a nanofluid past a stretching sheet. Int. J. Heat Mass. Tran. **53**, 2477–2483 (2010)
15. A.V. Kuznetsov, D.A. Nield, Natural convective boundary layer flow of a nanofluid past a vertical plate. Int. J. Thermal. Sci. **49**, 243–247 (2010)
16. Y.Y. Lok, A. Ishak, I. Pop, MHD stagnation-point flow towards a shrinking sheet. Int. J.. Numer. Meth. Heat Fluid Flow **21**, 61–72 (2011)
17. T.R. Mahapatra, A.S. Gupta, Heat transfer in stagnation-point flow towards a stretching sheet. Heat Mass Tran. **38**, 517–521 (2002)
18. J. Maxwell, C. James, *A Treatise on Electricity and Magnetism*, 1st edn. (Clarendon Press, Oxford, UK, 1873)
19. J. Maxwell, *A Treatise on Electricity and Magnetism*, 2nd edn. (Oxford University, UK, 1904).
20. D.A. Nield, A.V. Kuznetsov, The Cheng-Minkowycz problem for natural convective boundary layer flow in a porous medium saturated by a nanofluid. Int. J. Heat Mass Tran. **52**, 3187–3196 (2009)
21. H.F. Oztop, E. Abu-Nada, Numerical study of natural convection in partially heated rectangular enclosures filled with nanofluids. Int. J.. Heat Fluid Flow **29**, 1326–1336 (2008)
22. M.K. Partha, P. Murthy, G.P. Rajasekhar, Effect of viscous dissipation on the mixed convection heat transfer from an exponentially stretching surface. Heat Mass Transf. **41**(4), 360–366 (2005)
23. B.C. Sakiadis, Boundary-layer behaviour on continuous solid surface: I. Boundary-layer equations for two-dimensional and axisymmetric flow. Am. Inst. Chem. Eng. J. **7**, 26–28 (1961)
24. S.L. Soo, Effect of electrification on the dynamics of a particulate system. I and EC Fund **3**, 75–80 (1964)
25. S.L. Soo, *Particulates and Continuum- Multiphase Fluid Dynamics*. ISBN 9780471970767
26. R.K. Tiwari, M.K. Das, Heat transfer augmentation in a two-sided lid-driven differentially heated square cavity utilizing nanofluids. Int. J. Heat Mass Tran. **50**, 2002–2018 (2007)

27. K. Vajravelu, A. Hadjinicolaou, Heat transfer in a viscous fluid over a stretching sheet with viscous dissipation and internal heat generation. Int. Commun. Heat Mass Trans. **20**(3), 417–430 (1993)
28. C.Y. Wang, The three-dimensional flow due to a stretching flat surface. Phys. Fluids **27**(8), 1915–1917 (1984)

Effect of Variable Viscosity on Slow Rotation of a Porous Sphere in a Cavity

Madasu Krishna Prasad

Abstract In the limit of low Reynolds number, the effect of variable viscosity engendered by the rotation of a porous sphere in a closed spherical boundary is filled up with a viscous fluid. The fluid motion is steady and axisymmetric. Stokes equation holds in the liquid region and porous region obeys the Brinkman equation. The slip boundary condition is used on the surface of cavity. The continuity of velocity components and stress jump boundary condition on the porous-liquid surface are used. The couple and wall correction factor exerted by the surrounding viscous fluid on a rotating porous sphere are obtained. The effect of apparent viscosity, permeability parameter, slip parameter, and stress jump coefficient on the wall correction factor are discussed. Special cases are recovered.

Keywords Creeping flow · Porous sphere · Variable viscosity · Rotation · Couple

1 Introduction

The flow problems of the slow rotation of a cavity wrapping a concentric porous particle have attracted the attention of many investigators to examine due to its various applications in the areas of biomedical, environmental engineering, chemical, and science. These problems are modeled by using the Stokes version for the flow inside cavity and Brinkman's equation holds in the porous region. The objective is to determine the velocities in both liquid and porous regions. The torque is to be calculated.

Considerable amount of work has been done on rotation of a rigid sphere, porous sphere, composite sphere, and approximate sphere in a spherical cavity under slow motion. One can find useful literature in Happel and Brenner [1] and Kim and Karrila [2]. Kanwal [3] discussed the incompressible viscous fluid flow of a steadily rotating a torus, a spindle, and a lens. Keh and Chou [4] and Keh and Lu [5] investigated the effects of boundary wall that significantly the slow motion of a porous shell particles.

M. K. Prasad (✉)
Department of Mathematics, National Institute of Technology, Raipur 492010, India
e-mail: madaspra.maths@nitrr.ac.in; kpm973@gmail.com

S. R. Mishra et al. (eds.), *Recent Trends in Applied Mathematics*, Lecture Notes in Mechanical Engineering, https://doi.org/10.1007/978-981-15-9817-3_15

Srinivasacharya and Prasad [6–8] extended their work ([4, 5]) by using stress jump condition [9–13], respectively. Srivastava and Saxena [14] discussed the rotation of a rigid sphere in a viscous fluid immersed in spherical porous medium.

Ashmawy [15] studied the effect of slip condition for oscillatory rotational motion of a porous shell bounded by concentric spherical cavity. Prasad et al. [16] presented an application of steady rotation of solid spheroidal particle enclosed in a concentric spheroidal container. Awasthi et al. [17] investigated the effect of variable permeability on the torque exerted by porous sphere in a spherical container. Recently, Filippov and Koroleva [18] discussed uniform estimates for flow of porous cylindrical particles where porous medium is filled by liquid with varying viscosity. Filippov and Koroleva [19] studied the behavior of the flow velocity by modeling a set of porous particles immersed into liquid concentric cells. Ryzhikh and Filippov [20] studied the hydrodynamic permeability of a porous shell with solid impermeable core for variable viscosity.

In this work, the effect of variable viscosity is investigated for slow and steady rotation of a porous sphere fill up with a viscous fluid in a spherical cavity having common center. The flow in the fluid region is modeled by Stokes' equations and the flow in porous region obeys Brinkman's model. The slip condition at the cavity surface is employed. On the fluid-porous interface, continuity of the velocity and tangential stress jump conditions are used. The influences of the geometric parameters are discussed using graphs.

2 Mathematical Formulation

As depicted in Fig. 1, the problem of an axially symmetric steady rotational flow caused by a porous sphere of radius a bounded by a Newtonian fluid having viscosity μ in a concentric spherical cavity of radius b is studied. The inner sphere is rotating about z-axis with angular velocity Ω. The angular velocity of the outer sphere is $-\Omega$. Porous region has uniform porosity and variable liquid viscosity. The viscosity of liquid region μ is constant $(a < r < b)$, and the effective viscosity is assumed to depend on the viscosity of liquid region by the power law $\mu_e = \mu \left(\frac{a}{r}\right)^{\alpha}$, i.e., the μ_e is a monotonically along the depth of the porous sphere from the value of μ at the liquid region and porous region. The parameter α is chosen on the basis of the specific value of effective viscosity. The fluid dynamic pressure is constant everywhere.

Let us assume the flow outside the porous sphere is ruled by steady Stokes flow

$$\nabla \cdot \mathbf{v}^{(1)} = 0, \tag{1}$$

$$\nabla p^{(1)} + \mu \nabla \times \nabla \times \mathbf{v}^{(1)} = 0, \tag{2}$$

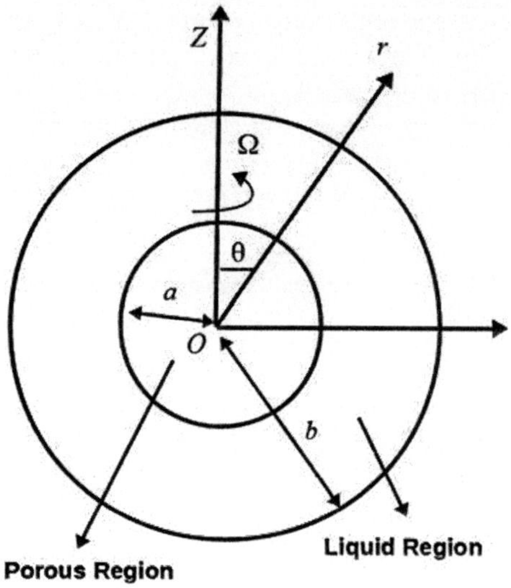

Fig. 1 The physical situation of the problem

where $p^{(1)}$ is the pressure, μ is the coefficient of viscosity, and $\mathbf{v}^{(1)}$ is the velocity. and the flow inside the porous region is described by Brinkman's equation [21]

$$\nabla \cdot \mathbf{v}^{(2)} = 0, \tag{3}$$

$$\nabla p^{(2)} + \frac{\mu}{k}\mathbf{v}^{(2)} + \mu_e \nabla \times \nabla \times \mathbf{v}^{(2)} = 0, \tag{4}$$

where $p^{(2)}$ is the pressure, k is permeability of the porous medium, μ_e is the effective viscosity , and $\mathbf{v}^{(2)}$ is the velocity.

According to the geometry of the problem and nature of the flow, the velocity vectors can be represented as

$$\mathbf{v}^{(i)} = v_\phi^{(i)}(r, \theta)\, \mathbf{e}_\phi, \; i = 1, 2, \tag{5}$$

where $(\mathbf{e}_r, \mathbf{e}_\theta, \mathbf{e}_\phi)$ are the unit vectors along the increasing directions of the spherical coordinate system (r, θ, ϕ)

The suitable boundary conditions which describe the present problem [6, 7, 15, 16] are given as

$$v_\phi^{(1)} = v_\phi^{(2)} \quad \text{on} \quad r = a, \tag{6}$$

$$\frac{\partial v_\phi^{(2)}}{\partial r} - \frac{\partial v_\phi^{(1)}}{\partial r} = \frac{\sigma}{\sqrt{k}} v_\phi^{(2)} \quad \text{on} \quad r = a, \tag{7}$$

where σ is the stress jump coefficient. If this parameter $\sigma = 0$, the continuity of tangential stresses occurs at the interface. Experimentally, it has been asserted σ lies between -1 and 1.

$$\lambda_1 \left(v_\phi^{(1)} + \Omega\, r\, \sin\theta \right) = -\left(\frac{\partial v_\phi^{(1)}}{\partial r} - \frac{v_\phi^{(1)}}{r} \right) \quad \text{on } r = b. \tag{8}$$

If $\lambda_1 \to \infty$, the special case of slow motion generated by rotation of a no-slip spherical container about a porous sphere is obtained.

The following dimensionless equations are obtained by introducing non-dimensional variables $r = a\tilde{r}, v_\phi^{(i)} = \Omega a \tilde{v}_\phi^{(i)},\ E^2 = \frac{1}{a^2}\tilde{E}^2$ in the Eqs. (2) and (4) and dropping the tildes.

$$E^2\,(r\,\sin\theta\, v_\phi^{(1)}) = 0, \tag{9}$$

$$(E^2 - r^\alpha \beta^2)\,(r\,\sin\theta\, v_\phi^{(2)}) = 0, \tag{10}$$

where $E^2 = \dfrac{\partial^2}{\partial r^2} + \dfrac{(1-\zeta^2)}{r^2}\dfrac{\partial^2}{\partial \zeta^2}$ is the Stokesian operator, $\zeta = \cos\theta$, and $\beta^2 = a^2/k$.

3 Solution of the Problem

Assuming the solution of second order partial differential equations (9) and (10) in the form $w^{(1)} = f(r)\sin\theta$ and $w^{(2)} = g(r)\sin\theta$. The transformed ordinary differential equations are

$$r^2 f''(r) + 2rf'(r) - 2f(r) = 0, \tag{11}$$

$$v_\phi^{(1)} = \left(A\,r + B\,r^{-2}\right)\sin\theta, \tag{12}$$

$$r^2 g''(r) + 2rg'(r) - (2 + r^{\alpha+2}\beta^2)g(r) = 0, \tag{13}$$

Let $p = \alpha + 2$,

$$r^2 g''(r) + 2rg'(r) - (2 + r^p\beta^2)g(r) = 0. \tag{14}$$

This differential equation of the second order posses two kinds of solutions when $p \geq 0$ and $p < 0$. For $p < 0$, this equation shows that two linearly independent solutions are singular at $r = 0$. Hence, we discuss cases for $p > 0$ and $p = 0$. If $p > 0$ [17],

$$g(r) = C\frac{1}{\sqrt{r}}I_{\frac{3}{p}}\left(\frac{2\sqrt{r^p}\beta}{p}\right), \tag{15}$$

$$v_\phi^{(2)} = C\frac{1}{\sqrt{r}}I_{\frac{3}{p}}\left(\frac{2\sqrt{r^p}\beta}{p}\right)\sin\theta. \tag{16}$$

Applying the boundary conditions (6)–(8), we get

$$A + B - C\,I_{\frac{3}{p}}\left(\frac{2\beta}{p}\right) = 0, \tag{17}$$

$$A - 2B + C\left((\beta\sigma + 2)I_{\frac{3}{p}}\left(\frac{2\beta}{p}\right) - \beta I_{\frac{3}{p}-1}\left(\frac{2\beta}{p}\right)\right) = 0, \tag{18}$$

$$A + B\eta^3(1 - 3\eta\lambda^{-1}) + 1 = 0. \tag{19}$$

The unknowns A, B, and C are given as

$$A = (W_3 - 2W_1)\lambda\Delta_1$$
$$, B = -(W_3 + W_1)\lambda\Delta_1$$
$$, C = -3\lambda\Delta_1$$
$$,$$

$$\Delta_1 = \left(\left((W_3 + W_1)\eta^3 - W_3 + 2W_1\right)\lambda - 3(W_3 + W_1)\eta^4\right)^{-1},$$

$$W_1 = I_{\frac{3}{p}}\left(\frac{2\beta}{p}\right)$$
$$, W_2 = I_{\frac{3}{p}-1}\left(\frac{2\beta}{p}\right)$$
$$, W_3 = (\beta\sigma + 2)W_1 - \beta W_2$$
$$.$$

If $p = 0(\alpha = -2)$, we have

$$v_\phi^{(2)} = \left(C\,r^\delta + D\,r^\chi\right)\sin\theta, \tag{20}$$

where

$$\delta = \frac{-1 + \sqrt{4\beta^2 + 9}}{2} \quad \text{and} \quad \chi = \frac{-1 - \sqrt{4\beta^2 + 9}}{2}.$$

As $r \to 0$, $v_\phi^{(2)}$ must be finite for that $D = 0$. Hence,

$$v_\phi^{(2)} = C\,r^\delta\sin\theta \tag{21}$$

the system of equations satisfying the boundary conditions (6)–(8) are

$$A + B - C = 0, \tag{22}$$

$$A - 2B + C(\beta\sigma - \delta) = 0, \tag{23}$$

$$A + B\eta^3(1 - 3\eta\lambda^{-1}) + 1 = 0. \tag{24}$$

The constants involved in above system are

$$A = \lambda(\beta\sigma - \delta - 2)\,\Delta_2$$
$$, B = -\lambda(\beta\sigma - \delta + 1)\,\Delta_2$$
$$, C = -3\lambda\Delta_2$$
,

where

$$\Delta_2 = \left(\beta\sigma((\eta^3 - 1)\lambda - 3\eta^4) + ((1 - \delta)\eta^3 + \delta + 2)\lambda + 3(\delta - 3)\eta^4\right)^{-1}.$$

4 Torque

The torque experienced by the fluid on the particle is given by

$$T_z = 2\pi a^3 \int_0^{\pi} t_{r\phi}^{(1)}\Big|_{r=1} \sin^2\theta d\theta,$$

where $t_{r\phi}^{(1)}$ is the tangential stress tensor of the liquid region.

$$T_z = -8\pi a^3 \mu \Omega B. \tag{25}$$

Case $p > 0$:

$$T_z = -8\pi a^3 \mu \Omega (W_3 + W_1)\lambda\Delta_1. \tag{26}$$

Case $p = 2$:

$$T_z = -8\pi a^3 \mu \Omega (W_4\sigma + W_5)\lambda\Delta_3, \tag{27}$$

where

$$W_4 = \beta\sinh\beta - \beta^2\cosh\beta$$
$$W_5 = (\beta^2 + 3)\sinh\beta - 3\beta\cosh\beta$$
$$\Delta_3 = \left(((\eta^3 - 1)\lambda - 3\eta^4)\,W_4\sigma + W_5\eta^3(1 - 3\lambda\eta) - \beta^2\lambda\sinh\beta\right)^{-1}$$

Case $p = 0$:

$$T_z = -8\,\pi\,a^3\,\mu\,\Omega\,\lambda\,(\beta\sigma - \delta + 1)\,\Delta_2. \tag{28}$$

The torque exerted on the particle in an infinite expense
Case $p > 0$:

$$T_\infty = -8\,\pi\,a^3\,\mu\,\Omega\left(\frac{W_3 + W_1}{W_3 - 2W_1}\right). \tag{29}$$

Case $p = 2$:

$$T_\infty = -8\,\pi\,a^3\,\mu\,\Omega\left[\frac{(3 + \beta\,(\beta + \sigma))\,\sinh\beta - \beta\,(3 + \beta\,\sigma)\,\cosh\beta}{\beta\,((\beta + \sigma)\,\sinh\beta - \beta\,\sigma\,\cosh\beta)}\right].$$

This result has been obtained previously by Srinivasacharya and Prasad [6, 7].
Case $p = 0$:

$$T_\infty = -8\,\pi\,a^3\,\mu\,\Omega\left[\frac{(\beta\sigma - \delta + 1)}{\beta\sigma - \delta - 2}\right]. \tag{30}$$

The effect of wall can be calculated as the ratio of the torque exerted by the fluid on the particle in the envelope to the torque experienced by the fluid in an open space. It is defined as correction factor W. Using Eqs. (26) and (29), this becomes

$$W = \frac{T_z}{T_\infty}.$$

When $\sigma = 0$ and $b \to \infty$ (or $\eta = 0$), the torque is

$$T_\infty = -8\,\pi\,a^3\,\mu\,\Omega\left[\frac{3 + \beta^2 - 3\,\beta\,\coth\beta}{\beta^2}\right] \tag{31}$$

this well-known result for porous sphere was found by Keh and Chou [4], and Keh and Lu [5].

If $\beta \to \infty$, we get well-known result of the torque exerted by the fluid on a solid sphere [22]

$$T_\infty = -8\,\pi\,\mu\,\Omega\,a^3. \tag{32}$$

The correction factor in case of solid sphere is given by

$$W = \frac{1}{1 - \eta^3}. \tag{33}$$

The wall correction parameter W depends on the following parameters :

(i) The separation parameter, $\eta = \frac{a}{b}$.
(ii) The slip parameter $\lambda = \frac{\lambda_1 a}{\mu}$.
(iii) The stress jump coefficient σ.
(iv) The permeability parameter $k_1 = \frac{1}{\beta^2}$.
(v) The parameter p.

The effect of separation parameter η on the correction factor W is plotted in Figs. 2, 3, 4, and 5. The following observations are found:

(i) Increasing parameter p decreases the W.
(ii) W increases monotonically with an increase in the value of η.

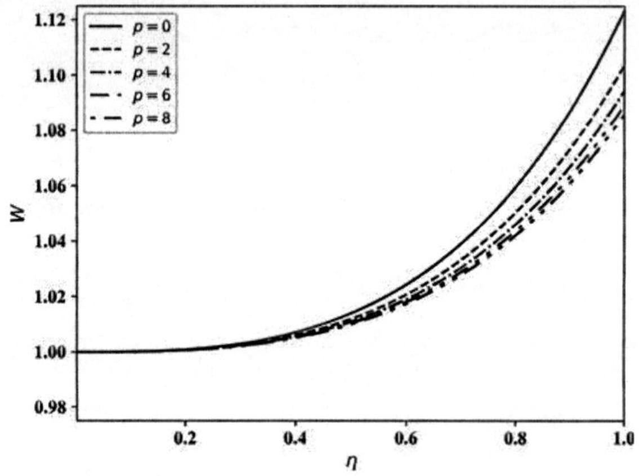

Fig. 2 Variations of W_c versus η for different values of the p, $\lambda \to \infty$, $\sigma = -0.3$, $k_1 = 2$

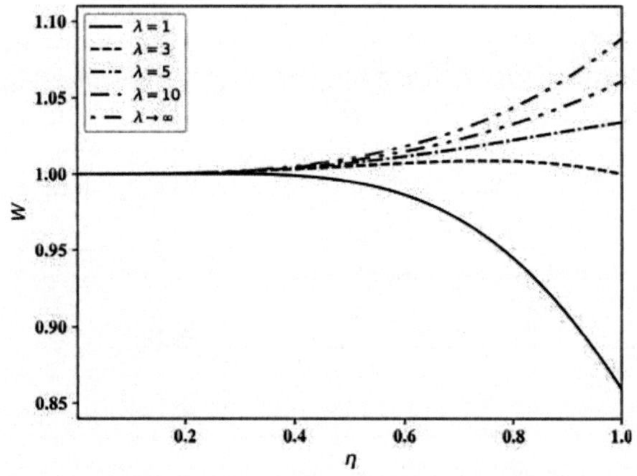

Fig. 3 Variations of W_c versus η for different values of λ, $p = 6$, $\sigma = -0.3$, $k_1 = 2$

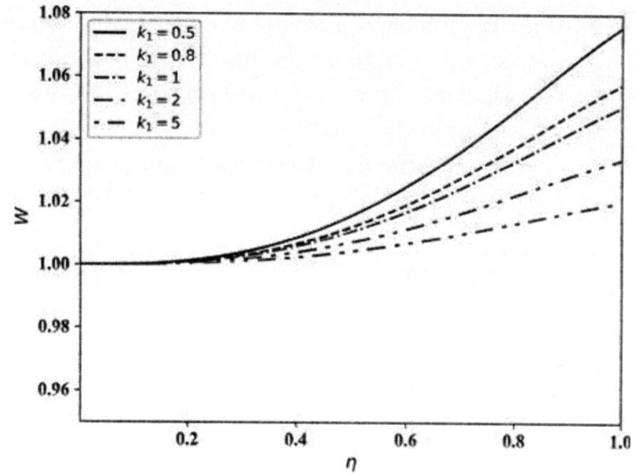

Fig. 4 Variations of W_c versus η for different values of k_1, $\lambda = 5$, $p = 6$, $\sigma = -0.3$

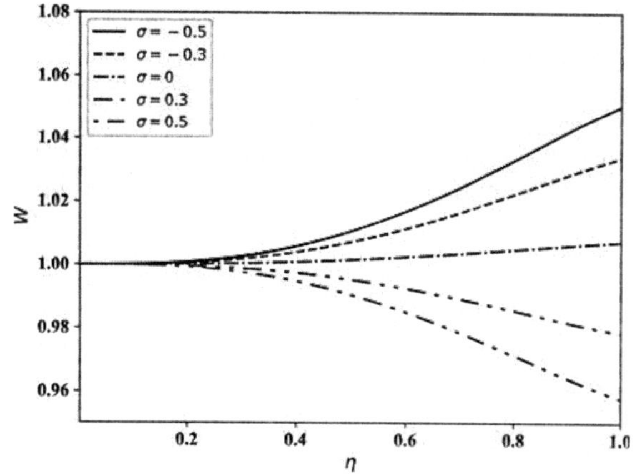

Fig. 5 Variations of W_c versus η for different values of σ, $\lambda = 5$, $p = 6$,$k_1 = 2$

(iii) W increases with an increase in the value of λ. The value of W is less than 1 for $\lambda = 1$ after particular value of η which is not physically possible.
(iv) Increasing permeability k_1 decreases the W.
(v) W is monotonically increasing function for non-positive values of σ. For positive values of σ, $W < 1$ after particular value of separation parameter η. There is significant torque or wall correction factor generated by rotation of a spherical cavity because the shear stress of cavity region is more than that of a inner region.
(vi) The boundary effect on the correction factor is stronger when the permeability k_1 is smaller, slip parameter $\lambda \to \infty$, non-positive value of stress jump coefficient σ, $p = 0$.

(vii) If the parameter α is positive, the effective viscosity decreases in the direction of viscosity of fluid. Hence, the viscosity inside the porous region is greater than the viscosity of the fluid. If the effective viscosity is less than the fluid viscosity, the situation describes the superhydrophobic surface [19]. If the parameter $\alpha = 0$, the effective viscosity is equal to fluid viscosity and the results are reported earlier in Srinivasacharya and Prasad [6, 7].

Hence, the value of parameter p has a significant effect on rotatory motion of a porous particle.

5 Conclusion

The effect of variable viscosity is examined by providing an analytical solution of rotational motion of a porous sphere located at a concentric position in a spherical cavity. The torque exerted by the fluid on particle is calculated. Hence, the effect of wall is computed as a correction factor. It is seen that the torque decreases for an increasing value of the permeability parameter or the stress jump coefficient. The results show that torque exerted by the fluid on the porous sphere of superhydrophobic surface is less than that of the torque experienced by the fluid on the porous sphere of variable viscosity $\alpha > 0$.

References

1. J. Happel, H. Brenner, *Low Reynolds Number Hydrodynamics* (Prentice-Hall, Englewood Cliffs, N.J., 1965)
2. S. Kim, S.J. Karrila, *Microhydrodynamics : Principles and Selected Applications* (Butterworth-Heinemann, Bostan, 1991)
3. R.P. Kanwal, Slow steady rotation of axially symmetric bodies in a viscous fluid. J. Fluid Mech. **10**, 17–24 (1961). https://doi.org/10.1017/S0022112061000020
4. H.J. Keh, J. Chou, Creeping motion of a composite sphere in a concentric spherical cavity. Chem. Eng. Sci. **59**, 407–415 (2004). https://doi.org/10.1016/j.ces.2003.10.006
5. H.J. Keh, Y.S. Lu, Creeping motions of a porous spherical shell in a concentric spherical cavity. J. Fluid Struct. **20**, 735–747 (2005). https://doi.org/10.1016/j.jfluidstructs.2005.03.005
6. D. Srinivasacharya, M.K. Prasad, Slow steady rotation of a porous sphere in a spherical container. J. Porous Media **15**(12), 1105–1110 (2012). https://doi.org/10.1615/JPorMedia.v15.i12.30
7. D. Srinivasacharya, M.K. Prasad, Slow rotation of a composite sphere in a concentric spherical cavity. Acta Mech. Sin. **28**(3), 653–658 (2012). https://doi.org/10.1007/s10409-012-0057-z
8. D. Srinivasacharya, M.K. Prasad, Rotation of a porous approximate sphere in an approximate spherical container. Lat. Am. Appl. Res. **45**, 107–112 (2015)
9. J.A. Ochoa-Tapia, S. Whitaker, Momentum transfer at the boundary between a porous medium and a homogeneous fluid-I, theoretical development. Int. J. Heat Mass Tran. **38**, 2635–2646 (1995a). https://doi.org/10.1016/0017-9310(94)00346-W
10. J.A. Ochoa-Tapia, S. Whitaker, Momentum transfer at the boundary between a porous medium and a homogeneous fluid-II, comparison with experiment. Int. J. Heat Mass Tran. **38**, 2647–2655 (1995b). https://doi.org/10.1016/0017-9310(94)00347-X

11. A.V. Kuznetsov, Analytical investigation of the fluid flow in the interface region between a porous media and a clear fluid in channels partially filled with porous medium. Appl. Sci. Res. **56**, 53–67 (1996). https://doi.org/10.1007/BF02282922
12. A.V. Kuznetsov, Analytical investigation of Couette flow in a composite channel partially filled with a porous media and partially with a clear fluid. Int. J. Heat Mass Tran. **41**, 2556–2560 (1998). https://doi.org/10.1016/S0017-9310(97)00296-2
13. G.P.R. Sekhar, A. Bhattacharyya, Stokes flow inside a porous spherical shell: Stress jump boundary condition. Z. Angew. Math. Phys. **56**, 475–496 (2005). https://doi.org/10.1007/s00033-004-2115-2
14. A.C. Srivastava, P. Saxena, Rotation of a solid sphere in a viscous fluid bounded by a concentric spherical porous medium. J. Porous Media **11**(6), 575–588 (2008). https://doi.org/10.1615/JPorMedia.v11.i6.50
15. E.A. Ashmawy, Rotary oscillation of a composite sphere in a concentric spherical cavity using slip and stress jump conidtions. Eur. Phys. J. Plus **130**, 163 (2015). https://doi.org/10.1140/epjp/i2015-15163-8
16. M.K. Prasad, M. Kaur, D. Srinivasacharya, Slow steady rotation of an approximate sphere in an approximate spherical container with slip surfaces. Int. J. Appl. Comput. Math **3**, 987–999 (2017). https://doi.org/10.1007/s40819-016-0151-1
17. M. Awasthi, S. Datta, N. Pandya, Slow rotation of a spherical cavity enclosing a concentric porous sphere of variable permeability. Jnanabha **48**(1), 92–101 (2018)
18. A. Filippov, Y. Koroleva, Viscous flow through a porous medium filled by liquid with varying viscosity. Bul. Acad. Stiinte Repub. Mold. Mat. **85**(3), 74–87 (2017)
19. A. Filippov, Y. Koroleva, On a hydrodynamic permeability of a system of coaxial partly porous cylinders with superhydrophobic surfaces. Appl. Math. Comp. **338**, 363–375 (2018). https://doi.org/10.1016/j.amc.2018.06.034
20. P.O. Ryzhikh, A.N. Filippov, A cell model of a membrane with allowance for variable viscosity of liquid in porous shells of spherical grains. Colloid J. **80**(2), 199–206 (2018). https://doi.org/10.1134/S1061933X18020072
21. H.C. Brinkman, A calculation of viscous force exerted by a flowing fluid on dense swarm of particles. Appl. Sci. Res. **A1**, 27–34 (1957). https://doi.org/10.1007/BF02120313
22. H. Lamb, *Hydrodynamics* (Cambridge University Press, Bentley House, London, 1933)

Arbitrary Amplitude Double Layers in Dust Kinetic Alfvén Wave Plasmas with κ-Distributed Electrons

Latika Kalita, Ranjit Kumar Kalita, and Jnanjyoti Sarma

Abstract An investigation has been done to study the effects of κ-distributed electrons on the waves with arbitrary amplitude of magnetized plasma in the presence of warm ions and negatively charged dust particles. The kappa distribution ($\kappa >$ 3/2), which represents a velocity distribution, contains high energy tail but nears to the Maxwell–Boltzmann distribution as $\kappa \to \infty$. In this work, by using the non-perturbation technique (i.e., Sagdeev pseudopotential analysis), a suitable mathematical expression for the arbitrary amplitude double layers in dust kinetic Alfvén waves is formulated. Using the standard numerical values for different plasma parameter related to it, the Sagdeev Potential (SP) has been calculated for the plasma waves. It is observed that the spatial index κ plays a substantial role in finding the size and shape of double layers. Depth of double layers increases with decreasing k_z, keeping ω fixed. For this parameter set, the double layer does not exist when $M \leq 1.1$. This theoretical study may be helpful for explaining some of recent in situ observations (e.g., Freja, Cassini) in space plasmas, where one of its constituents is kappa distributed electron.

Keywords Solitary waves · Double layers

L. Kalita (✉)
Department of Mathematics, Kamrup Polytechnic, Baihata Chariali 781381, Assam, India
e-mail: latika84k@rediffmail.com

R. K. Kalita
Department of Mathematics, Morigaon College, Morigaon 782105, Assam, India
e-mail: kalitaranjit@yahoo.com

J. Sarma
Department of Mathematics, R. G. Baruah College, Guwahati 781025, Assam, India
e-mail: jsarma_2001@yahoo.com

S. R. Mishra et al. (eds.), *Recent Trends in Applied Mathematics*, Lecture Notes in Mechanical Engineering, https://doi.org/10.1007/978-981-15-9817-3_16

1 Introduction

The well-known Alfvén waves are a basic physical phenomenon in magnetized plasma that leads to a different kind of physical process in the space plasmas. For example, the turbulence, plasma heating, acceleration along the field lines, wave particle interactions, and generation of geometric perturbations. Many theoretical studies as well as experimental works have been thus dedicated to the study of Alfvén waves and brought out many aspects of Alfvén waves [1–4]. The observational data obtained from the artificial satellites like Freja have showed the formation and propagation of low-frequency auroral electromagnetic waves similar to Kinetic Alfvén waves (KAWs) ensuing in density pulses [5]. The structures so obtained, can be distinguished as Solitary waves (SWs) or Double layers (DLs). These waves propagate in fluid plasmas and can be studied with the help of Reductive Perturbation Technique (RPT) [6] or Sagdeev's Pseudopotential method (SP) [7–9]. For first time, Hasegawa and Mima [10] and Yu and Shukula [11] were studied analytically to explain their numerical behavior of SKAWs by using Sagdeev's potential [7] approach. Double layer [12] is a mechanism which is exclusively responsible for particles energization both in space plasmas and laboratory plasmas. Double layers are associated with currents and are of interest in astrophysics as a direct intends to accelerate the particles. They can hold a local area of parallel electric field leading to the magneto-hydrodynamic (MHD) relief to a large scale magnetic field. The studies linked to double layers are becoming impetus due to their contributions in any system of plasma, which includes discharge tubes, space plasmas, etc. to the Birkeland currents supplying the Earth's aurora. Because of the potential drop-down across a DL, the acceleration of electrons and positive ions take place in opposite directions (regardless of the width of the DL). The acceleration of charged particles may result in beams or jets of the charged particles. Investigations as well as studies on double layers had improved in a big way with the launching of the triple plasma device [13]. Temerin et al. [14] were observed double layers in space plasma for the first time. The astrophysical and space plasmas are mostly found as non-Maxwellian with high energy tail. The measurements of such profiles of electron energy spectra, done by the spacecraft, have been fruitfully modeled with superthermal electrons. The kappa (κ) distribution, as it has an empirical fit to the observed particle distributions, was suggested by Vasyliunas [15] for the first to model solar wind data which was obtained from OGO 1 and OGO 2 satellites. Kappa distribution is not only useful to analyze and interpret spacecraft data on the Earth's magneto-sphere plasma sheet but also for the solar wind, Jupiter, Saturn, and planetary magnetospheres. The value of the κ (spectral index) provides the slope of the energy spectrum of the superthermal particles, which form the tail of the velocity distribution function. Low values of κ represent distributions with a huge part of superthermal particles which are generally named as a "hard" spectrum [15]. For very big values of κ ($\kappa \to \infty$), the velocity distribution function nears to Maxwellian distribution. Remarkable uses of the κ distribution function admit, for example, an interpretation of the observations in foreshock [16] of the Earth (where $3 < \kappa_e < 6$) and the models for solar wind having coronal electrons [17–19] fulfilling the condition $2 < \kappa_e < 6$. The natures of ion acoustic solitary waves having arbitrary amplitude, obliquely propagation in magnetized plasma with the electrons following κ distribution is investigated by Sultana et al. [20]. They have considered those plasmas which have two components collisionless but embedded in a uniform magnetic field comprised of cold ions describes by the fluid moment equations and electrons following κ velocity distribution. Yet, to the authors' knowledge, the formation of double layers is not investigated. Data obtained from Freja satellite [21] depicted that the low-frequency auroral electromagnetic fluctuations ensuing on strong electric spikes that show to interpreted as density pulses displaying kinetic theory study of waves in space plasmas. The significance of kinetic Alfvén waves for the study of coupling has been seen between the ionosphere and magnetosphere [22]. An observable fact related to the Alfvén variation waves in auroral region can also be analyzed by in situ measurements, where the variation of electric as well as magnetic fields and the correlations between them, for the propagation of the Alfvén waves can be determined.

Moreover it is observed that in space plasmas, the presence of charged dust particles is very common with sizes from 10 nm to 10 μm and masses of them can vary from 10^{-11} to 10^{-16} gm in background plasma of neutral gas particles, ions, electrons, etc. Such plasma is known as dusty plasma, a naturalistic medium in the studies associated with interplanetary space dust, comets, planetary rings, planetary magnetospheres, interstellar clouds, etc. A significant phenomenon has been developed in the process of dust particles' charging in plasmas and its effects, not only on dusty plasma dynamics in astrophysical surroundings but also in the processing of laboratory plasma and semiconductor manufacturing [23]. The charges on the dust particles may change because of the wave motion-induced electrons and ions currents flowing onto the grain surface. Consequently the charge of dust grain turns into a new dynamical variable, passing to certain new results in dusty plasmas, which are generally missing in the general multicomponent plasmas. Many heuristic works are found to study salient properties of coherent structures occurring in Earth's space region and auororal ionosphere filled up with dusty plasmas.

2 Basic Equations

In the present work, we consider a set of relevant fluid model equations, which contain relativistic effects as the part of the dynamics of superthermal electron. For the sake of in-depth investigations of arbitrary amplitude double layers in dust kinetic Alfven waves in the environment of space and astrophysical plasma, the universal Newton's equation for $j = i, e$, respectively, for the ions and electrons goes as $m_j n_j \left[\partial_t \vec{u}_j + \vec{u}_j \cdot \nabla \vec{u}_j + \nabla \vec{p}\right] = q_j n_j \left[\vec{E} + \vec{u}_j \times \vec{B}\right]$. This equation shows the variations of relativistic *momentum* and thermal *pressure* of the plasma. We take a plasma model in the presence of uniform external magnetic field $\boldsymbol{B_0} = B0\hat{z}$ along z-axis and with $\beta << m_e/m_i$, where the fluctuations are presumed in the (x, z) plane.

The introductory equations governing the dynamics of nonlinear Alfvén waves in low-β plasma are given by

$$\frac{\partial n_e}{\partial t} + \frac{\partial}{\partial z}(n_e v_{ez}) = 0 \tag{1}$$

$$\frac{\partial n_i}{\partial t} + \frac{\partial}{\partial x}(n_i v_{ix}) + \frac{\partial}{\partial z}(n_i v_{iz}) = 0 \tag{2}$$

$$\frac{\partial v_{ix}}{\partial t} + v_{ix}\frac{\partial v_{ix}}{\partial x} + v_{iz}\frac{\partial v_{ix}}{\partial z} = \frac{\beta}{2}\left(-\frac{\partial \phi}{\partial x} - \frac{\sigma}{n_i}\frac{\partial n_i}{\partial x}\right) + v_{iy} \tag{3}$$

$$\frac{\partial v_{iy}}{\partial t} + v_{ix}\frac{\partial v_{iy}}{\partial x} + v_{iz}\frac{\partial v_{iy}}{\partial z} = -v_{ix} \tag{4}$$

$$\frac{\partial v_{iz}}{\partial t} + v_{ix}\frac{\partial v_{iz}}{\partial x} + v_{iz}\frac{\partial v_{iz}}{\partial z} = \frac{\beta}{2}\left(-\frac{\partial \psi}{\partial z} - \frac{\sigma}{n_i}\frac{\partial n_i}{\partial z}\right) \tag{5}$$

$$\frac{\partial^4(\phi - \psi)}{\partial x^2 \partial z^2} = \frac{2}{\beta}\left[\frac{\partial^2}{\partial t^2}(n_e) + \frac{\partial^2}{\partial t \partial z}(n_i v_{iz})\right] \tag{6}$$

The simplest form of electrons having κ distribution is given by

$$n_e = \left(1 - \frac{\psi}{\kappa - \frac{3}{2}}\right)^{-\kappa + \frac{1}{2}} \tag{7}$$

As the dust grains have heavy mass and carry negative charges, so they are static. In this case we consider the quasi-neutrality condition [24] as

$$\delta_e n_e + z\delta_d - n_i = 0 \tag{8}$$

And the charge neutrality at equilibrium reads

$$\delta d = 1 - \delta e \tag{9}$$

where n_e and n_i represent the densities of electron and ion, respectively. In the derivation of Eq. (6) we have used continuity equation for the electron (n_e), i.e., Eq. (1). The normalization for densities is done by the equilibrium plasma density n_O, distance by gyroradius $\rho_s = c\omega_{p_i}^{-1}$, the potential by $\frac{KT_e}{e}$, the velocities by Alfvén velocity $v_A = \frac{cB_0}{(4\pi n_0 m_i)^{\frac{1}{2}}}$, and time t by ion gyroperiod Ω_i^{-1}. Here Ω_i represent the gyrofrequency of ion and T_e symbolizes for the electron temperature, σ is the temperature ratio (ion to electrons) and $\beta = \frac{8\pi n_0 T_i}{c^2 B_0^2}$, (is the ratio of kinetic to the magnetic pressure), is a significant parameter for examining the efficiency of magnetic fields in close proximity of plasma in tokomak's investigation. In low-β supposition in case of high magnetic pressure, the electric field can be expressed with the help of two potential ϕ and ψ. z (dust charge number), $\delta e = \frac{ne0}{ni0}$, $\delta d = \frac{znd0}{ni0}$, $Q = m_e/m_i$ (mass ratio of electron to ion), *the* subscripts i and e represent ions and electrons, respectively. To obtain a dispersion relation we have utilized Poisson's equation as

$$\frac{\partial^2\phi}{\partial x^2} + \frac{\partial^2\psi}{\partial z^2} = -4\pi e(n_i - n_e) \tag{10}$$

Using linearized Eqs. (1)–(7) and (10) and after some algebraic calculations, by employing the following variables:

$$V_A = \frac{eB_0}{\sqrt{4\pi n_{i0} m_i}}, \Omega_c = \frac{eB_0}{m_i}, C_s^2 = \frac{T_i}{m_i}, Q = \frac{m_e}{m_i}, \omega_{pi}^2 = \frac{4\pi n_{i0} e^2}{m_i}, \omega_{pe}^2 = \frac{4\pi n_{e0} e^2}{m_e}, \frac{\omega_{pi}^2}{\Omega_c^2} = \frac{c^2}{V_A^2}, \frac{\omega_{pe}^2}{\omega_{pi}^2} = Q$$

We have found the dispersion relation as given below:

$$\left(1 + \frac{\Omega_c^2}{V_A^2}\left(\frac{1}{k_x^2 k_z^2} + \frac{\sigma k_z^2 C_s^2}{\omega}\right) + \frac{Q\omega_{pe}^2}{C_s^2}\left(\frac{2\kappa - 1}{2\kappa - 3}\right)\left(1 - \frac{\omega^2}{c^2}\right)\right)$$
$$\left(1 - \frac{\sigma k_z^2 C_s^2}{\omega} - \frac{\sigma\omega C_s^2}{\omega^2 - \Omega_c^2}\left(\frac{\Omega_c^2}{V_A^2} + \frac{k_x^2}{\omega}\right)\right)$$
$$= \frac{\omega_{pi}^2 k_z^2}{\omega^2} + \frac{\left(B - \sigma C_s^2\right)}{\left(\omega^2 - \Omega_c^2\right)k_z^2}\frac{\Omega_c^2}{V_A^2}\left(k_x^2 k_z^2 + \frac{\Omega_c^2}{k_x^2 V_A^2}\right) - \left(\frac{2\kappa - 1}{2\kappa - 3}\right)\frac{\omega^2 Q\omega_{pe}^2}{c^2 C_s^2} \tag{11}$$

where $B = \frac{1}{m_i c^2}$, $\sigma = \frac{T_i}{T_e}$. From Eq. (11), we notice that the dispersion relation (11) depends on parameters k_x, k_z, m_i, c,ω,σ and other plasma parameters.

The object of our study is directly concerned to the formation of kinetic Alfvén waves as in the dispersion curves we find out the wave number k_z with larger values pointing to the dispersion created.

3 Formulation of the Sagdeev Potential Equation

In order to obtain one-dimensional time-stationary planar solutions, the dependence on the quantity

$$\eta = xk_x + zk_z - Mt \tag{12}$$

where $M = \frac{v}{v_A}$ = ratio of wave's phase velocity in the unit of Alfven velocity which is also called normalized Mach number and k_x, k_z are direction cosines related by

$$k_x^2 + k_z^2 = 1 \tag{13}$$

Employing the coordinate transformation to the Eqs. (2–6), the set fluid equations becomes

$$k_x v_{ix} + k_z v_{iz} = M\left(1 - \frac{1}{n_i}\right) \tag{14}$$

$$-\frac{M}{n_i}\frac{\partial v_{ix}}{\partial \eta} + \frac{\sigma k_x \beta}{2n_i}\frac{\partial n_i}{\partial \eta} = -\frac{k_x \beta}{2}\frac{\partial \phi}{\partial \eta} + v_{iy} \tag{15}$$

$$\frac{M}{n_i}\frac{\partial v_{iy}}{\partial \eta} = v_{ix} \tag{16}$$

$$-\frac{M}{n_i}\frac{\partial v_{iz}}{\partial \eta} + \frac{\sigma k_z \beta}{2n_i}\frac{\partial n_i}{\partial \eta} = -\frac{k_z \beta}{2}\frac{\partial \psi}{\partial \eta} \tag{17}$$

$$k_x^2 k_z^2 \frac{\partial^4}{\partial \eta^4}(\phi - \psi) = \frac{2}{\beta}\left[M^2 \frac{\partial^2 n_e}{\partial \eta^2} - k_z M \frac{\partial^2}{\partial \eta^2}(n_i v_{iz})\right] \tag{18}$$

Now (15) $\times\, k_x$ + (17) $\times\, k_z$ and using (14) we get

$$-\frac{M^2}{n_i^3}\frac{\partial n_i}{\partial \eta} + \frac{\sigma\beta}{2n_i}\frac{\partial n_i}{\partial \eta} = -\frac{\beta}{2}\left[k_x^2 \frac{\partial \phi}{\partial \eta} + k_z^2 \frac{\partial \psi}{\partial \eta}\right] + k_x v_{iy} \tag{19}$$

Diff. w. r. t. "η" we get

$$\frac{\partial}{\partial \eta}\left[\left(\frac{M^2}{n_i^3} - \frac{\beta}{2}\frac{\sigma}{n_i}\right)\frac{\partial n_i}{\partial \eta}\right] = \frac{\beta}{2}\left[k_x^2 \frac{\partial^2 \phi}{\partial \eta^2} + k_z^2 \frac{\partial^2 \psi}{\partial \eta^2}\right] - \frac{n_i}{M}\left[M\left(1 - \frac{1}{n_i}\right) - k_z v_{iz}\right] \tag{20}$$

From Eq. (7)

$$\frac{\partial \psi}{\partial \eta} = \left(\frac{\kappa - \frac{3}{2}}{\kappa - \frac{1}{2}}\right) n_e^{\frac{\kappa + \frac{1}{2}}{-\kappa + \frac{1}{2}}} \frac{\partial n_e}{\partial \eta} \tag{21}$$

Using (21), (17) becomes

$$\Rightarrow \frac{M}{n_i}\frac{\partial v_{iz}}{\partial \eta} = \frac{\sigma k_z \beta}{2n_i}\frac{\partial n_i}{\partial \eta} + \frac{k_z \beta}{2}\left(\frac{\kappa - \frac{3}{2}}{\kappa - \frac{1}{2}}\right) n_e^{\frac{\kappa + \frac{1}{2}}{-\kappa + \frac{1}{2}}} \frac{\partial n_e}{\partial \eta} \tag{22}$$

From now we shall employ the following quasi-neutrality condition,

$$\delta ene + z\delta d - ni = 0$$

$$\Rightarrow ni = \delta ene + z\delta d \text{ and } \delta d = 1 - \delta_e$$

Equation (22) becomes

$$v_{iz} = \frac{k_z \sigma \beta \delta e}{2\omega}(ne - 1) + \frac{\beta k_z}{2\omega}\left[\left(n_e^{\frac{\frac{3}{2}-\kappa}{\frac{1}{2}-\kappa}} - 1\right) - \left(\kappa - \frac{3}{2}\right) z(1 - \delta e)\left(n_e^{\frac{1}{\frac{1}{2}-\kappa}} - 1\right)\right] \tag{23}$$

Using (23) in Eq. (20) we get

$$\Rightarrow \frac{\beta}{2}\left[k_x^2 \frac{\partial^2 \phi}{\partial \eta^2} + k_z^2 \frac{\partial^2 \psi}{\partial \eta^2}\right]$$
$$= \frac{\partial}{\partial \eta}\left[\left(\frac{M^2 \delta e}{(\delta ene + z(1 - \delta e))^3} - \frac{\sigma \beta \delta e}{2(\delta ene + z(1 - \delta e))}\right)\frac{\partial ne}{\partial \eta}\right] + (\delta ene + z(1 - \delta e) - 1)$$
$$- \frac{(\delta ene + z(1 - \delta e))}{M^2}$$
$$k_z^2\left[\frac{\sigma \beta \delta e}{2}(ne - 1) + \frac{\beta}{2}\left[\left(n_e^{\frac{\frac{3}{2}-\kappa}{\frac{1}{2}-\kappa}} - 1\right) - \left(\kappa - \frac{3}{2}\right) z(1 - \delta e)\left(n_e^{\frac{1}{\frac{1}{2}-\kappa}} - 1\right)\right]\right] \tag{24}$$

Equation (18), integrating twice and using boundary conditions $\phi = \psi = v_{iz} \to 0$ and $ne \to 1$ we get

$$k_x^2 k_z^2\left(\frac{\partial 2\phi}{\partial \eta 2} - \frac{\partial 2\psi}{\partial \eta 2}\right) = \frac{2}{\beta}\left[M^2(ne - 1) - \frac{k_z^2 \sigma \beta \delta e}{2}(\delta ene + z(1 - \delta e))(ne - 1) - \right.$$
$$\left.\frac{\beta k_z^2(\delta ene + z(1 - \delta e))}{2}\left[\left(n_e^{\frac{\frac{3}{2}-\kappa}{\frac{1}{2}-\kappa}} - 1\right) - \left(\kappa - \frac{3}{2}\right) z(1 - \delta e)\left(n_e^{\frac{1}{\frac{1}{2}-\kappa}} - 1\right)\right]\right] \tag{25}$$

Now (24) $\times\ k_z^2 -$ (25) $\times\ \frac{\beta}{2}$ and using (21) we get

$$\Rightarrow \frac{\partial}{\partial \eta}$$
$$\left[\left(\frac{\beta k_z^2}{2}\left(\frac{\kappa - \frac{3}{2}}{\kappa - \frac{1}{2}}\right) n_e^{\frac{\kappa + \frac{1}{2}}{-\kappa + \frac{1}{2}}} - k_z^2\left(\frac{M^2 \delta e}{(\delta ene + z(1 - \delta e))^3} - \frac{\sigma \beta \delta e}{2(\delta ene + z(1 - \delta e))}\right)\right)\frac{\partial n_e}{\partial \eta}\right]$$

$$
\begin{aligned}
&= \frac{\beta}{2} k_z^2 (\delta ene + z(1-\delta e))\left(1 - \frac{k_z^2}{M^2}\right) \\
&\left[\left(n_e^{\frac{\frac{3}{2}-\kappa}{\frac{1}{2}-\kappa}} - 1\right) - \left(\kappa - \frac{3}{2}\right) z(1-\delta e)\left(n_e^{\frac{1}{\frac{1}{2}-\kappa}} - 1\right)\right] + \\
&\quad \frac{\sigma\beta\delta e}{2} k_z^2 (\delta ene + z(1-\delta e))(ne-1)\left(1 - \frac{k_z^2}{M^2}\right) \\
&\quad + k_z^2(\delta ene + z(1-\delta e) - 1) - M^2(ne-1)
\end{aligned}
\tag{26}
$$

We now multiply both sides of Eq. (26) by the term of the left hand within the parentheses, which gives

$$
\frac{1}{2}\left(\frac{dn_e}{d\eta}\right)^2 + K(n_e) = 0 \tag{27}
$$

where $K(n_e) = \lambda(n_e)\mu(n_e)$

$$
\lambda(n_e) = \frac{-1}{\left[\frac{\beta}{2} k_z^2 \left(\frac{\kappa - \frac{3}{2}}{\kappa - \frac{1}{2}}\right) n_e^{\frac{\frac{1}{2}+\kappa}{\frac{1}{2}-\kappa}} - k_z^2\left(\frac{M^2\delta e}{(\delta en_e + z(1-n_e))^3} - \frac{\sigma\beta\delta e}{2(\delta en_e + z(1-n_e))}\right)\right]^2}
$$

$$
\begin{aligned}
\mu(n_e) = &-\frac{k_z^4\beta^2(k_z^2 - M^2)}{8(-5+2\kappa)M^2} \\
&\left[2(-3+2\kappa)\left(-1 + ne^{1+\frac{4}{1-2\kappa}}\right) z(1-\delta e) + (-5+2\kappa)\left(-1 + ne^{2+\frac{4}{1-2\kappa}}\right)\delta e\right] \\
&+ \frac{k_z^4\beta^2(M^2 - k_z^2)}{8M^2}\left[(-3+2\kappa)\left(-1 + ne^{\frac{2}{1-2\kappa}}\right) z(1-\delta e) - 2\left(-1 + ne^{1+\frac{1}{1-2\kappa}}\right)\delta e\right] \\
&\quad + \frac{(3-2\kappa)^2 k_z^4\beta^2 z(k_z^2 - M^2)(1-\delta e)}{32(-5+2\kappa)M^2} \\
&\quad \left[(-5+2\kappa)z(1-\delta e) - 4\delta e + ne^{\frac{4}{1-2\kappa}}\{(5-2\kappa)z(1-\delta e) + 4ne\delta e\}\right] \\
&\quad + \frac{(-3+2\kappa)k_z^4\beta^2 z(k_z^2 - M^2)(1-\delta e)}{16M^2} \\
&\quad \left[(-3+2\kappa)\left(-1 + ne^{\frac{2}{1-2\kappa}}\right) z(1-\delta e) - 2\left(-1 + ne^{1+\frac{2}{1-2\kappa}}\right)\delta e\right] \\
&- \frac{k_z^4\beta\left(1 - \frac{k_z^2}{M^2}\right)M^2}{4(\delta ene + z(1-\delta e))} \\
&\left[2 - (-1+2\kappa)ne^{\frac{3-2\kappa}{1-2\kappa}}\, Hypergeometric\right. \\
&\left. 2F1\left[\frac{2}{-1+2\kappa}, \frac{3-2\kappa}{-1+2\kappa}, \frac{1+2\kappa}{-1+2\kappa}, \frac{z(1-\delta e)}{z(1-\delta e) + ne\delta e}\right]\left(\frac{ne\delta e}{z(1-\delta e) + ne\delta e}\right)^{\frac{3-2\kappa}{-1+2\kappa}}\right] \\
&+ \frac{k_z^4\beta\left(1 - \frac{k_z^2}{M^2}\right)M^2}{4(\delta e + z(1-\delta e))}
\end{aligned}
$$

$$[2-(-1+2\kappa)Hypergeometric$$

$$2F1\left[\frac{2}{-1+2\kappa},\frac{3-2\kappa}{-1+2\kappa},\frac{1+2\kappa}{-1+2\kappa},\frac{z(1-\delta e)}{z(1-\delta e)+\delta e}\right]\left(\frac{\delta e}{z(1-\delta e)+\delta e}\right)^{\frac{3-2\kappa}{-1+2\kappa}}\Bigg]$$

$$+\frac{k_z^4\sigma\beta^2\delta e}{16}\left(1-\frac{k_z^2}{M^2}\right)$$

$$\left[4+\frac{1-2\kappa}{-1+\kappa}-4ne+\frac{(-1+2\kappa)ne^{\frac{4(-1+\kappa)}{-1+2\kappa}}}{-1+\kappa}\right]$$

$$+\frac{(-3+2\kappa)k_z^4\beta z(k_z^2-M^2)(1-\delta e)}{4(1+2\kappa)}\left[\frac{1}{(\delta e+z(1-\delta e))}\{1+2\kappa-(-1+2\kappa)\times\right.$$

$$Hypergeometric\,2F1\left[\frac{2}{-1+2\kappa},1+\frac{2}{-1+2\kappa},\frac{4\kappa}{-1+2\kappa},\frac{z(1-\delta e)}{z(1-\delta e)+\delta e}\right]$$

$$\left(\frac{\delta e}{z(1-\delta e)+\delta e}\right)^{\frac{2}{-1+2\kappa}}+\frac{1}{(ne\delta e+z(1-\delta e))}\times$$

$$\Bigg\{-1-2\kappa+(-1+2\kappa)ne^{\frac{2}{-1+2\kappa}}Hypergeometric\,2F1$$

$$\left[\frac{2}{-1+2\kappa},1+\frac{2}{-1+2\kappa},\frac{4\kappa}{-1+2\kappa},\frac{z(1-\delta e)}{z(1-\delta e)+\delta ene}\right]\left(\frac{ne\delta e}{z(1-\delta e)+ne\delta e}\right)^{\frac{2}{-1+2\kappa}}\Bigg\}$$

$$\left.+\frac{k_z^4\sigma\beta^2 z\delta e(1-\delta e)}{8M^2}\left(k_z^2-M^2\right)\left[-2+(3-2\kappa)ne+(-1+2\kappa)ne^{1+\frac{2}{1-2\kappa}}\right]\Bigg\}\right]$$

$$+\frac{k_z^4\sigma\beta^2\delta e}{16(-1+\kappa)M^2}\left(k_z^2-M^2\right)$$

$$\Big[2(-1+\kappa)(-1+2\kappa)z(1-\delta e)+(1-2\kappa)\delta e-ne^{\frac{2}{1-2\kappa}}[2(-1+\kappa)\times$$

$$(-3+2\kappa+2ne)z(1-ne)+(4+2\kappa(-2+ne)-3ne)ne\delta e]]$$

$$+\frac{k_z^4\sigma\beta}{2}\left(k_z^2-M^2\right)$$

$$\left[-1-\mathrm{Log}[z(1-\delta e)+\delta e]+\mathrm{Log}[z(1-\delta e)+ne\delta e]+\frac{z(1-\delta e)+\delta e}{z(1-\delta e)+ne\delta e}\right]$$

$$-\frac{k_z^4\sigma^2\beta^2\delta_e^2}{8M^2}(-1+ne)^2\left(k_z^2-M^2\right)$$

$$-\frac{k_z^2\beta M^2}{4}\left[1+2ne^{1+\frac{2}{1-2\kappa}}-3ne^{\frac{2}{1-2\kappa}}+2\kappa\left(-1+ne^{\frac{2}{1-2\kappa}}\right)\right]$$

$$+\frac{k_z^4\beta}{4}\Big[3-2\kappa+(-3+2\kappa)z(1-\delta e)-2\delta e+ne^{\frac{2}{1-2\kappa}}$$

$$(-3+2\kappa+(3-2\kappa)z(1-\delta e)+2ne\delta e)]$$

$$+\frac{k_z^4}{2}\Bigg[\sigma\beta(\text{Log}[z(1-\delta e)+\delta e]-z(1-\delta e)-\delta e)+\frac{M^2(1-2z(1-\delta e)-2\delta e)}{(z(1-\delta e)+\delta e)^2}+$$

$$\sigma\beta(-\text{Log}[z(1-\delta e)+ne\delta e]+z(1-\delta e)+ne\delta e)+\frac{M^2(-1+2z(1-\delta e)+2ne\delta e)}{(z(1-\delta e)+ne\delta e)^2}\Bigg]$$

$$+\frac{k_z^2 M^4\delta e(-1+ne)^2}{2(z(1-\delta e)+\delta e)(z(1-\delta e)+ne\delta e)^2}$$

$$-\frac{k_z^2\sigma\beta M^2}{2\delta e}\Big[z(\text{Log}[z(1-\delta e)+\delta e]-\text{Log}[z(1-\delta e)+ne\delta e])(1-\delta e)$$

$$+(-1+ne+\text{Log}[z(1-\delta e)+\delta e]-\text{Log}[z(1-\delta e)+ne\delta e])\delta e\Big]$$

Equation (27) can be interpreted as the analog "energy-balance" equation in mechanics for the different dynamical problem of motion of an unit mass turned up at "position" ne and developing in "time" η with "velocity" $\frac{dne}{d\eta}$ in a potential $K(ne)$. The hypergeometric 2F1 can be obtained as a solution of linear ordinary differential equation of the second order, like $\left(\frac{d^2w}{dz^2}+p\frac{dw}{dz}+qw\right)=0$.

4 Conditions for the Existence of Solitary Waves and Double Layers

The solutions of the Sagdeev potential equation are generally found with reference to favored frame of references for investigations of solitary waves as well as double layers. The Sagdeev method is a general one, for the propagating-wave solutions of arbitrary amplitude are reachable through macroscopic plasma models [25] for nonlinear waves. Seeking for exact solutions for propagating-wave of nonlinear plasma waves is the widely known way for theoretical plasma research in recent times related to contemporaneous laboratory experimental studies of coherent plasma. The dynamical conditions for propagating SWs and DLs demand a procedure like reflection of a pseudo particle from a point of maximum n in the potential well. Equation (27) is the main tool in our research works and it could describe many salient features of plasma-acoustic mode. For existence of solitary waves, the following conditions must be satisfied

(i) $K(n_e)=0$ at $n_e=1$ and $n_e=N$ (maximum variation of n_e).
(ii) $\left(\frac{dK}{dn_e}\right)_{n_e=1}=0$ but $\left(\frac{dK}{dn_e}\right)_{n_e=N}\neq 0$, Additional requirement is that the double root at $n_e=1$ is a local maximum, which gives us
(iii) $\left(\frac{d^2K}{dn_e^2}\right)_{n_e=1}<0$

For the double layer $K(n_e)$ should have a maximum at $n_e = N(\neq 1)$ which entails $K(n_e) = 0$ for $n_e = N$

$$\left(\frac{dK}{dn_e}\right)_{n_e=N} = 0 \text{ and } \left(\frac{d^2K}{dn_e^2}\right)_{n_e=N} < 0$$

5 Results and Discussions

Interpretations of observed results in satellite experiments are often ambiguous and disputed among researchers. The changes that the space vehicles may miss observation of vital areas of space, lead to hidden causes of space phenomena. Such are the situations where mathematical modeling gives a high degree of reliability towards explaining space phenomena in plasma. Hence consideration of the nonlinearity of the plasma models is of prime importance. Different types of plasma waves, their interactions and related formation of leading structures which are basically nonlinear with coherence to be studied in this pursuit. Alfvén waves in space and astrophysical dusty plasmas by introducing some basic physical concepts of the dust environments. It was proven that the dispersion relation of Alfven waves is changed importantly in the dusty plasma system because the charged grains in a magnetized plasma are extremely coupled to the waves because of cyclotron resonances. The computations show the existence of rarefactive DLs under the effect of superthermal electrons. Here the supersonic propagation of the structures of the dust kinetic Alfvén DLs under the effect of superthermality of the electrons with spectral index kappa to the relevant in space plasma and astrophysical plasmas. The energy enhancement can be encountered by computing the widths of the structures. Use of the charge neutrality condition Eq. (8) is being made during our mathematical calculation to arrive at this potential in terms of $z = 1$ for the appearance of dust particles in the plasma. Without the effect of dust consideration [18, 19], some results show the formation of compressive double layers with similar ranges of kappa values between 4 and 6. We would spotlight our outcomes on the formation of DLs only.

Depth of DLs increases with decreasing k_z keeping M fixed (Fig. 1). For this parameter set, double layer does not exist when $M \leq 1.1$. The widths of the double layers become bigger for higher concentration of dust density δ_e in the plasma. The propagation of the electrostatic double layers is found to be directed significantly close to the direction of the ambient magnetic field. It is comfortably understood that DLs' amplitude depends on the external magnetic field. In some events, to obtained the existence of DLs for various values of M, it is need to exchange k_z values also that can be distinctly observed from Fig. 2 for $M = 1.4$, $k_z = 0.2$; $M = 1.6$, $k_z = 0.3$; $M = 1.8$, $k_z = 0.4$ and other parameters are $z = 1, \delta_e = 0.97$, $\sigma = 0.03$, $\beta = 0.002$, $\kappa = 4$. This depicts the nature of double layers for various set of values of M and k_z. To examine the consequence of κ parameter on the formation of nonlinear wave

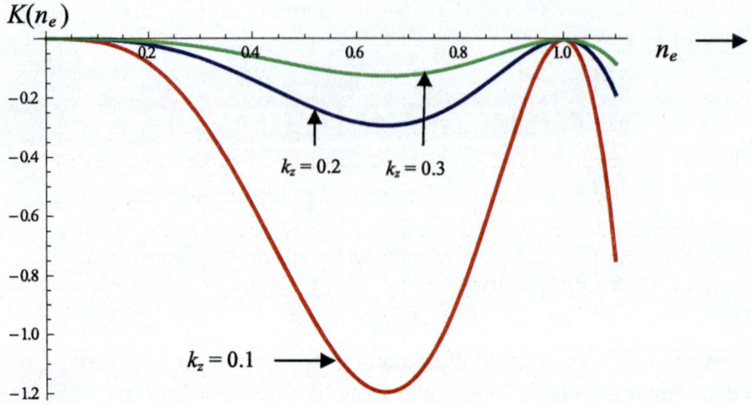

Fig. 1 Variation of $K(ne)$ against density ne showing double layers formation for several values of $kz = 0.1$, $kz = 0.2$, $kz = 0.3$ and other parameters are, $M = 1.2$, $z = 1$, $\delta e = 0.97$, $\sigma = 0.03$, $\beta = 0.002$, $\kappa = 4$

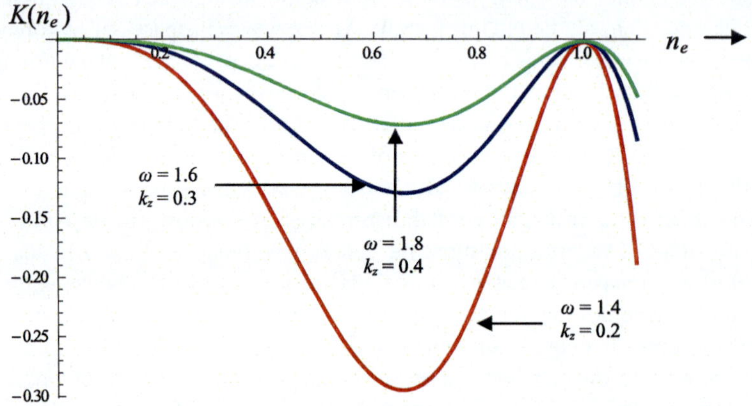

Fig. 2 Variation of $K(ne)$ against density ne showing double layers formation for several values of $M = 1.4$, $kz = 0.2$; $M = 1.6$, $kz = 0.3$; $M = 1.8$, $kz = 0.4$ and other parameters are $z = 1$, $\delta e = 0.97$, $\sigma = 0.03$, $\beta = 0.002$, $\kappa = 4$

structures, we would vary one parameter at a time, where the remaining parameters remain fixed. A graph for K (n_e) is plotted for a bigger value of $\kappa = 50$ (i.e., Maxwellian) for a comparison with kappa distribution $\kappa = 4$. As the spectral index κ increases it is shows that the curve ($\kappa = 50$) stands for neither a solitary wave nor a double layer (Fig. 3). This is because in presence of magnetic field, the higher energetic electrons have a tendency to be magnetized easily compared to the lower energetic electrons. The kappa index of superthermality of the non-Maxwellian electrons is in good agreement with the double layer formations in space and astrophysical plasmas. Because in our computations for double layers kappa values are seen

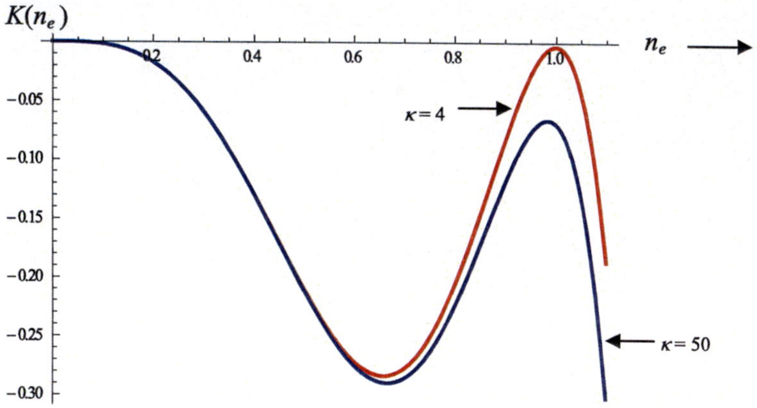

Fig. 3 Variation of $K(ne)$ against density ne showing double layers formation for κ = & κ = 50 (Maxwellian) and other parameters are, $M = 1.2$, $z = 1$, $\delta e = 0.98$, $\sigma = 0.03$, $\beta = 0.002$, $kz = 0.2$

ranging between 4 and 5. 3D Variation of $K(n_e)$ with density n_e and Mach number M showing double layers formation (Fig. 4). From the study of nonlinear dispersion relations (Eq. 11), the variation of the wave frequency ω with k_z for $\kappa = 5, \sigma = 0.02$ is shown in Fig. 5. The present theoretical outcomes could be of interest and it is anticipated to describe some of the new in situ observations (e.g., Freja, Viking) as spiky structures and the upcoming electric signatures as auroral radiation (rarefactive double layer) in space plasma where a kappa distributed electron component is observed. As the meteoric dust particles are not detected by in situ investigations in

Fig. 4 3D Variation of $K(ne)$ with density ne and Mach number M showing double layers formation for κ $= 4$, $z = 1$, $\delta e = 0.98$, $\sigma = 0.03$, $\beta = 0.002$, $kz = 0.2$

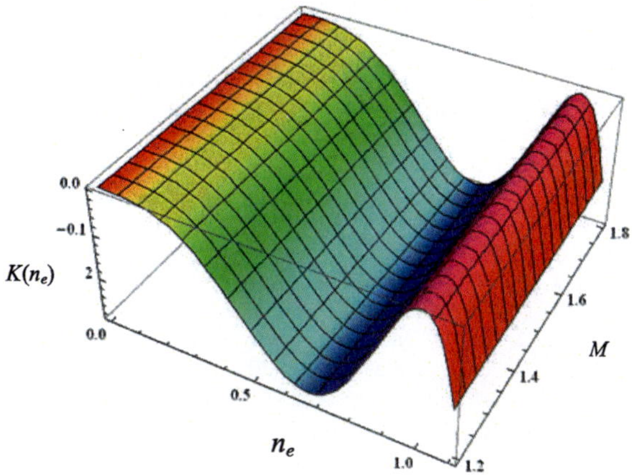

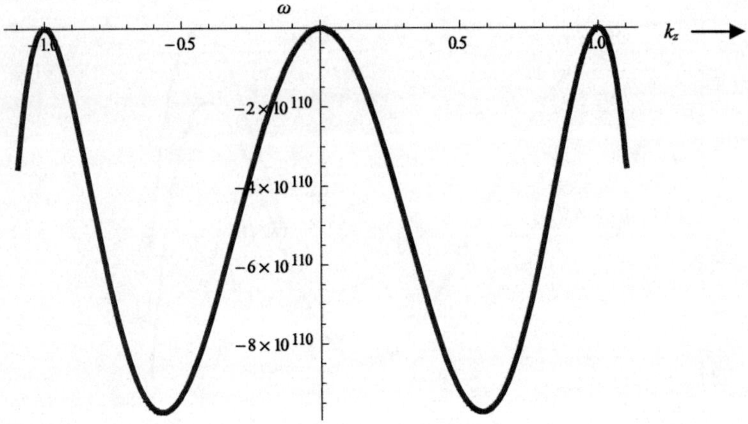

Fig. 5 Variation of the wave frequency ω (given in Eq. (11)) with kz for $\kappa = 5$, $\sigma = 0.02$

the mesosphere, so our findings are relevant to space as well as astrophysical plasmas in connection to noctilucent clouds (NLC) linked to dusty plasma prevailing in the mesosphere (~80–110 km attitude, i.e., some part between D and E region of Earth's ionosphere).

References

1. W. Gekelman, S. Vincena, B. Van Compernolle, G.J. Morales, J.E. Maggs, P. Pribyl, T.A. Carter, Phys. Plasmas **18**, 055501 (2011)
2. S. Houshmandyar, E.E. Scime, Phys. Plasmas **18**, 112111 (2011)
3. D. Leneman, W. Gekelman, J. Maggs, Phys. Rev. Lett. **82**, 2673 (1999)
4. C. Watts, J. Hanna, Phys. Plasmas **11**, 1358 (2004)
5. P.O. Dovner, A.I. Eriksson, G. Holmgreen, Geophys. Res. Lett. **21**, 1847 (1994)
6. H. Washimi, T. Tanuity, Phys. Rev. Lett. **17**, 996 (1966)
7. R.Z. Sagdeev, Rev. Plasma Phys. New York Consul. Bureau **4**, 52 (1966)
8. L. Kalita, N. Devi, R. Chaharia, Int. J. Math. Sci. Engg. Appls. (IJMSEA) **9**(III), 165–178. ISSN: 0973-9424 (2015)
9. L. Kalita, *Plasma and Fusion Science* (Apple Academic Press, Oakville, ON L6L 0A2 Canada, 2018). Hard ISBN: 9781771884532, E-Book ISBN: 9781315365947
10. A. Hasegawa, K. Mima, Phys. Rev. Lett. **37**, 690 (1976)
11. M.Y. Yu, P.K. Shukla, Phys. Fluids **21**, 1457 (1978)
12. M.A. Raadu, Phys. Rep. **178**, 25 (1989)
13. P. Coakley, N. Hershkowitz, Phys. Lett. **83A**, 131 (1981)
14. M. Temerin, K. Cerny, W. Lotko, F.S. Mozer, Phys. Rev. Lett. **48**, 1175 (1982)
15. V.M. Vasyliunas, J. Geophys. Res. (Space Physics) **73**, 2839 (1968)
16. W.C. Feldman, R.C. Anderson, J.R. Asbridge, S.J. Bame, J.T. Gosling, R.D. Zwickl, J. Geophys. Res. **87**, 632 (1982)
17. M. Maksimovic, V. Pierrard, J.F. Lemaire, Astron. Astrophys., **324**, 725 (19971)
18. R. Gogoi, M. Khan, Phys. Plasmas **17**, 112311 (2010)
19. L. Kalita, M.K. Deka, Int. J. Eng. Sci. Technol. (IJEST) **10**(02S). ISSN (Print): 2278–9510, ISSN (Online): 0975–5462, https://doi.org/10.21817/ijest/2018/v10i2s/181002s024 (2018)

20. S. Sultana, I. Kourakis, N.S. Saini, M.A. Hellberg, Phys. Plasmas **17**, 032310 (2010)
21. J.D. Wu, G.-L. Huang, D.-Y. Wang, C.G. Falthammar, Phys. Plasmas, **3**, 2879 (1996)
22. R. Roychoudhury, J. Plasma Phys. **67**, 199 (2002)
23. P.K Shukla, A.A. Mamun, *Introduction to Dusty Plasma Physics* (IOP Publishing Ltd., 2002)
24. N. Devi, R. Chaharia, L. Kalita, Int. J. Phys. Math. Sci. **8**(4), 13–21. ISSN: 2277-2111 (Online), October–December (2018)
25. R.C. Davidson, *Methods in Nonlinear Plasma Theory* (Academic Press, New York, 1972)

Mathematical Modeling for an Optimal Order Inventory with Demand Dependent Selling Price, Nonlinear Stock, and Nonlinear Holding Cost

Mamta Kumari and P. K. De

Abstract In the present paper an EOQ model has been developed with nonlinear holding cost, where demand is found to be linearly dependent on selling price and nonlinearly on inventory level. In this model the prevailing assumption of zero ending inventory level has been changed into a nonzero ending inventory level. Here an inventory model with shortages is analyzed which is partially backlogged. The main purpose of the inventory model is to find out the optimal order quantity along with ending inventory level so as to maximize retailer's total profit per unit time and also to determine the best-selling price of a given product. The trade credit policy is also introduced in the model. To demonstrate our model a numerical example has been presented and a sensitivity analysis is incorporated to highlight the findings of the suggested inventory model.

Keywords Inventory · Stock dependent demand · Selling price · Shortages · Nonlinear holding cost

1 Introduction

From past few years, a lot of attention has been paid towards the inventory management policy. Price is an important factor on which the demand of a product depends. A common question is what should be the selling price of a product? Although the ability to vary price in an inventory cycle is appreciable but sometimes the retailer may choose to keep constant price for administrative convenience. A system of inventory for non-instantaneous deteriorating items with price-dependent demand was formulated. According to many researchers, it has been observed that a display of large quantities of a product increases the product demand within the customers. Demand is also found to be dependent on the stock in hand. An inventory production model

M. Kumari (✉) · P. K. De
Department of Mathematics, NIT Silchar, Assam 788010, India
e-mail: mk656769@gmail.com

P. K. De
e-mail: pijusde@gmail.com

S. R. Mishra et al. (eds.), *Recent Trends in Applied Mathematics*, Lecture Notes in Mechanical Engineering, https://doi.org/10.1007/978-981-15-9817-3_17

in an imperfect production process where demand rate is dependent on both selling price as well as time was developed by Sarkar et al. [7]. A deteriorating item inventory model was formulated by Hsieh and Dye [4] considering the displayed stock level and sales price-dependent demand. Leopoldo Eduardo Cárdenas—Barrón et al. [8] formulated an EOQ inventory model considering nonlinear demand dependent on stock level, nonlinear stock dependent holding cost, and trade credit. In this paper, an inventory model has been proposed taking shortages into consideration. This paper demonstrates an inventory model where demand is nonlinear with nonlinear holding cost along with trade credit policy. Demand is considered to be a function of the amount of stock in hand and price, when the stock is nonzero, whereas demand is considered constant during the backlog period. A nonzero ending inventory level instead of zero ending inventory level has been developed. Backlogging gives an idea about the quantity of product to be ordered. Backlogging is useful for the retailer to have an idea of the order quantity on one side but it also comes with loss of sales due to shortage on the other side. In this paper we are calculating the optimal selling price, order quantity as well as the ending inventory level.

This paper is sorted out as follows. Section 2 states the assumptions and describes the notations which are necessary to depict the proposed inventory model. Section 3 builds up the inventory model considering nonlinear demand, nonlinear holding cost, along with trade credit policy. Section 4 presents theoretical results and optimization methods for optimizing the total profit. Section 5 solves few numerical examples. Section 6 depicts sensitivity analysis as well as discusses few observations. At the end, Sect. 7 provides future research directions and few conclusions.

2 Notation and Assumptions

The following notations and assumptions have been used as described below.

2.1 Assumptions

Demand is considered to be the function of price and stock level, given by

$$\begin{aligned} \mathrm{D(t)} &= \alpha(a - bp)[q(t)]^{\beta} && \text{when } q(t) > 0 \\ &= \alpha && \text{when } q(t) \le 0 \end{aligned}$$

It is deterministic in nature and a > 0, b > 0.

1. Holding cost of the inventory is considered to be a nonlinear function of stock level formulated as

$$H(t) = c_h[q(t)]^{\gamma} \quad \text{where } \gamma > 0$$

With the decline of stock level, it is found that holding cost also decreases. For $\gamma = 1$ we obtain an inventory model where holding cost is linearly dependent on stock.
2. Instantaneous replenishment rate, with negligible lead time.
3. Inventory system planning horizon is considered to be infinite.
4. In this case we have considered a single-level policy of trade credit where the retailer/manufacturer/supplier grants a credit policy to his or her customers for a given slot of time with well-defined terms and conditions.
5. Shortages are permitted in this model, and it is partially backlogged with backlogging parameter δ.

2.2 Notation

Notation	Description
Parameters	
c_o	Replenishment price per order
c	Cost of purchasing per unit
c_h	Holding cost per unit per unit time
c_b	Shortage cost per unit per unit time
c_l	Cost of lost sale per unit
γ	Holding cost elasticity; $\gamma > 0$
	Elasticity of demand; $0 \leq \beta < 1$
δ	Partial backlogging parameter;
	Fraction of the demand within the
	Stock out period which is backlogged,
	$\in [0,1]$
α	Demand rate scale parameter
t_1	Time at which inventory level position
	Reaches to zero
T	Length of the replenishment cycle
M	Trade credit period granted by the
	Supplier to the retailer
I_e	Interest percentage per unit time
	Gained by the retailer
I_p	Interest percentage per unit time paid
	By the retailer
Functions	
$q(t)$	Inventory level at a given time t where
	$0 \leq t \leq T$

(continued)

(continued)

Notation	Description
TP (Q, B, p)	Total profit per unit time
Decision Variables	
Q	Lot size per cycle
B	Ending inventory level at time *T*
p	Selling price per unit

3 Mathematical Modeling of the Inventory Model with Nonlinear Stock Dependent Holding Cost and Nonlinear Demand with Trade Credit

Inventory model having a nonzero ending level of inventory has been developed where holding cost is found to have a nonlinear dependence on stock, demand is a function of selling price and nonlinear stock level. Initially in the beginning of the inventory cycle Q units of a product exist. A replenishment order is placed when the level of inventory reaches to B units, then an order quantity of $Q - B$ units is placed which brings back the stock level again to the height of Q units at the beginning of the next cycle. The supplier grants a trade credit period M to his or her retailer additionally. For the inventory model with shortage ($B \leq 0$), the total profit per unit time has been derived.

3.1 An Inventory Model with Shortage

Initially Q units of an item are purchased by the retailer. After that the lot size of Q units decreases due to demand during the interval [0, T]. At t = t_1 the level of inventory reaches to zero. After that shortages occur and it is partially backlogged at the rate δ, the inventory level drops further down than zero. The inventory situation can be best explained by the following differential equations:

$$\frac{dq(t)}{dt} = -\alpha(a - bp)[q(t)]^{\beta}, \quad 0 < t \leq t_1 \tag{1}$$

$$\frac{dq(t)}{dt} = -\alpha\delta, \quad t_1 < t \leq T \tag{2}$$

With the following boundary conditions: $q(T) = B \leq 0$, $q(0) = Q$.
Solving the differential Eqs. (1) and (2) we get

$$q(t) = [Q^{1-\beta} - \alpha(a - bp)(1 - \beta)t]^{\frac{1}{1-\beta}} \quad 0 < t \le t_1 \tag{3}$$

$$q(t) = B + \alpha\delta(T - t) \quad t_1 < t \le T \tag{4}$$

Using the condition $q(t_1) = 0$ in Eq. (4) we get

$$t_1 = T + \frac{B}{\alpha\delta} \tag{5}$$

Using the continuity conditions from Eqs. (4) and (3) at the point t $= t_1$, the cycle time is determined as follows:

$$T = \frac{Q^{1-\beta}}{\alpha(a - bp)(1 - \beta)} - \frac{B}{\alpha\delta} \tag{6}$$

Different costs associated with this inventory model with shortage are as follows:

1. Ordering cost of the item per order

$$= C_0 \tag{7}$$

2. Inventory holding cost per cycle (C_{hol}) $= c_h \int_0^{t_1} [q(t)]^{\gamma} dt$

$$C_{hol} = \int_0^{t_1} [Q^{1-\beta} - \alpha(a - bp)(1 - \beta)t]^{\frac{\gamma}{1-\beta}} dt$$

$$C_{hol} = \frac{c_h}{\alpha(a - bp)(\gamma + 1 - \beta)} \left[Q^{\gamma+1-\beta} - \left\{ Q^{1-\beta} - \alpha(a - bp)(1 - \beta)t_1 \right\}^{\frac{\gamma+1-\beta}{1-\beta}} \right] \tag{8}$$

Substituting the value of t_1 from Eq. (5) into the above expression we get.

$$C_{hol} = \frac{c_h}{\alpha(a - bp)(\gamma + 1 - \beta)} Q^{\gamma+1-\beta} \tag{9}$$

3. Cost of purchasing

$$= c(Q - B) \tag{10}$$

4. Revenue collected from sales during the given period (SR)

$$= p(Q - B) \tag{11}$$

5. Shortage cost (C_{sho}) during the inventory cycle $= c_b \int_{t_1}^{T} [-q(t)] dt$

$$C_{sho} = - c_b \int_{t_1}^{T} [B + \alpha\delta(T - t)]dt$$

$$= - c_b \left[B(T - t_1) + \frac{\alpha\delta}{2}(T - t_1)^2 \right] \quad (12)$$

Putting the value of t_1 and T in the above expression shortage cost reduces to

$$\mathrm{C_{sho}} = \frac{c_b B^2}{2\alpha\delta} \quad (13)$$

6. The opportunity cost ($\mathrm{OC_{ls}}$) due to loss of sales during the inventory cycle

$$= c_l \int_{t_1}^{T} \alpha(1 - \delta)dt$$

$$= c_l\alpha(1 - \delta)(T - t_1)$$

Substituting the value of t_1 and T from Eqs. (5) and (6) into the above expression, the cost of opportunity further simplifies to

$$\mathrm{OC_{ls}} = -\frac{c_l(1 - \delta)B}{\delta} \quad (14)$$

In accordance with the policy of granting trade credit, the manufacturer/supplier offers a specific time period (M) to his/her retailer. Depending on the trade credit period the following cases occur:

CASE-1: $0 < M \le t_1$
CASE-2: $t_1 < M \le T$
CASE-3: $M > T$

Case-1:

Trade credit period is less than or equal to the time period when the inventory level reaches to zero.

Here the supplier provides a trade credit period to his or her retailer but it is less than or equal to t_1. It is to be observed that at the end of credit time M, the retailer faces interest charges and he or she must pay the interest during the time interval [M,T]. Therefore, the interest paid is determined as follows:

$$\mathrm{IP} = cI_p \left[\frac{[Q^{1-\beta} - \alpha(a - bp)(1 - \beta)M]^{\frac{2-\beta}{1-\beta}}}{\alpha(a - bp)(2 - \beta)} \right] \quad (15)$$

Since the retailer has a credit time period M, the retailer earns interest up to time $t = M$. Therefore, the interest earned (IE) is calculated as shown below:

$$\text{IE} = pI_e \int_0^M \int_0^t \alpha(a - bp)[q(u)]^\beta du dt + pI_e \int_0^M [-B] dt$$

$$= pI_e \left[(Q - B)M + \frac{[Q^{1-\beta} - \alpha(a - bp)(1 - \beta)M]^{\frac{2-\beta}{1-\beta}} - Q^{2-\beta}}{\alpha(a - bp)(2 - \beta)} \right] \quad (16)$$

Total profit per unit time,

$$TP_1(Q, B, p) = \frac{X_1}{T} \quad (17)$$

X_1 = revenue collected from sales + interest earned − interest paid−cost of lost sales − shortage cost − ordering cost − holding cost − cost of purchasing

$$X_1 = \text{SR} + \text{IE} - c(Q - B) - c_o - \text{C}_{\text{hol}} - \text{C}_{\text{sho}} - \text{IP} - \text{OC}_{\text{ls}} \quad (18)$$

Problem 1

$$\text{Maximize } TP_1(Q, B, p) = \frac{X_1}{T}$$
$$\text{Subject to } \quad 0 < M \le t_1 \quad (19)$$

<u>Case-2:</u>

Trade credit time is greater than the time at which inventory level reaches to zero but it is less than or equal to the cycle length.

In this case, the trade credit time period M is greater than t_1 but less than or equal to T. Here the retailer does not need to pay interest since the trade credit period M is found to be greater than t_1.

$$\text{IP} = 0 \quad (20)$$

The retailer does not need to pay interest but earns interest. The interest earned is calculated as shown below:

$$\text{IE} = pI_e \left[\int_0^M -B dt + \int_0^{t_1} \int_0^t \alpha(a - bp)(q(u))^\beta du dt \right]$$

$$+\int_0^{t_1}\left(\alpha(a-bp)(q(t))^{\beta}dt\right)[M-t_1]$$

$$= pI_e\left[-BM+\left[\frac{Q^{2-\beta}}{\alpha(a-bp)(1-\beta)(2-\beta)}\right]+Q\left[M-\frac{Q^{1-\beta}}{\alpha(a-bp)(1-\beta)}\right]\right] \quad (21)$$

The total profit per unit time is given by

$$TP_2(Q,B,p)=\frac{X_2}{T} \quad (22)$$

X_2 = revenue collected from sales + interest earned − cost of lost sales − ordering cost − cost of purchasing − shortage cost − holding cost − interest paid

$$X_2 = \mathrm{SR}+\mathrm{IE}-\mathrm{C_{hol}}-c_o-\mathrm{C_{sho}}-\mathrm{IP}-\mathrm{OC_{ls}} - c(Q-B) \quad (23)$$

Problem 2

$$\text{Maximize } TP_2(Q,B,p)=\frac{X_2}{T}$$
$$\text{Subject to } \ t_1 < M \le T \quad (24)$$

Case-3:
Trade credit time is greater than the cycle time.
Here, the trade credit period M is greater than T. In this case the retailer does not need to pay interest since the trade credit period M is greater than T.

$$\mathrm{IP}=0 \quad (25)$$

There is no requirement of paying interest by the retailer but he earns interest in this phase. The interest earned is given by

$$\mathrm{IE} = pI_e\left[\int_0^{M}-Bdt+\int_0^{t_1}\int_0^{t}\alpha(a-bp)(q(u))^{\beta}dudt + \int_0^{t_1}\left(\alpha(a-bp)(q(t))^{\beta}dt\right)[M-t_1]\right]$$

$$= pI_e\left[-BM + \left[\frac{Q^{2-\beta}}{\alpha(a-bp)(1-\beta)(2-\beta)}\right] + Q\left[M - \frac{Q^{1-\beta}}{\alpha(a-bp)(1-\beta)}\right]\right] \tag{26}$$

Total profit per unit time is given by

$$TP_3(Q, B, p) = \frac{X_3}{T} \tag{27}$$

X_3 = interest earned + revenue collected from sales − cost of lost sales − ordering cost − cost of purchasing − shortage cost − holding cost − interest paid

$$\begin{aligned} X_3 &= \mathrm{SR} + \mathrm{IE} - \mathrm{C_{hol}} - c_o - \mathrm{C_{sho}} - \mathrm{IP} - \mathrm{OC_{ls}} \\ &\quad - c(Q - B) \end{aligned} \tag{28}$$

Problem 3

$$\begin{aligned} &\text{Maximize } TP_3(Q, B, p) = \frac{X_3}{T} \\ &\text{Subject to } \quad T < M \end{aligned} \tag{29}$$

4 Theoretical Results and Optimization Procedures

The total profit functions formulated below are very complex in nature. It is not always easy to find out a closed form solution of the decision variables. Additionally, mathematically sometimes it is hard to present the concavity property of the total profit gained per unit time. So, in order to optimize the total profit earned per unit time, a search algorithm is used.

4.1 An Inventory Model with Shortage

Case-1:

$$TP_1(Q, B, p) = \left[\frac{1}{\frac{Q^{1-\beta}}{\alpha(a-bp)(1-\beta)} - \frac{B}{\alpha\delta}}\right]$$

$$\left[(p-c)(Q-B) - c_o - \frac{B^2 c_b}{2\alpha\delta} + \frac{c_l(1-\delta)B}{\delta} - \frac{c_h Q^{\gamma+1-\beta}}{\alpha(a-bp)(\gamma+1-\beta)}\right.$$

$$+pI_e\left[(Q-B)M+\frac{\left[\left\{Q^{1-\beta}-\alpha(a-bp)(1-\beta)M\right\}^{\frac{2-\beta}{1-\beta}}-Q^{2-\beta}\right]}{\alpha(a-bp)(2-\beta)}\right]$$
$$-\frac{cI_P}{\alpha(a-bp)(2-\beta)}\left[Q^{1-\beta}-\alpha(a-bp)(1-\beta)M\right]^{\frac{2-\beta}{1-\beta}}\Bigg] \tag{30}$$

Differentiating partially Eq. (30) with respect to Q we get

$$H_1=\frac{\partial TP_1(Q,B,p)}{\partial Q}$$
$$=\Bigg[\frac{1}{\frac{Q^{1-\beta}}{\alpha(a-bp)(1-\beta)}-\frac{B}{\alpha\delta}}\Bigg\{(p-c)-\frac{c_hQ^{\gamma-\beta}}{\alpha(a-bp)}$$
$$-\frac{cI_pQ^{-\beta}\left\{Q^{1-\beta}-\alpha(a-bp)(1-\beta)M\right\}^{\frac{1}{1-\beta}}}{\alpha(a-bp)}+pI_eM$$
$$+\frac{pI_e\left\{\left\{Q^{-\beta}(Q^{1-\beta}-\alpha(a-bp)(1-\beta)M)^{\frac{1}{1-\beta}}\right\}-Q^{1-\beta}\right\}}{\alpha(a-bp)}\Bigg\}\Bigg]$$
$$-\frac{Q^{-\beta}}{\alpha(a-bp)\left(\frac{Q^{1-\beta}}{\alpha(a-bp)(1-\beta)}-\frac{B}{\alpha\delta}\right)^2}$$
$$\Bigg[(p-c)(Q-B)-c_o-\frac{B^2c_b}{2\alpha\delta}+\frac{c_l(1-\delta)B}{\delta}-\frac{c_hQ^{\gamma+1-\beta}}{\alpha(a-bp)(\gamma+1-\beta)}$$
$$+pI_e\left[(Q-B)M+\frac{\left[\left\{Q^{1-\beta}-\alpha(a-bp)(1-\beta)M\right\}^{\frac{2-\beta}{1-\beta}}-Q^{2-\beta}\right]}{\alpha(a-bp)(2-\beta)}\right]$$
$$-\frac{cI_P}{\alpha(a-bp)(2-\beta)}\left[Q^{1-\beta}-\alpha(a-bp)(1-\beta)M\right]^{\frac{2-\beta}{1-\beta}}\Bigg]=0 \tag{31}$$

Next differentiating Eq. (30) partially with respect to B we get

$$H_2=\frac{\partial TP_1(Q,B,p)}{\partial B}$$
$$=\left[\frac{1}{\frac{Q^{1-\beta}}{\alpha(a-bp)(1-\beta)}-\frac{B}{\alpha\delta}}\right]\left[c-p-\frac{c_bB}{\alpha\delta}+\frac{C_l(1-\delta)}{\delta}-pMI_e\right]$$
$$+\frac{1}{\alpha\delta\left(\frac{Q^{1-\beta}}{\alpha(a-bp)(1-\beta)}-\frac{B}{\alpha\delta}\right)^2}\Bigg[(p-c)(Q-B)-c_o-\frac{B^2c_b}{2\alpha\delta}$$
$$+\frac{c_l(1-\delta)B}{\delta}-\frac{c_hQ^{\gamma+1-\beta}}{\alpha(a-bp)(\gamma+1-\beta)}$$

$$+pI_e\left[(Q-B)M+\frac{\left[\left\{Q^{1-\beta}-\alpha(a-bp)(1-\beta)M\right\}^{\frac{2-\beta}{1-\beta}}-Q^{2-\beta}\right]}{\alpha(a-bp)(2-\beta)}\right]$$

$$-\frac{cI_P}{\alpha(a-bp)(2-\beta)}\left[Q^{1-\beta}-\alpha(a-bp)(1-\beta)M\right]^{\frac{2-\beta}{1-\beta}} \quad (32)$$

Differentiating Eq. (30) with respect to p we get

$$H_3=\frac{\partial TP_1(Q,B,p)}{\partial p}$$

$$=\left[\frac{1}{\frac{Q^{1-\beta}}{\alpha(a-bp)(1-\beta)}-\frac{B}{\alpha\delta}}\left[Q-B-\frac{bcI_pM(Q^{1-\beta}-\alpha M(a-bp)(1-\beta))^{\frac{1}{1-\beta}}}{a-bp}\right.\right.$$

$$+pI_e\left\{\frac{bM(Q^{1-\beta}-\alpha M(a-bp)(1-\beta))^{\frac{1}{1-\beta}}}{a-bp}\right.$$

$$\left.+\frac{b\left((Q^{1-\beta}-\alpha M(a-bp)(1-\beta))^{\frac{2-\beta}{1-\beta}}-Q^{2-\beta}\right)}{\alpha(a-bp)^2(1-\beta)}\right\}$$

$$+I_e\left\{(Q-B)M+\frac{\left((Q^{1-\beta}-\alpha M(a-bp)(1-\beta))^{\frac{2-\beta}{1-\beta}}-Q^{2-\beta}\right)}{\alpha(a-bp)(1-\beta)}\right\}$$

$$\left.\left.-\frac{bcI_p\left((Q^{1-\beta}-\alpha M(a-bp)(1-\beta))^{\frac{2-\beta}{1-\beta}}\right)}{\alpha(a-bp)^2(1-\beta)}-\frac{bc_hQ^{\gamma+1-\beta}}{\alpha(a-bp)^2(\gamma+1-\beta)}\right]\right]$$

$$-\frac{bQ^{1-\beta}}{\alpha(a-bp)^2(1-\beta)\left(\frac{Q^{1-\beta}}{\alpha(a-bp)(1-\beta)}-\frac{B}{\alpha\delta}\right)^2}$$

$$\left[(p-c)(Q-B)-c_o-\frac{B^2c_b}{2\alpha\delta}+\frac{c_l(1-\delta)B}{\delta}-\frac{c_hQ^{\gamma+1-\beta}}{\alpha(a-bp)(\gamma+1-\beta)}\right.$$

$$+pI_e\left[(Q-B)M+\frac{\left[\left\{Q^{1-\beta}-\alpha(a-bp)(1-\beta)M\right\}^{\frac{2-\beta}{1-\beta}}-Q^{2-\beta}\right]}{\alpha(a-bp)(2-\beta)}\right]$$

$$\left.-\frac{cI_P}{\alpha(a-bp)(2-\beta)}\left[Q^{1-\beta}-\alpha(a-bp)(1-\beta)M\right]^{\frac{2-\beta}{1-\beta}}\right] \quad (33)$$

Differentiating Eq. (31) with respect to Q we get

$$\frac{\partial H_1}{\partial Q} = \frac{-2Q^{-\beta}}{\alpha(a-bp))\left(\frac{Q^{1-\beta}}{\alpha(a-bp)(1-\beta)} - \frac{B}{\alpha\delta}\right)^2}$$

$$\left[(p-c) - c_o - \frac{c_h Q^{\gamma-\beta}}{\alpha(a-bp)}\right.$$

$$+pI_e\left[M + \frac{\left[Q^{-\beta}\left\{Q^{1-\beta} - \alpha(a-bp)(1-\beta)M\right\}^{\frac{1}{1-\beta}} - Q^{1-\beta}\right]}{\alpha(a-bp)}\right]$$

$$\left.-\frac{cI_P Q^{-\beta}}{\alpha(a-bp)}\left[Q^{1-\beta} - \alpha(a-bp)(1-\beta)M\right]^{\frac{1}{1-\beta}}\right]$$

$$+\left[\frac{1}{\frac{Q^{1-\beta}}{\alpha(a-bp)(1-\beta)} - \frac{B}{\alpha\delta}}\left[-\frac{cI_p(Q^{-2\beta}(Q^{1-\beta} - \alpha M(a-bp)(1-\beta))^{\frac{\beta}{1-\beta}}}{a-bp}\right.\right.$$

$$+\frac{\beta Q^{-1-\beta}(Q^{1-\beta} - \alpha M(a-bp)(1-\beta))^{\frac{1}{1-\beta}}}{a-bp}$$

$$+\frac{pI_e}{\alpha(a-bp)}\left\{-Q^{-\beta}(1-\beta) + Q^{-2\beta}(Q^{1-\beta} - \alpha M(a-bp)(1-\beta))^{\frac{\beta}{1-\beta}}\right.$$

$$\left.\left.\left.-Q^{-1-\beta}(Q^{1-\beta} - \alpha M(a-bp)(1-\beta))^{\frac{1}{1-\beta}}\right\} - \frac{bc_h Q^{\gamma+1-\beta}}{\alpha(a-bp)^2(\gamma+1-\beta)}\right]\right]$$

$$+\left[\frac{2Q^{-2\beta}}{\alpha^2(a-bp)^2\left(\frac{Q^{1-\beta}}{\alpha(a-bp)(1-\beta)} - \frac{B}{\alpha\delta}\right)^3} + \frac{\beta Q^{-1-\beta}}{\alpha(a-bp)\left(\frac{Q^{1-\beta}}{\alpha(a-bp)(1-\beta)} - \frac{B}{\alpha\delta}\right)^2}\right]$$

$$\left[(p-c)(Q-B) - c_o - \frac{B^2 c_b}{2\alpha\delta} + \frac{c_l(1-\delta)B}{\delta} - \frac{c_h Q^{\gamma+1-\beta}}{\alpha(a-bp)(\gamma+1-\beta)}\right.$$

$$+pI_e\left[(Q-B)M + \frac{\left[\left\{Q^{1-\beta} - \alpha(a-bp)(1-\beta)M\right\}^{\frac{2-\beta}{1-\beta}} - Q^{2-\beta}\right]}{\alpha(a-bp)(2-\beta)}\right]$$

$$\left.-\frac{cI_P}{\alpha(a-bp)(2-\beta)}\left[Q^{1-\beta} - \alpha(a-bp)(1-\beta)M\right]^{\frac{2-\beta}{1-\beta}}\right]$$

Differentiating Eq. (32) with respect to B we get

$$\frac{\partial H_2}{\partial B} = \left[\frac{2}{\alpha^2\delta^2\left(\frac{Q^{1-\beta}}{\alpha(a-bp)(1-\beta)} - \frac{B}{\alpha\delta}\right)^3}\right]$$

$$\left[(p-c)(Q-B) - c_o - \frac{B^2 c_b}{2\alpha\delta} + \frac{c_l(1-\delta)B}{\delta} - \frac{c_h Q^{\gamma+1-\beta}}{\alpha(a-bp)(\gamma+1-\beta)}\right.$$

$$+pI_e\left[(Q-B)M+\frac{\left[\left\{Q^{1-\beta}-\alpha(a-bp)(1-\beta)M\right\}^{\frac{2-\beta}{1-\beta}}-Q^{2-\beta}\right]}{\alpha(a-bp)(2-\beta)}\right]$$
$$\left.-\frac{cI_P}{\alpha(a-bp)(2-\beta)}\left[Q^{1-\beta}-\alpha(a-bp)(1-\beta)M\right]^{\frac{2-\beta}{1-\beta}}\right]$$
$$-\frac{c_b}{\alpha\delta\left(\frac{Q^{1-\beta}}{\alpha(a-bp)(1-\beta)}-\frac{B}{\alpha\delta}\right)}+\frac{2}{\alpha\delta\left(\frac{Q^{1-\beta}}{\alpha(a-bp)(1-\beta)}-\frac{B}{\alpha\delta}\right)^2}$$
$$\left[c-p-pMI_e-\frac{Bc_b}{\alpha\delta}+\frac{Bc_l(1-\delta)}{\delta}\right] \quad (35)$$

Differentiating Eq. (33) with respect to p, we get

$$\frac{\partial H_3}{\partial p}=\frac{2bQ^{1-\beta}}{\alpha(1-\beta)(a-bp)^2\left(\frac{Q^{1-\beta}}{\alpha(a-bp)(1-\beta)}-\frac{B}{\alpha\delta}\right)^2}$$
$$\left[Q-B-\frac{bcI_pM(Q^{1-\beta}-\alpha M(a-bp)(1-\beta))^{\frac{1}{1-\beta}}}{\alpha(a-bp)}\right.$$
$$+pI_e\left[\frac{bM(Q^{1-\beta}-\alpha M(a-bp)(1-\beta))^{\frac{1}{1-\beta}}}{a-bp}\right.$$
$$\left.+\frac{b\left((Q^{1-\beta}-\alpha M(a-bp)(1-\beta))^{\frac{2-\beta}{1-\beta}}-Q^{2-\beta}\right)}{\alpha(a-bp)^2(1-\beta)}\right]$$
$$+I_e\left[M(Q-B)+\frac{\left((Q^{1-\beta}-\alpha M(a-bp)(1-\beta))^{\frac{2-\beta}{1-\beta}}-Q^{2-\beta}\right)}{\alpha(a-bp)(1-\beta)}\right]$$
$$\left.-\frac{bcI_P\left((Q^{1-\beta}-\alpha M(a-bp)(1-\beta))^{\frac{2-\beta}{1-\beta}}\right)}{\alpha(a-bp)^2(1-\beta)}-\frac{bc_hQ^{\gamma+1-\beta}}{\alpha(a-bp)^2(\gamma+1-\beta)}\right]$$
$$+\left[\frac{1}{\frac{Q^{1-\beta}}{\alpha(a-bp)(1-\beta)}-\frac{B}{\alpha\delta}}\right]\left[-\frac{2b^2cI_pM\left((Q^{1-\beta}-\alpha M(a-bp)(1-\beta))^{\frac{1}{1-\beta}}\right)}{(a-bp)^2}\right.$$
$$+pI_e\left[\frac{2b^2M\left((Q^{1-\beta}-\alpha M(a-bp)(1-\beta))^{\frac{1}{1-\beta}}\right)}{\alpha(a-bp)^2(1-\beta)}\right.$$
$$+\frac{b^2M^2\alpha\left((Q^{1-\beta}-\alpha M(a-bp)(1-\beta))^{\frac{\beta}{1-\beta}}\right)}{(a-bp)}$$

$$+\frac{2b^2\left(\left\{Q^{1-\beta}-\alpha(a-bp)(1-\beta)M\right\}^{\frac{2-\beta}{1-\beta}}-Q^{2-\beta}\right)}{\alpha(a-bp)^3(2-\beta)}\Bigg]$$

$$+2I_e\left[\frac{bM\left((Q^{1-\beta}-\alpha M(a-bp)(1-\beta))^{\frac{\beta}{1-\beta}}\right)}{(a-bp)}\right.$$

$$\left.+\frac{b\left(\left\{Q^{1-\beta}-\alpha(a-bp)(1-\beta)M\right\}^{\frac{2-\beta}{1-\beta}}-Q^{2-\beta}\right)}{\alpha(a-bp)^2(2-\beta)}\right]$$

$$-\frac{b^2M^2\alpha cI_p\left((Q^{1-\beta}-\alpha M(a-bp)(1-\beta))^{\frac{\beta}{1-\beta}}\right)}{(a-bp)}$$

$$-\frac{2b^2cI_p\left(\left\{Q^{1-\beta}-\alpha(a-bp)(1-\beta)M\right\}^{\frac{2-\beta}{1-\beta}}\right)}{\alpha(a-bp)^3(2-\beta)}-\frac{2b^2c_hQ^{\gamma+1-\beta}}{\alpha(a-bp)^3(\gamma+1-\beta)}\Bigg]\Bigg]$$

$$-\left[\frac{2b^2Q^{2-2\beta}}{\alpha^2(a-bp)^4(1-\beta)^2\left(\frac{Q^{1-\beta}}{\alpha(a-bp)(1-\beta)}-\frac{B}{\alpha\delta}\right)^3}\right.$$

$$\left.-\frac{2b^2Q^{1-\beta}}{\alpha(a-bp)^3(1-\beta)\left(\frac{Q^{1-\beta}}{\alpha(a-bp)(1-\beta)}-\frac{B}{\alpha\delta}\right)^2}\right]$$

$$\Bigg[(p-c)(Q-B)-c_o-\frac{B^2c_b}{2\alpha\delta}+\frac{c_l(1-\delta)B}{\delta}-\frac{c_hQ^{\gamma+1-\beta}}{\alpha(a-bp)(\gamma+1-\beta)}$$

$$+pI_e\left[(Q-B)M+\frac{\left[\left\{Q^{1-\beta}-\alpha(a-bp)(1-\beta)M\right\}^{\frac{2-\beta}{1-\beta}}-Q^{2-\beta}\right]}{\alpha(a-bp)(2-\beta)}\right]$$

$$-\frac{cI_P}{\alpha(a-bp)(2-\beta)}\left[Q^{1-\beta}-\alpha(a-bp)(1-\beta)M\right]^{\frac{2-\beta}{1-\beta}}\Bigg] \tag{36}$$

Case-2:

$$TP_2(Q,B,p)$$

$$=\left[\frac{1}{\frac{Q^{1-\beta}}{\alpha(a-bp)(1-\beta)}-\frac{B}{\alpha\delta}}\right]$$

$$\Bigg[(p-c)(Q-B)-c_o-\frac{B^2c_b}{2\alpha\delta}+\frac{c_l(1-\delta)B}{\delta}-\frac{c_hQ^{\gamma+1-\beta}}{\alpha(a-bp)(\gamma+1-\beta)}$$

$$+pI_e\left[-BM+\left(\frac{Q^{2-\beta}}{\alpha(a-bp)(1-\beta)(2-\beta)}\right)\right.$$
$$\left.\left.+Q\left(M-\frac{Q^{1-\beta}}{\alpha(a-bp)(1-\beta)}\right)\right]\right] \tag{37}$$

Differentiating partially Eq. (37) with respect to Q we get

$$G_1=\frac{\partial TP_2(Q,B,p)}{\partial Q}$$
$$=\left[\frac{1}{\frac{Q^{1-\beta}}{\alpha(a-bp)(1-\beta)}-\frac{B}{\alpha\delta}}\left\{(p-c)-\frac{c_hQ^{\gamma-\beta}}{\alpha(a-bp)}+pI_e\left(M-\frac{Q^{1-\beta}}{\alpha(a-bp)}\right)\right\}\right]$$
$$-\frac{Q^{-\beta}}{\alpha(a-bp)\left(\frac{Q^{1-\beta}}{\alpha(a-bp)(1-\beta)}-\frac{B}{\alpha\delta}\right)^2}$$
$$-\left[(p-c)(Q-B)-c_o-\frac{B^2c_b}{2\alpha\delta}+\frac{c_l(1-\delta)B}{\delta}-\frac{c_hQ^{\gamma+1-\beta}}{\alpha(a-bp)(\gamma+1-\beta)}\right.$$
$$+pI_e\left[-BM+\left(\frac{Q^{2-\beta}}{\alpha(a-bp)(1-\beta)(2-\beta)}\right)\right.$$
$$\left.\left.+Q\left(M-\frac{Q^{1-\beta}}{\alpha(a-bp)(1-\beta)}\right)\right]\right]=0 \tag{38}$$

Next differentiating Eq. (37) partially with respect to B we get

$$G_2=\frac{\partial TP_2(Q,B,p)}{\partial B}$$
$$=\left[\frac{1}{\frac{Q^{1-\beta}}{\alpha(a-bp)(1-\beta)}-\frac{B}{\alpha\delta}}\right]\left[c-p-\frac{c_bB}{\alpha\delta}+\frac{C_l(1-\delta)}{\delta}-pMI_e\right]$$
$$+\frac{1}{\alpha\delta\left(\frac{Q^{1-\beta}}{\alpha(a-bp)(1-\beta)}-\frac{B}{\alpha\delta}\right)^2}\left[(p-c)(Q-B)-c_o-\frac{B^2c_b}{2\alpha\delta}\right.$$
$$+\frac{c_l(1-\delta)B}{\delta}-\frac{c_hQ^{\gamma+1-\beta}}{\alpha(a-bp)(\gamma+1-\beta)}$$
$$+pI_e\left[-BM+\left(\frac{Q^{2-\beta}}{\alpha(a-bp)(1-\beta)(2-\beta)}\right)\right.$$
$$\left.\left.+Q\left(M-\frac{Q^{1-\beta}}{\alpha(a-bp)(1-\beta)}\right)\right]\right]=0 \tag{39}$$

Differentiating Eq. (37) with respect to p we get

$$G_3 = \frac{\partial TP_2(Q,B,p)}{\partial p}$$

$$= \left[\frac{1}{\frac{Q^{1-\beta}}{\alpha(a-bp)(1-\beta)} - \frac{B}{\alpha\delta}} \left[Q - B - pI_e\left(\frac{bQ^{2-\beta}}{\alpha(a-bp)^2(1-\beta)(2-\beta)}\right) \right.\right.$$

$$+I_e\left\{ -BM + \left(\frac{Q^{2-\beta}}{\alpha(a-bp)(1-\beta)(2-\beta)}\right) + Q\left(M - \frac{Q^{1-\beta}}{\alpha(a-bp)(1-\beta)}\right)\right\}$$

$$\left.\left. - \frac{bc_h Q^{\gamma+1-\beta}}{\alpha(a-bp)^2(\gamma+1-\beta)} \right]\right] - \frac{bQ^{1-\beta}}{\alpha(a-bp)^2(1-\beta)\left(\frac{Q^{1-\beta}}{\alpha(a-bp)(1-\beta)} - \frac{B}{\alpha\delta}\right)^2}$$

$$\left[(p-c)(Q-B) - c_o - \frac{B^2 c_b}{2\alpha\delta} + \frac{c_l(1-\delta)B}{\delta} - \frac{c_h Q^{\gamma+1-\beta}}{\alpha(a-bp)(\gamma+1-\beta)} \right.$$

$$+pI_e\left[-BM + \left(\frac{Q^{2-\beta}}{\alpha(a-bp)(1-\beta)(2-\beta)}\right) \right.$$

$$\left.\left. +Q\left(M - \frac{Q^{1-\beta}}{\alpha(a-bp)(1-\beta)}\right)\right]\right] = 0 \tag{40}$$

Differentiating Eq. (38) with respect to Q we get

$$\frac{\partial G_1}{\partial Q} = \left[\frac{-2Q^{-\beta}}{\alpha(a-bp)\left(\frac{Q^{1-\beta}}{\alpha(a-bp)(1-\beta)} - \frac{B}{\alpha\delta}\right)^2} \right]$$

$$\left[(p-c) - \frac{c_h Q^{\gamma-\beta}}{\alpha(a-bp)} + pI_e\left(M - \frac{Q^{1-\beta}}{\alpha(a-bp)}\right)\right]$$

$$+ \left[\frac{1}{\left(\frac{Q^{1-\beta}}{\alpha(a-bp)(1-\beta)} - \frac{B}{\alpha\delta}\right)} \left[-pI_e\left\{\frac{(1-\beta)Q^{-\beta}}{\alpha(a-bp)}\right\} - \frac{c_h(\gamma-\beta)Q^{\gamma-\beta}}{\alpha(a-bp)} \right]\right]$$

$$+ \left[\frac{2Q^{-2\beta}}{\alpha^2(a-bp)^2\left(\frac{Q^{1-\beta}}{\alpha(a-bp)(1-\beta)} - \frac{B}{\alpha\delta}\right)^3} + \frac{\beta Q^{-1-\beta}}{\alpha(a-bp)\left(\frac{Q^{1-\beta}}{\alpha(a-bp)(1-\beta)} - \frac{B}{\alpha\delta}\right)^2} \right]$$

$$\left[(p-c)(Q-B) - c_o - \frac{B^2 c_b}{2\alpha\delta} + \frac{c_l(1-\delta)B}{\delta} - \frac{c_h Q^{\gamma+1-\beta}}{\alpha(a-bp)(\gamma+1-\beta)} \right.$$

$$+pI_e\left[-BM + \left(\frac{Q^{2-\beta}}{\alpha(a-bp)(1-\beta)(2-\beta)}\right) \right.$$

$$\left.\left. +Q\left(M - \frac{Q^{1-\beta}}{\alpha(a-bp)(1-\beta)}\right)\right]\right] \tag{41}$$

Differentiating Eq. (39) with respect to B we get

$$
\begin{aligned}
\frac{\partial G_2}{\partial B} = &\left[-\frac{c_b}{\alpha\delta\left(\frac{Q^{1-\beta}}{\alpha(a-bp)(1-\beta)}-\frac{B}{\alpha\delta}\right)}\right] \\
&+\frac{2}{\alpha\delta\left(\frac{Q^{1-\beta}}{\alpha(a-bp)(1-\beta)}-\frac{B}{\alpha\delta}\right)^2}\left[c-p-pMI_e-\frac{Bc_b}{\alpha\delta}+\frac{c_l(1-\delta)}{\delta}\right] \\
&+\left[\frac{2}{\alpha^2\delta^2\left(\frac{Q^{1-\beta}}{\alpha(a-bp)(1-\beta)}-\frac{B}{\alpha\delta}\right)^3}\right]\left[(p-c)(Q-B)-c_o-\frac{B^2c_b}{2\alpha\delta}\right. \\
&+\frac{c_l(1-\delta)B}{\delta}-\frac{c_hQ^{\gamma+1-\beta}}{\alpha(a-bp)(\gamma+1-\beta)} \\
&+pI_e\left[-BM+\left(\frac{Q^{2-\beta}}{\alpha(a-bp)(1-\beta)(2-\beta)}\right)\right. \\
&\left.\left.+Q\left(M-\frac{Q^{1-\beta}}{\alpha(a-bp)(1-\beta)}\right)\right]\right]
\end{aligned} \tag{42}
$$

Differentiating Eq. (40) with respect to p we get

$$
\begin{aligned}
\frac{\partial G_3}{\partial p} = &\left[-\frac{2bQ^{1-\beta}}{\alpha(1-\beta)(a-bp)^2\left(\frac{Q^{1-\beta}}{\alpha(a-bp)(1-\beta)}-\frac{B}{\alpha\delta}\right)^2}\right] \\
&\left[Q-B-pI_e\left[\frac{bQ^{2-\beta}}{\alpha(a-bp)^2(2-\beta)}\right]\right. \\
&+I_e\left[-BM+\left(\frac{Q^{2-\beta}}{\alpha(a-bp)(1-\beta)(2-\beta)}\right)\right. \\
&\left.\left.+Q\left(M-\frac{Q^{1-\beta}}{\alpha(a-bp)(1-\beta)}\right)\right]-\frac{bc_hQ^{\gamma+1-\beta}}{\alpha(a-bp)^2(\gamma+1-\beta)}\right] \\
&+\left[\frac{1}{\left(\frac{Q^{1-\beta}}{\alpha(a-bp)(1-\beta)}-\frac{B}{\alpha\delta}\right)}\right]\left[pI_e\left[\frac{-2b^2Q^{2-\beta}}{\alpha(a-bp)^3(2-\beta)}\right]\right. \\
&\left.+2I_e\left[-\frac{bQ^{2-\beta}}{\alpha(a-bp)^2(2-\beta)}\right]-\frac{2b^2c_hQ^{\gamma+1-\beta}}{\alpha(a-bp)^3(\gamma+1-\beta)}\right]\Bigg] \\
&+\left[\frac{2b^2Q^{2-2\beta}}{\alpha^2(a-bp)^4(1-\beta)^2\left(\frac{Q^{1-\beta}}{\alpha(a-bp)(1-\beta)}-\frac{B}{\alpha\delta}\right)^3}\right. \\
&\left.-\frac{2b^2Q^{1-\beta}}{\alpha(a-bp)^3(1-\beta)\left(\frac{Q^{1-\beta}}{\alpha(a-bp)(1-\beta)}-\frac{B}{\alpha\delta}\right)^2}\right]
\end{aligned}
$$

$$
* \left[(p-c)(Q-B) - c_o - \frac{B^2 c_b}{2\alpha\delta} + \frac{c_l(1-\delta)B}{\delta} - \frac{c_h Q^{\gamma+1-\beta}}{\alpha(a-bp)(\gamma+1-\beta)} + pI_e\left[-BM + \left(\frac{Q^{2-\beta}}{\alpha(a-bp)(1-\beta)(2-\beta)} \right) + Q\left(M - \frac{Q^{1-\beta}}{\alpha(a-bp)(1-\beta)} \right) \right] \right] \tag{43}
$$

Case-3:

$$
TP_3(Q,B,p) = \left[(p-c)(Q-B) - c_o - \frac{B^2 c_b}{2\alpha\delta} + \frac{c_l(1-\delta)B}{\delta} - \frac{c_h Q^{\gamma+1-\beta}}{\alpha(a-bp)(\gamma+1-\beta)} + pI_e\left[-BM + \left(\frac{Q^{2-\beta}}{\alpha(a-bp)(1-\beta)(2-\beta)} \right) + Q\left(M - \frac{Q^{1-\beta}}{\alpha(a-bp)(1-\beta)} \right) \right] \right] \tag{44}
$$

The total profit function is same as Case-2; therefore, all the derivatives and corresponding results will be the same.

5 Numerical Example

The following values has been considered for the input parameters: $c_o = \$\ 40$/order, $c = \$\ 58$/unit, $c_h = \$\ 11$/unit/year, $\gamma = 1.1$, $\beta = 0.4$, $\alpha = 120$, $\delta = 0.83$, $I_p =$ 12%/year, $I_e = 10\%$/year, $M = 0.04$ year, $c_b = \$\ 20$/unit/year, $c_l = \$\ 6$/unit/year, $a = 200$, $b = 3$.

The optimal solution is given by $t_1^* = 0.0324315$, $T^* = 0.985543$, $Q^* = 3.27527$, $B^* = -65.8583$, $p^* = \$66.6564$/unit.

It corresponds to case-2, therefore, $TP\ (Q^*, B^*, p^*) = \$\ 25.184$.

6 Sensitivity Analysis

This section does an analysis of the preceding numerical example to study the effect of underestimation or overestimation of the input parameters on the optimal values of Q, p, t_1, B, T and the total profit $TP\ (Q, B, p)$. It is done by varying the parametric values between 20% and −20%. The following results are drawn from the table mentioned below:

1. It is clear from the data that with the increasing vales of α, the optimal order quantity (Q) increases and the ending level of inventory quantity (B) decreases. The total profit decreases.
2. It is observed that with the increase in value of (c_o), optimal order quantity decreases. Optimal ending level of inventory (B), t_1, T, and optimal price (p) increases. Hence, total profit increases.
3. It is to be pointed that with the increment in values of holding cost (c_h), the optimal order quantity (Q) decreases whereas the ending inventory level (B) increases. Optimal price (p) does not vary much but total profit increases.
4. With the increasing values of (c_l), it is observed that the optimal order quantity (Q) increases for some time and then start decreasing. Optimal ending inventory level (B), optimal price (p), and total profit start decreasing in the beginning and then start rising up.
5. With the increment in value of γ, ending inventory level (B) increases. Optimal stock (Q) decreases but total profit increases (Table 1).

7 Conclusion

This paper highlights the following important facts: (i) display of product in large quantities enhances its demand within the customers, (ii) price is also an important factor on which demand of a product depends, (iii) holding cost is not always linearly dependent on stock. The retailer's optimal strategy is examined in this paper based on nonlinear demand and nonlinear holding cost of his or her product when the supplier uses a policy of trade credit for its customers. An inventory model is portrayed with nonlinear demand as well as nonlinear holding cost relaxing the ending-zero inventory level condition, in which shortages occur and demand during that time is partially backlogged which helps the retailer in running his or her business. The main objective is to determine the optimal order quantity (Q), ending level of inventory (B), and optimal rice (p) in order to maximize the total profit earned per unit time by the retailer.

Table. 1 Sensitivity analysis by changing the input parameters

Parameters	% change in parameters	$TP^*(Q,B,p)$	% change in				
			Q^*	B*	p^*	t_1^*	T^*
α	−20	−4.37586	0.010123	−56.31029	66.6651	0.234782	0.941488
	−10	−4.95253	0.0104676	−65.4329	66.6665	2.00154	2.73149
	10	−39.3965	3.67087	−88.3008	66.6511	0.589956	1.39591
	20	−52.1037	5.14655	−99.0546	66.6509	0.653935	1.48271
c_o	−20	−15.1336	2.16136	−76.5598	66.6567	0.73762	1.50629
	−10	−12.3277	1.78224	−75.2828	66.6583	0.782661	1.5389
	10	−8.14765	1.20847	−73.0026	66.6608	0.88409	1.61705
	20	−5.0935	0.778559	−71.5169	66.6627	1.00437	1.72242
c_h	−20	−7.72601	1.29576	−73.6571	66.6612	0.989324	1.7288
	−10	−6.72488	1.06964	−73.2569	66.6617	0.970563	1.70607
	10	−0.58305	0.0053467	−71.1429	66.6606	0.033071	0.747357
	20	−0.0229076	0.00508875	−71.1424	66.6666	2.92149	3.63577
δ	−20	−62.0047	7.92441	−57.6059	66.6416	0.639484	1.36245
	−10	−65.8165	8.40825	−72.8153	66.639	0.600361	1.41267
	10	−48.6975	6.46064	−98.3363	66.6433	0.606897	1.50445
	20	−36.5466	4.99855	−110.489	66.6476	0.637643	1.56208
c_b	−20	−158.404	20.4196	−128.725	66.6128	0.52115	1.81753
	−10	−46.4611	6.20918	−92.3453	66.6525	0.97746	1.90462
	10	−86.1187	11.2951	−83.9905	66.6368	0.663886	1.50716
	20	−88.5031	11.5872	−77.9307	66.6355	0.646016	1.42845

(continued)

Table. 1 (continued)

Parameters	% change in parameters	$TP^*(Q,B,p)$	% change in				
			Q^*	B*	p^*	t_1^*	T^*
c_l	−20	−96.1152	12.5839	−96.03	66.6324	0.617396	1.58155
	−10	−108.402	14.058	−97.2673	66.0629	0.600265	1.57684
	10	−74.9873	9.84529	−88.3002	66.6398	0.679625	1.56617
	20	−12.4022	1.81025	−72.2315	66.66	0.991472	1.71669
c	−20	−27.115	3.81426	−196.876	66.6582	1.2209	3.19757
	−10	−27.046	3.74507	−138.273	66.6563	0.986241	2.37452
	10	−288.747	4.03877	−68.9485	66.6507	0.67001	1.36226
	20	−1606.52	19.2747	−175.476	66.6285	0.715904	2.47771
β	−20	−27.7307	3.7961	−78.5236	66.6529	0.735037	1.52343
	−10	−40.1303	5.55112	−81.6435	66.649	0.735811	1.55553
	10	−53.7532	7.85169	−85.3352	66.6468	0.791706	1.64848
	20	−4.5234	0.79847	−72.6253	66.6619	0.996905	1.72607
γ	−20	−225.375	43.2169	−125.654	66.6149	0.856817	2.1184
	−10	−221.008	35.0901	−120.339	66.6123	0.719986	1.92821
	10	−77.0521	9.44888	−89.0063	66.6447	0.81098	1.70462
	20	−0.520498	0.132264	−71.35	66.6657	1.42277	2.13913
a	−20	8259.57	0.0293595	−491.742	190.838	−0.000004	4.93716
	−10	−21,464.2	2015.03	−2175.14	59.7982	2.20391	24.0427
	10	−78.7311	17.1638	−96.0064	66.6379	0.003806	0.967726

(continued)

Table. 1 (continued)

Parameters	% change in parameters	$TP^*(Q,B,p)$	% change in				
			Q^*	B*	p^*	$t_1{}^*$	T^*
	20	−488.513	17.4153	−133.816	66.6291	0.0019228	1.34546
b	−20	−329.104	5.935	−110.679	66.651	0.0010098	1.11224
	−10	−375.102	9.8307	−118.1	66.6437	0.002728	1.18847
	10	−69.3142	9.89997	−11.3825	60.5581	0.347256	0.461538
	20	5387.53	0.0166383	−311.582	144.151	−0.000003	3.12833
M	−20	−213.364	28.2477	−114.607	66.6083	0.588805	1.73948
	−10	−213.71	28.287	−114.95	66.6083	0.589296	1.74341
	10	−294.79	38.4192	−128.427	66.5944	0.571923	1.86135
	20	−156.352	20.365	−105.846	66.6329	0.83635	1.89906

Acknowledgements First author is thankful to MHRD for giving financial support for carrying research at NIT Silchar. Also, she is thankful to TEQIP-III for providing financial support to present this paper in AMSE 2019 at Bhubaneshwar.

References

1. H.K. Alfares, Inventory model with stock-level dependent demand rate and variable holding cost. Int. J. Prod. Econ. **108**(1–2), 259–265 (2007)
2. H.K. Alfares, A.M. Ghaithan, Inventory and pricing model with price-dependent, time-varying holding cost, and quantity discounts. Comput. Ind. Eng. **94**, 170–177 (2016)
3. S.C. Chen, C. Barrón, L. E., & Teng, J. T. , Retailer's economic order quantity when the suppliers offers conditionally permissible delay in payments link to order quantity. Int. J. Prod. Econ. **155**, 284–291 (2014)
4. T.P. Hsieh, C.Y. Dye, Optimal dynamic pricing for deteriorating items with reference price effects when inventories stimulate demand. Eur. J. Oper. Res. **262**(1), 136–150 (2017)
5. S. Mukhopadhyay, R.N. Mukherjee, K.S. Chaudhuri, Joint pricing and ordering policy for a deteriorating inventory. Comput. Ind. Eng. **47**(4), 339–349 (2004)
6. R. Maihami, I.N. Kamalabadi, Joint pricing and inventory control for non-instantaneous deteriorating items with partial backlogging and time and price dependent demand. Int. J. Prod. Econ. **136**(1), 116–122 (2012)
7. B. Sarkar, P. Mandal, S. Sarkar, An EMQ model with price and time dependent demand under the effect of reliability and inflation. Appl. Math. Comput. **231**, 414–421 (2014)
8. L.E.C. Barrón et al., An EOQ inventory model with nonlinear stock dependent holding cost, nonlinear stock dependent demand and trade credit. Comput. Indus. Eng. (2018) (in Press)

An Inventory Model for Seasonal Deteriorating Items with Price Dependent Demand and Preservation Technology Investment in Crisp and Fuzzy Environments

Swagatika Sahoo and Milu Acharya

Abstract The main objective of this research is to establish an inventory model for seasonal deteriorating items, in which demand for the product is price dependent and dealer invests in preservation technology to reduce the rate of deterioration of a product in crisp and fuzzy environment. So, this work considers both deterioration and seasonal properties of the inventory simultaneously. Our main aim is to take the optimal decisions about selling price, preservation technology investment, frequency of orders and the total profit. A numerical example along with its sensitivity analysis is included to illustrate the model. This work ends with a conclusion and possible future directions.

Keywords Inventory · Fuzzy · Triangular fuzzy numbers · Signed distance method · Preservation technology · Defuzzification

1 Introduction

Over the last few years, inventory problems for deteriorating items have been broadly investigated. Deterioration is otherwise known as decompose, change, or spoilage of items that are not in a condition of being used for their actual purpose. Various researchers have explored this problem over time.

Ghare and Scharder [7] were the first who introduced the idea of deterioration in their proposed work. Covert and Philip [10] and Mishra [11] introduced variable rate of deterioration in their proposed models. Sana et al. [15] developed a production inventory model for deteriorating items with shortages. Ghosh et al. [13] explored an optimal price of a finite production model for perishable products. Sahoo et al. [16] considered three rates of EOQ/EPQ model for deteriorating items under shortages.

S. Sahoo · M. Acharya (✉)
Department of Mathematics, Institute of Technical Education and Research, Siksha 'O' Anusandhan (Deemed To Be University), Bhubaneswar, Odisha, India
e-mail: miluacharya@soa.ac.in

S. Sahoo
e-mail: swagatika.lucy@gmail.com

S. R. Mishra et al. (eds.), *Recent Trends in Applied Mathematics*, Lecture Notes in Mechanical Engineering, https://doi.org/10.1007/978-981-15-9817-3_18

An inventory model to minimize the deterioration rate with the expenditure on preservation technology was demonstrated by Hsu et al. [8]. Dye and Hsieh [2] extended the work of Hsu et al. by incorporating a time varying rate of deterioration. Dye [1] studied on a non-instantaneous deteriorating inventory model with preservation technology investment. Again, increase in preservation cost leads to decrease in the rate of deterioration of an inventory model was discussed in [5].

Shah et al. [6] discussed on the inventory model for fashion goods with a quadratic time-dependent demand. He and Huang [17] proposed on pricing strategies of a seasonal model by including preservation technology investment.

In crisp models, some of the parameters are known and have definite values. In case of rapidly changing market scenario, mostly the parameters involved in inventory problems cannot have known and definite values. In such cases the models might be formulated either in fuzzy or in probabilistic considerations.

To meet the unstable market situations, researchers have proposed different fuzzy inventory models, which is just one part of the problem. In case of crisp models, different types of cost parameters or coefficients associated in demand function, deterioration function, are assumed either constants or time dependent. However, these assumptions create problems while taking decisions. Therefore, formulations of inventory models with different fuzzy parameters were become the prime objective of researchers. Researchers related to this area are Zimmermann [3, 4], Bellman and Zadeh [9], etc. The other part of the problem is to defuzzify and get the result in crisp sense. In defuzzification analysis, particularly on ranking fuzzy numbers, the benefaction of Yager [12] put up conclusive termination. After some years, a good number of researchers took initiative to study on the ranking methods and finally derived many simulated formulae on the subject.

Novelty behind the formulation of this production inventory model: (i) production model for deteriorating items; (ii) demand is taken as linearly price dependent; (iii) fuzzy model is presented using triangular fuzzy for the disposal cost; (iv) fuzzy total cost is defuzzified by using signed distance method via Yager's ranking index method.

The research work is organized in the following order: Sect. 1 establishes introduction and literature review. Section 2 presents preliminary discussion and Sect. 3 illustrates the notations and assumptions. Section 4 presents model formulation in both crisp and fuzzy sense. Section 5 describes numerical examples and sensitivity analysis and managerial insights. Conclusion and future work are given in Sect. 6.

2 Preliminaries

2.1 *Normalized General Triangular Fuzzy Number (NGTFN) [12]*

Let D be a NGTFN having the form $\tilde{D} = (a_1, a_2, a_3)$. Then its membership function is defined by

$$\mu\left(\tilde{D}\right) = \begin{cases} \frac{D-a_1}{a_2-a_1}\ if\ a_1 \le D \le a_2 \\ \frac{a_3-D}{a_3-a_2}\ if\ a_2 \le D \le a_3 \\ 0\ if\ D < a_1\ and\ D > a_2 \end{cases} \tag{1}$$

Now, the left and right cuts of $\mu\left(\tilde{D}\right)$ are given by

$$L(\alpha) = a_1 + \alpha(a_2 - a_1)\ and\ R(\alpha) = a_3 - \alpha(a_3 - a_2) \tag{2}$$

Yager's Ranking Index [12]

If $L\alpha$ and $R\alpha$ are the left and right α-cuts of a fuzzy number then $\tilde{D}$ the defuzzification rule under Yager's ranking index is given by

$$I\left(\tilde{D}\right) = \frac{1}{2}\int_0^1 [L(\alpha) + R(\alpha)]d\alpha = \frac{1}{4}[a_1 + 2a_2 + a_3].$$

3 Notations and Assumptions

The following are the subsections meant for the notations and the required assumptions for the model problem:

3.1 *Notations*

Decision Variables:

- n_1: frequency of orders
- β: Unit wise preservation technology cost
- p: Selling price.

Parameters:

- B_1: Unit wise buying cost
- h_c: Unit wise inventory holding cost in per unit time
- O_c: Unit wise ordering cost per order
- D_c: Unit wise disposal cost for the deteriorated items
- M_d: Deteriorated items (cycle wise)
- $D_1(p)$: Price dependent demand
- $T_{AP}(n_1, \beta, p)$: Total profit (per unit) in per unit time.

Notations for Fuzzy Model

- $\tilde{T}_{AP}(n_1, \beta, p)$: Fuzzy total profit in per unit time
- $\tilde{D}_c$: Disposal cost per unit time.

3.2 Assumptions

(1) $D_1(p) = l - mp$ (l = scale factor, m = index of price elasticity).
(2) $D_1(p)$ is considered in a limited time horizon T.
(3) Shortages are not allowed.
(4) Lead time is zero.
(5) Disposal cost, denoted by D_c, is added to the total cost incurred due to the disposal of deteriorating products.

For the present problem, there is an interdependence between deterioration rate ($\mu(\beta)$) and preservation technology investment (β) and which satisfies $\frac{\partial\mu(\beta)}{\partial\beta} < 0$, $\frac{\partial^2\mu(\beta)}{\partial^2\beta} > 0$. In the present research,$\mu(\beta) = \mu_0 e^{-\tau\beta}$ where μ_0 is the deterioration rate without preservation technology and the small change takes place for deterioration due to β is denoted as τ.

4 Mathematical Model Formulation

4.1 Model 1(Crisp)

(1) Deterioration rate is reduced with the application of preservation technology.
(2) Here, there are two arrangements; (i) with the increase in n_1, there can be a decrease in the deterioration cost and the ordering cost increases, (ii) with the increase in β, the deterioration cost decreases.
(3) The purchased materials are sent to the production center to produce the required product.
(4) To fulfill the demand of customers the product is stored as inventory.

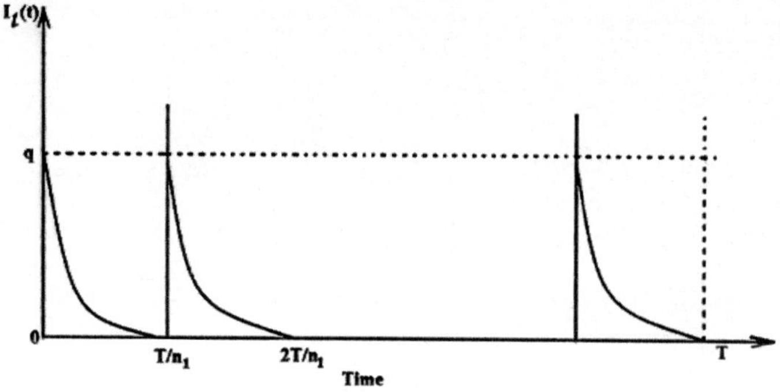

Fig. 1 The inventory systems

(5) Stored items from the inventory start deteriorating at a certain rate and are transferred to the disposal centers.
(6) Here for all the ordering periods the time length is equal. So we study on the first period (Fig. 1).

The rate of change of inventory at any instant of time t is mentioned as follows:

$$\frac{\partial I_l(t)}{\partial t} = -\mu(\beta)I_l(t) - D_1(p), \left(t \in \left[0, \frac{T}{n_1}\right]\right) \tag{3}$$

The boundary condition is

$$I_l\left(\frac{T}{n}\right) = 0 \tag{4}$$

By solving (1) we have

$$I_l(t) = \frac{D_1(p)}{\mu(\beta)}\left(e^{\mu(\beta)(T/n_1 - t)} - 1\right) \tag{5}$$

1. **Sales revenue**
The total sales revenue in time T is calculated as

$$R_s = p.D_1(p).T \tag{6}$$

2. **Buying cost**
The total buying cost can be calculated as

$$P_c = n_1.B_1.q = n_1.B_1.\frac{D_1(p)}{\mu(\beta)}\left(e^{\mu(\beta)(T/n_1 - t)} - 1\right) \tag{7}$$

3. **Holding cost**
   The total inventory holding cost can be calculated as

$$C_H = h_c.n_1.D_1(p).\int_0^{\frac{T}{n}} I_l(t)dt = h_c.n_1.\frac{D_1(p).\left(e^{\mu(\beta)(T/n_1)} - \mu(\beta)(T/n) - 1\right)}{\mu^2(\beta)} \tag{8}$$

4. **Ordering cost**
   The total ordering cost can be calculated as

$$C_O = n_1.o_c \tag{9}$$

5. **Cost for Preservation Technology**
   Preservation technology cost can be calculated as

$$C_p = \beta.T \tag{10}$$

6. **Disposal cost**

The disposal cost incurred due to the deterioration of products in spite of the application of preservation technology is calculated as follows:

$$M_d = \int_0^{\frac{T}{n_1}} \mu_0.I_l(t)dt = \frac{\left(\mu_0.T^2.D_1(p)\right)}{2n}$$

The disposal cost per cycle is calculated as

$$D_cM_d = \frac{D_c.\left(\mu_0.T^2.D_1(p)\right)}{2n} \tag{11}$$

So the profit function is defined as

$$T_{AP}(n_1, \beta, p) = R_s - P_c - C_H - C_O - C_p - D_cM_d \tag{12}$$

$$\begin{aligned} T_{AP}(n_1, \beta, p) &= R_s - P_c - C_H - C_O - C_p - D_cM_d \\ &= p.D_1(p).T - n_1.B_1.\frac{D_1(p)}{\mu(\beta)}\left(e^{\mu(\beta)(T/n_1-t)} - 1\right) \\ &-h_c.n_1.\frac{D_1(p).\left(e^{\mu(\beta)(T/n_1)} - \mu(\beta)(T/n) - 1\right)}{\mu^2(\beta)} - n_1.o_c - \beta.T - \frac{D_c.\left(\mu_0.T^2.D_1(p)\right)}{2n} \end{aligned} \tag{13}$$

From Taylor's series we know that for small value of μ and T/n

$$e^{\mu T/n_1} \approx 1 + \frac{\mu T}{n_1} + \left(\frac{\mu T}{n_1}\right)^2/2. \tag{14}$$

Substituting Eq. (14) in Eq. (13) we get

$$T_{AP}(n_1, \beta, p) = p.D_1(p).T - B_1.D_1(p)\left(T + \frac{T^2}{2n_1}\mu(\beta)\right) - h_c.D_1(p).\frac{T^2}{2n} - n_1.o_c$$
$$-\beta.T - \frac{D_c.\mu_0.T^2.D_1(p)}{2n}. \qquad (15)$$

4.2 *Model 2(Fuzzy)*

Here the disposal cost D_c follows the fuzzy extensibility for the inventory runtime.
Using (1)–(2) the index values of fuzzy total profit is given by

$$\tilde{T}_{AP}(n_1, \beta, p) = p.D_1(p).T - B_1.D_1(p)\left(T + \frac{T^2}{2n_1}\mu(\beta)\right) - h_c.D_1(p).\frac{T^2}{2n} - n_1.o_c$$
$$-\beta.T - \frac{(D_{c_1} + D_{c_2} + D_{c_3}).\left(\mu_0.T^2.D_1(p)\right)}{8n} \qquad (16)$$

5 Numerical Examples

This section includes a numerical study to show the applicability of the suggested model (Tables 1, 2 and 3).

Table 1 Values taken for different parameters

μ_0	β	h_c	O_c	l	m	B_1	D_c	D_{c_1}	D_{c_2}	D_{c_3}	τ
0.02	10	1	10	0.2	10	10	0.1	0.1	0.2	0.3	0.5

Table 2 Optimum solution of the present Crisp model for $D_c = 0.1$ and 0.5

$D_c = 0.1$					$D_c = 0.5$				
n_1	β	p	μ	$T_{AP}(n_1, \beta, p)$	n_1	β	p	μ	$T_{AP}(n_1, \beta, p)$
3	2.63	38.78	0.0053	687.09	3	2.63	38.78	0.0053	417.81
4	2.63	38.78	0.0053	1149.85	**4**	**2.63**	**38.78**	**0.0053**	**790.81**
5	**2.63**	**38.78**	**0.0053**	**1218.54**	5	2.63	38.78	0.0053	769.81
6	2.63	38.78	0.0053	1090.18	4	2.45	38.78	0.0058	794.63
5	2.45	38.78	0.0058	1225.19	4	2.25	38.78	0.0064	797.30
5	2.25	38.78	0.0064	1231.32	**4**	**2.05**	**38.78**	**0.0071**	**798.14**
5	2.05	38.78	0.0071	1236.04	4	1.85	38.78	0.0079	796.97
5	**1.85**	**38.78**	**0.0079**	**1239.06**	**4**	**2.05**	**36.78**	**0.0071**	**804.66**
5	1.25	38.78	0.0107	1236.80	4	2.05	34.78	0.0071	651.87
5	1.85	36.78	0.0079	1409.74					
5	**1.85**	**34.78**	**0.0079**	**1420.42**					
5	1.85	32.78	0.0079	1271.10					

Table 3 Optimum solution for the problem in fuzzy sense

n_1	β	p	μ	$T_{AP}(n_1, \beta, p)$
3	2.63	38.78	0.0053	653.43
4	2.63	38.78	0.0053	1149.77
5	**2.63**	**38.78**	**0.0053**	**1162.44**
6	2.63	38.78	0.0053	1022.86
5	2.45	38.78	0.0058	1169.09
5	2.25	38.78	0.0064	1175.22
5	2.05	38.78	0.0071	1179.99
5	1.85	38.78	0.0079	1182.69
5	**1.65**	**38.78**	**0.0087**	**1184.25**
5	1.45	38.78	0.0096	1183.35
5	1.65	36.78	0.0087	1341.59
5	1.65	34.78	0.0087	1338.93

5.1 Inputs for the Numerical Study

5.1.1 Solutions and Discussions

5.2 Sensitivity Analysis

Table 4 presents the sensitivity analysis of $\mu(\beta)$ and $T_{AP}(n_1, \beta, p)$ by changing the values of the parameters in a range from -50 to $+ 50\%$, taking one parameter at a time keeping the other parameters to be constant.

In model 1, $T_{AP}(n_1, \beta, p)$ first increases and then decreases with the increase in B_1/h_c and decreases in $O_c/D_c/\mu_0$.

Table 4 Sensitivity test in Crisp and Fuzzy sense respectively

Para-meter	Changes (%)	μ	$T_{AP}(n_1, \beta, p)$	Para-meter	Changes (%)	μ	$T_{AP}(n_1, \beta, p)$
B_1	−50	0.0079	2454.32	B_1	−50	0.0087	2779.46
	−25	0.0079	3275.67		−25	0.0087	2060.52
	+25	0.0079	−9728.40		+25	0.0087	6226.58
	+50	0.0079	−8310.80		+50	0.0087	−96.27
O_c	−50	0.0079	2016.60	O_c	−50	0.0087	2591.59
	−25	0.0079	1436.62		−25	0.0087	1966.59
	+25	0.0079	1866.23		+25	0.0087	7165.93
	+50	0.0079	−438.37		+50	0.0087	91.59
h_c	−50	0.0079	2333.62	h_c	−50	0.0087	2663.59
	−25	0.0079	3094.62		−25	0.0087	2002.59
	+25	0.0079	50.623		+25	0.0087	6805.93
	+50	0.0079	−710.37		+50	0.0087	19.59
D_c	−50	0.0079	1192.12	μ_0	−50	0.0043	1556.61
	−25	0.0079	1001.87		−25	0.0065	1449.10
	+25	0.0079	621.37		+25	0.0109	1234.08
	+50	0.0079	431.12		+50	0.0131	1226.57
μ_0	−50	0.0039	1312.82	τ	−50	0.0132	1223.26
	−25	0.0059	1062.22		−25	0.0107	1288.51
	+25	0.0099	561.02		+25	0.0071	1384.78
	+50	0.0118	310.41		+50	0.0058	1419.92
τ	−50	0.0125	669.66				
	−25	0.0099	748.81				
	+25	0.0062	861.46				
	+50	0.0049	901.01				

In model 2, $T_{AP}(n_1, \beta, p)$ first increases and then decreases with respect to the increase in $O_c/B_1/\mu_0/\tau$.

6 Conclusion

This model built up the connection between seasonal deteriorating products and the cost of the preservation technology through the cleaning of the environment with a disposal cost under price dependent demand. It is established that by applying β, reducing the deterioration rate.

Sensitivity test was exhibited to observe the effects of variations of the Parameters $\mu(\beta)$ and $T_{AP}(n_1, \beta, p)$. This work can still be studied by introducing shortages and backlogging of items.

References

1. C.Y. Dye, The effect of preservation technology investment on a non instantaneous deteriorating inventory model. Omega **41**(5), 872–880 (2013)
2. C.Y. Dye, T.P. Hsieh, An optimal replenishment policy for deteriorating items with effective investment in preservation technology. Eur. J. Oper. Res. **218**(1), 106–112 (2012)
3. H.J. Zimmerman, Fuzzy programming and linear programming with several objective functions. Fuzzy Sets Syst. **1**(1), 45–55 (1978)
4. H.J. Zimmermann, *Fuzzy Set Theory and Its Applications*, 3rd edn. (Kluwer Academic Publishers, Dordrecht, 1996).
5. N.H. Shah, U. Chaudhari, M.Y. Jani, Optimal policies for time varying deteriorating item with preservation technology under selling price and trade credit dependent quadratic demand in a supply chain. Int. J. Appl. Comput. Math. **3**, 363–379 (2017)
6. N.H. Shah, A.S. Gor, C.A. Jhaveri, Determination of optimal ordering and transfer policy for deteriorating inventory system when demand is quadratic. Int. J. Manag. Sci. Eng. Manag. **6**, 278–283 (2011)
7. P.M. Ghare, G.P. Schrader, A model for exponentially decaying inventories. J. Ind. Eng. **14**, 238–243 (1963)
8. P.H. Hsu, M. Wee, H.M. Teng, Preservation technology investment for deteriorating inventory. Int. J. Prod. Econ. **124**(2), 388–394 (2010)
9. R.E. Bellman, L.A. Zadeh, Decision-making in a fuzzy environment. Manage. Sci. **17**, B141–B164 (1970)
10. R.P. Covert, G.C. Philip, An EOQ model for items with Weibull distribution deterioration. AIIE Transit. **5**, 323–326 (1973)
11. R.B. Mishra, Optimum production lot size model for a system with deteriorating inventory. Int. J. Prod. Res. **13**(5), 495–505 (1975)
12. R.R. Yager, A procedure for ordering fuzzy subsets of the unit interval. Inf. Sci. **24**, 143–161 (1981)
13. S.K. Ghosh, S. Khanra, K.S. Chauduri, Optimal price and lot size determination for a perishable product under conditions of finite production, partial backlogging and lost sale. Appl. Math. Comput. **217**(13), 6047–6053 (2011)
14. S.K. Manna, C. Chiang, Economic production quantity models for deteriorating items with ramp type demand. Int. J. Oper. Res. **7**(4), 429 (2010)

15. S. Sana, S.K. Goyal, K.S. Chaudhuri, A production-inventory model for a deteriorating item with trended demand and shortages. Eur. J. Oper. Res. **157**(2), 357–371 (2004)
16. S. Sahoo, M. Acharya, M.M. Nayak, A three rates of EOQ/EPQ model for instantaneous deteriorating items involving fuzzy parameter under shortages. Int. J. Innov. Technol. Explor. Eng. **8**(8S2), 405–418 (2019)
17. Y. He, H. Huang, Optimizing inventory and pricing policy for seasonal deteriorating products with preservation technology investment. J. Ind. Eng. 1–7 (2013)

A Study of Propagation of Love Waves in an Anisotropic Porous Layer Under Initial Stresss

Pankaj, P. K. De, and Alok Singh

Abstract The objective of this paper is to study the propagation of Love waves in a pre-stressed anisotropic porous layer which is a sandwich between a rigid layer and a non-homogeneous elastic half-space. Dynamical equations of motion are based on Biot's incremental deformation theory. Dispersion equations of phase velocity of Love waves have been derived. The impact of anisotropy, pre-stresses, and porosity on the phase velocity are examined in details. It has been observed that the velocity of Love waves increases when the porosity of the layer decreases. Also, velocity of Love waves increases when anisotropy in the layer and compressive initial stresses both are increased. Numerical computations for the velocity of Love waves have been carried out and the results are plotted in different graphs.

1 Introduction

In both practical applications and theory, the Earth is supposed to be isotropic or consists of isotropic layers. However, some studies show the presence of anisotropy. Also, Earth is an initially stressed medium. The term "Initial Stress" is defined as stresses that exist in a medium due to its internal properties not subjected to the action of exterior forces. It is also called pre-stress. These initial stresses may be developed in a medium by any artificial process or due to natural phenomena. Due to the variation of temperature, pressure, external loading, inelastic deformations, etc., a huge quantity of stresses stored inside a medium which is called initial stress. In the recent years, the study of wave propagation has gained attention of the researchers due to its practical importance in underground water, geophysics, and exploration of

Pankaj (✉) · P. K. De · A. Singh
Department of Mathematics, National Institute of Technology Silchar, Silchar 788010, India
e-mail: pankajnarang830@gmail.com

P. K. De
e-mail: pijusde@gmail.com

A. Singh
e-mail: alok.rawat478@gmail.com

S. R. Mishra et al. (eds.), *Recent Trends in Applied Mathematics*, Lecture Notes in Mechanical Engineering, https://doi.org/10.1007/978-981-15-9817-3_19

oil. Love waves travel in the sub-crustal and crustal layers which are near to the earth surface. This crust and sub-crust of the earth may be anisotropic in nature. Several authors have discussed wave propagation through porous medium by using the theory given by Biot [1]. Dey and De [2] in 1984 discussed Love wave propagation in a non-homogenous anisotropic layer lying over a pre-stressed orthotropic medium. Dey et al. [3] in 1987 studied propagation of Love waves in an initially stressed anisotropic porous layer which is sandwiched between a non-homogenous elastic half-space and a rigid layer. Gupta et al. [4] studied the effect of pre-stress on the propagation of Love waves in an anisotropic porous layer. By using the theory given by Biot [1], several authors, viz., Khurana and Vashisth [5], Gupta et al. [6], Vaishnav et al. [7], and others have been discussed problems of propagation of Love waves in porous and pre-stressed medium. It has been observed that the presence of pre-stress, porosity, and irregularity affect the Love wave propagation.

In this paper, it is assumed that the layer is anisotropic, water saturated, and porous in nature. The attention has been paid to the effect of pre-stress, porosity, and non-homogeneity on the propagation of Love waves, and velocity equation is derived in a simplified form. In half-space, the inhomogeneity has been presumed to be $\rho = \rho_0 e^{bz}$ and $\mu = \mu_0 e^{bz}$ where b is a positive constant and z is depth. The phase velocity of Love waves has been obtained by using the values of material constants given by Biot [8]. It has been observed that the velocity of Love waves is affected by anisotropy, inhomogeneity, and porosity. All the graphs that are shown in this paper are generated by using MATLAB software. Full views of the graphs are plotted which are not shown by the Dey et al. [3]. There are many uses of the study of propagation of seismic waves like it provides guidance to the civil engineers in the construction of a building and it also helps in analysing the earthquake in the hilly areas.

2 Formulation and Geometry of the Problem

We consider a sandwiched earth model with a half space at the bottom and a rigid layer at the top. It is assume that the layer is anisotropic, water saturated, and porous in nature which is under compressive pre-stress $P' = -S_{11}^{'}$ along the direction of x. The lower half space is non-homogeneous and elastic in nature which is under compressive pre-stress $P = -S_{11}$ in the direction of x. The plane of contact between the porous layer and half space is at $z = 0$ and $z = -H$ is the plane of contact between rigid boundary and porous layer. Z-axis directing vertically downwards. The wave propagation is performed along the direction of x (Fig. 1).

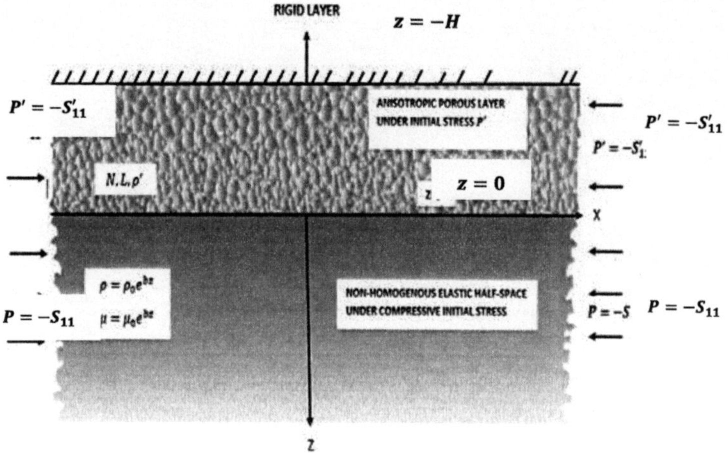

Fig. 1 Geometry of the problem

3 Porous Layer

Neglecting the body forces and viscosity of water, the equations of motions in the porous layer under pre-stress can be written as (Biot [1, 9])

$$
\begin{gathered}
\frac{\partial s_{11}'}{\partial x}+\frac{\partial s_{12}'}{\partial y}+\frac{\partial s_{13}'}{\partial z}-P'\frac{\partial w_z'}{\partial y}+P'\frac{\partial w_y'}{\partial z}=\frac{\partial^2}{\partial t^2}\left(\rho_{11}u_x'+\rho_{12}U_x\right)\\
\frac{\partial s_{21}'}{\partial x}+\frac{\partial s_{22}'}{\partial y}+\frac{\partial s_{23}'}{\partial z}-P'\frac{\partial w_z'}{\partial x}=\frac{\partial^2}{\partial t^2}\left(\rho_{11}v_y'+\rho_{12}V_y\right),\\
\frac{\partial s_{31}'}{\partial x}+\frac{\partial s_{32}'}{\partial y}+\frac{\partial s_{33}'}{\partial z}-P'\frac{\partial w_y'}{\partial x}=\frac{\partial^2}{\partial t^2}\left(\rho_{11}w_z'+\rho_{12}W_z\right),\\
\frac{\partial S}{\partial x}=\frac{\partial^2}{\partial t^2}\left(\rho_{12}u_x'+\rho_{22}U_x\right),\ \frac{\partial S}{\partial y}=\frac{\partial^2}{\partial t^2}\left(\rho_{12}v_y'+\rho_{22}V_y\right),\ \frac{\partial S}{\partial z}=\frac{\partial^2}{\partial t^2}\left(\rho_{12}w_z'+\rho_{22}W_z\right),
\end{gathered}
\tag{1}
$$

where $s_{ij}'(i,j=1,2,3)$ = incremental stress components, $\left(u_x', v_y', w_z'\right)$ = Vector form of displacement components for the solid, $\left(U_x, U_y, U_z\right)$ = Vector form of displacement components for the liquid,

S = Vector form of stress due to the liquid

$$
\omega_x'=\frac{1}{2}\left(\frac{\partial w_z'}{\partial y}-\frac{\partial v_y'}{\partial z}\right),\ \omega_y'=\frac{1}{2}\left(\frac{\partial u_x'}{\partial z}-\frac{\partial w_z'}{\partial x}\right),\ \omega_z'=\frac{1}{2}\left(\frac{\partial v_y'}{\partial x}-\frac{\partial u_x'}{\partial y}\right)
\tag{2}
$$

where $\omega_x', \omega_y',$ and ω_z' are the rotational vector (ω') components.

Stress–strain relations for the upper layer are

$$
s_{11}'=\left(D+P'\right)e_{xx}+\left(D-2N+P'\right)e_{yy}+\left(G+P'\right)e_{zz}+Q\varepsilon,
$$

$$s'_{22} = (D - 2N)e_{xx} + De_{yy} + Ge_{zz} + Q\varepsilon,$$
$$s'_{33} = Ge_{xx} + Ge_{yy} + He_{zz} + Q\varepsilon,$$
$$s'_{12} = 2Ne_{xy}, s'_{23} = 2Le_{yz}, s'_{13} = 2Le_{zx} \tag{3}$$

where $H, G, D, L,$ and N represent the elastic constants of the medium. L and N are the shear moduli of the layer which is anisotropic in nature along the direction of z and x, respectively, and

$$e_{ij} = \frac{1}{2}\left(\frac{\partial u_i}{\partial x_j} + \frac{\partial u_j}{\partial x_i}\right), \varepsilon = \frac{\partial U_x}{\partial x} + \frac{\partial U_y}{\partial y} + \frac{\partial U_z}{\partial z} \tag{4}$$

Here, Q is the measure of coupling between the change in volume of the liquid and solid, p is the Fluid pressure, f = porosity of the layer and ρ', ρ_w, ρ_s are the densities of layer, water, and solid,

$\rho_{11}, \rho_{12,}$ and ρ_{22} are the mass coefficients related to ρ', ρ_s, ρ_w

$$-S = fp,$$
$$\rho_{11} + \rho_{12} = (1 - f)\rho_s,$$
$$\rho_{12} + \rho_{22} = f\rho_w \tag{5}$$

Aggregate mass density is

$$\rho' = \rho_{11} + 2\rho_{12} + \rho_{22} = \rho_s + f(\rho_w - \rho_s) \tag{6}$$

The following inequalities also obey these mass coefficients:

$$\rho_{11} > 0, \rho_{22} > 0, \rho_{12} < 0, \rho_{11}\rho_{22} - \rho_{12}^2 > 0 \tag{7}$$

4 Solution of the Upper Layer

The propagation of Love waves along x-axis, having the movement of particles along y-axis, we have

$$u'_x = 0, w'_z = 0, v'_y = v'(x, z, t)$$
$$U_x = 0, W_z = 0, V_y = V(x, z, t) \tag{8}$$

Only e_{yz} and e_{xy} strain components will be produced by the above displacements and the remaining strain components will become zero. Hence the strain–stress relations which are helpful in the problem are

$$s'_{23} = 2Le_{yz} \text{ and } s'_{12} = 2Ne_{xy} \tag{9}$$

Using Eq. (9) in Eq. (1), the equations of motion which are not automatically satisfied are

$$\frac{\partial s'_{21}}{\partial x} + \frac{\partial s'_{22}}{\partial y} + \frac{\partial s'_{23}}{\partial z} - P'\frac{\partial w'_z}{\partial x} = \frac{\partial^2}{\partial t^2}\left(\rho_{11}v'_y + \rho_{12}V_y\right),$$
$$\frac{\partial S}{\partial y} = \frac{\partial^2}{\partial t^2}\left(\rho_{12}v'_y + \rho_{22}V_y\right) \tag{10}$$

Using Eqs. (8) and (9), we get

$$\left(N - \frac{P'}{2}\right)\frac{\partial^2 v'}{\partial x^2} + L\frac{\partial^2 v'}{\partial z^2} = \frac{\partial^2}{\partial t^2}\left(\rho_{11}v' + \rho_{12}V\right) \tag{11}$$

$$\frac{\partial^2}{\partial t^2}\left(\rho_{11}v' + \rho_{12}V\right) = 0 \tag{12}$$

Eliminating V from the Eq. (11) with the help of Eq. (12)

$$\left(N - \frac{P'}{2}\right)\frac{\partial^2 v'}{\partial x^2} + L\frac{\partial^2 v'}{\partial z^2} = d'\frac{\partial^2 v'}{\partial t^2} \tag{13}$$

where $d' = \rho_{11} - \frac{\rho_{12}^2}{\rho_{22}}$, from Eq. (13) it is clear that $\sqrt{\frac{N-\frac{P'}{2}}{d'}}$ is the velocity of the shear wave along the direction of x, $\sqrt{\frac{L}{d'}}$ is the shear wave velocity in a permeable layer along the direction of z.

The shear wave velocity in a permeable medium in the direction of x-axis can be expressed as

$$\beta' = \beta_a.\sqrt{\frac{1-\xi'}{d}} \tag{14}$$

where $d = r_{11} - \frac{r_{12}^2}{r_{22}}$, $\beta_a = \sqrt{\frac{N}{\rho'}}$ is the shear wave velocity in the non-porous, elastic, pre-stress free, anisotropic medium in the direction of x-axis under rigid boundary, $\xi' = \frac{P'}{2N}$ is the dimensionless parameter because of the pre-stress P'

$$r_{11} = \frac{\rho_{11}}{\rho'}, r_{12} = \frac{\rho_{12}}{\rho'}, r_{22} = \frac{\rho_{22}}{\rho'} \tag{15}$$

where r_{11}, r_{12}, r_{22} = dimensionless parameters for the porous layer.

The propagation of Love wave along x-axis, the solution of the Eq. (13) is taken as

$$v'(x, z, t) = V(z)e^{ik(x-ct)}$$

Using this in Eq. (13)

$$\frac{\partial^2 V}{\partial z^2} + q'^2 V = 0 \quad \text{where} \quad q'^2 = \frac{k^2}{L}\left[c^2 d' - \left(N - \frac{P'}{2}\right)\right]$$

Solution of this equation is

$$V(z) = Ae^{iq'z} + Be^{-iq'z}$$

Hence the solution is

$$v' = \left(Ae^{iq'z} + Be^{-iq'z}\right)e^{ik(x-ct)}, -H \leq z \leq 0, \tag{16}$$

where $q' = k\sqrt{rd\left(\frac{c^2}{\beta_a^2} - \frac{(1-\xi')}{d}\right)}$, k = wave number, $r = \frac{N}{L}$.

5 Solution of the Half-Space

The equations of motion for the lower medium can be written as (Biot [9])

$$\frac{\partial s_{21}}{\partial x} + \frac{\partial s_{23}}{\partial z} - \frac{P}{2}\left(\frac{\partial^2 v}{\partial x^2}\right) = \frac{\partial^2}{\partial t^2}(\rho v) \tag{17}$$

where P = initial compressive stress in the direction of x-axis, s_{ij} are the components of the incremental stress in half-space, ρ = density for the half-space.

For the medium, the non-homogeneity is defined as

$$\mu = \mu_0 e^{bz}$$
$$\rho = \rho_0 e^{bz} \tag{18}$$

where μ_0 is the value of μ at z = 0, ρ_0 is the value of ρ at $z = 0$ and b is constant.

By the strain–stress relations, we have

$$s_{21} = 2\mu e_{xy}, s_{23} = 2\mu e_{yz} \tag{19}$$

using the relation (18) in the equation of motion (17) gives

$$(1-\xi)\frac{\partial^2 v}{\partial x^2}+\frac{\partial^2 v}{\partial z^2}+b\frac{\partial v}{\partial z}=\frac{1}{\beta_0^2}\frac{\partial^2 v}{\partial t^2} \tag{20}$$

where $\xi = \frac{P}{2\mu}$ and $\beta_0 = \sqrt{\frac{\mu_0}{\rho_0}}$

Let the solution of Eq. (20) be

$$v = g(z)e^{ik(x-ct)}$$

where $g(z)$ satisfies

$$\frac{\partial^2 g(z)}{\partial z^2}+b\frac{\partial g(z)}{\partial z}+k^2\left(\frac{c^2}{\beta_0^2}-(1-\xi)\right)g(z)=0 \tag{21}$$

$$\text{Let } q = k\sqrt{(1-\xi)-\frac{c^2}{\beta_0^2}}$$

$$\frac{\partial^2 g(z)}{\partial z^2}+b\frac{\partial g(z)}{\partial z}-q^2 g(z)=0$$

The above equation has solution

$$g(z) = Ce^{\eta z}+De^{\eta' z} \tag{22}$$

where

$$\eta=\frac{-b+\sqrt{b^2+4q^2}}{2},\ \eta'=\frac{-b-\sqrt{b^2+4q^2}}{2} \tag{23}$$

and $q = k\sqrt{(1-\xi)-\frac{c^2}{\beta_0^2}}$.

As the solution (22) is bounded when $z \to \infty$, the solution for the half-space may become

$$v = Ce^{\eta z}e^{ik(x-ct)},\ 0 \le z \le \infty \tag{24}$$

6 Boundary Conditions and Dispersive Equation

According to the formulation of the problem, the suitable boundary conditions are

$$\begin{aligned} v' &= 0 \text{ at } z = -H \\ L\frac{\partial v'}{\partial z} &= \mu\frac{\partial v}{\partial z} \text{ at } z = 0 \\ v' &= v \text{ at } z = 0 \end{aligned} \tag{25}$$

Now, using Eqs. (16) and (24) in the above boundary equations, we get

$$v' = 0 = \left(Ae^{-iq'H} + Be^{iq'H}\right)e^{ik(x-ct)} = 0 \text{ when } z = -H$$

$$Ae^{-iq'H} + Be^{iq'H} = 0 \qquad (a)$$

$$L\frac{\partial v'}{\partial z} = \mu\frac{\partial v}{\partial z} \text{ at } z = 0$$

$$L(Aiq' - Biq') = \mu_0 C\eta$$

$$iq'L(A - B) = \mu_0 C\eta \qquad (b)$$

$$v' = v \text{ at } z = 0 = (A + B)e^{ik(x-ct)} = Ce^{ik(x-ct)}$$

$$A + B = C \qquad (c)$$

Using the solutions (16) and (24) and the boundary conditions (25) give three homogenous equations in three unknowns A, B, and C

$$\begin{aligned} &Ae^{-iq'H} + Be^{iq'H} = 0 \\ &iq'L(A - B) = \mu_0 C\eta \\ &A + B = C \end{aligned} \tag{26}$$

Eliminating the unknown constants A, B, and C from Eq. (26) the following velocity equation of Love wave is derived.

$$\begin{vmatrix} e^{-iq'H} & e^{iq'H} & 0 \\ iq'L & -iq'L & -\mu_0\eta \\ 1 & 1 & -1 \end{vmatrix} = 0 \tag{27}$$

which on simplification gives the following result

$$\tan\left\{\sqrt{rd\left[\frac{c^2}{\beta_a^2} - \frac{(1-\xi')}{d}\right]}\right\}kH = \frac{2L\sqrt{rd\left[\frac{c^2}{\beta_a^2} - \frac{(1-\xi')}{d}\right]}}{\mu_0\left(\frac{-b}{k} + \sqrt{\frac{b^2}{k^2} + 4\left[(1-\xi) - \frac{c^2}{\beta_0^2}\right]}\right)} \tag{28}$$

where c = phase velocity of Love wave.

The above equation is called the dispersive equation of Love wave in pre-stressed, anisotropic, water saturated, permeable layer above the elastic half-space which is pre-stressed and non-homogeneous.

Condition	Value of d
Layer is non-porous	$d \to 1$
Layer tends to be fluid	$d \to 0$
Layer is porous	$0 < d < 1$

7 Particular Cases

Case 1

Consider both the medium are free from pre-stresses,

$$\text{i.e., } \xi' = \xi = 0$$

Then, Eq. (28) becomes

$$\tan\left\{\sqrt{rd\left[\frac{c^2}{\beta_a^2} - \frac{1}{d}\right]}\right\}.kH = \frac{2L\sqrt{rd\left[\frac{c^2}{\beta_a^2} - \frac{1}{d}\right]}}{\mu_0\left(\frac{-b}{k} + \sqrt{\frac{b^2}{k^2} + 4\left[1 - \frac{c^2}{\beta_0^2}\right]}\right)} \tag{29}$$

Case 2

Consider the upper layer as non-porous.

i.e, $f = 0$, then $d = 1$

Then, Eq. (28) becomes

$$\tan\left\{\sqrt{r\left[\frac{c^2}{\beta_a^2} - (1 - \xi')\right]}\right\}.kH = \frac{2L\sqrt{r\left[\frac{c^2}{\beta_a^2} - (1 - \xi')\right]}}{\mu_0\left(\frac{-b}{k} + \sqrt{\frac{b^2}{k^2} + 4\left[(1 - \xi) - \frac{c^2}{\beta_0^2}\right]}\right)} \tag{30}$$

The above Eq. (30) involving pre-stressed parameter ξ and ξ' represents the dispersive equation of Love wave in a medium which is anisotropic.

Case 3

The Eq. (30) reduces to Eq. (31), when the half-space is homogenous i.e., $b = 0$

$$\tan\left\{\sqrt{r\left[\frac{c^2}{\beta_a^2}-(1-\xi')\right]}\right\}.kH=\frac{L\sqrt{r\left[\frac{c^2}{\beta_a^2}-(1-\xi')\right]}}{\mu_0\left(\sqrt{\left[(1-\xi)-\frac{c^2}{\beta_0^2}\right]}\right)} \tag{31}$$

Case 4

Further, if we consider both the medium are isotropic, homogenous, and free from pre-stress, i.e., $\xi'=0, r=1, L=N=\mu'$ and $b=0$

Then the Eq. (30) changes to

$$\tan\left\{\sqrt{\left[\frac{c^2}{\beta_a^2}-1\right]}\right\}.kH=\frac{\mu'\sqrt{\left[\frac{c^2}{\beta_a^2}-1\right]}}{\mu_0\left(\sqrt{\left[1-\frac{c^2}{\beta_0^2}\right]}\right)} \tag{32}$$

The equation matches with the standard equation of Love wave.

8 Observations

By Eq. (28) we can say that Love waves can propagate in the permeable layer under rigid boundary, if

$$\sqrt{\frac{(1-\xi')}{d}}\beta_a<c<\beta_0\sqrt{\frac{b^2}{4k^2}+(1-\xi)} \tag{33}$$

Following observations are obtained from the condition (33)

Observation 1

When the pre-stresses in the half-space and in the layer are zero,

$$\text{i.e., } \xi'=\xi=0$$

Then, Eq. (33) becomes

$$\sqrt{\frac{1}{d}}\beta_a<c<\beta_0\sqrt{\frac{b^2}{4k^2}+1}$$

which is written as

$$\frac{1}{d}<\frac{c^2}{\beta_a^2}<\frac{\beta_0^2}{\beta_a^2}\left(\frac{b^2}{4k^2}+1\right) \tag{34}$$

The above results indicate that Love wave propagation depends on both the porosity of the medium and the ratio of shear velocities in the half-space and the layer.

From the above equation, we conclude that the phase velocity increases when the value of 'd' decreases, i.e., the porosity of layer increases. As to $\frac{b}{k}$ there is no limitation. The estimation of b/k might be taken positive or negative or zero.

Observation 2

When the porosity and $\frac{\beta_a}{\beta_0}$ are taken as constant and $\frac{b}{k} = 0$. The Eq. (33) becomes

$$\frac{(1-\xi')}{d} < \frac{c^2}{\beta_a^2} < \frac{\beta_0^2}{\beta_a^2}(1-\xi) \tag{35}$$

When $\frac{\beta_a}{\beta_0} = 0.7$ and $d = 0.6154$ are taken as a particular case, the above Eq. (35) changes to

$$(1-\xi')1.625 < \frac{c^2}{\beta_a^2} < 2.041(1-\xi) \tag{36}$$

Consequently the propagation is supported by positive value of ξ' and negative value of ξ, i.e., compressive pre-stress in the porous layer and tensile pre-stress in the half-space.

Observation 3

When we take $\xi' = 0, \xi = 0, d = 1$, and $b = 0$ then Eq. (33) becomes

$$\beta_a < c < \beta_0 \tag{37}$$

9 Numerical Calculations and Discussion

In this section, the graphical representation is provided to reflect the effect of porosity, pre-stress, non-homogeneity, and anisotropy on Love waves propagation under rigid boundary. In all these calculations the values of $\frac{\mu_0}{L}$ and $\frac{\beta_a}{\beta_0}$ have been taken as 2.5 and 0.7, respectively.

Figure 2 manifest the effect of porosity in a homogenous, isotropic, pre-stress free medium below rigid boundary. This figure shows that when the porosity decreases, i.e., the value of d $\left[r_{11} - \left(\frac{r_{12}^2}{r_{22}}\right)\right]$ increases, the velocity of Love waves increases for the same value of kH.

Figure 3 manifests the effect of rigidity and density in the half-space. The curves show that if the rigidity and density decreases with depth ($\frac{b}{k} = -1$), the Love wave

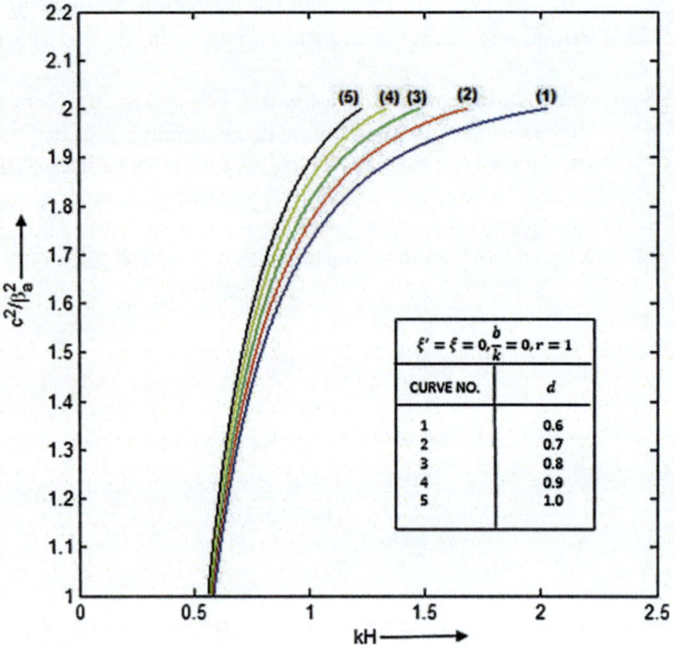

Fig. 2 Love wave dispersion curves in the porous layer for various porosity in terms of d

propagates in a small range of kH and when the rigidity and density increases Love wave propagates in a large range of kH.

Figure 4 manifests the effect of anisotropy in a porous medium. The figure shows when the anisotropy increases, the velocity of Love waves decreases but after a particular value it increases and when layer is non-porous ($d = 1$), the velocity of Love waves increases when the anisotropy increases.

Figure 5 manifests the effect of compressive pre-stresses (ξ') present in the permeable layer when half-space is free from pre-stress, i.e., $\xi = 0$. This figure shows that when the compressive initial stresses are increased in the porous layer, the velocity of Love waves is also increased for the same value of kH.

Figure 6 manifests the effect of tensile pre-stresses ($\xi \leq 0$) in half-space, when the pre-stress is absent in the permeable layer, i.e., $\xi' = 0$. It is analysed that as the magnitude of tensile initial stresses increases in half-space, the velocity of Love waves increases for the same value of kH.

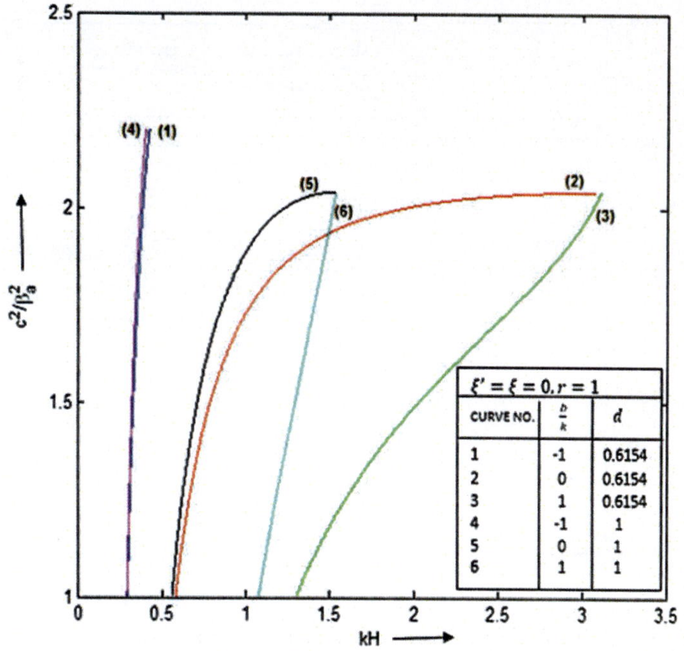

Fig. 3 Effect of non-homogeneity $\left(\frac{b}{k}\right)$ present in the half-space on the propagation of Love waves in the porous layer

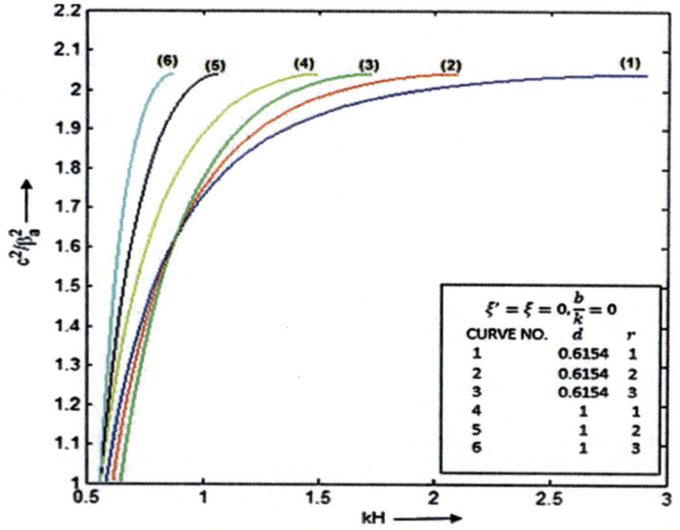

Fig. 4 Effect of anisotropy (r) in the porous layer on the propagation of Love waves

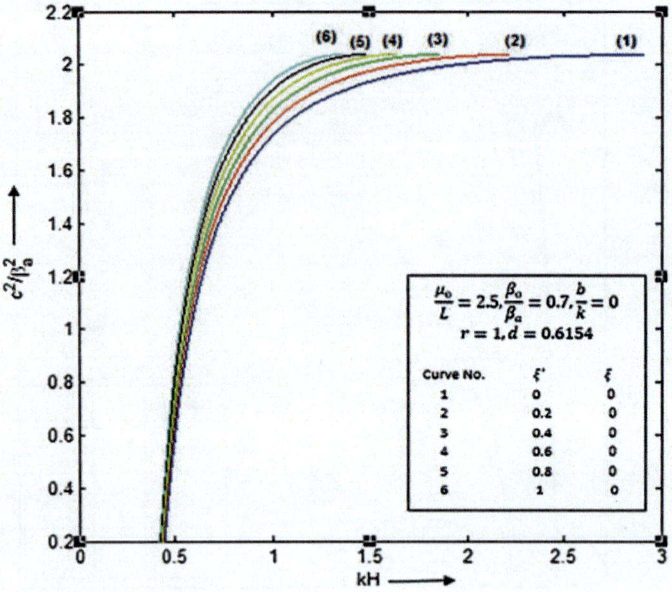

Fig. 5 Love wave dispersion curves when only the porous layer is pre-stressed

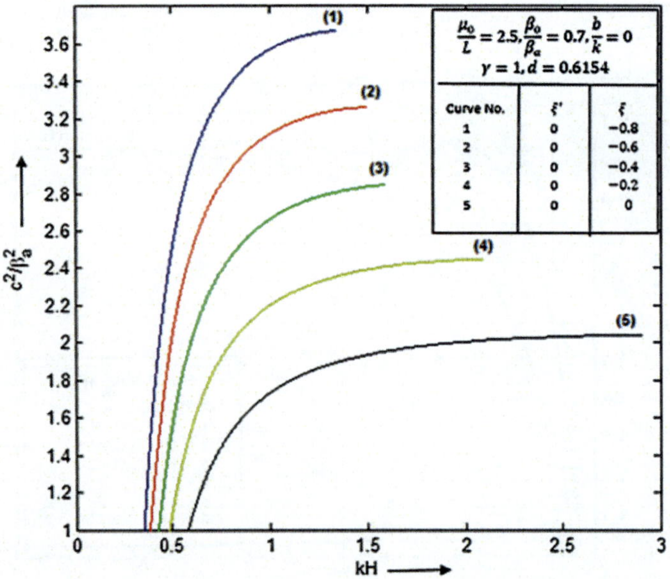

Fig. 6 Love waves dispersion curves when only the half space is under tensile initial stresses $(-0.8 \leq \xi \leq 0, \xi' = 0)$

10 Conclusion

A scientific method is used to examine Love wave propagation in a porous layer under pre-stress which is anisotropic in nature with a half-space at the bottom and a rigid layer at the top. The impact of anisotropy, porosity, and pre-stresses on the wave velocity is examined. It has been examined that when the porosity of the layer decreases, velocity of Love wave increases. The velocity increases with increases in the value of anisotropy and when the compressive pre-stresses increase in the layer the velocity of Love waves increases. As the effect of pre-stresses is concerned, it has been discovered that as the magnitude of tensile pre-stresses increases in half-space then the velocity of Love waves increases.

Acknowledgements First author is thankful to MHRD for giving financial support for carrying research at NIT Silchar. Also, He is thankful to TEQIP III for providing financial support to present this paper in AMSE 2019 at Bhubaneswar.

References

1. M.A. Biot, Theory of deformation of a porous viscoelastic anisotropic solid. J. Appl. Phys. **27**, 459–467 (1956)
2. S. Dey, R.K. De, Love waves in a non-homogenous anisotropic layer lying over a pre-stressed orthotropic medium. Geophys. Res. Bull. **22**(1) (1984)
3. S. Dey, N. Roy, S. Ghosh, Propagation of Love waves in an initially stressed anisotropic porous layer sandwiched between a rigid layer and a non-porous homogenous elastic half-space. Geophys. Res. Bull. **25**(3) (1987)
4. S. Gupta, A. Chattopadhyay, D.K. Majhi, Effect of initial stress on propagation of Love waves in an anisotropic porous layer. J. Solid Mech. **2**(1), 50–62 (2010)
5. P. Khurana, A.K. Vashisth, Love wave propagation in a pre-stressed medium. Indian J. Pure Appl. Math. **32**(8), 1201–1207 (2001)
6. S. Gupta, S.K. Vishwakarma, D.K. Majhi, S. Kundu, Possibility of Love wave propagation in a porous layer under the effect of linearly varying directional rigidities. Appl. Math. Model. **37**, 6652–6660 (2013)
7. P.K. Vaishnav, S. Kundu, S. Gupta, A. Saha, Propagation of Love-type wave in porous medium over an orthotropic semi-infinite medium with rectangular irregularity, in *Mathematical Problems in Engineering, 2016* (Hindawi Publishing Corporation, 2016), 9 pp. Article ID 2081505
8. M.A. Biot, Mechanics of deformation and acoustic propagation in porous media. J. Appl. Phys. **33**, 1482–1498 (1962)
9. M.A. Biot, *Mechanics of Incremental Deformation* (Wiley Incorporation, New York, 1965).

Deformation of an Elastic-Layer Overlying an Elastic Half-Space Caused by a Finite, Buried, Inclined, Locked Strike-Slip Fault

Piu Kundu and Seema Sarkar (Mondal)

Abstract The process of stress accumulation near earthquake faults during the aseismic period has become a subject of research during the last few decades. It is noted that seismic waves, generated by an earthquake, result in a considerable disturbance in a seismic region causing a movement of the free surface. Such ground movements are not observed during the aseismic period. But a slow quasi-static aseismic surface displacement of the order of few cms. per year or less can be observed during the aseismic period which indicates a slow subsurface process of stress–strain accumulation. Keeping this in view we here consider an aseismically locked, buried, finite strike-slip fault inclined to the vertical at an arbitrary angle. The fault is situated in an elastic layer over an elastic half-space representing the lithosphere-asthenosphere system. An analytical study for displacement, stress, and strain has been carried out for a buried, finite, inclined fault. The solutions for displacement, stress, and strain are then found before the onset of fault movement and then superpose the effect of fault movement using Laplace transform and suitable mathematical techniques of Green's function. The model is validated by numerical examples employing MATLAB. It is observed that the inclination of the fault and velocity of the fault movement has a noticeable effect on displacements, stresses, and strains.

Keywords Finite strike-slip fault · Locked · Aseismic period · Green's function technique · Laplace transform

1 Introduction

Study of dynamic process leading to an earthquake is one of the main concerns of seismologists. About 90% of all earthquakes which are the results of tectonic events are natural primarily due to movement across the fault. An earthquake of

P. Kundu (✉) · S. Sarkar (Mondal)
Department of Mathematics, National Institute of Technology Durgapur, Durgapur 713209, India
e-mail: piukundu91@gmail.com

S. Sarkar (Mondal)
e-mail: seemasarkarmondal17@gmail.com

S. R. Mishra et al. (eds.), *Recent Trends in Applied Mathematics*, Lecture Notes in Mechanical Engineering, https://doi.org/10.1007/978-981-15-9817-3_20

magnitude 7.0 occurred on October 10, 2018 in the east of Kimbe, Papua New Guinea was the result of thrust faulting on or near the plate boundary interface between the subducting Australia and overriding Pacific plates. Due to horizontal movement between the North Americal plate and Pacific plate, an earthquake of magnitude 6 occurred on August 24, 2014 in California. These are caused by the release of elastic strain energy accumulated over a long period of time within the subsurface rocks in the region. This stress and strain accumulate in the region due to various tectonic processes, such as mantle convection and plate movement. These tectonic processes ultimately lead to movements across the fault. The quasi-static period in between two seismic events is called aseismic period which may last for years or even for centuries. Though this period apparently seems to be static but a slow and continuous aseismic surface movements were observed with the help of sophisticated measuring instruments. Such aseismic surface movements indicate that slow aseismic change of stress and strain are occurring in the region. It may eventually lead to sudden or creeping movements across the seismic faults situated in the region. It, therefore, seems to be an essential feature to identify the nature of the stress and strain accumulation in the vicinity of seismic faults situated in the region by studying the observed ground deformations during the aseismic period. Such study and its analysis may give us some precursory information with respect to the impending earthquakes.

2 Literature Review

A pioneering work including static ground deformation in elastic media was started by Steketee [36, 37]. Chinnery [5–7] compared the theoretically calculated surface displacements due to static dislocation across plane faults with the observed residual surface displacements near the San Andreas fault due to the San Francisco earthquake of 1906. Maruyama [19, 20] considered both strike-slip and dip-slip faults to calculate displacements and stresses due to prescribed static dislocations characterized by a prescribed discontinuity of the displacement across the faults using Green's function technique. Chinnery et al. [8] discussed the displacement fields of a very long vertical strike-slip fault model consisting of two layers over a half-space. Later some hypothetical models towards this path were produced by various authors namely Rybicki [29, 30], Sato [31], Rosenman [28], Sen [35], Mukhopadhyay et al. [22–25]. Savage and Prescott [32] described a review of geodimeter measurements made along the "big-bend" section of the San Andreas fault in southern California and this indicates no significant increment in strain during the period of major uplift (late 1959 to mid-1963). Savage [33] predicted vertical displacements in a subsiding elastic layer. Matsu'ura and Sato [18] have constructed a kinematic model for the earthquake cycle at convergent plate boundaries on the basis of dislocation theory. Okada [26] presented an analytical expression for the surface displacements, strains, and tilts due to inclined shear and tensile faults in a half-space for both point and finite rectangular source. Further Okada [27] calculated the internal displacements, strains and tilts for the same model (1986). Debnath and Sen [9, 10] developed models of finite strike-slip fault and long dip-slip fault in viscoelastic half-space. Debnath and Sen [11, 14] have developed movement of different types of long dip-sip fault situated in a viscoelastic layer and in an elastic layer over a viscoelastic half-space. In another paper, Debnath [15] calculated displacements, stresses and strains due to a buried, vertical rectangular strike-slip fault in a viscoelastic layer over a viscoelastic half-space. He also described ground deformation due to movement of finite rectangular fault in an elastic layer over a viscoelastic half-space in [12, 13, 16]. The earthquake deformation has been described in Segall [34].

In most of these works, the medium was taken to be elastic and/or viscoelastic half-space or layer. The faults are taken to be too long compared to their depth so that the problem reduces to a two-dimensional model. For instance, the famous San Andreas Fault is very long, while the faults like Calaveras, Hayward, San Jacinmto are not so long compared to their depth. The faults comparatively short in length are called finite faults and a three-dimensional model is imminent for them. Keeping this in view, in the present paper we have considered a finite fault. This fault is situated in an elastic layer over an elastic half-space representing the lithosphere-asthenosphere. The fault undergoes a slipping movement when the stress in the region exceeds certain threshold values.

3 Formulation

We consider a locked, buried, rectangular strike-slip fault F of length 2L (L-finite) and width D. The fault is locked that is located at a particular position at time t = 0 and then suddenly displaced by an amount but thereafter locked. The inclination of the fault plane with the free surface is θ and the thickness of the elastic layer is h_1.The depth of the fault from the free surface is r_1.

A three-dimensional Cartesian co-ordinate system (y_1, y_2, y_3) is introduced with y_1 axis parallel to the strike of the fault, y_3 axis is pointing downwards, and y_2 axis is perpendicular to $y_1 - y_3$ plane. The free surface is given by $y_3 = 0$. The elastic layer is represented by $0 \leq y_3 \leq h_1$, in which the entire fault is situated and $y_3 \geq h_1$ represents the elastic half-space. Since depth of the fault is r_1 from the free surface and the fault makes an angle θ with the horizontal, we take another Cartesian co-ordinate system (y_1', y_2', y_3'), the origin of which is at O'(Fig. 1). y_1' is along the strike of the fault and is parallel to y_1 axis, y_3' axis is on the fault plane perpendicular to y_1 axis making an angle θ with the plane free surface and y_2' axis is perpendicular to the fault plane. Then the fault is given by F:$-L \leq y_1' \leq L$, $y_2' = 0, 0 \leq y_3' \leq D$ as shown in Fig. 1. The relation between (y_1, y_2, y_3) and (y_1', y_2', y_3') are given by

$$
\begin{aligned}
y_1' &= y_1 \\
y_2' &= y_2 \sin\theta - (y_3 - r_1)\cos\theta \\
y_3' &= y_2 \cos\theta + (y_3 - r_1)\sin\theta
\end{aligned}
\tag{1}
$$

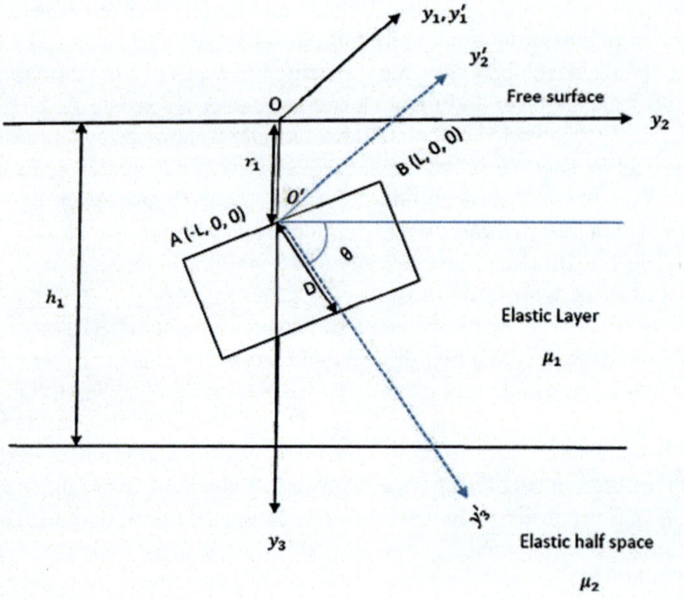

Fig. 1 A finite, buried, inclined, strike-slip fault in elastic layer over an elastic half-space

Let (u_1, u_2, u_3) be the components of the displacement and $\tau_{ij}(i, j = 1, 2, 3)$, $e_{ij}(i, j = 1, 2, 3)$ are, respectively, stress and strain components in the elastic layer $0 \leq y_3 \leq h_1$. Also let (u_1', u_2', u_3') be the components of displacement, $\tau_{ij}'(i, j = 1, 2, 3)$ and $e_{ij}'(i, j = 1, 2, 3)$ be stress and strain components, respectively, in the elastic half-space $(y_3 \geq h_1)$.

3.1 Constitutive Equations

Isotropic form of Hook's Law for homogeneous isotropic elastic material is

$$\tau_{ij} = \lambda e_{pp} \delta_{ij} + 2\mu e_{ij}$$

where $e_{ij} = \frac{1}{2}\left(\frac{\partial u_i}{\partial x_j} + \frac{\partial u_j}{\partial x_j}\right)$, λ and μ are Lame's constants, μ is shear modulus. For simple shear strain e_{ij} is non zero. Then Hooke's Law states

$$\tau_{ij} = 2\mu e_{ij}$$

So, the constitutive equations that relate the stresses acting on a material element to the resultant strains and/or rate of strain will be as follows:

For elastic layer $(0 \leq y_3 \leq h_1, |y_2| \leq \infty)$:

$$
\begin{aligned}
\tau_{11} &= 2\mu_1 \frac{\partial u_1}{\partial y_1} \\
\tau_{12} &= \mu_1 \left(\frac{\partial u_1}{\partial y_2} + \frac{\partial u_2}{\partial y_1} \right) \\
\tau_{13} &= \mu_1 \left(\frac{\partial u_1}{\partial y_3} + \frac{\partial u_3}{\partial y_1} \right) \\
\tau_{22} &= 2\mu_1 \frac{\partial u_2}{\partial y_2} \\
\tau_{23} &= \mu_1 \left(\frac{\partial u_2}{\partial y_3} + \frac{\partial u_3}{\partial y_2} \right) \\
\tau_{33} &= 2\mu_1 \frac{\partial u_3}{\partial y_3}
\end{aligned}
\tag{2}
$$

For elastic half-space ($y_3 \geq h_1, |y_2| < \infty$):

$$
\begin{aligned}
\tau'_{11} &= 2\mu_2 \frac{\partial u'_1}{\partial y_1} \\
\tau'_{12} &= \mu_2 \left(\frac{\partial u'_1}{\partial y_2} + \frac{\partial u'_2}{\partial y_1} \right) \\
\tau'_{13} &= \mu_2 \left(\frac{\partial u'_1}{\partial y_3} + \frac{\partial u'_3}{\partial y_1} \right) \\
\tau'_{22} &= 2\mu_2 \frac{\partial u'_2}{\partial y_2} \\
\tau'_{23} &= \mu_2 \left(\frac{\partial u'_2}{\partial y_3} + \frac{\partial u'_3}{\partial y_2} \right) \\
\tau'_{33} &= 2\mu_2 \frac{\partial u'_3}{\partial y_3}
\end{aligned}
\tag{3}
$$

where μ_1 and μ_2 are the rigidities of the elastic layer and elastic half-space, respectively.

3.2 *Stress Equations of Motion*

For a slow, aseismic, quasi-static deformation, the inertial terms are very small and can be neglected as explained by Mukhopadhyay et al. [24]. Then stress equations for elastic layer ($0 \leq y_3 \leq h_1, |y_2| < \infty$) are

$$\begin{aligned}
&\frac{\partial}{\partial y_1}(\tau_{11}) + \frac{\partial}{\partial y_2}(\tau_{12}) + \frac{\partial}{\partial y_3}(\tau_{13}) = 0\\
&\frac{\partial}{\partial y_1}(\tau_{21}) + \frac{\partial}{\partial y_2}(\tau_{22}) + \frac{\partial}{\partial y_3}(\tau_{23}) = 0\\
&\frac{\partial}{\partial y_1}(\tau_{31}) + \frac{\partial}{\partial y_2}(\tau_{32}) + \frac{\partial}{\partial y_3}(\tau_{33}) = 0
\end{aligned} \tag{4}$$

and for elastic half-space ($y_3 \geq h_1, |y_2| < \infty$) are

$$\begin{aligned}
&\frac{\partial}{\partial y_1}(\tau'_{11}) + \frac{\partial}{\partial y_2}(\tau'_{12}) + \frac{\partial}{\partial y_3}(\tau'_{13}) = 0\\
&\frac{\partial}{\partial y_1}(\tau'_{21}) + \frac{\partial}{\partial y_2}(\tau'_{22}) + \frac{\partial}{\partial y_3}(\tau'_{23}) = 0\\
&\frac{\partial}{\partial y_1}(\tau'_{31}) + \frac{\partial}{\partial y_2}(\tau'_{32}) + \frac{\partial}{\partial y_3}(\tau'_{33}) = 0
\end{aligned} \tag{5}$$

3.3 Boundary Conditions

The boundary conditions near the tips of the fault at time t = 0 that is at an instant when there is no fault movement are taken as.

$$\begin{aligned}
&\lim_{y'_1 \to L-} \tau_{11}(y_1, y_2, y_3, t) = \lim_{y'_1 \to L+} \tau_{11}(y_1, y_2, y_3, t) = \tau_L(\text{say})\\
&y'_2 = 0, 0 \leq y'_3 \leq D, t \geq 0
\end{aligned} \tag{6}$$

and $\lim\limits_{y'_1 \to (-L)-} \tau_{11}(y_1, y_2, y_3, t) = \lim\limits_{y'_1 \to (-L)-} \tau_{11}(y_1, y_2, y_3, t) = \tau_{-L}(\text{say})$

$$y'_2 = 0, 0 \leq y'_3 \leq D, t \geq 0 \tag{7}$$

assuming that the stresses maintain constant values τ_L and τ_{-L} at the tips of the fault along y'_1 axis [the value of this constant stress is likely to be small enough so that no further extension is possible along y'_1 axis].

$$\text{Also} \quad \lim_{|y_2| \to \infty} \tau_{12}(y_1, y_2, y_3, t) \to \tau_\infty(\text{t}) \tag{8}$$

for $-L \leq y_1 \leq L, 0 \leq y_3 \leq h_1, t \geq 0$, where $\tau_\infty(\text{t})$ is the shear stress. This is maintained by tectonic forces which arises due to mantle convection and other tectonic phenomena.

On the free surface $y_3 = 0$, $(-L \le y_1 \le L, 0 \le y_3 \le h_1, t \ge 0)$.

$$\tau_{13} = \tau_{23} = \tau_{22} = \tau_{33} = 0 \tag{9}$$

Also as $y_3 \to \infty$ $(-L \le y_1 \le L, 0 \le y_3 \le h_1, t \ge 0)$.

$$\tau'_{13} = \tau'_{23} = \tau'_{22} = \tau'_{33} = 0 \tag{10}$$

On the interface $y_3 = h_1$ $(-L \le y_1 \le L, 0 \le y_3 \le h_1, t \ge 0)$, since the layer and half-space are in welded contact, therefore, we have

$$\begin{aligned} u_1 &= u'_1 \\ u_2 &= u'_2 \\ u_3 &= u'_3 \\ \tau_{13} &= \tau'_{13} \\ \tau_{23} &= \tau'_{23} \\ \tau_{33} &= \tau'_{33} \end{aligned} \tag{11}$$

3.4 *Initial Conditions and Conditions at Infinity*

$(u_i)_0, \left(u'_i\right)_0 (i = 1, 2, 3), \left(\tau_{ij}\right)_0, \left(e_{ij}\right)_0, \left(\tau'_{ij}\right)_0, \left(e'_{ij}\right)_0 (i, j = 1, 2, 3)$ are the values of $u_i, u'_i, \tau_{ij}, e_{ij}, \tau'_{ij}, e'_{ij} (i, j = 1, 2, 3)$, respectively, at t = 0. We assume that tectonic forces result in a shear strain far away from the fault which may change with time. Then the following conditions are satisfied

$$\begin{aligned} e_{12} &\to (e_{12})_{0,\infty} + g(t) + h(t) \\ e'_{12} &\to \left(e'_{12}\right)_{0,\infty} + g(t) + h(t) \end{aligned} \tag{12}$$

where $(e_{12})_{0,\infty} = \lim_{|y_2| \to \infty} (e_{12})_0$, $\left(e'_{12}\right)_{0,\infty} = \lim_{|y_2| \to \infty} \left(e'_{12}\right)_0$ and $(e_{12})_0$, $\left(e'_{12}\right)_0$ are the values of e_{12}, e'_{12} at t = 0. $g(t)$, $h(t)$ are gradually increasing continuous function of t with $g(0) = 0$, $h(0) = 0$ at t = 0. Same $g(t)$ and $h(t)$ are taken since the layer and half-space are in welded contact and the strains are continuous in the boundary.

4 Method of Solution

From Eq. (2), we find

$$\tau_{11} = 2\mu_1 \frac{\partial u_1}{\partial y_1}, \tau_{12} = 2\mu_1 \frac{\partial u_1}{\partial y_2}, \tau_{13} = 2\mu_1 \frac{\partial u_1}{\partial y_3}$$

$$\tau_{21} = 2\mu_1 \frac{\partial u_2}{\partial y_1}, \tau_{22} = 2\mu_1 \frac{\partial u_2}{\partial y_2}, \tau_{23} = 2\mu_1 \frac{\partial u_2}{\partial y_3}$$

$$\tau_{31} = 2\mu_1 \frac{\partial u_3}{\partial y_1}, \tau_{32} = 2\mu_1 \frac{\partial u_3}{\partial y_2}, \tau_{33} = 2\mu_1 \frac{\partial u_3}{\partial y_3}$$

By using Eq. (4) we get, $\nabla^2 u_i = 0$ where $i = 1, 2, 3$
Similarly from Eqs. (3) and (5) we get

$$\nabla^2 u'_i = 0 \text{ where } i = 1, 2, 3$$

These are the general solutions.

For the model considered above, the solutions for displacements, stresses, and strains in the absence of fault movement and after the fault movement have been discussed in the Sects. 4.1 and 4.2.

4.1 Displacements, Stresses, and Strains in the Absence of Fault Movement

The boundary value problem given by (1)–(12) can be solved by taking Laplace transform with respect to time "t" of all the constitutive equations and the boundary conditions. On taking inverse Laplace transform we get solutions for displacements, stresses, and strains as follows:

For elastic layer ($0 \le y_3 \le h_1, |y_2| < \infty$)

$$\begin{aligned}
u_1 &= (u_1)_0 + 2h(t)y_2 \\
u_2 &= (u_2)_0 + \frac{1}{\mu_1}[\tau_\infty(t) - \tau_\infty(0)]y_1 - 2h(t)y_1 \\
u_3 &= (u_3)_0 \\
\tau_{11} &= (\tau_{11})_0 \\
\tau_{12} &= (\tau_{12})_0 + [\tau_\infty(t) - \tau_\infty(0)] \\
\tau_{13} &= (\tau_{13})_0 \\
\tau_{23} &= (\tau_{23})_0 \\
\tau_{22} &= (\tau_{22})_0 \\
\tau_{33} &= (\tau_{33})_0 \\
e_{12} &= (e_{12})_0 + \frac{1}{2\mu_1}[\tau_\infty(t) - \tau_\infty(0)]
\end{aligned} \tag{13}$$

For elastic half-space ($y_3 \geq h_1, |y_2| < \infty$).

$$\begin{aligned}
u'_1 &= \left(u'_1\right)_0 + 2h(t)y_2 \\
u'_2 &= \left(u'_2\right)_0 + \frac{1}{\mu_1}[\tau_\infty(t) - \tau_\infty(0)]y_1 - 2h(t)y_1 \\
u'_3 &= \left(u'_3\right)_0 \\
\tau'_{11} &= \left(\tau'_{11}\right)_0 \\
\tau'_{12} &= \left(\tau'_{12}\right)_0 + [\tau_\infty(t) - \tau_\infty(0)] \\
\tau'_{13} &= \left(\tau'_{13}\right)_0 \\
\tau'_{23} &= \left(\tau'_{23}\right)_0 \\
\tau'_{22} &= \left(\tau'_{22}\right)_0 \\
\tau'_{33} &= \left(\tau'_{33}\right)_0 \\
e'_{12} &= \left(e'_{12}\right)_0 + \frac{1}{2\mu_1}[\tau_\infty(t) - \tau_\infty(0)]
\end{aligned} \tag{14}$$

Here the solutions in the Eqs. (13) and (14) are trivial, this indicates that there are no effect on displacement, stress, and strain components before the fault movements for both the layer and half-space.

$$\begin{aligned}
\tau''_{12} &= \text{Stress component which tends to cause} \\
&\quad\ \text{strike - slip movement across the fault F} \\
&= \tau_{12}\sin\theta - \tau_{13}\cos\theta \\
\tau''_{12} &= \left(\tau''_{12}\right)_0 + [\tau_\infty(t) - \tau_\infty(0)]\sin\theta
\end{aligned} \tag{15}$$

where $\left(\tau''_{12}\right)_0 = (\tau_{12})_0 \sin\theta - (\tau_{13})_0 \cos\theta$.

If $\tau_\infty(t)$ does not have a constant value and it is a time-dependent function, then we assume that $\tau_\infty(t) = \tau_\infty(0)(1 + k_1 t)$. Here k_1 is a constant and is taken as 10^{-9} and $\tau_\infty(0) = 5 \times 10^6\ \text{N/m}^2$.

From the above solution it is clear that for the fault F τ''_{12} increase with time. The characteristic of the fault is such that the fault moves when the magnitude of stress τ''_{12} reaches some critical value τ_c. Here we consider $\tau_c = 210$ bar, i.e., $21 \times 10^6 \text{N/m}^2$(Pascal). From Fig. 2, it is found that τ''_{12} reaches the value 210 bar after time T $= 149$ years for $\theta = 60°$.

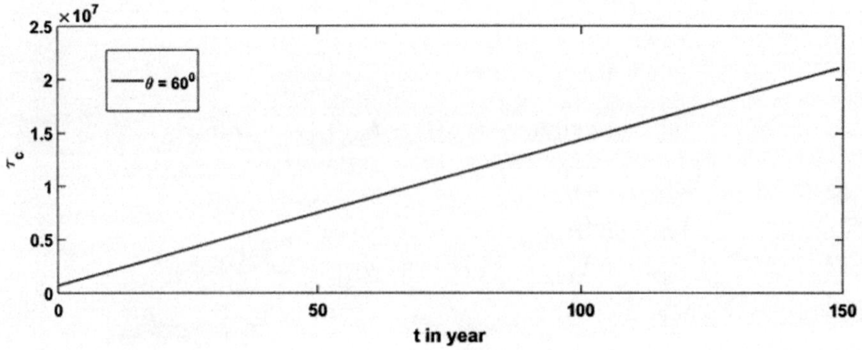

Fig. 2 Critical stress with time

4.2 Displacements, Stresses, Strains After the Commencement of Fault Movement

We consider that for a period $t_1 = t - T \geq 0$ (which now corresponds to the new phase of aseismic state of the model re-established after the sudden fault movement), the inertial forces again become very small and are therefore neglected. The displacements, stresses, and strains are continuous everywhere in the model except for the fault F across which the displacement component u_1 has a discontinuity which characterizes the sudden fault movement across the fault F given by the following conditions:

$$[(u_1)]_F = U_1 f\left(y_1', y_3'\right) H(t_1) \tag{16}$$

where U_1 is constant independence of y_1', y_3', $f\left(y_1', y_3'\right)$ is a continuous function of y_1', y_3' giving the dependence of the relative displacement across the fault F on the depth along the fault, $H(t_1)$ is the Heaviside function and $t_1 = t - T$.

$$[(u_1)]_F = \text{The discontinuity ofacross F is given by}$$

$$[(u_1)]_F = \lim_{y_2' \to 0^+} (u_1) - \lim_{y_2' \to 0^-} (u_1)\left[-L \leq y_1' \leq L, 0 \leq y_3' \leq D\right] \tag{17}$$

Taking Laplace transform on (16) we get $[(\overline{u_1})]_F = \frac{U_1}{p} f\left(y_1', y_3'\right)$.

After time T the fault F is locked, we try to obtain the displacements, stresses, and strains for $t \geq 0$ (with respect to new time origin, i.e., the instant at which this aseismic state has been restored in the system after the fault movement) in the following form:

For elastic layer $(0 \leq y_3 \leq h_1, |y_2| < \infty)$

$$u_1 = (u_1)_1 + (u_1)_2$$

$$
\begin{aligned}
u_2 &= (u_2)_1 + (u_2)_2 \\
u_3 &= (u_3)_1 + (u_3)_2 \\
\tau_{11} &= (\tau_{11})_1 + (\tau_{11})_2 \\
\tau_{12} &= (\tau_{12})_1 + (\tau_{12})_2 \\
\tau_{13} &= (\tau_{13})_1 + (\tau_{13})_2 \\
\tau_{22} &= (\tau_{22})_1 + (\tau_{22})_2 \\
\tau_{23} &= (\tau_{23})_1 + (\tau_{23})_2 \\
\tau_{33} &= (\tau_{33})_1 + (\tau_{33})_2
\end{aligned} \tag{18}
$$

For the half-space ($y_3 \geq h_1, |y_2| < \infty$).

$$
\begin{aligned}
u'_1 &= \left(u'_1\right)_1 + \left(u'_1\right)_2 \\
u'_2 &= \left(u'_2\right)_1 + \left(u'_2\right)_2 \\
u'_3 &= \left(u'_3\right)_1 + \left(u'_3\right)_2 \\
\tau'_{11} &= \left(\tau'_{11}\right)_1 + \left(\tau'_{11}\right)_2 \\
\tau'_{12} &= \left(\tau'_{12}\right)_1 + \left(\tau'_{12}\right)_2 \\
\tau'_{13} &= \left(\tau'_{13}\right)_1 + \left(\tau'_{13}\right)_2 \\
\tau'_{22} &= \left(\tau'_{22}\right)_1 + \left(\tau'_{22}\right)_2 \\
\tau'_{23} &= \left(\tau'_{23}\right)_1 + \left(\tau'_{23}\right)_2 \\
\tau'_{33} &= \left(\tau'_{33}\right)_1 + \left(\tau'_{33}\right)_2
\end{aligned} \tag{19}
$$

where $(u_i)_1, \left(\tau_{ij}\right)_1, (e_{12})_1, \left(u'_i\right)_1, \left(\tau'_{ij}\right)_1, \left(e'_{12}\right)_1$ $(i, j = 1, 2, 3)$ are continuous everywhere in the model and given in Eqs. (13) and (14). $(u_i)_2, \left(\tau_{ij}\right)_2, (e_{12})_2, \left(u'_i\right)_2, \left(\tau'_{ij}\right)_2, \left(e'_{12}\right)_2 (i, j = 1, 2, 3)$ are satisfied all the conditions from (2) to (12). These are obtained by using a modified form of Green's function technique developed by Maruyama [20] and Rybicki [29] explained in Appendix 1. We note that $(u_2)_2, (u_3)_2, \left(u'_2\right)_2, \left(u'_3\right)_2$ are continuous even after the fault movement, so that $[(u_2)]_2 = 0, [(u_3)]_2 = 0, \left[\left(u'_2\right)\right]_2 = 0, \left[\left(u'_3\right)\right]_2 = 0$, while $(u_1)_2$ satisfy the dislocation condition (17).

We get
$$
\begin{aligned}
(u_1)_2 &= \frac{U_1}{2\pi} H(t_1)\phi(y_1, y_2, y_3) \\
\left(u'_2\right)_2 &= \frac{U_1}{2\pi} H(t_1)\psi(y_1, y_2, y_3)
\end{aligned} \tag{20}
$$

ϕ and ψ are described in Appendix 1.
Now for elastic layer ($0 \leq y_3 \leq h_1, |y_2| < \infty$)

$$
\begin{aligned}
(\tau_{11})_2 &= 2\mu_1 H(t_1)\frac{U_1}{2\pi}\phi_1 \\
(\tau_{12})_2 &= \mu_1 H(t_1)\frac{U_1}{2\pi}\phi_2 \\
(\tau_{13})_2 &= \mu_1 H(t_1)\frac{U_1}{2\pi}\phi_3 \\
(\tau_{23})_2 &= 0 \\
(\tau_{22})_2 &= 0 \\
(\tau_{33})_2 &= 0 \\
(e_{12})_2 &= H(t_1)\frac{U_1}{4\pi}\phi_2
\end{aligned} \tag{21}
$$

where $\phi_1 = \frac{\partial\phi}{\partial y_1}$, $\phi_1 = \frac{\partial\phi}{\partial y_2}$, $\phi_1 = \frac{\partial\phi}{\partial y_3}$ and $t_1 = t - T$.

For the half-space $(y_3 \geq h_1, |y_2| < \infty)$

$$
\begin{aligned}
\left(\tau'_{11}\right)_2 &= 2\mu_2 H(t_1)\frac{U_1}{2\pi}\psi_1 \\
\left(\tau'_{12}\right)_2 &= \mu_2 H(t_1)\frac{U_1}{2\pi}\psi_2 \\
\left(\tau'_{13}\right)_2 &= \mu_2 H(t_1)\frac{U_1}{2\pi}\psi_3 \\
\left(\tau'_{23}\right)_2 &= 0 \\
\left(\tau'_{22}\right)_2 &= 0 \\
\left(\tau'_{33}\right)_2 &= 0 \\
\left(e'_{12}\right)_2 &= H(t_1)\frac{U_1}{4\pi}\psi_2
\end{aligned} \tag{22}
$$

where $\psi_1 = \frac{\partial\psi}{\partial y_1}$, $\psi_2 = \frac{\partial\psi}{\partial y_2}$, $\psi_3 = \frac{\partial\psi}{\partial y_3}$ and $t_1 = t - T$.

Finally the solutions for displacement, stress, and strain are obtained as follows: for elastic layer $(0 \leq y_3 \leq h_1, |y_2| < \infty)$.

$$
\begin{aligned}
u_1 &= (u_1)_0 + 2h(t)y_2 + \frac{U_1}{2\pi}H(t_1)\phi(y_1, y_2, y_3) \\
u_2 &= (u_2)_0 + \frac{1}{\mu_1}[\tau_\infty(t) - \tau_\infty(0)]y_1 - 2h(t)y_1 \\
u_3 &= (u_3)_0 \\
\tau_{11} &= (\tau_{11})_0 + 2\mu_1 H(t_1)\frac{U_1}{2\pi}\phi_1 \\
\tau_{12} &= (\tau_{12})_0 + [\tau_\infty(t) - \tau_\infty(0)] + \mu_1 H(t_1)\frac{U_1}{2\pi}\phi_2
\end{aligned}
$$

$$
\begin{aligned}
\tau_{13} &= (\tau_{13})_0 + \mu_1 H(t_1)\frac{U_1}{2\pi}\phi_3 \\
\tau_{23} &= (\tau_{23})_0 \\
\tau_{22} &= (\tau_{22})_0 \\
\tau_{33} &= (\tau_{33})_0 \\
e_{12} &= (e_{12})_0 + \frac{1}{2\mu_1}[\tau_\infty(t) - \tau_\infty(0)] + H(t_1)\frac{U_1}{4\pi}\phi_2
\end{aligned}
\tag{23}
$$

For elastic half-space ($y_3 \geq h_1, |y_2| < \infty$).

$$
\begin{aligned}
u'_1 &= \left(u'_1\right)_0 + 2h(t)y_2 + \frac{U_1}{2\pi}H(t_1)\psi(y_1, y_2, y_3) \\
u'_2 &= \left(u'_2\right)_0 + \frac{1}{\mu_1}[\tau_\infty(t) - \tau_\infty(0)]y_1 - 2h(t)y_1 \\
u'_3 &= \left(u'_3\right)_0 \\
\tau'_{11} &= (\tau_{\prime 11})_0 + 2\mu_2 H(t_1)\frac{U_1}{2\pi}\psi_1 \\
\tau'_{12} &= \left(\tau'_{12}\right)_0 + [\tau_\infty(t) - \tau_\infty(0)] + \mu_2 H(t_1)\frac{U_1}{2\pi}\psi_2 \\
\tau'_{13} &= \left(\tau'_{13}\right)_0 + \mu_2 H(t_1)\frac{U_1}{2\pi}\psi_3 \\
\tau'_{23} &= \left(\tau'_{23}\right)_0 \\
\tau'_{22} &= \left(\tau'_{22}\right)_0 \\
\tau'_{33} &= \left(\tau'_{33}\right)_0 \\
e'_{12} &= \left(e'_{12}\right)_0 + \frac{1}{2\mu_1}[\tau_\infty(t) - \tau_\infty(0)] + H(t_1)\frac{U_1}{4\pi}\psi_2
\end{aligned}
\tag{24}
$$

$$
\begin{aligned}
T''_{12} &= \text{Stress component which tends to cause strike} - \text{slip} \\
&\quad \text{movement across the fault F} \\
&= \tau_{12}\sin\theta - \tau_{13}\cos\theta \\
&= \left(T''_{12}\right)_0 + [\tau_\infty(t) - \tau_\infty(0)]\sin\theta \\
&\quad + \frac{\mu_1 H(t_1)U_1}{2\pi}[\phi_2\sin\theta - \phi_3\cos\theta]
\end{aligned}
\tag{25}
$$

where $\left(T''_{12}\right)_0 = (\tau_{12})_0\sin\theta - (\tau_{13})_0\cos\theta$.

5 Numerical Computations

In order to study in greater details, the nature of surface displacement, shear strain, and accumulation of shear stress near the fault due to the movement across the fault, we analyze the solution by assigning suitable numerical values to the model parameters involved.

It is assumed that as a result of some structural reason there is a slow but steady accumulation of shear strain at a distance far away from the fault. Keeping this in view we take g(t) and h(t) to be linearly increasing function with time and g(0) = 0, h(0) = 0 at t = 0. With this assumption we take h(t) = kt. We assume k = 3.2×10^{-14} [21], noting also that during the aseismic period the rate of strain accumulation observed in the seismically active region is of the order of 10^{-6} to 10^{-8} per year.

The rigidities μ_1, μ_2 of the layer and half-space, respectively, are taken as $\mu_1 = 3 \times 10^{10}\,\mathrm{N/m^2}$ and $\mu_2 = 3.5 \times 10^{10}\,\mathrm{N/m^2}$ as suggested by Aki [1], Bullen [2], Cathles [3], Chift [4], Karato [17] for lithosphere-asthenosphere system.

We denote $\gamma_1 = \frac{\mu_2}{\mu_1} = 1.667$.

λ = Lame's constant = $2.7290 \times 10^{10}\mathrm{N/m^2}$.
h_1 = thickness of the layer = 20 km.

Then we define $\Delta_1 = 1 - \frac{1-\gamma_1}{1+\gamma_1} e^{-2\lambda h_1} = 1$.

T = 150 years.
$t_1 = 1$ year.

The slip of fault U_1 is taken as constant, since the fault is locked.

$$U_1 = 0.01$$

We consider θ as the angle of inclination of the fault plane to the horizon. Noting that for values of $\theta > \frac{\pi}{2}$, say $\theta = \pi - \theta \left(0 \le \theta_1 < \frac{\pi}{2}\right)$, the nature of displacements, stresses, and strains in the medium will be similar to the case for which $\theta = \theta_1$. So in our model, the results are shown for $\theta = 30^{\circ}, 45^{\circ}, 60^{\circ}, 90^{\circ}$.

From Debnath [15] the values of the model parameters are taken as below:

The width of the fault D = 10 km = 10^4 m, noting that the San Andreas fault in North America, the value of D has been estimated to be in the range 5–15 km.

Length of the fault 2L = 20 km = 2×10^4 m.

$$\tau_{\infty}(t) = 2 \times 10^7\,\mathrm{N/m^2}$$

$$(\tau_{12})_0 = 2 \times 10^6\,\mathrm{N/m^2}$$

$$(\tau_{13})_0 = 2 \times 10^6\,\mathrm{N/m^2}$$

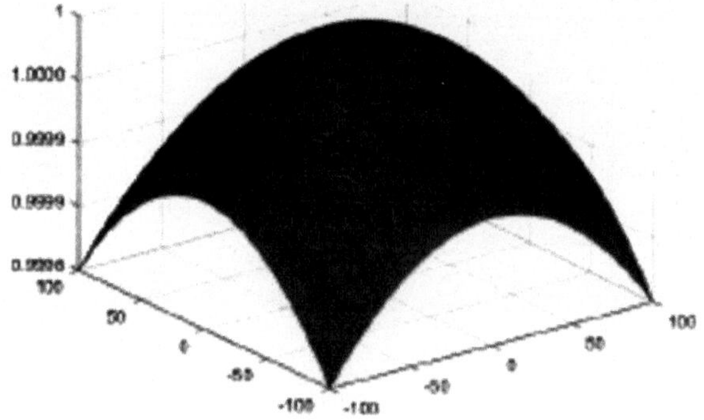

Fig. 3 Slip function

$$(e_{12})_0 = 5$$

$$(u_1)_0 = 5\,\text{m}$$

Slip function is $f(x_1', x_3') = U\left(1 - \frac{x_1'^2}{L^2} - \frac{x_2'^2}{D^2}\right)$, U = 1 cm in Fig. 3.

6 Result and Discussion

We have calculated the displacement, stress, and strain components of the fault movement. The results of the effect of this fault movement on displacements, stresses, and strains have been discussed below:

(a) ***Effect of displacement*:**

Let us consider the change of displacement on the free surface $y_3 = 0$ near the strike of the fault. The rate of surface displacement depends on the inclination of the fault and depth of the fault from the free surface. R_D = change of the displacement in the layer due to fault movement after restoration of aseismic state = $[u_1 - 2h(t)y_2] = \frac{U_1}{2\pi}H(t_1)\phi$.In Fig. 4, R_D has been plotted against y_2 for $y_1 = 10$ km (representing the distance of the point at the tip along the strike of the fault) and $y_3 = 0$.

The maximum value of residual surface displacement is attained near the fault for both $y_2 > 0$ and $y_2 < 0$. This falls off rapidly as we move away from the fault on the free surface and becomes very small for $|y_2| \ominus D$. For $y_2 > 0$ and $y_2 < 0$, it is in opposite directions. R_D depends on the inclination of the fault. For $y_2 > 0$, R_D attains its maximum value as $y_2 \to 0^+$ and this depends on θ. The higher values of θ give rise to higher magnitude of rate of displacement for $y_2 > 0$. For higher θ, lower value attains for $y_2 < 0$. It has been observed that residual surface displacement attains maximum and minimum value near $y_2 = 0$ at $\theta = 90^\circ$, i.e., for vertical fault. The maximum value of R_D is 1.2298×10^9 m and the minimum value is -9.45×10^8 m. For vertical fault change of displacement is symmetric w.r.t $y_2 = 0$.

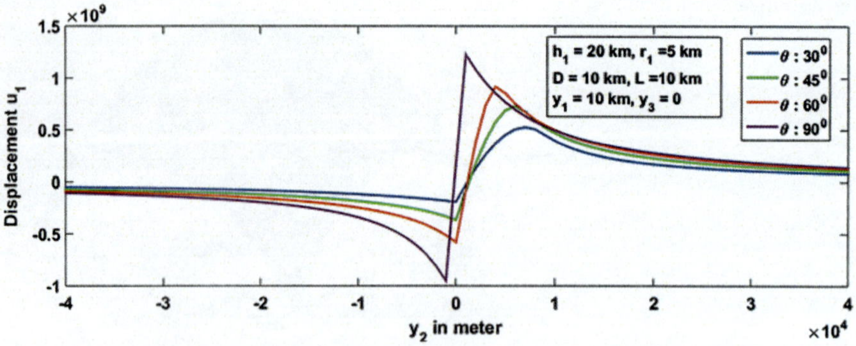

Fig. 4 Change of surface displacement with y_2 for different values of inclination θ

In Fig. 5, residual surface displacement has been plotted against y_2 for different values of depth of the fault (r_1) form the free surface. R_D attains maximum value near the fault for $r_1 = 0$, i.e., for surface breaking fault. It attains minimum value for $r_1 = 10\,\text{km}$ near the fault. It is in the opposite direction for $y_2 > 0$ and $y_2 < 0$. It falls off rapidly for $|y_2| > 0$. For $y_2 > 0$, surface displacement increases for r_1 decreases and for $y_2 < 0$, it increases with decreasing values of r_1.

(b) ***Effect of stress***: In Fig. 6, changes of stress component which tend to cause strike-slip movement across the fault F $\left(T_{12}''\right)$ has been plotted against time t. T_{12}'' given in Eq. (25). Figure 6 has explained the changes in T_{12}'' before the fault movement for different inclination of the fault. It shows that T_{12}'' increases with θ increase and attains maximum value at $\theta = 90^{\circ}$. The maximum value attains for different inclinations at t = 149 years, where T_{12}'' reaches some critical value. After attains critical value, the fault movement occurs.

T_{12}'' has been plotted against t for the different inclination of the fault after the fault movement in Fig. 7. It shows that after the fault movement T_{12}'' attains negative constant value. This shows that before the fault movement stress increases with time,

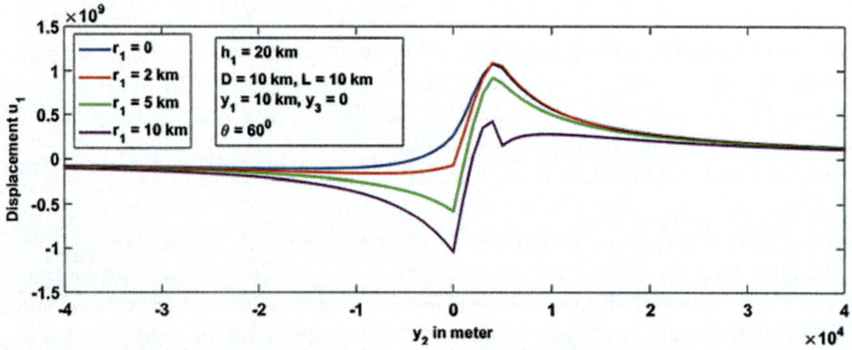

Fig. 5 Change of surface displacement with y_2 for different depth of the fault from the free surface

i.e., stress accumulates. After attaining maximum value the fault movement occurs and after the fault movement all the stress release with time. The maximum stress accumulates before the fault movement for $\theta = 90^\circ$ which is clear from Fig. 6 and maximum stress releases after the fault movement at $\theta = 90^\circ$, which is clear from Fig. 7.

(c) ***Effect of shear strain***: Surface shear strain due to fault movement near the fault at the time of restoration of aseismic state is $R_w = \left[e_{12} - (e_{12})_0 - \frac{1}{2\mu_1}(\tau_\infty(t) - \tau_\infty(0))\right] = \frac{H(t_1)U_1}{4\pi}\phi_2$.

Figure 8 shows the changes of residual surface shear strain with depth y_3 from the free surface for different inclination θ of the fault. For all different values of θ, shear strain increases with depth increase, i.e., shear strain accumulates with depth. After attaining maximum value it decreases. After releases a small amount of strain it again increases with depth. Maximum shear strain attains at $\theta = 30^\circ$ and its value is 1.522×10^4.

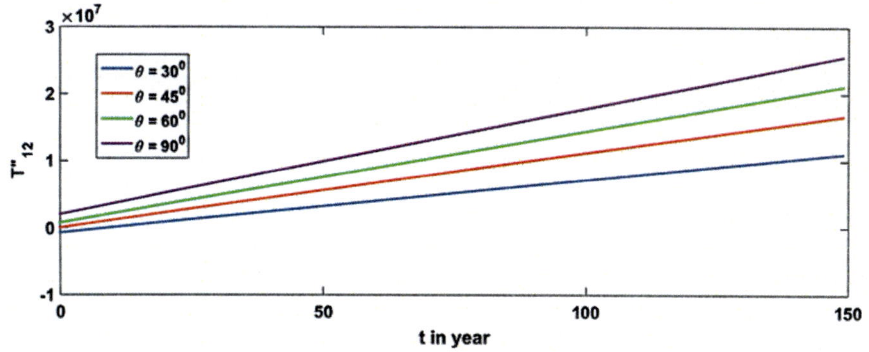

Fig. 6 T''_{12} with t before the fault movement for different values of inclination θ

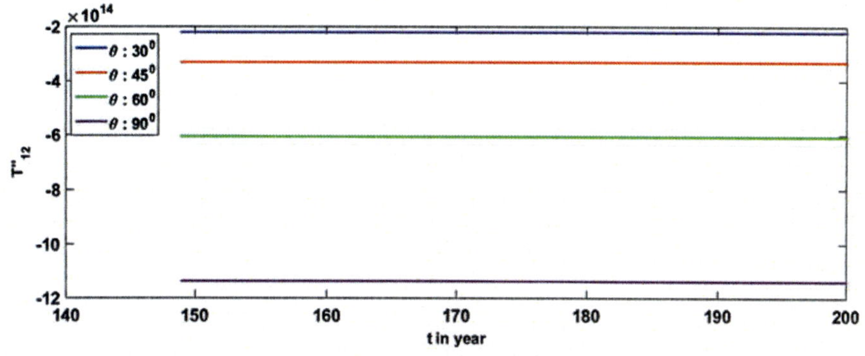

Fig. 7 T''_{12} with t after the fault movement for different values of inclination θ

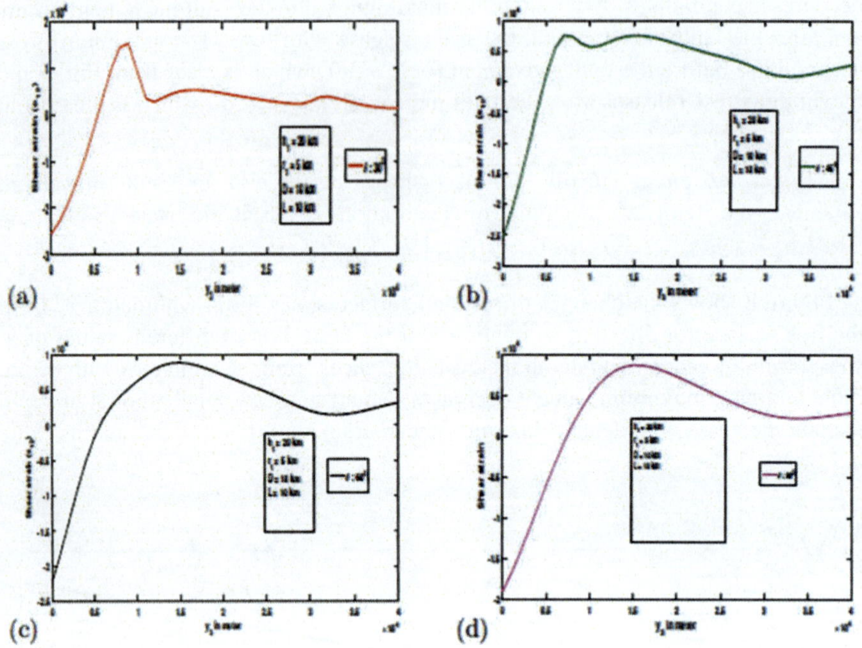

Fig. 8 Change of shear strain with y_3 for inclination **a** $\theta = 30^\circ$, **b** $\theta = 45^\circ$, **c** $\theta = 60^\circ$, **d** $\theta = 90^\circ$

In Fig. 9, shear strain has been plotted against y_3 for different values of thickness of the elastic layer, i.e., h_1. From this figure, it is clear that for all values of h_1, shear strain increases with depth, i.e., it accumulates with depth up to h_1 equal to near 16 km. Then stress releases a small amount and then further accumulates.

For the buried fault, the changes of strain with depth depending on the depth of the fault from the free surface, which shows in Fig. 10. When the fault is surface breaking, then the shear strain accumulates maximum value near $y_3 = 20$ km. For

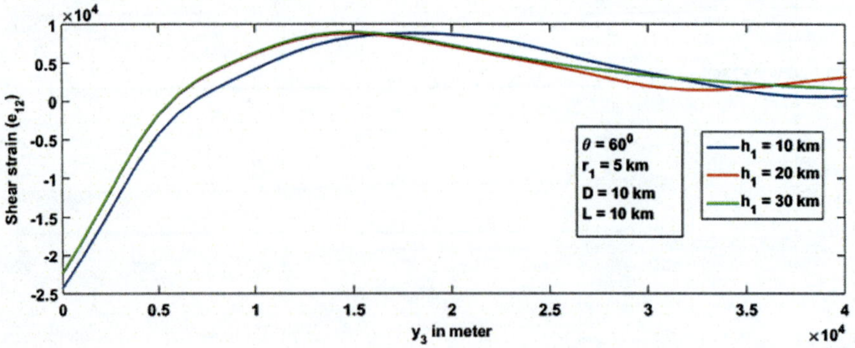

Fig. 9 Change of shear strain with y_3 for different thickness of the layer (h_1)

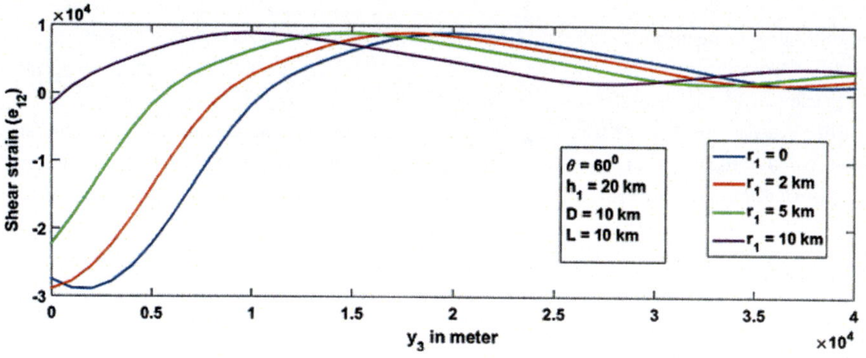

Fig. 10 Change of shear strain with y_3 for values of r_1

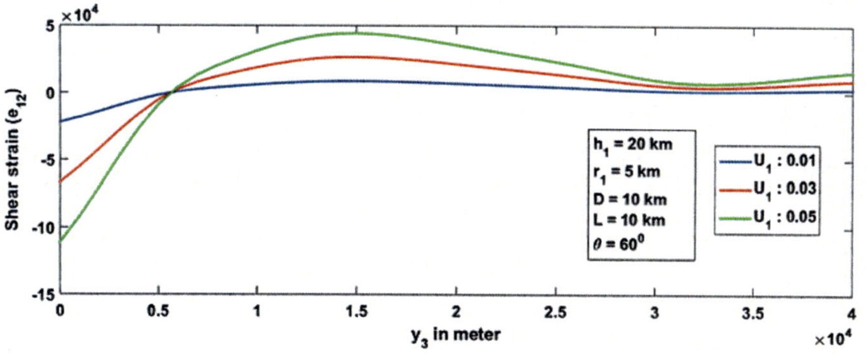

Fig. 11 Change of shear strain with y_3 for different values of U_1

$r_1 = 2, 5, 10$ km, respectively, then shear strain accumulates maximum value near depth 15, 18, 10 km, respectively. That is for a buried fault, shear strain attains maximum value on the fault always for all values of $r_{1.}$.

Figure 11 shows the changes of shear strain with depth y_3 for different values of slip function U_1. Strain accumulation increases with depth increases. For $y_3 < 5$ km, it attains some negative value and then increases with depth. At $y_3 = 5$ km all the stress releases that is earthquake occurs and after that stress accumulates with depth. For $y_3 > 5$ km, shear strain accumulation increases with U_1 increases and for $y_3 < 5$ km, it increases with U_1 decreases.

7 Conclusion

This model of the lithosphere-asthenosphere system is represented by a buried, inclined, locked, finite strike-slip fault. The model is validated by numerical results which are computed using suitable values of model parameters. The nature of displacements, stresses, and strains are analyzed by considering their graphical representation. It is observed that there are certain regions of strain accumulation in the elastic layer and other regions of strain release. The rate of strain accumulation/release in the layer near the fault depends on the various inclination of the fault, depth of the fault, and thickness of the layer. The analysis of accumulation/release of shear stress may help in making earthquake prediction program.

Appendix 1

The displacements, stresses, and strains after the commencement of fault movement are $(u_i)_2, (u_i')_2, (\tau_{ij})_2, \left(\tau_{ij}'\right)_2, (e_{12})_2, (e_{12}')_2 (i, j = 1, 2, 3)$. This satisfies all the Eqs. (2)–(12). This boundary value problem can be solved by using modified Green's function technique developed by Maruyama [20] and Rybicki [29] and correspondence principle. According to them we get

$(\overline{u_1})_2(Q_1) = \iint_F \left[(\overline{u_1})_2(p)\right] G(P, Q_1) dx_3 dx_1$ in the layer.

$(\overline{u_1}')_2(Q_1) = \iint_F \left[(\overline{u_1})_2(p)\right] G'(P, Q_2) dx_3 dx_1$ in the half-space.

where $Q_1(y_1, y_2, y_3)$ and $Q_2(y_1, y_2, y_3)$ are field points in the layer and half-space, respectively, and $P(x_1, x_2, x_3)$ is any point on the fault F and $\left[(u_1)_2(p)\right]$ is the magnitude of discontinuity of u_1 across F and G, G' are Green's function.

where $G(P, Q_1) = \frac{\partial}{\partial x_2}\left[G_{12(1)}(P, Q_1) - G_{13(1)}(P, Q_1)\right]$

$$G(P, Q_2) = \frac{\partial}{\partial x_2}\left[G_{12(2)}(P, Q_2) - G_{13(2)}(P, Q_2)\right]$$

$$\text{Here } G_{12(1)}(P, Q_1) = \int_0^\infty \left[A_1(\lambda)e^{-\lambda(y_3 - y_1)} + B_1(\lambda)e^{\lambda(y_3 - y_1)}\right] \sin[\lambda(x_2 - y_2)] \, d\lambda - \frac{1}{\left[(y_1 - x_1)^2 + (y_2 - x_2)^2 + (y_3 - x_3)^2\right]^{\frac{1}{2}}}$$

$$G_{13(1)}(P, Q_1) = \int_0^\infty \left[C_1(\lambda)e^{-\lambda(y_3 - y_1)} + D_1(\lambda)e^{\lambda(y_3 - y_1)}\right] \cos[\lambda(x_2 - y_2)] \, d\lambda - \frac{1}{\left[(y_1 + x_1)^2 + (y_2 - x_2)^2 + (y_3 - x_3)^2\right]^{\frac{1}{2}}}$$

$$G_{12(2)}(P, Q_2) = \int_0^\infty A_2(\lambda)e^{-\lambda(y_3 - y_1)} \sin[\lambda(x_2 - y_2)] d\lambda$$

$$G_{13(2)}(P, Q_2) = \int_0^\infty C_2(\lambda)e^{-\lambda(y_3 - y_1)} \cos[\lambda(x_2 - y_2)] d\lambda$$

where $A_1 = -\frac{1}{2\pi\Delta_1}\left[e^{-\lambda(x_3+x_1)} + \frac{1-\gamma_1}{1+\gamma_1}e^{-\lambda(2h_1-x_3+x_1)}\right]$

$$B_1 = -\frac{1}{2\pi\Delta_1}\frac{1-\gamma_1}{1+\gamma_1}[e^{-\lambda(2h_1+x_3-x_1)} + e^{-\lambda(2h_1-x_3-x_1)}]$$

$$C_1 = -\frac{1}{2\pi\Delta_1}\left[e^{-\lambda(x_3-x_1)} - \frac{1-\gamma_1}{1+\gamma_1}e^{-\lambda(2h_1-x_3-x_1)}\right]$$

$$D_1 = -\frac{1}{2\pi\Delta_1}\frac{1-\gamma_1}{1+\gamma_1}\left[e^{-\lambda(2h_1+x_3+x_1)} - e^{-\lambda(2h_1-x_3+x_1)}\right]$$

$$A_2 = -\frac{1}{\pi(\gamma_1+1)\Delta_1}\left[e^{\lambda(x_3-x_1)} + e^{-\lambda(x_3+x_1)}\right]$$

$$C_2 = \frac{1}{\pi(\gamma_1+1)\Delta_1}\left[e^{\lambda(x_3+x_1)} - e^{-\lambda(x_3-x_1)}\right]$$

where $\gamma_1 = \frac{\mu_1}{\mu_2}$, μ_1, μ_2 are rigidity of elastic layer and half-space, λ is Lame's constant and h_1 is the thickness of the layer.

$$\Delta_1 = 1 - \frac{1-\gamma_1}{1+\gamma_1}e^{-2\lambda h_1}$$

P (x_1, x_2, x_3) being a point on the fault F. Since the fault inclined at an angle θ and depth is r_1 from the free surface, then $0 \le x_2 \le D\cos\theta, 0 \le x_3 \le D\sin\theta$ and $x_2 = x_3\cot\theta$. A change in co-ordinate from (x_1, x_2, x_3) to $\left(x_1', x_2', x_3'\right)$ connected by the relations:

$$x_1 = x_1', x_2 = x_2'\sin\theta + x_3'\cos\theta, x_3 = -x_2'\cos\theta + x_3'\sin\theta + r_1$$

From $x_2 = x_3\cot\theta$ we get $x_2' = 0$.

Then $x_1 = x_1', x_2 = x_3'\cos\theta, x_3 = x_3'\sin\theta + r_1$ and $dx_1 = dx_1', dx_2' = 0, dx_3 = \sin\theta dx_3'$

$$\begin{aligned}(\overline{u_1})_2(Q_1) &= \iint_F (\overline{u}_1)_2(p)G(P, Q_1)dx_3dx_1\\ &= \int_{-L}^{L}\int_0^D U_1 f\left(x_1', x_3'\right)G(P, Q_1)\sin\theta d\,x'_1\,d\,x'_1\\ &= U_1\phi(y_1, y_2, y_3)\end{aligned}$$

Taking inverse Laplace transform

$(u_1)_2(Q_1) = \frac{U_1}{2\pi}\phi(y_1, y_2, y_3)H(t_1)$, where $H(t_1)$ is the Heaviside function, $t_1 = t - T$ and

$$\phi = \int_{-L}^{L}\int_{0}^{D} f(x_1', x_3')G(P, Q_1)\sin\theta d\,x'_3\,d\,x'_1$$

$$= \int_{-L}^{L}\int_{0}^{D} f(x_1', x_3')\left(\frac{\lambda}{2\pi\Delta_1}\left[-A_{11} - \frac{1-\gamma_1}{1+\gamma_1}A_{12} - \frac{1-\gamma_1}{1+\gamma_1}A_{13} - \frac{1-\gamma_1}{1+\gamma_1}A_{14}\right.\right.$$

$$\left.- B_{11} + \frac{1-\gamma_1}{1+\gamma_1}B_{12} - \frac{1-\gamma_1}{1+\gamma_1}B_{13} + \frac{1-\gamma_1}{1+\gamma_1}B_{14}\right] + A_{15} - B_{15})\sin\theta dx_3' dx_1'$$

where $A_{11} = \frac{(y_3 - y_1 + x_3' \sin\theta + r_1 + x_1')}{(y_3 - y_1 + x_3' \sin\theta + r_1 + x_1')^2 + (x_3' \cos\theta - y_2)^2}$

$$A_{12} = \frac{(2h_1 + y_3 - y_1 - x_3' \sin\theta - r_1 + x_1')}{(2h_1 + y_3 - y_1 - x_3' \sin\theta - r_1 + x_1')^2 + (x_3' \cos\theta - y_2)^2}$$

$$A_{13} = \frac{(2h_1 - y_3 + y_1 + x_3' \sin\theta + r_1 - x_1')}{(2h_1 - y_3 + y_1 + x_3' \sin\theta + r_1 - x_1')^2 + (x_3' \cos\theta - y_2)^2}$$

$$A_{14} = \frac{(2h_1 - y_3 + y_1 - x_3' \sin\theta - r_1 - x_1')}{(2h_1 - y_3 + y_1 - x_3' \sin\theta - r_1 - x_1')^2 + (x_3' \cos\theta - y_2)^2}$$

$$A_{15} = \frac{(x_3' \cos\theta - y_2)}{\left((x_1' - y_1)^2 + (x_3' \cos\theta - y_2)^2 + (x_3' \sin\theta + r_1 - y_3)^2\right)^{\frac{3}{2}}}$$

$$B_{11} = \frac{(x_3' \cos\theta - y_2)}{(x_3' \cos\theta - y_2)^2 + (x_3' \sin\theta + r_1 + y_3 - x_1' - y_1)^2}$$

$$B_{12} = \frac{(x_3' \cos\theta - y_2)}{(x_3' \cos\theta - y_2)^2 + (2h_1 - x_3' \sin\theta - r_1 + y_3 - x_1' - y_1)^2}$$

$$B_{13} = \frac{(x_3' \cos\theta - y_2)}{(x_3' \cos\theta - y_2)^2 + (2h_1 + x_3' \sin\theta + r_1 - y_3 + x_1' + y_1)^2}$$

$$B_{14} = \frac{(x_3' \cos\theta - y_2)}{(x_3' \cos\theta - y_2)^2 + (2h_1 - x_3' \sin\theta - r_1 - y_3 + x_1' + y_1)^2}$$

$$B_{15} = \frac{(x_3' \cos\theta - y_2)}{\left((x_1' + y_1)^2 + (x_3' \cos\theta - y_2)^2 + (x_3' \sin\theta + r_1 - y_3)^2\right)^{\frac{3}{2}}}$$

Similarly $\left(\overline{u_1'}\right)_2(Q_2) = U_1\psi(y_1, y_2, y_3)$.

Taking inverse Laplace transform $\left(u_1'\right)_2(Q_2) = \frac{U_1}{2\pi}\psi H(t_1)$.

where $H(t_1)$ is Heaviside step function, $t_1 = t - T$ and

$$\psi = \int_{-L}^{L}\int_{0}^{D} f(x_1', x_3')\left(\frac{\lambda}{\pi(\gamma_1 + 1)\Delta_1}\left[-A_{11}' - A_{12}' + B_{11}' - B_{12}'\right]\right)\sin\theta dx_3' dx_1'$$

where

$$A'_{11} = \frac{(y_3 - y_1 - x'_3 \sin\theta - r_1 + x'_1)}{\left(y_3 - y_1 - x'_3 \sin\theta - r_1 + x'_1\right)^2 + \left(x'_3 \cos\theta - y_2\right)^2}$$

$$A'_{12} = \frac{(y_3 - y_1 + x'_3 \sin\theta + r_1 + x'_1)}{\left(y_3 - y_1 + x'_3 \sin\theta + r_1 + x'_1\right)^2 + \left(x'_3 \cos\theta - y_2\right)^2}$$

$$B'_{11} = \frac{(x'_3 \cos\theta - y_2)}{\left(x'_3 \cos\theta - y_2\right)^2 + \left(x'_3 \sin\theta + r_1 - y_3 + x'_1 + y_1\right)^2}$$

$$B'_{12} = \frac{(x'_3 \cos\theta - y_2)}{\left(x'_3 \cos\theta - y_2\right)^2 + \left(x'_3 \sin\theta + r_1 + y_3 - x'_1 - y_1\right)^2}$$

Reference

1. K. Aki, Scaling law of seismic spectrum. JGR **72**(4) (1967)
2. K.E. Bullen, *An Introduction to the Theory of Seismology* (Cambridge Univ. Press, London, 1963), p. 381
3. L.M. Cathles III., *The Viscoelasticity of the Earth's Mantle* (Princeton University Press, Princeton, N.J., 1975).
4. P. Chift, J. Lin, U. Barcktiausen, Marine Pet. Geol. **19**, 951–970 (2002)
5. M.A. Chinnery, The deformation of the ground around surface faults. Bull. Seis. Soc. Am. **51**, 355–372 (1961)
6. M.A. Chinnery, The strength of the Earth's crust under horizontal shear stress. J. Geophys. Res. **69**, 2085–2089 (1964)
7. M.A. Chinnery, The vertical displacements associated with transcurrent faulting. J. Geophys. Res. **70**, 4627–4632 (1965)
8. M.A. Chinnery, D.B. Jovanovich, Effect of earth layering on earthquake displacement fields. Bull. Seis. Soc. Am. **62**, 1629–1646 (1972)
9. S.K. Debnath, S. Sen, A creeping vertical strike-slip fault of finite length in a viscoelastic half-space model of the Lithosphere. Int. J. Comput. **2**(3), 687–697 (2012)
10. S.K. Debnath, S. Sen, Long dip-slip fault in a viscoelastic half-space model of the Lithosphere. Am. J. Comput. Appl. Math. **2**(6), 249–256 (2012)
11. S.K. Debnath, S. Sen, Stress and strain accumulation due to a long dip-slip fault movement in an elastic layer over a viscoelastic half-space model of the lithosphere-asthenosphere system. Int. J. Geosci. **4**(3), 549–557 (2013)
12. S.K. Debnath, S. Sen, Aseismic ground deformation in a viscoelastic layer overlying a viscoelastic half-space model of the lithosphere-asthenosphere system. Geosciences **2**(3), 60–67 (2013)
13. S.K. Debnath, Nature of stress-strain accumulation due to a rectangular finite fault in a viscoelastic layer over a viscoelastic half-space. Int. J. Sci. Technol. Res. **2**(3), 254–265 (2013)
14. S.K. Debnath, A buried vertical long dip-slip fault in a viscoelastic half-space model of the lithosphere. J. Emerg. Trends Eng. Appl. Sci. **4**(1), 7–15 (2013)
15. S.K. Debnath, Nature of stress-strain accumulation due to a rectangular finite fault in a viscoelastic layer over a viscoelastic half-space. Int. J. Sci. Technol. Res. **2**(3) (2013)
16. S.K. Debnath, A buried vertical rectangular finite fault in an elastic layer over a viscoelastic half-space. Int. J. Curr. Res. **5**(6), 1407–1414 (2013)
17. S. Karato, Rheology of the Earth's mantle. A historical review. Gondwana Res. **18**(1) (2010)

18. M. Matsu'ura, T. Sato, A dislocation model for the earthquake cycle at convergent plate boundaries. Geophys. J. **96**, 23–32 (1989)
19. T. Maruyama, Statical elastic dislocations in an infinite and semi-infinite medium. Bull. Earthq. Res. Inst. Tokyo Univ. **42**, 289–368 (1964)
20. T. Maruyama, On two dimensional dislocations in an infinite and semi-infinite medium. Bull. Earthq. Res. Inst. Tokyo Univ. **44**(part 3), 811–871 (1966)
21. B. Mondal, S. Sen, Long vertical strike-slip fault in a multi-layered elastic model. Geosciences **6**(2), 29–40 (2016)
22. A. Mukhopadhyay, et al., On stress accumulation near finite rectangular fault. Indian J. Meteorol. Hydrol. Geophys. (Mausam) **30**, 347–352 (1979)
23. A. Mukhopadhyay, et al., On stress accumulation and fault slip in lithosphere. Indian J. Meteorol. Hydrol. Geophys. (Mausam) **30**, 353–358 (1979)
24. A. Mukhopadhyay, S. Sen, B.P. Paul, On stress accumulation in a viscoelastic lithosphere containing a continuously slipping fault. Bull. Soc. Earthq. Technol. **17**(1), 1–10 (1980)
25. A. Mukhopadhyay, S. Sen, B.P. Paul, On stress accumulation near a continuously slipping fault in a two layered model of lithosphere. Bull. Soc. Earthq. Technol. **17**(4), 29–38 (1980)
26. Y. Okada, Surface deformation due to shear and tensile fault in a half-space. Bull. Seis. Soc. Am. **75**(4), 1135–1154 (1986)
27. Y. Okada, Internal deformation due to shear and tensile fault in a half-space. Bull. Seis. Soc. Am. **82**(2), 1018–1040 (1992)
28. M. Rosenman, S.J. Singh, Quasi-static strains and tilts due to faulting in a viscoelastic half-space. Bull. Seis. Soc. Am. **63**(5), 1737–1752 (1973)
29. K. Rybicki, The elastic residual field of a very long strike-slip fault in the presence of a discontinuity. Bull. Seis. Soc. Am. **61**, 79–92 (1971)
30. K. Rybicki, Static deformation of a multilayered half-space by a very long strike-slip fault. Pure Appl. Geophys. **110**, 1955–1966 (1973)
31. R. Sato, Stress drop of finite fault. J. Phys. Earth **20**, 397–407 (1972)
32. J.C. Savage, W.H. Prescott, Geodimeter measurements of strain during the Southern California uplift. J. Geophys. Res. **84**(B1) (1979)
33. W.Z. Savage, Prediction of vertical displacements in a subsiding elastic layer. Geophys. Res. Lett. **8**(3), 195–198 (1981)
34. P. Segall, *Earthquake and Volcano Deformation* (Princeton University Press, 2010)
35. S. Sen, S. Sarker, A. Mukhopadhyay, A creeping and surface breaking long strike-slip fault inclined to the vertical in a viscoelastic half-space. Mausam **44**(4), 365–372 (1993)
36. J.A. Steketee, On Volterra's dislocations in a semi-infinite medium. Can. J. Phys. **36**, 192–205 (1958)
37. J.A. Steketee, Some geophysical applications of the theory of dislocations. Can. J. Phys. **36**, 1168–1198 (1958)

Influence of Velocity Slip on the MHD Flow of a Micropolar Fluid Over a Stretching Surface

P. K. Pattnaik, D. K. Moapatra, and S. R. Mishra

Abstract Free convection of an electrically conducting micropolar fluid past a permeable stretching surface is considered in the present analysis. The crux of the investigation is the study of velocity slip boundary condition that affects the flow behavior. In addition to that the temperature profile enhances with the inclusion of dissipative heat energy, thermal radiation and the heat generation/absorption parameter. Employing suitable similarity variables, the governing equations are transformed to nonlinear ODEs and numerical treatment such as fourth-order Runge-Kutta method in conjunction with shooting technique. Physical behavior of several contributing parameters for the flow phenomena, local skin-friction coefficient, the wall couple stress, and the local Nusselt number are presented via graphs and further described in the results and discussion section.

Keywords MHD · Micropolar fluid · Slip velocity · Stretching surface · Heat generation

Nomenclature

a, b	Constants
B_0	External uniform magnetic field
C_{fx}	Local skin friction coefficient
Cp	Specific heat at constant pressure
Ec	Eckert number
f	Dimensionless stream function
f_w	Suction/injection parameter

P. K. Pattnaik (✉)
Department of Mathematics, College of Engineering and Technology, BBSR, 751029 Bhubaneswar, Odisha, India
e-mail: papun.pattnaik@gmail.com

D. K. Moapatra · S. R. Mishra
Department of Mathematics, Siksha 'O' Anusandhan Deemed to be University, Khandagiri, Bhubaneswar 751030, Odisha, India

S. R. Mishra et al. (eds.), *Recent Trends in Applied Mathematics*, Lecture Notes in Mechanical Engineering, https://doi.org/10.1007/978-981-15-9817-3_21

g	Acceleration due to gravity
G	Micro-rotation parameter
G_1	Micro-rotation constant
j	Micro-inertia density
k	Thermal conductivity
K	Material parameter
M	Magnetic parameter
m	Heat flux exponent
m_w	Wall couple stress
M_x	Dimensionless wall couple stress
N	Micro-rotation/angular velocity
n	Micro-rotation boundary condition
Nu_x	Local Nusselt number
Pr	Prandtl number
Q_0	Heat generation/absorption constant
q_r	Radiative heat flux
q_s	Variable surface heat flux
q_w	Heat transfer rate
R	Radiation absorption parameter
Re_x	Reynold's number
S	Heat generation/absorption parameter
T	Temperature of the fluid
T_∞	Onset temperature
(u, v)	Velocity components along x-, y-axes
v_w	Suction/injection velocity
(x, y)	Horizontal and vertical co-ordinate axes

Greek Symbols

μ	Dynamic viscosity
ρ	Fluid density
β_T	Coefficient of thermal expansion
σ	Electrical conductivity
γ	Spin gradient parameter
α	Velocity slip parameter
α^*	Velocity slip coefficient
ω	Dimensionless micro-rotation velocity
η	Scaled boundary layer coordinate
θ	Dimensionless temperature
λ	Thermal buoyancy parameter
τ_w	Local wall shear stress
υ	Kinematic viscosity
ψ	Stream function

1 Introduction

In recent days, a considerable interest among the researchers is found for the study of flow phenomena through a stretching sheet. The fact is the extensive application in both engineering and industries. As a pioneer work, Crane [1] presented his study for the laminar boundary flow of an incompressible, time-independent flow through a stretching surface. Further, Gupta and Gupta [2] extended the work of [1] for the influence of suction/injection in the boundary layer over a stretching surface. Vajravelu and Rollins [3] and Pavlov [4] have investigated the heat transfer properties in an electrically conducting fluid in conjunction with internal heat generation or absorption. Baag et al. [5] studied numerically by using the fourth-order Runge-Kutta method with shooting technique to compare their result with previous study. They confirmed the accuracy of their study. Ayano et al. [6] reported that the flow of micro-rotation components will be in opposite direction and one of these components is not rotating.

Das [7] investigated the chemical reaction and thermal radiation effect of MHD micropolar fluid by considering a rotating frame of reference. An analytical treatment by using least square method (LSM) has been carried out to investigate the effects of Reynolds number and Peclet number on a micropolar fluid flow by Fakour et al. [8]. Shamshuddin and Narayana [9] have considered an unsteady case of MHD micropolar fluid whose flow past an inclined plate with reference to a rotating system. They observed the regular behavior of micropolar fluid in their study. Ishak et al. [10] considered the MHD micropolar fluid flow in presence of magnetic field which is applied normal to the plate and thermal buouyancy in their study. They observed the dual behavior of solutions which exist for the assisting flow. Nazeer et al. [11] have considered a micropolar fluid in porous medium with uniform and non-uniform heated bottom wall. A study of micropolar fluid flow in porous medium over a stretchable disk by considering all the profiles like axial velocity, radial velocity, micro-rotation, temperature, and concentrations profiles have been carried out by Rauf et al. [12]. Sheikholeslami et al. [13] in their study of micropolar fluid used an analytical method (Homotopy Analysis Method) to investigate the behavior of Reynolds number and Pecelet number on all the used profiles. They observed the inter link of both these said numbers with Nusselt and Sherwood numbers.Viscous dissipation taken into consideration on the study of a MHD micropolar fluid flow is to investigate the behavior of translation velocity, micro-rotation, and temperature profiles. It has been observed that all these profiles showed the decreasing behavior for increasing values of viscous dissipation (see [14]). Srivastava [15] in his research paper considered the flow of MHD micropolar fluid in between two eccentrically rotating disks to study the effects of the micropolar parameter and Hartmann number on the velocity and micro-rotation profiles. Mishra et al. [16, 17] used the uniform magnetic field strengths along the flow direction to check the behavior of all the profiles considered in the work in presence of heat source and radiation parameter. Ashmawy [18] considered a convective micropolar fluid in between two vertical uniformly heated channels with velocity slip condition applied. Ferdows and Liu

[19] used magnetic field and thermal buoyancy parameters in momentum equation, non-uniform heat source parameter in energy equation to investigate the behavior of magneto-micropolar fluid flow in a vertical plate.

In view of aforesaid discussion it is important to describe the physical significance of heat generation and absorption. Though it is difficult to model the exact internal heat generation or absorption, a mathematical model, following Foraboschi and Federico [20], can be expressed as $S = \begin{cases} Q_0(T - T_\infty), & T \geq T_\infty \\ 0, & T < T_\infty \end{cases}$ which is valid for the state of some exothermic processes. We have extended the work of Mahmoud et al. [21] by incorporating thermal buoyancy parameter in momentum equation, thermal radiation, and viscous dissipation term in energy equation and also boundary condition of micropolar profile has been modified.

2 Mathematical Formulation

Two-dimensional free convective flow of an electrically conducting micropolar fluid past a porous stretching surface is considered in the present investigation. The plate is along the plane $y = 0$, the flow takes place in the region $y > 0$. Applied uniform magnetic field of strength B_0 is imposed along the normal direction of the flow. Variable surface heat flux $q_s(x) = bx^m$ (where b, m are constants) as well as the slip velocity boundary conditions are also assumed. Based upon the aforesaid assumptions the basic governing equations for the flow are

$$\frac{\partial u}{\partial x} + \frac{\partial v}{\partial y} = 0 \tag{1}$$

$$u\frac{\partial u}{\partial x} + v\frac{\partial u}{\partial y} = \left(\frac{\mu + k}{\rho}\right)\frac{\partial^2 u}{\partial y^2} + \frac{k}{\rho}\frac{\partial N}{\partial y} + g\beta_T(T - T_\infty) - \frac{\sigma B_0^2}{\rho}u \tag{2}$$

$$u\frac{\partial N}{\partial x} + v\frac{\partial N}{\partial y} = \frac{\gamma}{\rho j}\frac{\partial^2 N}{\partial y^2} - \frac{k}{\rho j}\left(2N + \frac{\partial u}{\partial y}\right) \tag{3}$$

$$u\frac{\partial T}{\partial x} + v\frac{\partial T}{\partial y} = \frac{k}{\rho Cp}\frac{\partial^2 T}{\partial y^2} + \frac{Q_0}{\rho Cp}(T - T_\infty) - \frac{1}{\rho Cp}\frac{\partial q_r}{\partial y} + \frac{\mu}{\rho Cp}\left(\frac{\partial u}{\partial y}\right)^2 \tag{4}$$

The boundary conditions are

$$\left.\begin{array}{llll} u = ax + \alpha^*\left[(\mu + k)\dfrac{\partial u}{\partial y} + kN\right], & v = v_w, & N = -n\dfrac{\partial u}{\partial y}, \dfrac{\partial T}{\partial y} = -\dfrac{bx^m}{k}, & at\ y = 0 \\ u \to 0, & N \to 0, & T \to T_\infty & as\ y \to \infty \end{array}\right\} \tag{5}$$

Jena and Matkur [22] considered the case ($n = 0$) of concentrated particle flows in which they observed that micro-elements close to the wall are unable to rotate. But Ahmadi [23] examined the case ($n = 1/2$) of weak concentrations and indicates the vanishing of antisymmetric part where as the case for ($n = 1$) turbulent boundary layer flows suggested by Peddieson [24]. The radiative heat flux term by using the Rosseland approximation [25] is given by

$$q_r = -\frac{4\sigma^*}{3k^*}\frac{\partial T^4}{\partial y} \tag{6}$$

where σ^* Stefan–Boltzmann constant and k^* mean absorption coefficient. We have assumed that the temperature differences are very small within the fluid. We have expanded T^4 by Taylor series expansion about T_∞ and neglecting higher order terms to express as a linear function. So q_r can be written as

$$q_r = -\frac{16\sigma^* T_\infty^3}{3k^*}\frac{\partial T}{\partial y}$$

Equation (4) takes the form:

$$u\frac{\partial T}{\partial x} + v\frac{\partial T}{\partial y} = \left(\frac{k}{\rho Cp} + \frac{1}{\rho Cp}\frac{16\sigma^* T_\infty^3}{3k^*}\right)\frac{\partial^2 T}{\partial y^2} + \frac{Q_0}{\rho Cp}(T - T_\infty) + \frac{\mu}{\rho Cp}\left(\frac{\partial u}{\partial y}\right)^2 \tag{7}$$

3 Method of Solution

The equation of continuity (1) is satisfied by introducing the stream function ψ such that

$$u = \frac{\partial \psi}{\partial y}, \quad v = -\frac{\partial \psi}{\partial x} \tag{8}$$

and with the following dimensionless variables:

$$\eta = \sqrt{\frac{a}{v}}\,y,\ \psi = \sqrt{av}\,xf(\eta),\ N = ax\sqrt{\frac{a}{v}}\,\omega(\eta),\ T = T_\infty + \frac{q_s(x)}{k}\sqrt{\frac{v}{a}}\theta(\eta) \tag{9}$$

So the modified equations of the flow can be written as

$$(1 + K)f''' + ff'' - \left(f'\right)^2 + K\omega' - Mf' + \lambda\theta = 0 \tag{10}$$

$$G\omega'' + f\omega' - f'\omega - K\left(2\omega + f''\right) = 0 \quad (11)$$

$$\frac{1}{\Pr}\left(1 + \frac{4}{3}R\right)\theta'' + f\theta' - mf'\theta + S\theta + Ec\left(f''\right)^2 = 0 \quad (12)$$

$$\left.\begin{array}{l} f = f_w,\ f' = 1 + \alpha(1+K)f'',\ \omega = -nf'', \theta' = -1 \text{ at } \eta = 0 \\ f' \to 0, \qquad \omega \to 0, \qquad \theta \to 0 \text{ as } \eta \to \infty \end{array}\right\} \quad (13)$$

$$\begin{array}{l} K = \dfrac{k}{\mu},\ M = \dfrac{\sigma B_0^2}{a\rho},\ \lambda = \dfrac{g\beta_T q_s\sqrt{\upsilon}}{ka^{3/2}},\ G = \dfrac{\gamma}{j\mu},\ j = \dfrac{\upsilon}{a},\ \Pr = \dfrac{\mu c_p}{k} \\ R = \dfrac{4\sigma^* T_\infty^3}{kk^*},\ S = \dfrac{Q_0}{a\rho c_p},\ Ec = \dfrac{\sqrt{\upsilon}ka^{3/2}x^2}{c_p q_s},\ \alpha = \mu\alpha^*\sqrt{\dfrac{a}{\upsilon}} \end{array} \quad (14)$$

The physical quantities of interest are the local skin-friction coefficient C_{fx}, the dimensionless wall couple stress M_x, and the local Nusselt number Nu_x, which are defined as

$$C_{fx} = \frac{2\tau_w}{\rho(ax)^2},\ M_x = \frac{m_w}{\rho a \upsilon x^3},\ Nu_x = \frac{xq_w}{\kappa(T_w - T_\infty)} \quad (15)$$

where the local wall shear stress τ_W, the wall couple stress m_w, and the heat transfer from the plate q_w are defined by

$$\tau_W = \left[(\mu + k)\frac{\partial u}{\partial y} + kN\right]_{y=0},\ m_w = \gamma_0\left[\frac{\partial N}{\partial y}\right]_{y=0},\ q_w = -k\left(\frac{\partial T}{\partial y}\right)_{y=0} \quad (16)$$

Using the similarity variables (10), we get

$$C_{fx}\mathrm{Re}_x^{1/2} = -2(1+K)f''(0),\ M_x\mathrm{Re}_x = KG\omega'(0),\ Nu_x\mathrm{Re}_x^{-1/2} = -\theta'(0) \quad (17)$$

where $\mathrm{Re}_x = \frac{ax^2}{\upsilon}$ is the local Reynolds number.

4 Results and Discussion

Free convection of an electrically conducting micropolar fluid past a stretching surface is considered in the current investigation. The characteristics of the energy equation are enhanced by incorporating the heat generation/absorption parameter as well as the viscous dissipation. In an addition, the slip boundary condition for the velocity is considered which affects the flow phenomena. The physical significance

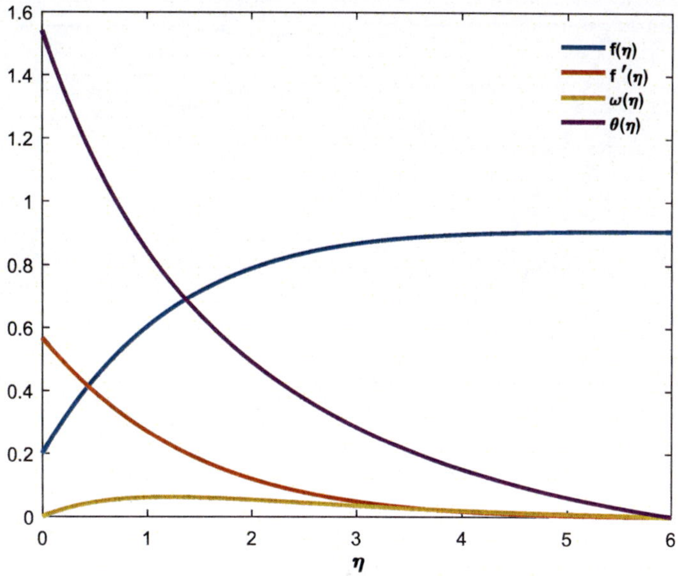

Fig. 1 Validation of stream function $f(\eta)$, Velocity $f'(\eta)$, Micro-rotation $\omega(\eta)$, and Temperature $\theta(\eta)$ profile

of the contributing parameters are obtained and presented via graphs. The rate coefficients for all the profiles are also displayed through graphs. The variation of several parameters on the profiles is presented in the corresponding figures. Figure 1 depicts the validation of the transverse velocity, longitudinal velocity, micro-rotation, and the temperature profiles in the absence of magnetic field, thermal buoyancy, and the thermal radiation. However, the result coincides with the work of Mahmoud et al. [21] showing the conformity of the convergence procedure of the current methodology. Figure 2 exhibits the behavior of the suction/injection parameter for various values of slip factor on the velocity distribution. The partial vacuum exerts upon a liquid is caused by the suction. Reduction in pressure is marked due to the removal of air from the space resulted to enter the fluid into the space. Therefore, the fluid exerts from the higher pressure region to lower pressure region. In comparison to suction and injection, it is seen that the suction lowers down the velocity profiles irrespective of the slip or no slip region. However, in case of no slip condition, the maximum velocity is rendered within the boundary layer and reduction in the profile is observed with increasing slip. Pick in the micro-rotation profiles is marked near the surface up to the region $\eta \leq 1$ and afterwards sudden fall is marked in Fig. 3. Moreover, suction produces higher pressure to reduce the profiles than that of injection. Similar observation is rendered in case of slip parameter as described in the Fig. 2. Figure 4 exhibits the distribution of temperature profiles for the variation of the suction/injection and slip parameters. It is observed that increasing slip enhances the fluid temperature in the entire region of the thermal boundary layer

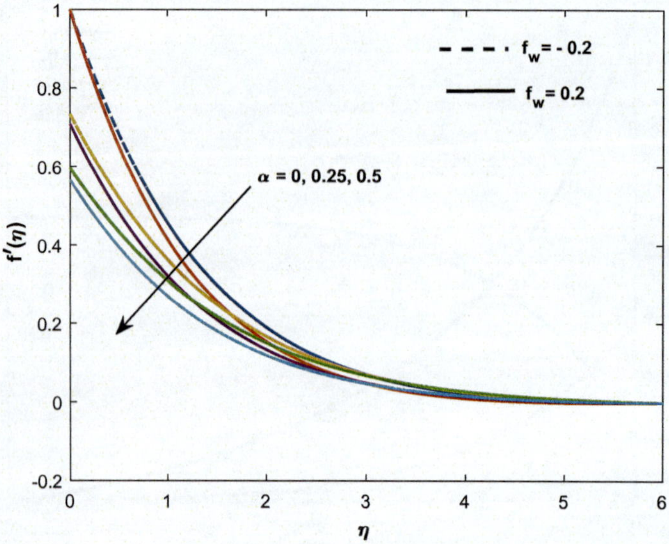

Fig. 2 Variation of Velocity Profile with α and f_w

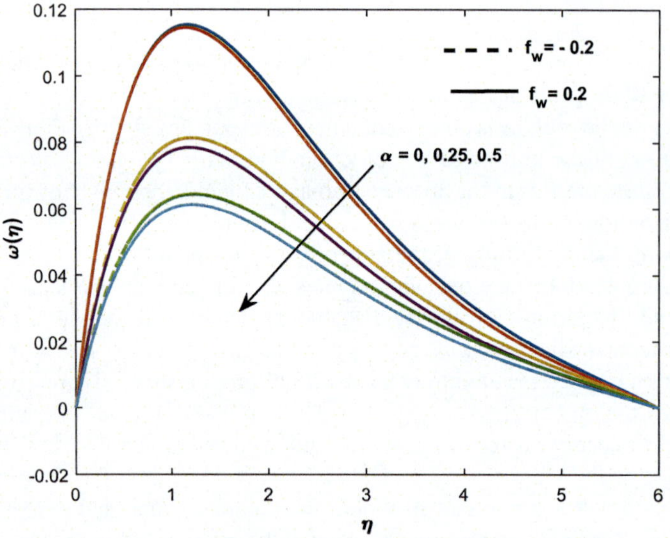

Fig. 3 Variation of Micro-rotation Profile with α and f_w

and injection also favorable to increase the temperature as well. The values of the material parameter (K) indicates the Newtonian and non-Newtonian characteristics of the fluid. $K = 0$ represents the Newtonian case and the $K \neq 0$ shows the

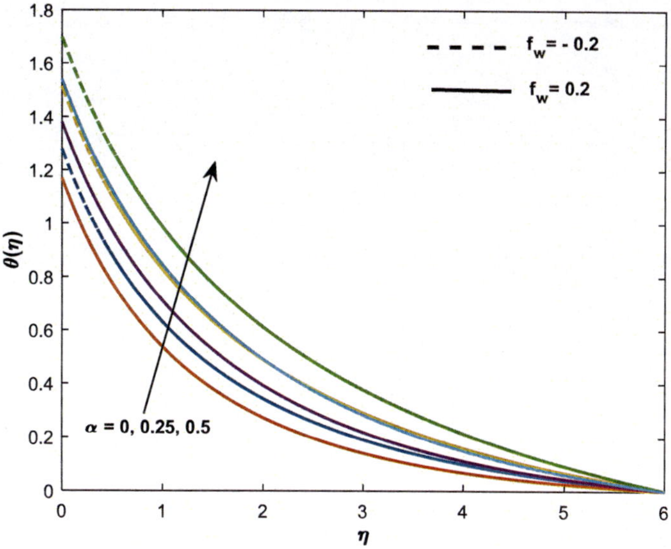

Fig. 4 Variation of Temperature Profile with α and f_w

non-Newtonian nature. However, in the present case we have considered the non-Newtonian behavior of the fluid. Figure 5 illustrates the profiles of micro-rotation in conjunction with suction/injection. An increase in the material parameter enhances the micro-rotation profiles with pick near the surface and further it decelerates. The

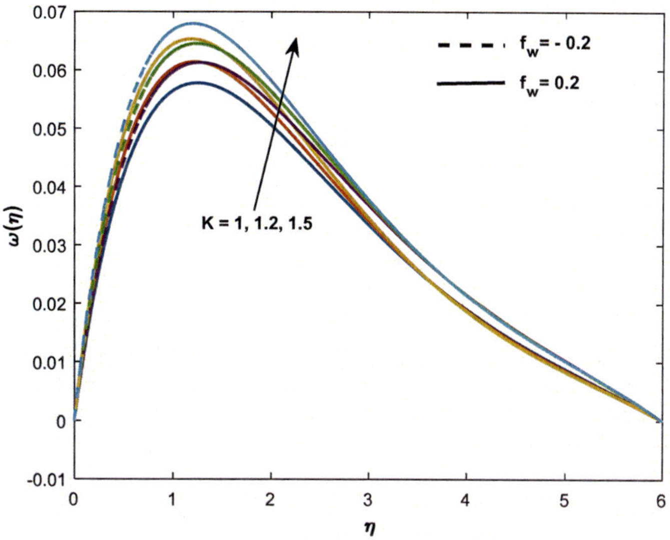

Fig. 5 Variation of Micro-rotation Profile with K and f_w

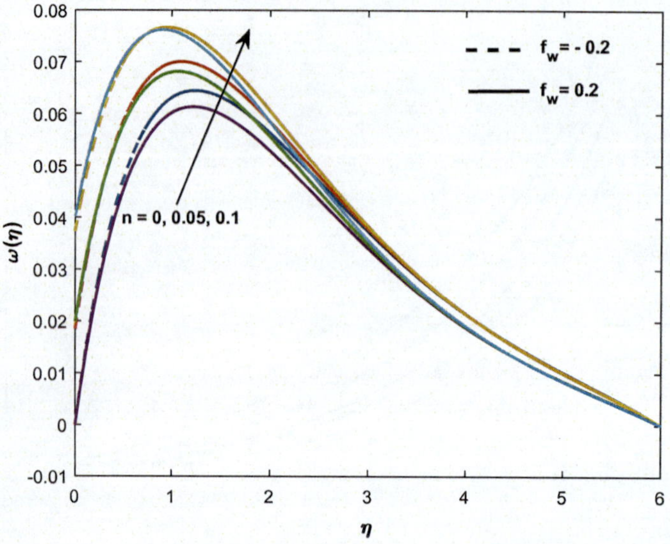

Fig. 6 Variation of Micro-rotation Profile with n and f_w

case of injection is also favorable to enhance it significantly. Figure 6 portrays the wall surface condition parameter (n) with suction/injection parameter on the micro-rotation distribution. The micro-rotation profile enhances rapidly near the surface with increasing the wall surface condition parameter. However, the injection is now favorable to enhance the profile for lower values of n but effect is reversed for higher values. Irrespective of values of suction/injection parameter, buoyancy parameter enriches the velocity profiles that exhibit in Fig. 7. The pressure difference results in a net upward force on the object. Figures 8 and 9 describe the impact of thermal radiation and heat source on the temperature profiles in conjunction with suction/injection parameter. Thermal radiation is one of the characteristics that depends on the various properties of the surface. Thermal enhancement occurs in the entire domain due to increase in the thermal radiation and heat source parameter. The coupling of temperature and velocity profile occurs due to the inclusion of coupling parameter, i.e., the Eckert number. Figure 10 describes the temperature distribution for the various values of Eckert number. From the mathematical definition, it is clear that increasing Eckert number enhances the fluid temperature. Finally, Figs. 11, 12, 13 display the computational results of shear stress, rate of heat transfer, and the couple stress for various values of suction/injection versus the slip parameter. The trend of the graph shows the decelerating effect, whereas increasing suction increases the rate coefficients with increasing slip.

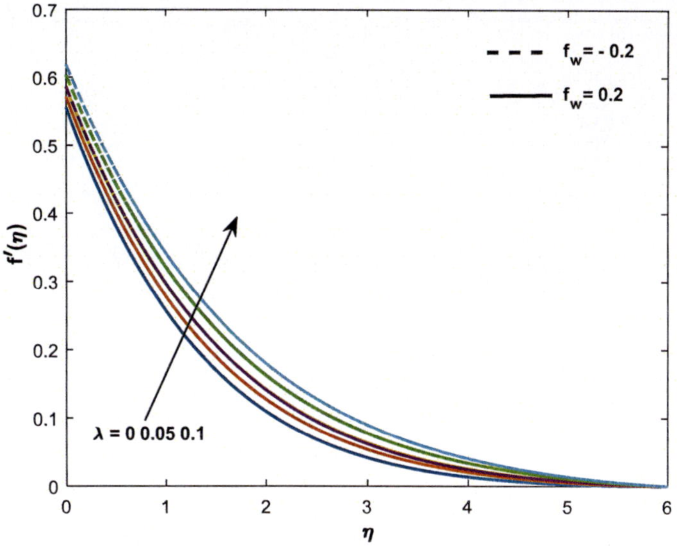

Fig. 7 Variation of Velocity Profile with λ and f_w

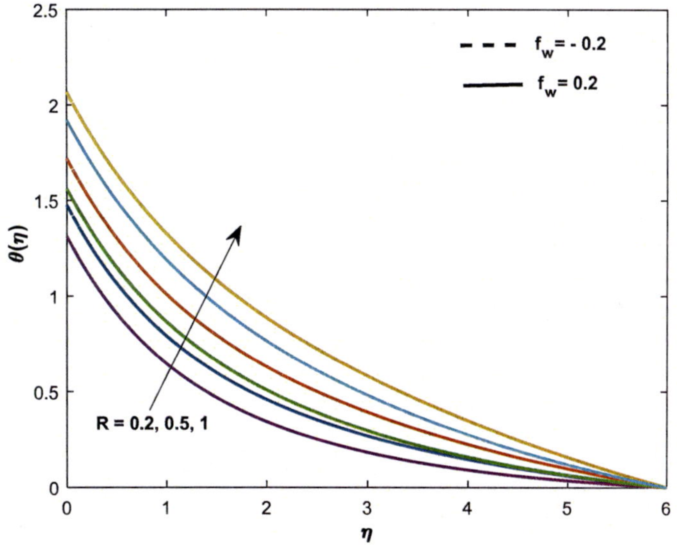

Fig. 8 Variation of Temperature Profile with *R* and f_w

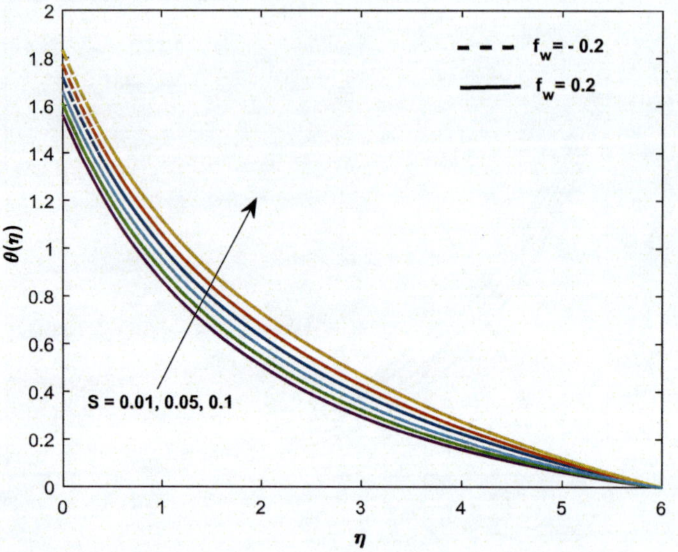

Fig. 9 Variation of Temperature Profile with *S* and f_w

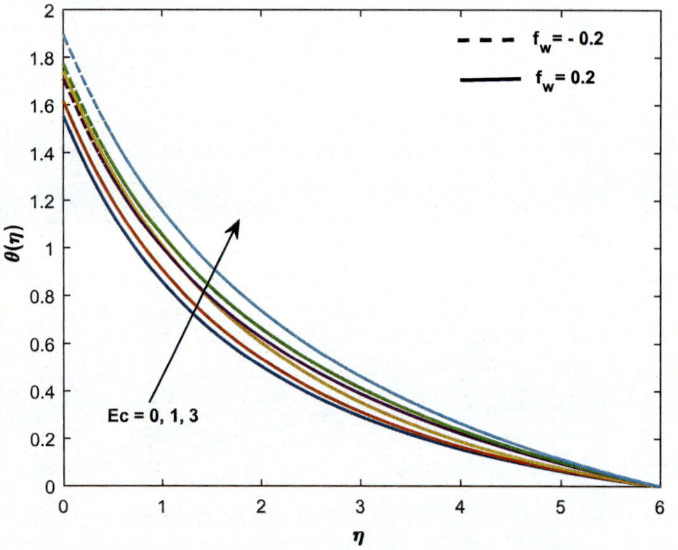

Fig. 10 Variation of Temperature Profile with *Ec* and f_w

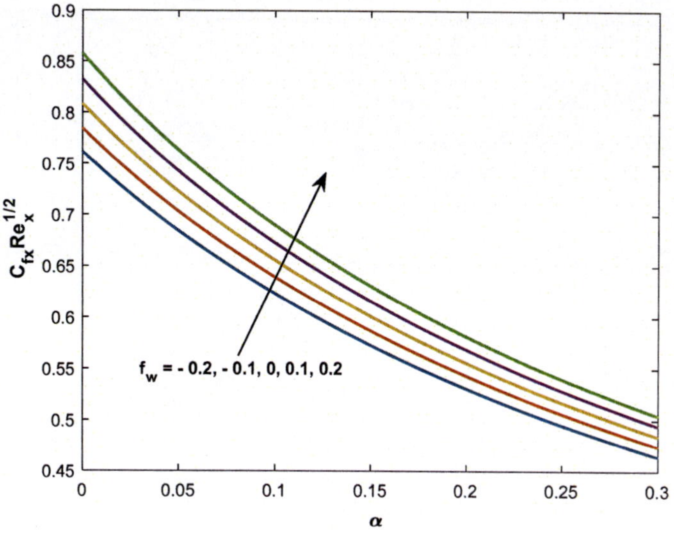

Fig. 11 Variation of Skin Friction Coefficient α and f_w

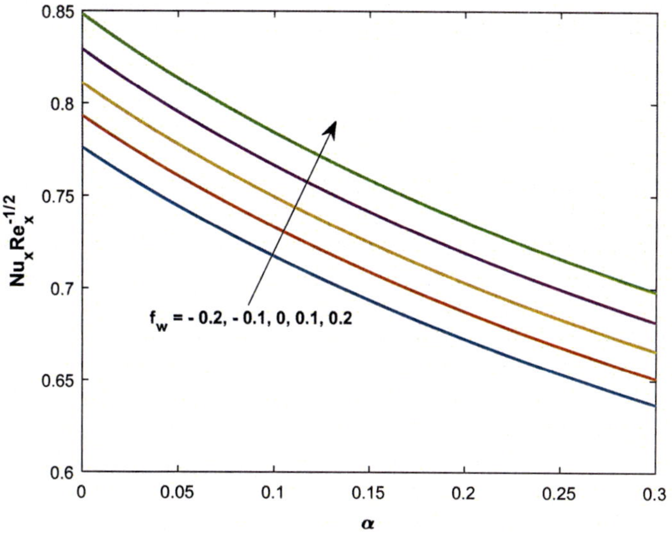

Fig. 12 Variation of Nusselt Number α and f_w

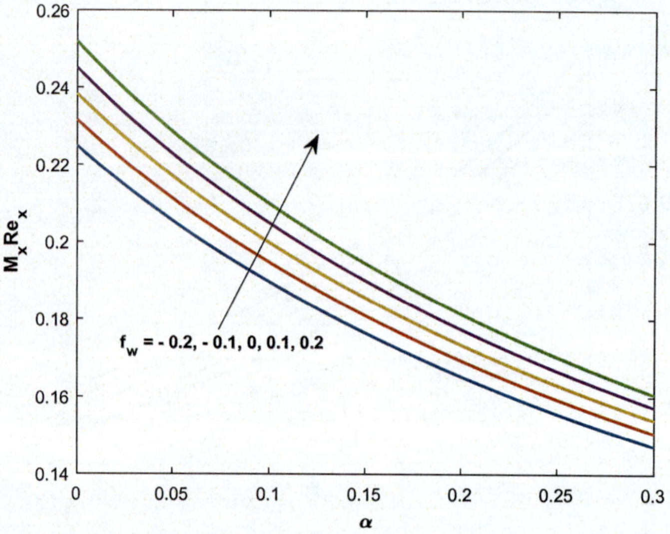

Fig. 13 Variation of Wall Couple Stress Coefficient α and f_w

5 Conclusive Remarks

Free convection of micropolar fluid in conjunction with slip parameter and the effect of heat source are exhibited in the present investigation. The behavior of characterizing parameter on the flow phenomena is displayed and discussed. However, the major contributions are laid down as

- The validation of present result with earlier established result shows the conformity of the convergence procedure of the methodology employed.
- Retardation in the velocity profiles is marked due to increase in suction regardless with the increase of slip parameter.
- The rate coefficients enhance with increasing suction with respect to the slip parameter.

References

1. L. Crane, Flow past a stretching plate. Z. Angew. Math. Phys. **21**, 645–647 (1970)
2. P.S. Gupta, A.S. Gupta, Heat and mass transfer on a stretching sheet with suction or blowing. Can. J. Chem. Eng. **55**, 744–746 (1977)
3. K. Vajravelu, D. Rollins, Heat transfer in electrically conducting fluid over a stretching sheet. Int. J. Non-Linear Mech. **27**, 265–277 (1992)
4. K.B. Pavlov, Magnetohydrodynamic flow of an incompressible viscous fluid caused by deformation of a surface. Magn. Gidrodin. **4**, 146–147 (1974)

5. S. Baag, S.R. Mishra, G.C. Dash, M.R. Acharya, Numerical investigation on MHD micropolar fluid flow toward a stagnation point on a vertical surface with heat source and chemical reaction. J. King Saud Univ. – Eng. Sci. **29**, 75–83 (2017)
6. M.S. Ayano, S.T. Sikwila, S. Shateyi, MHD mixed convection micropolar fluid flow through a rectangular duct. Hindawi Mathematical Problems in Engineering, ID 9840862, 8 p (2018). https://doi.org/10.1155/2018/9840862
7. K. Das, Effect of chemical reaction and thermal radiation on heat and mass transfer flow of MHD micropolar fluid in a rotating frame of reference. Int. J. Heat Mass Transf. **54**, 3505–3513 (2011)
8. M. Fakour, A. Vahabzadeha, D.D. Ganji, M. Hatami, Analytical study of micropolar fluid flow and heat transfer in a channel with permeable walls. J. Mol. Liq. **204**, 198–204 (2015)
9. M.D. Shamshuddin1, P.V. SatyaNarayana, Primary and secondary flows on unsteady mhd free convective micropolar fluid flow past an inclined plate in a rotating system: a finite element analysis. FDMP**14** (1), 57–86 (2018)
10. A. Ishak, R. Nazar, I. Pop, Magnetohydrodynamic (MHD) flow of a micropolar fluid towards a stagnation point on a vertical surface. Comput. Math. Appl. **56**, 3188–3194 (2008)
11. M. Nazeer, N. Ali and T. Javed, Numerical simulation of MHD flow of micropolar fluid inside a porous inclined cavity with uniform/non-uniform heated bottom wall. Can. J. Phys. 1–37 (2017)
12. A. Rauf, M. Ashraf, K. Batool, M. Hussain, M.A. Meraj, MHD flow of a micropolar fluid over a stretchable disk in a porous medium with heat and mass transfer. AIP Adv. **5**, 077156 (2015). https://doi.org/10.1063/1.4927501
13. M. Sheikholeslami, M. Hatami, D.D. Ganji, Micropolar fluid flow and heat transfer in a permeable channel using analytical method. J. Mol. Liq. **194**, 30–36 (2014)
14. S.R. Sheri, Md Shamshuddin, Heat and mass transfer on MHD flow of micropolar fluid in the presence of viscous dissipation and chemical reaction. Proceedia Eng. **127**, 885–892 (2015)
15. N. Srivastava, MHD flow of the micropolar fluid between eccentrically rotating disks. Hindawi Publishing Corporation International Scholarly Research Notices, 2014, ID 317075, 7 p (2014). https://doi.org/10.1155/2014/317075
16. S.R. Mishra, M.M. Hoque, B. Mohanty, N.N. Anika, Heat transfer effect on MHD flow of a micropolar fluid through porous medium with uniform heat source and radiation. (2018). https://doi.org/10.1515/nleng-2017-0126
17. S.R. Mishra, P.K. Pattnaik, G.C. Dash, Effect of heat source and double stratification on MHD free convection in a micropolar fluid. Alex. Eng. J. **54**, 681–689 (2015)
18. E.A. Ashmawy, Fully developed natural convective micropolar fluid flow in a vertical channel with slip. J. Egypt. Math. Soc. **23**, 563–567 (2015)
19. M. Ferdows, D. Liu, Natural convective flow of a magneto-micropolar fluid along a vertical plate. Propuls. Power Res. **7**(1), 43–51 (2018)
20. F.P. Foraboschi, I.D. Federico, Heat transfer in a laminar flow of non- Newtonian heat generating fluids. Int. J. Heat Mass Transf. **7**, 315–318 (1964)
21. M.A.A. Mahmoud, S.E. Waheed, MHD flow and heat transfer of a micropolar fluid over a stretching surface with heat generation (absorption) and slip velocity. J. Egypt. Math. Soc. **20**, 20–27 (2012)
22. S.K. Jena, M.N. Mathur, Similarity solutions for laminar free convective flow of a thermomicropolar fluid past a nonisother-mal vertical plate. Int. J. Eng. Sci. **19**, 1431–1439 (1981)
23. G. Ahmadi, Self-similar solution of incompressible micro-polar boundary layer flow over a semi-infinite plate. Int. J. Eng. Sci. **14**, 639–646 (1976)
24. J. Peddieson, An application of the micropolar fluid model to the calculation of turbulent shear flow. Int. J. Eng. Sci. **10**, 23–32 (1972)
25. M.Q. Brewster (Wiley, Inc., New York1992)

A Spatially Dependent Vaccination Model with Therapeutic Impact and Non-linear Incidence

Md. Shahriar Mahmud, Md. Kamrujjaman, and Md. Shafiqul Islam

Abstract This study presents a spatially dependent vaccination model considering therapeutic impact with non-linear incidence rate where the spatial habitat is a subset of $\mathbb{R}^n$ with smooth boundary. Auxiliary results such as disease-free and disease equilibrium states' basic reproduction number are calculated. This study also includes both local and global stability constraints, uniform persistence condition and existence of the unique solution of the model. Our study showed that the global stability of the model depends on the threshold level $\mathcal{R}_0$ in the way that $\mathcal{R}_0 \leq 1$ refers to disease-free equilibrium E_0 where $\mathcal{R}_0 > 1$ indicates unique disease equilibrium E^*.

Keywords Spatial vaccination model · Non-linear incidence · Threshold value · Local stability · Global stability · Uniform persistence

AMS Subject Classification 2010 92D30 · 92D25 · 93D05 · 93D20 · 93C20 · 76E30

Md. S. Mahmud (✉) · Md. Kamrujjaman
Department of Mathematics, University of Dhaka, Dhaka 1000, Bangladesh
e-mail: prism.shahriar@gmail.com

Md. Kamrujjaman
e-mail: kamrujjaman@du.ac.bd; mkamrujj@ucalgary.ca

Md. Kamrujjaman
Department of Mathematics and Statistics, University of Calgary,
Calgary, AB, Canada

Md. S. Islam
School of Mathematical and Computational Sciences, University of Prince
Edward Island, 550 University Ave., Charlottetown, PE, Canada
e-mail: sislam@upei.ca

S. R. Mishra et al. (eds.), *Recent Trends in Applied Mathematics*, Lecture Notes
in Mechanical Engineering, https://doi.org/10.1007/978-981-15-9817-3_22

1 Introduction

The terms "Vaccine" and "vaccinology" came into use soon after Edward Jenner discovered the smallpox vaccine called "variola vaccinae". The word "vaccine" originated from vacca, a Latin term for the cow. The credit for the first use of the term "vaccine" goes to Swiss physician Louis Odier (1748–1817), and the terms "vaccination" and "to vaccinate" were first used by Richard Dunning (1710–1797) [1].

Physicians from the times of Hippocrates (460–370 BC) tried to understand the pattern of diseases in the community, though the word "epidemiology" was first used in 1802 to describe the study of epidemics by Villalba, a Spanish physician, in the Epidemiologìa Española [2]. In modern times, John Snow (1813–1858) and William Farr (1807–1883) pioneered the work on epidemiology [3]. Epidemiology, though practiced from earlier times than vaccinology, gained attention and prominence in the nineteenth century. Now, the practice of vaccinology has become closely linked with that of epidemiology.

As most of the diseases have a recovered/immune stage for which vaccination is successful and some other bacteria can remain in host without causing any disease results in carriage, an SIS model is popular in epidemiology to observe the dynamical behaviors of the infectious disease. A model where recovery is short lived, i.e., brings the populations return to the compartment of susceptiblity is considerable in this action with vaccination as an SIS model [4].

Periodic fluctuations in abundance are common issues in many infectious diseases. Such periodicities may be shifted by outer factors, as contemplated in repeated transmission defeats, e.g., seasonality [5–7], or may be caused by time delays [8], age structure [9], or non-linearity of incidence rates.

The incidence rate means, the measure of the probability of occurrence of a given medical circumstance within a specific time period or in brief, the number of renewed briefs within a specific time period divided by the initial risk population size, i.e.,

$$\text{Incidence rate} = \frac{\text{New cases}}{\text{Population at risk}}.$$

It also can be explained as the inverse of waiting time for an individual to be infected in a specific disease condition as

$$\text{Incidence rate} = \frac{1}{\text{Waiting time}}.$$

Wilson and Worcester [10] were the first who introduced the general incidence rate S^p. They underlined that they did so primarily "to investigate the consequences of various assumptions when the laws are not known". For their motivations, it is also recognized that the suggested model does not fit the data well. In 1969, Severo [11] considered the incidence rate of the form kI^pS^q while $q < 1$; but the details of investigation was not considered in the study. Capasso and Serio [12] updated another incidence rate function such as $kh(I)S$ and again the study was conditional; the term

$h'(0)$ must be finite and positive. If $p \neq 1$ then the form kI^pS excludes the previous pattern. Later, Wangersky and Cunningham [13] noted that periodic solutions in a model may exist with an incidence rate $k(IS)^p$, provided $p > 1$. In 1986 and 1987, Liu et al. [14, 15] took some general incidence rates. They also established the conditions for which a durable periodic solution is possible in case of Hopf bifurcation and discussed feasible appointments underlying incidence rates $H(I)$, which is non-linear. They also declared that it yet remained to consider two important factors: various form of incidence rate and the outcomes of disease-persuaded mortality.

Since then, a lot of mathematical models have been studied with different types of interesting incidence rates. In 2003, Gumel and Moghadas' [16] introduced the non-linear incidence rate $H(I) = \frac{I}{1+I}$ and the proposed mathematical epidemic is defined as

$$\begin{cases} S_t(t) = a - bq_1H(I)S - (m+n)S + rI & \text{for } t \in (0,\infty), \quad \text{with } S(0) = S_0; \\ V_t(t) = nS - bq_2H(I)V - mV & \text{for } t \in (0,\infty), \quad \text{with } V(0) = V_0; \\ I_t(t) = b(q_1S + q_2V)H(I) - (m+r)I & \text{for } t \in (0,\infty), \quad \text{with } I(0) = I_0. \end{cases} \tag{1.1}$$

where S, V, I are the number of individuals in susceptible, vaccinated, and infectious compartments at time t, respectively. The susceptible individuals recruitment rate is a, q_1 and q_2 are the probabilities that the susceptible and vaccinated individuals will be transmitted. The average number of contact partners is notified by the parameter b, n is defined for vaccination coverage of susceptible populations, and m is the rate of natural death. Since system (1.1) monitors the dynamics of population with spatial heterogeneity, it concludes that all the parameters and variables are assumed to be non-negative. Further, it is assumed that the prevalent disease does not kill infected individuals and treatment does not offer permanent immunity.

They also added r as the therapeutic treatment coverage parameter of infected individuals $I(t)$ removed to $S(t)$ compartment. Since, it is an SIS model, it refers that the effectively treated infected individuals return to the susceptible compartments and behaves the same and as vaccination therapy can either diminish or eliminate the incidence of disease transmission, realistically it is observed that $q_2 \leq q_1$.

Gumel and Moghadas [16] analyzed the characteristic equation of the model and studied locally, its disease and disease-free equilibria and the optimal vaccine coverage threshold needed for disease control and eradication analytically. Lately, Buonomo et al. [17] constructed suitable Lyapunov functions and established global results of both disease and disease-free steady state of system (1.1) by using LaSalle's invariance principle [18]. It is also remark that in this study they also considered the treatment strategies and optimal vaccination to minimize the intervention and disease compulsion.

It has been established that a more realistic biological–mathematical model includes spatial variation as one of the principal factors that affect the spatial spreading of disease [19, 20]. Considering the individual movements, we are interested here to consider diffusive version of the deterministic model (1.1) and improving the incidence rate to a more general one.

We introduce α as the multiplicity of new cases of infection and β as the risk factor coefficient, where $\alpha, \beta > 0$ and surely $\alpha \leq \beta$. Then we form the incidence rate function as following:

$$f(I) = \frac{\alpha I}{1 + \beta I} \tag{1.2}$$

We see here, $f(I)$ coincides with $H(I)$ when $\alpha = \beta = 1$. And $f(I)$ gives a bilinear incidence when $\beta = 0$.

Before considering the spatially dependent vaccination model, for simplicity let us define the notations as $\mathcal{A} = \Omega \times (0, \infty)$ and $\partial\mathcal{A} = \partial\Omega \times (0, \infty)$. Now, in this study, we propose the following spatially dependent vaccination problem with a more general incidence function (1.2) as

$$\begin{cases} S_t(x,t) = \nu\Delta S + a - bq_1 f(I(x,t))S(x,t) - (m+n)\,S(x,t) + rI(x,t) & \text{in } \mathcal{A}, \\ V_t(x,t) = \upsilon\Delta V + nS(x,t) - bq_2 f(I(x,t))V(x,t) - mV(x,t) & \text{in } \mathcal{A}, \\ I_t(x,t) = \vartheta\Delta I + b\,(q_1 S(x,t) + q_2 V(x,t))\, f(I(x,t)) - (m+r)I(x,t) & \text{in } \mathcal{A}. \end{cases}$$

where $S(x,t)$, $V(x,t)$, $I(x,t)$ are the number of individuals in the susceptible, vaccinated, and infectious compartments at any time $t > 0$ and in location $x \in \Omega$, respectively, and the other parameters are same as in system (1.1). The symbolic notion Ω is a spatial niche/domain in $\mathbb{R}^n$ with a respective smooth boundary $\partial\Omega$, and the well-known Laplacian Operator Δ, and ν, υ and ϑ are the diffusion rates. The model represents the dynamical behaviors of population, all of its parametric values and variables, for example, a, b, m, n, q_1, q_2 and r must be non-negative as in the non-spatial model (1.1).

We know that in epidemiology, one of the fundamental issue is to find the stability of the two constant steady states, i.e., disease-free equilibrium and disease equilibrium. Considering all notations, we have studied the model briefly and presenting the results in this paper.

We have organized this paper with our results in the following manner.

In the primary phase, we check the well-posedness of the model verifying the model flow scheme and the mathematical reasoning of the model system in Sect. 2.

Here, we calculated the analytic expressions for disease-free and disease equilibrium positions in the Sect. 3.1 in Sect. 3. We also found the basic reproduction number in a subsection of this Sect. 3.2.

The must proving theorems such as existence and uniqueness of the solution of the model system is proved in Sect. 4.

Then in two Sects. 5.1 and 5.2 of Sect. 5, we showed the local and global steady states along with responsible constraints.

Uniform persistence theorems for the model are also highlighted as an interplay of our study in Sect. 6.

Finally, Sect. 7 discloses the summary of the results.

2 The Well-Posedness

We proposed the model system of partial differential equation for this manuscript as

$$\begin{cases} S_t(x,t) = \nu\Delta S + a - bq_1 f(I(x,t))S(x,t) - (m+n)\,S(x,t) + rI(x,t) & \text{in } \mathcal{A}, \\ V_t(x,t) = \upsilon\Delta V + nS(x,t) - bq_2 f(I(x,t))V(x,t) - mV(x,t) & \text{in } \mathcal{A}, \\ I_t(x,t) = \vartheta\Delta I + b\,(q_1 S(x,t) + q_2 V(x,t))\, f(I(x,t)) - (m+r)I(x,t) & \text{in } \mathcal{A}. \end{cases} \tag{2.1}$$

along with the initial values,

$$\begin{cases} S(x,0) = S^0(x) \geq 0 & \text{in } \Omega, \\ V(x,0) = V^0(x) \geq 0 & \text{in } \Omega, \\ I(x,0) = I^0(x) \geq 0 & \text{in } \Omega. \end{cases} \tag{2.2}$$

and the zero-flux Neumann boundary conditions,

$$\frac{\partial S}{\partial \omega}(x,t) = \frac{\partial V}{\partial \omega}(x,t) = \frac{\partial I}{\partial \omega}(x,t) = 0 \quad \text{on } \partial\mathcal{A}. \tag{2.3}$$

where the operator $\frac{\partial}{\partial \omega}$ is the outward normal to the boundary, $\partial\Omega$. The homogeneous Neumann boundary conditions indicate that there is no movement of populations on the boundary $\partial\Omega$ or the population going out and coming in are equal in number on the boundary.

A schematic representation of the model (2.1) is shown in Fig. 1.

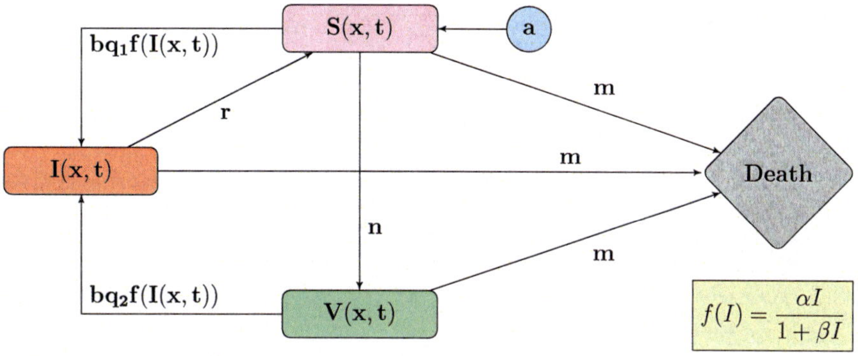

Fig. 1 Modeling scheme

3 Auxiliary Results

3.1 Disease-Free and Disease Equilibrium

To define the disease-free equilibrium (S_0, V_0, I_0) of system (2.1), we write $\nu = \upsilon = \vartheta = 0$, since at equilibrium state diffusion is always absent; then

$$\begin{cases} a - bq_1 f(I_0)S_0 - (m+n)\, S_0 + r I_0 = 0, \\ nS_0 - bq_2 f(I_0)V_0 - mV_0 = 0, \\ b\,(q_1 S_0 + q_2 V_0)\, f(I_0) - (m+r)I_0 = 0. \end{cases} \tag{3.1}$$

It is noted that for the disease-free equilibrium, we consider the count of compartments of infectious individuals $I_0 = 0$. Then we find,

$$\begin{cases} a - (m+n)S_0 = 0, \\ nS_0 - mV_0 = 0, \\ I_0 = 0. \end{cases}$$

This gives the disease-free equilibrium as

$$\begin{cases} S_0 = \dfrac{a}{m+n}, \\ V_0 = \dfrac{an}{m(m+n)}, \\ I_0 = 0. \end{cases} \tag{3.2}$$

Now, in the case of disease equilibrium (S^*, V^*, I^*), we write (2.1) as

$$\begin{cases} a - bq_1 f(I^*)S^* - (m+n)S^* + rI^* = 0, \\ nS^* - bq_2 f(I^*)V^* - mV^* = 0, \\ b\,(q_1 S^* + q_2 V^*)\, f(I^*) - (m+r)I^* = 0. \end{cases} \tag{3.3}$$

Here, the number of compartments of infectious individuals $I^* \neq 0$. Then, we find the count of susceptible individuals in the form

$$S^* = \frac{(a + rI^*)(1 + \beta I^*)}{bq_1\alpha I^* + (m+n)(1+\beta I^*)}, \tag{3.4}$$

and the vaccinated individuals

$$V^* = \frac{n(1+\beta I^*)^2(a + rI^*)}{(bq_1\alpha I^* + (m+n)(1+\beta I^*))(bq_2\alpha I^* + m(1+\beta I^*))}. \tag{3.5}$$

Then, for the count of infectious individuals, we get the following polynomial of degree two

$$\alpha_2(I^*)^2 + \alpha_1 I^* + \alpha_0 = 0, \tag{3.6}$$

where

$$\alpha_2 = (ab^2 q_1 q_2 - m(m + bq_2)(m + n + bq_1)),$$
$$\alpha_1 = (b(nq_2 + amq_1 + abq_1q_2) - m^2(m + n + bq_1) - m(m + n)(m + bq_2)),$$
$$\alpha_0 = b(a + nq_2 + amq_1) - m^2(m + n).$$

The real positive roots of (3.6) define the count of infectious individuals I^*. Thereby, we get the disease steady state $E^*(S^*, V^*, I^*)$ of the model (2.1).

Now, from (3.4) and (3.5) we claim that

$$0 < S^* < \frac{a}{m+n}, \quad 0 < V^* < \frac{an}{m(m+n)},$$

and similarly for I^*

$$0 < I^* < \frac{ab\alpha}{(m+n)(m+r)} \left(\frac{mq_1 + nq_2}{m} \right).$$

3.2 *Basic Reproduction Number*

The basic reproduction number of a model, the expected number of secondary cases reproduced by one infected individual during its entire infectious life, plays a numerous role in epidemiology.

The Jacobian matrix of the linearized model (2.1) at E_0 is

$$\mathcal{J} = \begin{pmatrix} -(m+n) & 0 & -\frac{abq_1\alpha}{m+n} + r \\ n & -m & -\frac{abnq_2\alpha}{m(m+n)} \\ 0 & 0 & \frac{ab\alpha(mq_1 + nq_2)}{m(m+n)} - (m+r) \end{pmatrix}.$$

with eigenvalues $\lambda_1 = -(m+n)$, $\lambda_2 = -m$ and $\lambda_3 = \frac{abq_1\alpha}{m+n} + \frac{abnq_2\alpha}{m(m+n)} - (m+r)$. Since all the model parameters are positive, it can be easily observed that $\lambda_1, \lambda_2 < 0$. Thus, the equilibrium E_0 that is locally asymptotically stable provides $\lambda_3 < 0$. Hence, by the definition of basic reproduction number [21], $\mathcal{R}_0$ of (2.1) is

$$\mathcal{R}_0 = \frac{ab\alpha(mq_1 + nq_2)}{m(m+n)(m+r)}. \tag{3.7}$$

4 Existence and Uniqueness of the Solution

Now, we are going to prove the existence and uniqueness of the solution of system (2.1) by learning the algorithm from a similar study of Xu et al. [22].

Let us denote the subset of $\mathbb{R}^3$ with vectors $x \geq 0$ as $\mathbb{R}^3_+$ and suppose $\mathbb{X} := \mathcal{C}(\Omega, \mathbb{R})$ is a Banach space with the supremum norm $\| \cdot \|_{\mathbb{X}}$. Also we consider $\mathbb{X}^+ := \mathcal{C}(\Omega, \mathbb{R}^3_+)$ then $(\mathbb{X}, \mathbb{X}^+)$ as a strongly ordered space. Suppose that $(T_1(t), T_2(t), T_3(t)) : \mathcal{C}(\Omega, \mathbb{R}) \to \mathcal{C}(\Omega, \mathbb{R})$ is the $\mathcal{C}_0$ semigroups associated with $\nu\Delta - (m+n)$, $\upsilon\Delta - m$ and $\vartheta\Delta - (m+r)$ subject to the respective boundary conditions, respectively; no-flux boundary conditions. It then follows that for any $\rho \in \mathcal{C}(\Omega, \mathbb{R})$ and $t \geq 0$

$$(T_1(t)\rho(x)) = e^{-(m+n)t} \int_\Omega \Gamma_1(\bar{x}, t)\rho(y)\mathrm{d}x$$

$$(T_2(t)\rho(x)) = e^{-mt} \int_\Omega \Gamma_2(\bar{x}, t)\rho(y)\mathrm{d}x$$

$$(T_3(t)\rho(x)) = e^{-(m+r)t} \int_\Omega \Gamma_3(\bar{x}, t)\rho(y)\mathrm{d}x,$$

where $\bar{x} = (x, y)$; Γ_i, $i = 1, 2, 3$ are the Green functions associated with ν, υ, and ϑ, subject to the Neumann boundary conditions, respectively. It then follows from [23] that the function

$$T_i(t) : \mathcal{C}(\Omega, \mathbb{R}) \to \mathcal{C}(\Omega, \mathbb{R}), \ i = 1, 2, 3, \ \text{for all } t > 0$$

is strongly positive and compact. Particularly,

$$T(t) = (T_1(t), T_2(t), T_3(t)) : \mathcal{C}(\Omega, \mathbb{R}) \to \mathcal{C}(\Omega, \mathbb{R}), \forall\, t \geq 0$$

is a strongly continuous semigroup.

If $A_i : \mathrm{G}(A_i) \to \mathbb{X}$ is the generated operator of T_i, $i = 1, 2, 3$, then $T(t) = (T_1(t), T_2(t), T_3(t)) : \mathbb{X} \to \mathbb{X}$ is a semigroup with operator $A = (A_1, A_2, A_3)$ which is defined on $\mathrm{G}(A) := \mathrm{G}(A_1) \times \mathrm{G}(A_2) \times \mathrm{G}(A_3)$. Now for any $\rho = (\rho_1, \rho_2, \rho_3) \in \mathbb{X}$, let us define $\mathcal{F} = (\mathcal{F}_1, \mathcal{F}_2, \mathcal{F}_3) : \mathbb{X}^+ \to \mathbb{X}$ by

$$\begin{aligned}
\mathcal{F}_1(\rho)(x) &= a - bq_1 f(\rho_3)\rho_1(x) - (m+n)\rho_1(x) + r\rho_3(x), && \forall\, x \in \Omega \\
\mathcal{F}_2(\rho)(x) &= n\rho_1(x) - bq_2 f(\rho_3)\rho_2(x) - m\rho_1(x), && \forall\, x \in \Omega \\
\mathcal{F}_3(\rho)(x) &= bq_1 f(\rho_3)\rho_1(x) + bq_2 f(\rho_3)\rho_2(x) - (m+r)\rho_3(x), && \forall\, x \in \Omega.
\end{aligned}$$

Using these operators, we can write (2.1)–(2.3) as the following integral equation

$$u(t) = T(t)\rho + \int_0^t T(t-s)\mathcal{F}(u(s))\mathrm{d}s,$$

where

$$u(t) = \begin{pmatrix} S(t) \\ V(t) \\ I(t) \end{pmatrix}, T(t) = \begin{pmatrix} T_1(t) & 0 & 0 \\ 0 & T_2(t) & 0 \\ 0 & 0 & T_3(t) \end{pmatrix}.$$

By writing $\mathbf{u} = (S, V, I)$ and $\rho = (S_0, V_0, I_0)$ it can also be rewritten as the following abstract differential equation

$$\begin{cases} \mathbf{u}_t = A\mathbf{u} + F(\mathbf{u}), & t > 0, \\ \mathbf{u}_0 = \rho \in \mathbb{X}^+, \end{cases} \tag{4.1}$$

As long as $F(\rho)$ is Lipschitz continuous locally on $\mathbb{X}^+$, then for any $\rho \in \mathbb{X}^+$, (4.1) obeys a unique non-continuous mild function $u(\cdot, t, \rho)$ for which $u(\cdot, t, \rho) \in \mathbb{X}$ for every t in its maximum medium of existence. In addition, $u(\cdot, t, \rho)$ is a class of solution of (2.1) with Neumann boundary conditions (2.3) for all $t > 0$ as followed from ([24], Corollary 2.2.5). Moreover, using the maximum principle of scalar parabolic, it is observed from (2.1) that all compartments $S(x, t)$, $V(x, t)$, and $I(x, t)$ are non-negative. Consequently, we get the basic result on solution of the governing Eqs. (2.1)–(2.3).

Lemma 1 *Define the initial value function* $\rho = (\rho_1, \rho_2, \rho_3) \in \mathbb{X}^+$. *Then for any* ρ, *system (2.1)–(2.3) has a unique solution* $u(x, t, \rho)$ *on* $[0, \sigma_\rho)$ *with* $u(x, t, \rho) = \rho$ *and* $u(\cdot, t, \rho) \in \mathbb{X}^+,\ \forall\, t \in [0, \sigma_\rho)$, *while* $\sigma_\rho \leq \infty$.

Now its time to show that the solution of the problem (2.1)–(2.3) indeed exists globally with the initial function $\rho \in \mathbb{X}^+$, that is, $\sigma = \infty$. To this conclusion, we need Lemma 1 as prescribed in ([25]).

Consider the single variable reaction-diffusion equation

$$\begin{cases} v_t(x, t) = D\Delta v(x, t) + A - dv(x, t), & \text{in } \mathcal{A}, \\ \dfrac{\partial v}{\partial \omega}(x, t) = 0, & \text{on } \partial\mathcal{A}, \end{cases} \tag{4.2}$$

where D, A, and d are all strictly positive and constants.

Lemma 2 *Eq. (4.2) has a unique positive equilibrium* $v^* = \frac{A}{d}$ *which is globally attractive in* $C(\Omega, \mathbb{R})$.

Now we are ready to produce the proof of the following theorem

Theorem 1 *For any initial value* $\rho \in \mathbb{X}^+$, *Eqs. (2.1)–(2.3) has a unique positive solution* $u(\cdot, t, \rho)$ *on* $[0, \infty)$. *In addition, the solution semiflow* $\Phi(t) := u(\cdot, t) : \mathbb{X}^+ \to \mathbb{X}^+,\ t \geq 0$ *has an attractor in* $\mathbb{X}^+$ *which is globally compact.*

Proof By Lemma 1, system (2.1)–(2.3) has a solution $u(\cdot, t, \rho)$ which is unique on $[0, \sigma_\rho)$ and $u(x, t, \rho) \geq 0$ for any $x \in \Omega$ while $t \in [0, \sigma_\rho)$. It is seen from Lemma 2 that the global attractor is $\frac{a}{m+n}$ for the following spatial equation

$$\begin{cases} v_t(x, t) = \nu \Delta v(x, t) + a - (m + n)v(x, t) + rv(x, t), & \text{in } \mathcal{A}, \\ \dfrac{\partial v}{\partial \omega}(x, t) = 0, & \text{on } \partial \mathcal{A}. \end{cases}$$

It is noted that since

$$S_t(x, t) \leq \nu \Delta S(x, t) + a - (m + n)v(x, t) + rv(x, t), \ \forall\, t \in [0, \sigma_\rho), \ x \in \Omega, \tag{4.3}$$

for parabolic equation, the standard comparison theorem [23] implies that $S(\cdot, t, \rho)$ is bounded on $[0, \sigma_\rho)$; so, there exist $S_0 > 0$ such that $S(\cdot, t, \rho) \leq S_0, \forall\, t \in [0, \sigma_\rho)$. Thus we obtain the following inequalities from the second equation of (2.1) and hence

$$V_t(x, t) \leq \upsilon \Delta V(x, t) + nS_0 - mV(x, t), \ \forall\, t \in [0, \sigma_\rho), \ x \in \Omega.$$

Again, $V(\cdot, t, \rho)$ is bounded on $[0, \sigma_\rho)$ using the comparison theorem and Lemma 2, that is, there is a $V_0 > 0$ such that $V(\cdot, t, \rho) \leq V_0$ for all $t \in [0, \sigma_\rho)$. Therefore, the third equation of (2.1) due to

$$I_t(x, t) \leq \vartheta \Delta I(x, t) + bq_1 S_0 + bq_2 V_0 - (m + r)I(x, t), \ \forall\, t \in [0, \sigma_\rho), \ x \in \Omega.$$

As a consequence, we have a bounded solution $u(\cdot, t, \rho)$ on $[0, \rho)$ and hence $\sigma_\rho = \infty$ for each $\rho \in \mathbb{X}^+$.

By theorem (4.3), for any $\rho \in \mathbb{X}^+$, it is clear that there exist some $t_1 = t_1(\rho) > 0$ such that

$$S(x, t) \leq \frac{a}{m + n} + 1 := M_1, \ \forall\, t \geq t_1, \ x \in \Omega.$$

Using the similar arguments as mentioned above, it can be shown that there are $M_i > 0$, independent of the choice of $\rho \in \mathbb{X}^+$ and $t_i = t_i(\rho) > 0$, $(i = 1, 2, 3)$, such that

$$V(x, t) \leq M_2, \ I(x, t) \leq M_3; \ \forall\, t \geq t_1, \ x \in \Omega.$$

As a result, the non-negative solution of (2.1)–(2.3) are finally bounded in case of maximum norm. This influence the semiflow of solution $\Phi(t) : \mathbb{X}^+ \to \mathbb{X}^+$ as designed by $(\Phi(t)\rho)(x) = u(x, t, \rho), \ x \in \Omega$, is point dissipative. However, $\Phi(t)$ is compact in view of [24] for any $t > 0$. Therefore, the solution $\Phi(t) : \mathbb{X}^+ \to \mathbb{X}^+, \ t \geq 0$, has a global compact attractor in $\mathbb{X}^+$ [26]. Hence the proof. □

5 Analysis of Steady States

5.1 *Local Steady States*

In this subsection, the local stability of the steady states for system (2.1) is explained. Thus we consider the proof of our second result:

Theorem 2 *The disease-free equilibrium E_0 of system (2.1) is locally asymptotically stable when $\mathcal{R}_0 < 1$.*

Proof By linearizing system (2.1) at E_0, we get

$$\mathbf{u}_t = \delta \Delta \mathbf{u}(x, t) + \mathcal{F}_1 \mathbf{u}(x, t),$$

where

$$\delta = \begin{pmatrix} \nu & 0 & 0 \\ 0 & \upsilon & 0 \\ 0 & 0 & \vartheta \end{pmatrix},$$

$$\mathcal{F}_1 = \begin{pmatrix} -(m+n) & 0 & -bq_1 S_0 + r \\ n & -m & -bq_2 V_0 \\ 0 & 0 & bq_1 S_0 + bq_2 V_0 - (m+r) \end{pmatrix}.$$

Then, one can find the characteristic polynomial, and they are defined as follows:

$$|\lambda \mathcal{I} + \delta \mathcal{L}^2 - \mathcal{F}_1| = 0,$$

where the eigenvalue, λ, determines temporal growth, $\mathcal{L}$ is the wave-number and identity matrix is $\mathcal{I}$ with size 3×3 [20]. Then, we get

$$(\lambda + \nu \mathcal{L}^2 + m + n)(\lambda + \upsilon \mathcal{L}^2 + m)(\lambda + \vartheta \mathcal{L}^2 + m + r - bq_1 S_0 - bq_2 V_0) = 0. \tag{5.1}$$

Now, it is clear that

$$\begin{aligned} \lambda_1 &= -(\nu \mathcal{L}^2 + m + n) < 0, \\ \lambda_2 &= -(\nu \mathcal{L}^3 + m) < 0, \\ \text{and } \lambda_3 &= -(\vartheta \mathcal{L}^2 + m + r - bq_1 S_0 - bq_2 V_0) \\ &= -(\vartheta \mathcal{L}^2 + (m + r)(1 - \mathcal{R}_0)). \end{aligned}$$

It follows from $\mathcal{R}_0 < 1$ that E_0 is locally asymptotically stable. □

Theorem 3 *When $\mathcal{R}_0 > 1$, the disease equilibrium E^* of system (2.1) is locally asymptotically stable.*

Proof Linearizing system (2.1) at E^*, we obtain

$$\mathbf{u}_t = \delta\Delta\mathbf{u}(x,t) + \digamma_2\mathbf{u}(x,t),$$

where

$$\digamma_2 = \begin{pmatrix} -\left(m+n+\frac{bq_1\alpha I^*}{1+\beta I^*}\right) & 0 & -\frac{bq_1\alpha S^*}{(1+\beta I^*)^2}+r \\ n & -\left(m+\frac{bq_2\alpha I^*}{1+\beta I^*}\right) & -\frac{bq_2\alpha V^*}{(1+\beta I^*)^2} \\ \frac{bq_1\alpha I^*}{1+\beta I^*} & \frac{bq_2\alpha I^*}{1+\beta I^*} & \frac{b(q_1S^*+q_2V^*)\alpha}{(1+\beta I^*)^2}-(m+r) \end{pmatrix}.$$

Then, we obtain the following characteristic equation

$$\lambda^3 + \mathcal{G}_1(\mathcal{L}^2)\lambda^2 + \mathcal{G}_2(\mathcal{L}^2)\lambda + \mathcal{G}_3(\mathcal{L}^2) = 0 \tag{5.2}$$

where

$$\begin{aligned}
\mathcal{G}_1(\mathcal{L}^2) =& \nu\mathcal{L}^2 + m + n + bq_1 f(I^*) + \upsilon\mathcal{L}^2 + m + bq_2 f(I^*) \\
&+ \vartheta\mathcal{L}^2 + m + r - b\left(q_1S^* + q_2V^*\right) f(I^*), \\
\mathcal{G}_2(\mathcal{L}^2) =& \left(\upsilon\mathcal{L}^2 + m + bq_2 f(I^*)\right)\left(\vartheta\mathcal{L}^2 + m + r\right) + bq_1S^*\frac{\alpha}{(1+\beta I^*)^2}bq_1 f(I^*) \\
&+ \left(\nu\mathcal{L}^2 + m + n + bq_1 f(I^*)\right)\left(\upsilon\mathcal{L}^2 + m + bq_2 f(I^*) + \vartheta\mathcal{L}^2 + m + r\right) \\
&+ bq_2V^*\frac{\alpha}{(1+\beta I^*)^2}bq_2 f(I^*) - \left(\nu\mathcal{L}^2 + \upsilon\mathcal{L}^2 + 2m + n + b(q_1+q_2) f(I^*)\right) \\
&\times\left(bq_1S^*\frac{\alpha}{(1+\beta I^*)^2} + bq_2V^*\frac{\alpha}{(1+\beta I^*)^2}\right), \\
\mathcal{G}_3(\mathcal{L}^2) =& \left(\nu\mathcal{L}^2 + m + n + bq_1 f(I^*)\right)\left(\upsilon\mathcal{L}^2 + m + bq_2 f(I^*)\right)\left(\vartheta\mathcal{L}^2 + m + r\right) \\
&+ bq_2V^*\frac{\alpha}{(1+\beta I^*)^2}bq_2 f(I^*)(\vartheta\mathcal{L}^2 + m + r) + nbq_1S^*\frac{\alpha}{(1+\beta I^*)^2}bq_2 f(I^*) \\
&+ bq_1S^*\frac{\alpha}{(1+\beta I^*)^2}bq_1 f(I^*)\left(\upsilon\mathcal{L}^2 + m + bq_2 f(I^*)\right) \\
&- \left(\nu\mathcal{L}^2 + m + n + bq_1 f(I^*)\right)\left(\upsilon\mathcal{L}^2 + m + bq_2 f(I^*)\right) \\
&\times\left(bq_1S^*\frac{\alpha}{(1+\beta I^*)^2} + bq_2V^*\frac{\alpha}{(1+\beta I^*)^2}\right) \\
&- bq_2 f(I^*)bq_2\frac{\alpha}{(1+\beta I^*)^2}\left(bq_1S^*\frac{\alpha}{(1+\beta I^*)^2} + bq_2V^*\frac{\alpha}{(1+\beta I^*)^2}\right).
\end{aligned}$$

Now, let us take

$$bq_1S^*\frac{\alpha}{(1+\beta I^*)^2} + bq_2V^*\frac{\alpha}{(1+\beta I^*)^2} \le b(q_1S^* + q_2V^*)\frac{\alpha}{1+\beta I^*} = m + r\,,$$

then we can get

$$\mathcal{G}_1(\mathcal{L}^2) \geq \nu\mathcal{L}^2 + m + n + bq_1 f(I^*) + \upsilon\mathcal{L}^2 + m + bq_2 f(I^*) + \vartheta\mathcal{L}^2 > 0\,,$$
$$\mathcal{G}_2(\mathcal{L}^2) > \vartheta\mathcal{L}^2\left(\upsilon\mathcal{L}^2 + m + bq_2 f(I^*)\right) > 0\,,$$
$$\mathcal{G}_3(\mathcal{L}^2) > \left(\nu\mathcal{L}^2 + m + n + bq_1 f(I^*)\right)\left(\upsilon\mathcal{L}^2 + m + bq_2 f(I^*)\right)\vartheta\mathcal{L}^2 > 0\,.$$

These lead us to the following conclusion:

$$\mathcal{G}_1(\mathcal{L}^2)\mathcal{G}_2(\mathcal{L}^2) - \mathcal{G}_3(\mathcal{L}^2) > bq_2 f(I^*) bq_2 V^* \frac{\alpha}{(1+\beta I^*)^2} b(q_1 S^* + q_2 V^*) \frac{\alpha}{(1+\beta I^*)^2} > 0\,.$$

Introducing the Routh–Hurwitz criterion, it is known that all eigenvalues of (5.2) have negative real parts. It concludes that as long as $\mathcal{R}_0 > 1$, the disease equilibrium E^* of (5.2) is locally asymptotically stable. □

5.2 Global Steady States

In this section, we investigate the global stability of the two constant equilibria E_0 and E^* in the case of a bounded domain Ω in which $E \equiv (S(x,t),\ V(x,t),\ I(x,t))$ is an arbitrary solution of (2.1) and is positive. For convenience, let us first consider the following shortcuts:

$$S \equiv S(x,t),\ V \equiv V(x,t),\ I \equiv I(x,t).$$

(h1) Let, $\mathcal{Q}(x,t,S) = \int_\Omega \left(\frac{S_0}{S} - 1\right) dx$ and $\mathcal{Q}(x,t,S)$ is monotonically increasing in S.

In case of global analysis, we consider a Lyapunov function and the results varies with basic reproduction number.

At this phase, we are in stable setting to establish the global stability theorems as the following:

Theorem 4 *If $\mathcal{R}_0 \leq 1$ and $\mathcal{Q}(x,t,S)$ satisfies 5.2 then the disease-free equilibrium $E_0(S_0, V_0, I_0)$ of system (2.1) is globally asymptotically stable.*

Proof Let define a Lyapunov function as

$$L_1(t) = \int_\Omega W_1(x,t)dx,$$

where

$$W_1(x,t) = S_0\left(\frac{S}{S_0} - 1 - \ln\frac{S}{S_0}\right) + V_0\left(\frac{V}{V_0} - 1 - \ln\frac{V}{V_0}\right) + I.$$

Differentiating $W_1(x,t)$ with respect to time, t, along the solution of (2.1) gives

$$\frac{\partial W_1}{\partial t} = \left(1 - \frac{S_0}{S}\right) S_t(x,t) + \left(1 - \frac{V_0}{V}\right) V_t(x,t) + I_t(x,t).$$

Then from (2.1), we can write

$$\begin{aligned}\frac{\partial W_1}{\partial t} &= \left(1 - \frac{S_0}{S}\right)(\nu\Delta S + a - bq_1 f(I)S - (m+n)S + rI) \\ &+ \left(1 - \frac{V_0}{V}\right)(\upsilon\Delta V + nS - bq_2 f(I)V - mV) \\ &+ (\vartheta\Delta I + bq_1 f(I)S + bq_2 f(I)V - (m+r)I).\end{aligned}$$

But, as $a = (m+n)S_0$ and $mV_0 = nS_0$, we can write

$$\begin{aligned}\frac{\partial W_1}{\partial t} &= \left(1 - \frac{S_0}{S}\right)\nu\Delta S + \left(1 - \frac{V_0}{V}\right)\upsilon\Delta V + \vartheta\Delta I + mS_0\left(2 - \frac{S}{S_0} - \frac{S_0}{S}\right) \\ &+ nS_0\left(3 - \frac{S_0}{S} - \frac{V}{V_0} - \frac{S}{S_0}\frac{V_0}{V}\right) - (m+r)(1+\beta I - \mathcal{R}_0)\frac{I}{1+\beta I} + \left(1 - \frac{S_0}{S}\right) rI.\end{aligned}$$

Applying Green's formula and no-flux boundary conditions (2.3), we have

$$\int_\Omega \Delta S \mathrm{d}x = \int_{\partial\Omega} \frac{\partial S}{\partial \omega} \mathrm{d}S = 0. \tag{5.3}$$

Similarly,

$$\int_\Omega \Delta V \mathrm{d}x = \int_\Omega \Delta I \mathrm{d}x = 0. \tag{5.4}$$

Again, by homogeneous Neumann boundary conditions (2.3) and Green's formula, we get the Green's first identity as

$$\int_\Omega \left(\frac{\Delta S}{S} - \frac{\|\nabla S\|^2}{S^2}\right) \mathrm{d}x = \int_{\partial\Omega} \frac{1}{S}(\nabla S \cdot \omega)\mathrm{d}S = 0,$$

which implies

$$\int_\Omega \frac{\Delta S}{S} \mathrm{d}x = \int_\Omega \frac{\|\nabla S\|^2}{S^2} \mathrm{d}x. \tag{5.5}$$

By the same arguments, we also can write

$$\int_\Omega \frac{\Delta V}{V} \mathrm{d}x = \int_\Omega \frac{\|\nabla V\|^2}{V^2} \mathrm{d}x, \tag{5.6}$$

$$\text{and} \int_\Omega \frac{\Delta I}{I} \mathrm{d}x = \int_\Omega \frac{\|\nabla I\|^2}{I^2} \mathrm{d}x. \tag{5.7}$$

Then using the above arguments, we have

$$\begin{aligned}\frac{dL_1}{dt} =& -\nu S_0 \int_\Omega \frac{\|\nabla S\|^2}{S^2} dx - \upsilon V_0 \int_\Omega \frac{\|\nabla V\|^2}{V^2} dx + m S_0 \int_\Omega \left(2 - \frac{S}{S_0} - \frac{S_0}{S}\right) dx \\ &+ n S_0 \int_\Omega \left(3 - \frac{S_0}{S} - \frac{V}{V_0} - \frac{S}{S_0}\frac{V_0}{V}\right) dx - (m+r) \int_\Omega \left((1+\beta I - \mathcal{R}_0)\frac{I}{1+\beta I}\right) dx \\ &+ r \int_\Omega I \left(1 - \frac{S_0}{S}\right) dx, \\ =& -\nu S_0 \int_\Omega \frac{\|\nabla S\|^2}{S^2} dx - \upsilon V_0 \int_\Omega \frac{\|\nabla V\|^2}{V^2} dx - m S_0 \int_\Omega \frac{(S-S_0)^2}{S_0 S} dx \\ &- n S_0 \int_\Omega \left(\frac{S_0}{S} + \frac{V}{V_0} + \frac{S}{S_0}\frac{V_0}{V} - 3\right) dx - (m+r) \int_\Omega \left((1+\beta I - \mathcal{R}_0)\frac{I}{1+\beta I}\right) dx \\ &- r \int_\Omega I \left(\frac{S_0}{S} - 1\right) dx. \\ \leq& -\nu S_0 \int_\Omega \frac{\|\nabla S\|^2}{S^2} dx - \upsilon V_0 \int_\Omega \frac{\|\nabla V\|^2}{V^2} dx - m S_0 \int_\Omega \frac{(S-S_0)^2}{S_0 S} dx \\ &- n S_0 \int_\Omega \left(\frac{S_0}{S} + \frac{V}{V_0} + \frac{S}{S_0}\frac{V_0}{V} - 3\right) dx - (m+r) \int_\Omega \left((1+\beta I - \mathcal{R}_0)\frac{I}{1+\beta I}\right) dx \\ &- r \min_{x\in\Omega} I\, Q(x,t,S)\end{aligned}$$

where $Q(x,t,S) = \int_\Omega \left(\frac{S_0}{S} - 1\right) dx$. Thus, using the property (h1) and whenever $\mathcal{R}_0 \leq 1$, we get $\frac{dL_1}{dt} \leq 0$.

When $S = S_0, V = V_0, I = 0$; we calculate, $\frac{dL_1}{dt} = 0$ and vice-versa. Consequently, the singleton E_0 is the greatest compact invariant set in $\{(S, V, I) \in C(\Omega, \mathbb{R}^3_+) : \frac{dL_1}{dt} = 0\}$. Then, LaSalle's invariance principle [27] refers to $\lim_{t\to\infty} (S, V, I) \to E_0$; which means, whenever $\mathcal{R}_0 \leq 1$, the disease-free equilibrium $E_0 = (S_0, V_0, 0)$ is globally asymptotically stable. This establishes Theorem 4. □

In a similar manner, the disease equilibrium of (2.1) is globally asymptotically stable and the proof is prescribed as follows:

Theorem 5 *The disease equilibrium $E^*(S^*, V^*, I^*)$ of (2.1) is globally asymptotically stable while $\mathcal{R}_0 > 1$.*

Proof Let us define a Lyapunov function as

$$L_2(t) = \int_\Omega W_2(x,t) dx,$$

where

$$W_2(x,t) = S^* \left(\frac{S}{S^*} - 1 - \ln \frac{S}{S^*}\right) + V^* \left(\frac{V}{V^*} - 1 - \ln \frac{V}{V^*}\right) + I^* \left(\frac{I}{I^*} - 1 - \ln \frac{I}{I^*}\right).$$

Calculating $W_2(x,t)$ with respect to the time derivative and the solution of (2.1) gives

$$\frac{\partial W_2}{\partial t} = \left(1-\frac{S^*}{S}\right)S_t(x,t) + \left(1-\frac{V^*}{V}\right)V_t(x,t) + \left(1-\frac{I^*}{I}\right)I_t(x,t).$$

Then from (2.1), it can be written as

$$\begin{aligned}\frac{\partial W_2}{\partial t} &= \left(1-\frac{S^*}{S}\right)(\nu\Delta S + a - bq_1 f(I)S - (m+n)S + rI)\\ &+ \left(1-\frac{V^*}{V}\right)(\upsilon\Delta V + nS - bq_2 f(I)V - mV)\\ &+ (I-I^*)\left(\frac{\vartheta\Delta I}{I} + bq_1\frac{\alpha}{1+\beta I}S + bq_2\frac{\alpha}{1+\beta I}V - (m+r)\right).\end{aligned} \tag{5.8}$$

Note that from (3.3), we have

$$\begin{aligned} a &= bq_1 f(I^*)S^* + (m+n)S^* - rI^*,\\ nS^* &= bq_2 f(I^*)V^* + mV^*,\\ (m+r)I^* &= b(q_1S^* + q_2V^*)f(I^*).\end{aligned}$$

and by substituting these in (5.8) yields

$$\begin{aligned}\frac{\partial W_2}{\partial t} &= \left(1-\frac{S^*}{S}\right)\nu\Delta S + \left(1-\frac{V^*}{V}\right)\upsilon\Delta V + \left(1-\frac{I^*}{I}\right)\vartheta\Delta I\\ &+ \left(1-\frac{S^*}{S}\right)(bq_1 f(I^*)S^* + (m+n)S^* - rI^* - bq_1 f(I)S - (m+n)S + rI)\\ &+ \left(1-\frac{V^*}{V}\right)\left(nS^*\left(\frac{S}{S^*}-\frac{V}{V^*}\right) + nS^*\frac{V}{V^*} - bq_2 f(I)V - mV\right)\\ &+ \left(\frac{I}{I^*}-1\right)\left(bq_1\frac{\alpha I^*}{1+\beta I}S + bq_2\frac{\alpha I^*}{1+\beta I}V - (m+r)I^*\right),\\ &= \left(1-\frac{S^*}{S}\right)\nu\Delta S + \left(1-\frac{V^*}{V}\right)\upsilon\Delta V + \left(1-\frac{I^*}{I}\right)\vartheta\Delta I\\ &+ \left(1-\frac{S^*}{S}\right)\left((m+n)S^*\left(1-\frac{S}{S^*}\right) + bq_1 f(I^*)S^*\left(1-\frac{S}{S^*}\frac{f(I)}{f(I^*)}\right) - rI^*\left(1-\frac{I}{I^*}\right)\right)\\ &+ \left(1-\frac{V^*}{V}\right)\left(nS^*\left(\frac{S}{S^*}-\frac{V}{V^*}\right) + (bq_2 f(I^*)V^* + mV^*)\frac{V}{V^*} - bq_2 f(I)V - mV\right)\\ &+ \left(\frac{I}{I^*}-1\right)\left(bq_1\frac{\alpha I^*}{1+\beta I}S + bq_2\frac{\alpha I^*}{1+\beta I}V - b(q_1S^* + q_2V^*)f(I^*)\right),\\ &= \left(1-\frac{S^*}{S}\right)\nu\Delta S + \left(1-\frac{V^*}{V}\right)\upsilon\Delta V + \left(1-\frac{I^*}{I}\right)\vartheta\Delta I\\ &+ \left(1-\frac{S^*}{S}\right)\left((m+n)S^*\left(1-\frac{S}{S^*}\right) + bq_1 f(I)S^*\left(1-\frac{S}{S^*}\frac{f(I)}{f(I^*)}\right)\right)\\ &+ \left(1-\frac{V^*}{V}\right)\left(nS^*\left(\frac{S}{S^*}-\frac{V}{V^*}\right) + bq_2 f(I^*)V\left(1-\frac{f(I)}{f(I^*)}\right)\right)\end{aligned}$$

$$+\left(\frac{I}{I^*}-1\right)\left(bq_1f(I^*)S^*\left(\frac{S}{S^*}\frac{1+\beta I^*}{1+\beta I}-1\right)+bq_2f(I^*)V^*\left(\frac{V}{V^*}\frac{1+\beta I^*}{1+\beta I}-1\right)\right)$$
$$-\left(1-\frac{S^*}{S}\right)rI^*\left(1-\frac{I}{I^*}\right).$$
$$=\left(1-\frac{S^*}{S}\right)\nu\Delta S+\left(1-\frac{V^*}{V}\right)\upsilon\Delta V+\left(1-\frac{I^*}{I}\right)\vartheta\Delta I$$
$$+mS^*\left(2-\frac{S}{S^*}-\frac{S^*}{S}\right)+mV^*\left(3-\frac{S^*}{S}-\frac{V}{V^*}-\frac{S}{S^*}\frac{V^*}{V}\right)$$
$$+bq_1f(I^*)S^*\left(3-\frac{S^*}{S}-\frac{S}{S^*}\frac{1+\beta I^*}{1+\beta I}-\frac{1+\beta I}{1+\beta I^*}\right)-b(q_1S^*+q_2V^*)\frac{\alpha\beta(I-I^*)^2}{(1+\beta I)(1+\beta I^*)^2}$$
$$+bq_2f(I^*)V^*\left(4-\frac{S^*}{S}-\frac{S}{S^*}\frac{V^*}{V}-\frac{1+\beta I}{1+\beta I^*}-\frac{V}{V^*}\frac{1+\beta I^*}{1+\beta I}\right)$$
$$-\left(1-\frac{S^*}{S}\right)rI^*\left(1-\frac{I}{I^*}\right).$$

Applying the Green's formula and no-flux (natural) boundary conditions, we obtain

$$\begin{aligned}\frac{\mathrm{d}L_2}{\mathrm{d}t}=&-\nu S^*\int_\Omega\frac{\|\nabla S\|^2}{S^2}\mathrm{d}x-\upsilon V^*\int_\Omega\frac{\|\nabla V\|^2}{V^2}\mathrm{d}x-\vartheta I^*\int_\Omega\frac{\|\nabla I\|^2}{I^2}\mathrm{d}x\\&+mS^*\int_\Omega\left(2-\frac{S}{S^*}-\frac{S^*}{S}\right)\mathrm{d}x+mV^*\int_\Omega\left(3-\frac{S^*}{S}-\frac{V}{V^*}-\frac{S}{S^*}\frac{V^*}{V}\right)\mathrm{d}x\\&+bq_1f(I^*)S^*\int_\Omega\left(3-\frac{S^*}{S}-\frac{S}{S^*}\frac{1+\beta I^*}{1+\beta I}-\frac{1+\beta I}{1+\beta I^*}\right)\mathrm{d}x\\&-b(q_1S^*+q_2V^*)\alpha\beta\int_\Omega\frac{(I-I^*)^2}{(1+\beta I)(1+\beta I^*)^2}\mathrm{d}x\\&+bq_2f(I^*)V^*\int_\Omega\left(4-\frac{S^*}{S}-\frac{S}{S^*}\frac{V^*}{V}-\frac{1+\beta I}{1+\beta I^*}-\frac{V}{V^*}\frac{1+\beta I^*}{1+\beta I}\right)\mathrm{d}x\\&+rI^*\int_\Omega\left(\frac{S^*}{S}+\frac{I}{I^*}-\frac{S^*}{S}\frac{I}{I^*}-1\right)\mathrm{d}x.\end{aligned}\tag{5.9}$$

We know that the geometric mean is less than or equal to the arithmetic mean. Consequently, for all strictly positive $S,\ V$ and I, we find

$$2-\frac{S}{S^*}-\frac{S^*}{S}\le 0,$$

$$3-\frac{S^*}{S}-\frac{V}{V^*}-\frac{S}{S^*}\frac{V^*}{V}\le 0,$$

$$3-\frac{S^*}{S}-\frac{S}{S^*}\frac{1+\beta I^*}{1+\beta I}-\frac{1+\beta I}{1+\beta I^*}\le 0,$$

$$\text{and } 4-\frac{S^*}{S}-\frac{S}{S^*}\frac{V^*}{V}-\frac{1+\beta I}{1+\beta I^*}-\frac{V}{V^*}\frac{1+\beta I^*}{1+\beta I}\le 0.$$

Let,

$$\mathcal{Z}_{\text{sup}} = rI^* \int_{\Omega} \left(\frac{S^*}{S} + \frac{I}{I^*} - \frac{S^*}{S}\frac{I}{I^*} - 1 \right) \mathrm{d}x.$$

Now, we need $\mathcal{Z}_{\text{sup}}$ to be zero or negative.

If either $r = 0$ or $S = S^*$ and $I = I^*$ then $\mathcal{Z}_{\text{sup}} = 0$.

But, if r is positive, then we can find a simply connected and positively invariant set containing no closed orbits by Bendixon's negative criterion, and the integral $\mathcal{Z}_{\text{sup}}$ is dominated by the negativity of other integrals in (5.9).

Thus, Eq. (5.9) reveals that $\frac{\mathrm{d}L_2}{\mathrm{d}t} \leq 0$ for $S, V, I > 0$. Since the above inequalities become equalities whenever $S \equiv S^*$, $V \equiv V^*$ and $I \equiv I^*$ and hence $\frac{\mathrm{d}L_2}{\mathrm{d}t} = 0$ for $(S, V, I) = (S^*, V^*, I^*)$. Now, LaSalle's invariance principle [27] refers to

$$\lim_{t\to\infty} (S(x,t), V(x,t), I(x,t)) \to E^*;$$

which means, when $\mathcal{R}_0 > 1$, the disease equilibrium $E^* = (S^*, V^*, I^*)$ is globally asymptotically stable.

This concludes the proof. □

6 Uniform Persistence

We first linearize the last equation of system (2.1) at the disease-free equilibrium E_0 and get the following:

$$\begin{cases} I_t(x,t) = \vartheta \Delta I + b(q_1 S_0 + q_2 V_0)I - (m+r)I & \text{in } \mathcal{A}, \\ \dfrac{\partial I}{\partial \omega} = 0 & \text{in } \partial\mathcal{A}. \end{cases} \tag{6.1}$$

Then referring the strategy followed in ([19], Theorem 2.2), ([20], Theorem 2), ([22], Theorem 3.2), ([23], Theorem 4.2), ([25], Theorem 2.11), ([27], Theorem 4.2), ([28], Theorem 3.4); Y. Yang et al. [29] established the uniform persistence result for their respective system through the following procedure.

Setting $I(x,t) = e^{\lambda t}\hat{p}(x)$, we get

$$\begin{cases} \lambda\hat{p}(x) = \vartheta\Delta\hat{p}(x) + (bq_1 S_0 + bq_2 V_0)\hat{p}(x) - (m+r)\hat{p}(x) & \text{for } x \in \Omega, \\ \dfrac{\partial\hat{p}(x)}{\partial\omega} = 0 & \text{for } x \in \partial\Omega. \end{cases} \tag{6.2}$$

Now substituting $\hat{p}(x) \equiv 1$ and the values of S_0, V_0 into (6.2), the principal eigenvalue of (6.1) is

$$\lambda(S_0, V_0) = b(q_1 S_0 + q_2 V_0) - (m+r) \ = (m+r)(\mathcal{R}_0 - 1).$$

Corresponding to this eigenvalue, there is the unique positive eigenfunction $\hat{\rho}(x) \equiv 1$.
Thus, observing this equation we can claim the following lemma:

Lemma 3 *The principal eigenvalue,* $\lambda(S_0, V_0)$ *has the same sign as* $(\mathcal{R}_0 - 1)$.

To claim the uniform persistence of (2.1)–(2.3), we now establish the following lemma and theorem using the similar arguments from [29].

Lemma 4 *If* $u(x,t,\rho)$ *is the solution of (2.1)–(2.3) such that* $u(\cdot, 0, \rho) = \rho \in X^+$, *then*

(i) for any $\rho \in X^+$, *always there exist* $S(x,t,\rho) > 0$ *and* $V(x,t,\rho) > 0$ *in* $\mathcal{A}$. *Furthermore, we have*

$$\liminf_{t\to\infty} S(x,t) \geq \frac{a}{m+n+bq_1}, \quad \textit{uniformly for } x \in \Omega,$$

and

$$\liminf_{t\to\infty} V(x,t) \geq \frac{an}{2(m+n+bq_1)(m+bq_2)}, \quad \textit{uniformly for } x \in \Omega,$$

(ii) if there exists some $t_0 \geq 0$ *such that* $I(\cdot, t_0, \rho) \not\equiv 0$ *is not true, then* $I(x,t,\rho) > 0,\ \forall\, x \in \Omega,\ t > t_0$.

Proof From the system of equations in (2.1), clearly it concludes that $S(x,t,\rho) > 0$ and $V(x,t,\rho) > 0$ in $\mathcal{A}$ for any $\rho \in \mathbb{X}^+$. Then,

$$S_t(x,t) \geq \nu\Delta S + a - (bq_1 + m + n - r)S \quad \text{in } \mathcal{A}.$$

Now applying (Lemma 1, [25]) and the comparison principle, we get

$$\liminf_{t\to\infty} S(x,t) \geq \frac{a}{m+n+bq_1}, \quad \text{uniformly for } x \in \Omega.$$

Then there exists a $t_1 > 0$ such that

$$S(x,t) \geq \frac{1}{2}\frac{a}{m+n+bq_1}, \qquad \forall\, t \geq t_1.$$

Consequently, the second equation of (2.1) follows that

$$\liminf_{t\to\infty} V(x,t) \geq \frac{an}{2(m+n+bq_1)(m+bq_2)},$$

uniformly for $x \in \Omega$. Finally, from the third equation of system (2.1), we can write

$$\begin{cases} I_t(x,t) \geq \vartheta\Delta I - (m+r)I & \text{in } \mathcal{A}, \\ \dfrac{\partial I}{\partial \omega} = 0 & \text{in } \partial\mathcal{A}. \end{cases}$$

By the strong maximum principle and the Hopf boundary Lemma [30], this result is valid. □

After the completion of the above arguments, we get the solutions for disease persistence as described in the following theorem:

Theorem 6 *If $\mathcal{R}_0 > 1$ then a constant $\eta > 0$ exists such that for any $\rho \in \mathbb{X}^+$ with $\rho_3(\cdot) \not\equiv 0$, we obtain the following inequalities*

$$\liminf_{t\to\infty} S(x,t) \geq \eta, \quad \liminf_{t\to\infty} V(x,t) \geq \eta, \quad \liminf_{t\to\infty} I(x,t) \geq \eta, \quad \textit{uniformly for } x \in \Omega.$$

Proof Let

$$\mathcal{X}_0 := \{\rho \in \mathbb{X}^+ : \rho_3(\cdot) \neq 0\},$$

and

$$\partial\mathcal{X}_0 := \mathbb{X}^+ \setminus \mathcal{X}_0 = \{\rho \in \mathbb{X}^+ : \rho_3(\cdot) = 0\},$$

By Lemma 4 and for any $\rho \in \mathcal{X}_0$, we get $I(x,t,\rho) > 0$, in $\mathcal{A}$, that is, $\Theta_t\mathcal{X}_0 \subseteq \mathcal{X}_0$, $\forall\, t \geq 0$.

Let define $R_\partial := \{\theta \in \mathcal{X}_0 : \Theta_t(\theta) \in \partial\mathcal{X}_0, \ \forall\, t \geq 0\}$, and $\omega(\theta)$ be the omega limit set of the orbit $\mathcal{O}^+(\theta) := \{\Theta_t(\theta) : t \geq 0\}$. Now, first, let us claim that $\omega(\rho) = \{(S_0, V_0, 0)\}$, $\forall\, \theta \in R_\partial$.

Since $\rho \in R_\partial$, we have $\Theta_t(\rho) \in \partial\mathcal{X}_0$, $\forall\, t \geq 0$. Hence, $I(\cdot, t, \rho) \equiv 0$. From the first equation of (2.1), it is clear that $\lim_{t\to\infty} S(x,t,\rho) = S_0$ uniformly for $x \in \Omega$. Hence $\omega(\rho) = \{(S_0, V_0, 0)\}$, $\forall\, \rho \in R_\partial$. It follows from Lemma 3 that $\lambda(S_0, V_0) > 0$ when $\mathcal{R}_0 > 1$. By the continuity of $\lambda(S_0, V_0)$, there exists a sufficiently small positive number $\delta_0 > 0$ such that $\lambda\left(\frac{S_0-\alpha\delta_0}{1+\beta\delta_0}, \frac{V_0-\alpha\delta_0}{1+\beta\delta_0}\right) > 0$.

Let us now claim that $(S_0, V_0, 0)$ is a uniform weak repeller for $\mathcal{X}_0$ in the sense that

$$\limsup_{t\to\infty} |\Theta_t(\rho) - (S_0, V_0, 0)| \geq \delta_0, \quad \forall\, \rho \in \mathcal{X}_0.$$

Suppose, by contradiction, there exists $\rho_0 \in \mathcal{X}_0$ such that

$$\limsup_{t\to\infty} |\Theta_t(\rho_0) - (S_0, V_0, 0)| < \delta_0.$$

Then there exists $t_2 > 0$ such that $S(x,t,\rho_0) > S_0 - \delta_0$, $V(x,t,\rho_0) > V_0 - \delta_0$ and $0 < I(x,t,\rho_0) < \delta_0$, for all $x \in \Omega$ and $t \geq t_2$. Therefore, $I(x,t,\rho_0)$ satisfies

$$\begin{cases} I_t(x,t) \geq \vartheta\Delta I + \dfrac{b(q_1(S_0-\alpha\delta_0)+q_2(V_0-\alpha\delta_0))}{1+\beta\delta_0} I - (m+r)I & \text{for } x \in \Omega \quad \text{and } t \geq t_2, \\ \dfrac{\partial I}{\partial\omega} = 0 & \text{for } x \in \partial\Omega \ \text{ and } t \geq t_2. \end{cases}$$

By Lemma 3, it is noted that the eigenfunction $\hat{\rho}$ is strongly positive corresponding to the eigenvalue $\lambda\left(\frac{S_0-\alpha\delta_0}{1+\beta\delta_0}, \frac{V_0-\alpha\delta_0}{1+\beta\delta_0}\right)$. For all $x \in \Omega$ and $t > 0$, the positive function $I(x, t, \rho_0) > 0$ ensures that there exists $\epsilon > 0$ such that $I(x, t, \rho_0) \geq \epsilon\hat{\rho}$. Clearly, $u(x, t) = \epsilon \exp\left(\lambda\left(\frac{S_0-\alpha\delta_0}{1+\beta\delta_0}, \frac{V_0-\alpha\delta_0}{1+\beta\delta_0}\right)(t - t_2)\right)\hat{\rho}$ is a solution of the system as described

$$\begin{cases} u_t \geq \vartheta\Delta u + \dfrac{b(q_1(S_0 - \alpha\delta_0) + q_2(V_0 - \alpha\delta_0))}{1 + \beta\delta_0}u - (m + r)u & \text{for } x \in \Omega \quad \text{and} \quad t \geq t_2, \\ \dfrac{\partial u}{\partial \omega} = 0 & \text{for } x \in \partial\Omega \quad \text{and} \quad t \geq t_2. \end{cases}$$

Thus the comparison principle yields

$$I(x, t, \rho_0) \geq \epsilon \exp\left(\lambda\left(\frac{S_0 - \alpha\delta_0}{1 + \beta\delta_0}, \frac{V_0 - \alpha\delta_0}{1 + \beta\delta_0}\right)(t - t_2)\right)\hat{\rho}, \quad \text{for } x \in \Omega \quad \text{and} \quad t \geq t_2.$$

This concludes that $I(x, t, \rho_0)$ is bounded, a contradiction.

Define a map $\mathcal{P} : \mathbb{X}^+ \to [0, \infty)$ by

$$\mathcal{P}(\rho) = \min_{x\in\Omega} \rho_3(x), \quad \forall\, \rho \in \mathbb{X}^+.$$

From this continuous function, it is clear that $\mathcal{P}^{-1}(0, \infty) \subseteq \mathcal{X}_0$. Additionally, it is derived that if $\mathcal{P}(\rho) > 0$ or $\mathcal{P}(\rho) = 0$ and $\rho \in \mathcal{X}_0$, then $\mathcal{P}(\Theta_t(\rho)) > 0$ for all $t > 0$. Therefore, for the semiflow $\Theta_t : \mathbb{X}^+ \to \mathbb{X}^+$, the distance function $\mathcal{P}$ is generalized. By observing the above discussion it is remarked that any forward orbit of Θ_t in R_∂ converges to $\{(S_0, V_0, 0)\}$. Also it is explicitly obvious that $\{(S_0, V_0, 0)\}$ is far away in $\mathbb{X}^+$ and $W^s(S_0, V_0, 0) \cap \mathcal{X}_0 = \varnothing$. Further, there is no cycle in R_∂ from $\{(S_0, V_0, 0)\}$ to $\{(S_0, V_0, 0)\}$. Applying the result (Theorem 3) as prescribed in [31] that there exists a $\varrho > 0$ such that

$$\min_{\psi\in\omega(\rho)} \mathcal{P}(\psi) > \varrho, \quad \forall\, \rho \in \mathcal{X}_0.$$

Thus,

$$\liminf_{t\to\infty} I(\cdot, t, \rho) \geq \varrho, \quad \forall\, \rho \in \mathcal{X}_0.$$

Then by Lemma 4(i), the result is established. □

7 Conclusion

In this paper, a spatially dependent vaccination model considering a general non-linear incidence rate in epidemiology is established with all its perspective aspects such as, analytic inter-locution of disease-free equilibrium, disease equilibrium, basic reproduction number, existence and uniqueness of the solution of the corresponding system, stability of local and global steady states and uniform persistence theorem for the system. Our study indicate that the dynamics of the model whether local or global is completely determined by the threshold value $\mathcal{R}_0$. This study will help to predict the upcoming probable results of treatments via vaccination and therapy against malignant diseases considering general incidence cases.

Conflict of Interest

The author(s) declare(s) that there is no conflict of interest regarding the publication of this paper.

Acknowledgements The authors are grateful to anonymous reviewers for their valuable comments that significantly contributed to the presentation of the paper. The research was partially supported by TWAS grant: 2019_19−169 RG/MATHS/AS_I.

References

1. S. Bhattacharya, M. Harrison, M. Worboys, *Fractured States: Smallpox, Public Health and Vaccination Policy in British India 1800–1947* (Educa Books, 2006), pp. 18–23. https://www.amazon.com/Fractured-States-Vaccination-1800-1947-Perspectives/dp/8125028668
2. A. Morabia, *A History of Epidemiologic Methods and Concepts* (Springer, 2004), pp. 346–347. https://www.springer.com/gp/book/9783764368180#aboutBook
3. D.E. Lilienfeld, Celebration: William Farr (1807–1883) an appreciation on the 200 th anniversary of his birth. Int. J. Epidemiol. **36**, 985–987 (2007). https://academic.oup.com/ije/article/36/5/985/775018
4. M. Martcheva, *An Introduction to Mathematical Epidemiology*, vol. 61 (Springer, 2015), pp. 217–220. https://books.google.com.bd/books/about/An_Introduction_to_Mathematical_Epidemio.html?id=tt7HCgAAQBAJ&printsec=frontcover&source=kp_read_button&redir_esc=y#v=onepage&q&f=false
5. J.L. Aron, I.B. Schwartz, Seasonality and period-doubling bifurcations in an epidemic model. J. Theoret. Biol., Springer **110**, 665–679 (1984). https://www.sciencedirect.com/science/article/pii/S0022519384801502
6. W.P. London, J.A. Yorke, Recurrent outbreak of measles, chickenpox and mumps: I. Seasonal variation in contact rates. Am. J. Epidemiol. **98**, 453–468 (1973). https://academic.oup.com/aje/article-abstract/98/6/453/116877
7. I.B. Schwartz, Multiple stable recurrent outbreaks and predictability in seasonally forced nonlinear epidemic models. J. Math. Biol., Springer **21**, 347–361 (1985). https://link.springer.com/article/10.1007/BF00276232
8. H.W. Hethcote, H.W. Stech, P. Van Den Driessche, Nonlinear oscillations in epidemic models. SIAM J. Appl. Math **40** (1981). https://doi.org/10.1137/0140001
9. D. Schenzle, An age-structured model of pre- and post-vaccination measles transmission. IMA J. Math. Appl. Med. Biol. **1**(2), 169–191 (1984). https://www.ncbi.nlm.nih.gov/pubmed/6600102
10. E.B. Wilson, J. Worcester, The law of mass action in epidemiology. PNAS (1945). https://www.pnas.org/content/31/4/109
11. N.C. Severo, Generalizations of some stochastic epidemic models. Math. Biosci. **4**, 395–402 (1969). https://www.sciencedirect.com/science/article/pii/0025556469900194
12. V. Capasso, G. Serio, A generalization of the Kermack-McKendrick deterministic epidemic model. Math. Biosci. **42**, 41–61 (1978). https://www.sciencedirect.com/science/article/pii/0025556478900068
13. P.J. Wangersky, W.J. Cunningham, Time lag in prey' predator population models. Ecology **38**, 136–139 (1979). https://esajournals.onlinelibrary.wiley.com/doi/abs/10.2307/1932137

14. W.M. Liu, H.W. Hetchote, S.A. Levin, Dynamical behavior of epidemiological models with non-linear incidence rates. J. Math. Biol. **25**, 359–380 (1987). https://link.springer.com/article/10.1007/BF00277162
15. W. Liu, S.A. Levin, Y. Iwasa, Influence of nonlinear incidence rates upon the behavior of SIRS epidemiological models. J. Math. Biol., Springer **23**, 187–204 (1986). https://link.springer.com/article/10.1007/BF00276956
16. A.B. Gumel, S.M. Moghadas, A qualitative study of a vaccination model with non-linear incidence. App. Math. Comput. **143**, 409–419 (2003). https://www.sciencedirect.com/science/article/abs/pii/S0096300302003727
17. B. Buonomo, D. Lacitignola, C. Vargar-De-Leon, Qualitative analysis and optimal control of an epidemic model with vaccination and treatment. Math. Comput. Simulat. **100**, 88–120 (2014). https://www.sciencedirect.com/science/article/abs/pii/S0378475414000391
18. J.P. LaSalle, *The Stability of Dynamical System*. Regional Conference Series in Applied Mathematics (SIAM, Philadelphia, 1976). https://books.google.com/books?hl=en&lr=&id=Mhm8sqYRRzsC&oi=fnd&pg=PP2&dq=J.+P.+LaSalle,+The+stability+of+dynamical+system&ots=epGVwv1XFy&sig=OB-LNX7_HYnCgMFHuSipJPA6wbc
19. C. Cosner, R.S. Cantrell, *Spatial Ecology Via Reaction-Diffusion Equations* (Wiley, Ltd., 2003). https://onlinelibrary.wiley.com/doi/book/10.1002/0470871296
20. J.D. Murray, *Mathematical Biology, I and II*, 3rd edn. (Springer, New York, 2002). http://www.ift.unesp.br/users/mmenezes/mathbio.pdf
21. R.M. Anderson, R.M. May, *Infectious Diseases of Humans* (Oxford University Press, London, 1991). https://global.oup.com/academic/product/infectious-diseases-of-humans-9780198540403?cc=us&lang=en&
22. Z. Xu, Y. Xu, Y. Huang, Stability and traveling waves of a vaccination model with nonlinear incidence. Comput. Math. App. **75**, 561–581 (2018). https://www.sciencedirect.com/science/article/pii/S0898122117306223
23. H.L. Smith, *Monotone Dynamical Systems: An Introduction to the Theory of Competitive and Cooperative Systems: Mathematical Surveys and Monographs*, vol. 41 (American Mathematical Society, 1995). https://bookstore.ams.org/surv-41-s
24. J. Wu, *Theory and Applications of Partial Functional Differential Equations* (Springer, New York, 1996). https://www.springer.com/gp/book/9780387947716
25. Y. Lou, X.-Q. Zhao, A reaction-diffusion malaria model with incubation period in the vector population. J. Math. Biol. **62**, 543–568 (2011). https://link.springer.com/article/10.1007/s00285-010-0346-8
26. J. Hale, *Asymptotic Behavior of Dissipative Systems* (American Mathematical Society, Providence, 1988). https://www.ams.org/books/surv/025/surv025-endmatter.pdf
27. D. Henry, *Geometric Theory of Semilinear Parabolic Equations* (Springer, New York, 1981). https://www.springer.com/gp/book/9783540105572
28. W. Wang, X.-Q. Zhao, A nonlocal and time-delayed reaction-diffusion model of dengue transmission. SIAM. Appl. Math. **71**, 147–168 (2011). https://www.researchgate.net/publication/220223186_A_Nonlocal_and_Time-Delayed_Reaction-Diffusion_Model_of_Dengue_Transmission
29. Y. Yang, S. Zhang, Dynamics of a diffusive vaccination model with nonlinear incidence. Comput. Math. Appl. **75**(12), 4355–4360 (2018). https://www.sciencedirect.com/science/article/pii/S0898122118301706
30. M.H. Protter, H.F. Weinberger, *Maximum Principles in Differential Equations* (Springer, New York, 1984). https://www.springer.com/gp/book/9780387960685
31. H.L. Smith, X. Zhao, Robust persistence for semi-dynamical systems. Nonlinear Anal. TMA **47**, 6169–6179 (2001). https://asu.pure.elsevier.com/en/publications/robust-persistence-for-semidynamical-systems

Robust Approach for Uncertain Portfolio Allocation Problems Under Box Uncertainty

Pulak Swain and A. K. Ojha

Abstract Portfolio Optimization is the process of investing the total wealth among different assets to get maximum return out of it with a least possible risk. Several risk-return optimization models can be used to determine the weights which should be given to each asset for the optimal profit. But there is a high possibility of the results being affected by the uncertainty in input parameters, namely, expected return and risk. A small perturbation in the input parameters can mislead the investor to invest in an inefficient portfolio. In uncertainty-based optimization problems, the uncertain parameters are assumed to lie in some specific uncertainty structure like box, ellipsoidal, polyhedral, etc. In the last two decades such problems are dealt with Robust Optimization approach, where the worst-case scenario problem is solved to get "immunized against uncertainty" solutions. This paper provides a discussion on robust mean–variance and robust mean-semi-variance problems under box uncertainty.

Keywords Portfolio optimization · Robust optimization · Box uncertainty

1 Introduction

Markowitz in 1952 formed an optimization model to solve investment related problems [1, 2]. His model is based on taking mean and variance of return as the input parameters for reward and risk, respectively. However, his mean–variance model does not seem to be appropriate when the return distribution is asymmetrical. Therefore, in many studies alternative measures of risk such as semi-variance, semi-absolute deviation, value at risk, and conditional value at risk have been used [3–5]. These measures are called downside risk measures as they consider only the lower side of variation from mean. By solving these portfolio models, we can get the optimal

P. Swain (✉) · A. K. Ojha
Indian Institute of Technology Bhubaneswar, Bhubaneswar, India
e-mail: ps28@iitbbs.ac.in

A. K. Ojha
e-mail: akojha@iitbbs.ac.in

S. R. Mishra et al. (eds.), *Recent Trends in Applied Mathematics*, Lecture Notes in Mechanical Engineering, https://doi.org/10.1007/978-981-15-9817-3_23

weights for allocating the assets. However, these models are very much sensitive to uncertainty in the input parameters. The uncertainty is due to the estimation of expected return and the covariance of return from the past historical return data.

Since the last two decades, the uncertainty-based optimization problems are being handled well by Robust Optimization approach. In this approach, we optimize the worst-case scenario problem to get a robust solution. Ben-Tal and Nemirovski [6] first introduced robust optimization for linear optimization problems and later it was applied in several other types of optimization problems. Robust optimization got tremendous success because of its diverse application in several fields [7, 8]. There have been many studies on the robust approach for portfolio optimization problems [9–11].

The goal of this paper is

(i) To derive the robust counterpart of uncertain mean–variance and mean-semi-variance models under box uncertainty.
(ii) To solve a portfolio problem with different orders of perturbation for input parameters in box uncertainty.
(iii) To compare the results of both mean–variance and mean-semi-variance models and to analyze which model can be more affected by uncertainty.

The organization of the paper is as follows: Sect. 2 presents a preliminary discussion on mean–variance and mean-semi-variance models and box uncertainty. In Sect. 3, we derive the robust counterpart of both the portfolio models for box uncertainty. In Sect. 4, a portfolio problem has been solved for both the models under box uncertainty and the result has been analyzed for different orders of perturbation. And finally, some concluding remarks have been incorporated in Sect. 5.

2 Preliminaries

2.1 *Mean–Variance and Mean-Semi-variance Portfolio Optimization*

Suppose a portfolio contains n number of assets with their returns at time t are given by $r_{it}(i = 1, \ldots, n)$. Markowitz portfolio model was based by taking mean of return as the reward and the variance of portfolio return as the risk factor. To calculate these, first we need to find expected return of each asset and covariance of return between each pair of assets, which are given by

$$\mu_i = E(r_i) = \frac{1}{T}\sum_{t=1}^{T} r_{it} \quad \text{for i} = 1, \ldots, \text{n}$$

$$\sigma_{ij} = E[(r_i - \mu_i)(r_j - \mu_j)]$$

$$=\frac{1}{T}\sum_{t=1}^{T}(r_{it}-\mu_i)\left(r_{jt}-\mu_j\right)\quad \text{for } \mathrm{i}=1,\ldots,\mathrm{n} \text{ and } \mathrm{j}=1,\ldots,\mathrm{n}.$$

The aim is to form a portfolio which will give our desired return with a minimum risk associated with it. Let the weight given to ith asset be x_i. Then the expected return and variance of return for the resulting portfolio are, respectively, given by

$$\mu_P=\sum_i \mu_i x_i,\ \sigma_P^2=\sum_{i,j}\sigma_{ij}x_i x_j$$

Markowitz Mean–Variance Model can be formulated as

$$\min\ \frac{1}{2}\sum_{\mathrm{i,j}}\sigma_{\mathrm{ij}}\mathrm{x_i}\mathrm{x_j}$$

$$s.t.\ \sum_i \mu_i x_i \geq \tau,\ \sum_i x_i = 1,\quad x_i \geq 0 \tag{1}$$

This model minimizes the variance of portfolio return at a fixed lower level of expected return (say τ). But variance measures the variation in both the sides of the mean which includes the potential losses as well as the gains. So this may not be a true representative of risk when the return distribution is not symmetric. In regard to that many downside risk measures have been proposed in the literature. Semi-variance is such a downside measure which takes the lower partial side of variance. This measure of risk gives more sense as it does not include the upside variation. So we do not have to minimize the potential gain.

To use semi-variance, first we have to calculate semi-covariance between each pair of assets which is defined by [4, 5] as

$$\sigma_{ij}^{-}=E\left[\min\{\mathrm{r_i}-\mathrm{B},0\}\cdot\min\left\{\mathrm{r_j}-\mathrm{B},0\right\}\right]$$
$$=\frac{1}{T}\sum_{t=1}^{T}\left[\min\{\mathrm{r_i}-\mathrm{B},0\}\cdot\min\left\{\mathrm{r_j}-\mathrm{B},0\right\}\right]$$

where B is any benchmark return chosen by the investor. Let it be equal to the expected portfolio return, i.e., $B=\mu_P$.

Now the Mean-Semi-variance model can be formulated as

$$\min\frac{1}{2}\sum_{\mathrm{i,j}}\sigma_{\mathrm{ij}}^{-}\mathrm{x_i}\mathrm{x_j}$$

$$s.t.:\ \sum_i \mu_i x_i \geq \tau,\ \sum_i x_i = 1,\quad x_i \geq 0 \tag{2}$$

2.2 Uncertainty in Optimization Problems and the Robust Counterpart

Suppose an optimization problem is given by

$$\min \ f(\boldsymbol{x}, \boldsymbol{u})$$

$$s.t.: \ c(\boldsymbol{x}, \boldsymbol{u}) \le 0$$

where $\boldsymbol{x} \in \mathbb{R}^n$ is a vector of decision variables and the elements of the vector $\boldsymbol{u} \in \mathbb{R}^n$ are some input parameters whose values are not known precisely at the time of decision-making. So they are called uncertain parameters and their values perturb around some approximated nominal values. Let $\mathcal{U}(\boldsymbol{x}) = \{\boldsymbol{u} : g(\boldsymbol{x}, \boldsymbol{u}) \le 0\}$ be the uncertainty set containing those uncertain parameters. So the general form of an uncertain optimization problem can be written as

$$\min \ f(\boldsymbol{x}, \boldsymbol{u})$$

$$s.t.: \quad c(\boldsymbol{x}, \boldsymbol{u}) \le 0, \ \forall \boldsymbol{u} \in \mathcal{U}(\boldsymbol{x}) = \{\boldsymbol{u} : g(\boldsymbol{x}, \boldsymbol{u}) \le 0\} \tag{3}$$

Let $\boldsymbol{u}^0$ be the vector of nominal values of the uncertain parameters. Then the box uncertainty set can be mathematically interpreted as

$$\mathcal{U}_{box} = \{\boldsymbol{u} : \|\boldsymbol{u} - \boldsymbol{u}^0\|_\infty \le \boldsymbol{\delta}\}$$

where $\|\cdot\|_\infty$ is the supremum norm and $\boldsymbol{\delta}$ is the maximum perturbation of the uncertain parameters.

In Robust Optimization approach, we get the solutions which are completely "immunized against uncertainty". That means here we solve the problem for the worst-case realization of the uncertain parameters, so that the solution will be feasible for any realization of the uncertain parameters, so that the solution will be feasible for any realization of uncertain parameters. Then the robust counterpart of the problem (3) is given by

$$\min_x \left\{ \max_{u \in \mathcal{U}(x)} f(\boldsymbol{x}, \boldsymbol{u}) \right\}$$

$$s.t.: \left\{ \max_{u \in \mathcal{U}(x)} c(\boldsymbol{x}, \boldsymbol{u}) \right\} \le 0 \tag{4}$$

Here, since it is a minimization problem, so the worst-case scenario can be obtained by taking those parameters from the uncertainty set which will give maximum values for the constraints. Similarly, for a maximization problem, we would have taken the minimum value parameter in the problem.

Now robust counterpart of the problem for box uncertainty can be written as

$$\min_{x}\left\{\max_{u\in\mathcal{U}(x)} \; f(\boldsymbol{x},\boldsymbol{u})\right\}$$

$$s.t.:\left\{\max_{u} \; c(\boldsymbol{x},\boldsymbol{u}) : \|\boldsymbol{u}-\boldsymbol{u}^{0}\|_{\infty}\le\delta\right\}\le 0. \tag{5}$$

3 Robust Counterparts of Uncertain Mean–Variance and Mean-Semi-variance Models Under Box Uncertainty

As in portfolio optimization, a small perturbation in data can mislead the investor to invest in an inefficient portfolio. So the uncertainty factor needs to be taken care in portfolio models.

3.1 *Uncertain Mean–Variance Model*

In mean–variance optimization, the perturbation may occur either in the mean return data or in the covariance matrix of the asset returns. The uncertain mean–variance portfolio problem is defined as

$$\min \; \frac{1}{2}\sum_{i,j}\sigma_{ij}x_ix_j$$

$$s.t.: \; \sum_{i}\mu_i x_i \ge \tau, \quad \sum_{i} x_i = 1, \quad x_i \ge 0$$

$$\forall \mu_i \in \mathcal{U}_{\mu} \text{ and } \sigma_{ij} \in \mathcal{U}_{\sigma} \tag{6}$$

where $\mathcal{U}_{\mu}$ and $\mathcal{U}_{\sigma}$ are the uncertain sets associated with mean and covariance terms, respectively.

The robust counterpart of the above problem is given by taking the worst-case realization of the uncertainty set,

$$\min_{x_i}\left\{\max_{\sigma_{ij}\in\mathcal{U}_{\sigma}}\sum_{i,j}\frac{1}{2}\sigma_{ij}x_ix_j\right\}$$

$$s.t.: \left\{ \min_{\mu_i \in \mathcal{U}_\mu} \sum_i \mu_i x_i \right\} \geq \tau, \quad \sum_i x_i = 1, \quad x_i \geq 0 \tag{7}$$

Now the box uncertainty set associated with the mean return can be represented by

$$\mathcal{U}_\mu = \{\mu_i : |\mu_i - \mu_i{}^0| \leq \delta_i, \; i = 1, 2, \cdots, n\}$$

where $\mu_i{}^0$ is the nominal expected return of ith asset.

That means the exact value of each component μ_i can be anyone from the interval $[\mu_i{}^0 - \delta_i, \mu_i{}^0 + \delta_i]$.

Since all the x_i's are non-negative so the robust counterpart for the expected portfolio return is

$$\min_{\mu_i \in \mathcal{U}_\mu} \sum_i \mu_i x_i = \sum_i (\mu_i{}^0 - \delta_i) x_i$$

Now let the box uncertainty set associated with covariance terms of return is given by

$$\mathcal{U}_\sigma = \{\sigma_{ij} : |\sigma_{ij} - \sigma_{ij}{}^0| \leq \delta_{ij}, \; i = 1, 2, \cdots, n, \; j = 1, 2, \cdots, n\}$$

where, $\sigma_{ij}{}^0$ is the nominal covariance of return between i^{th} and j^{th} asset. So the worst-case variance of the portfolio is given by,

$$\max_{\sigma_{ij} \in \mathcal{U}_\sigma} \sum_i \sigma_{ij} x_i x_j = \sum_i \sum_j (\sigma_{ij}{}^0 + \delta_{ij}) x_i x_j$$

Then the robust counterpart of the mean–variance problem under box uncertainty can be written as

$$\min_{x_i} \; \frac{1}{2} \sum_i \sum_j (\sigma_{ij}{}^0 + \delta_{ij}) x_i x_j$$

$$s.t.: \sum_i (\mu_i{}^0 - \delta_i) x_i \geq \tau, \quad \sum_i x_i = 1, \quad x_i \geq 0 \tag{8}$$

3.2 *Uncertain Mean-Semi-variance Model*

The uncertain mean-semi-variance model can be defined as

$$\min \ \frac{1}{2}\sum_{i,j} \sigma^{-}_{ij} x_i x_j$$

$$s.t.: \ \sum_i \mu_i x_i \geq \tau, \quad \sum_i x_i = 1, \quad x_i \geq 0$$

$$\forall \mu_i \in \mathcal{U}_\mu \text{ and } \sigma^{-}_{ij} \in \mathcal{U}_{\sigma^-} \tag{9}$$

where $\mathcal{U}_\mu$ and $\mathcal{U}_{\sigma^-}$ are the uncertain sets associated with mean and semi-covariance terms, respectively.

Let the box uncertainty set associated with semi-covariance of return is given by

$$\mathcal{U}_{\sigma^-} = \{\sigma^{-}_{ij} : |\sigma^{-}_{ij} - \sigma^{-0}_{ij}| \leq \delta^{-}_{ij}, i = 1, 2, \ldots, n, j = 1, 2, \ldots, n\}$$

Then the robust counterpart of the mean-semi-variance model can be written as

$$\min_{x_i} \ \frac{1}{2}\sum_i \sum_j (\sigma^{-0}_{ij} + \delta^{-}_{ij}) x_i x_j$$

$$s.t.: \ \sum_i (\mu_i{}^0 - \delta_i) x_i \geq \tau, \quad \sum_i x_i = 1, \quad x_i \geq 0. \tag{10}$$

4 Numerical Example

To understand how robust optimization approach works for portfolio problems, we solve a practical problem. The quarterly return data is taken for four assets: AMZN, DPZ, BTC, and NFLX from July 2013 to May 2019. We calculate the expected returns of each asset and the covariance and semi-covariance of return between each pair of assets. Since we have to deal with uncertainty, so those values of expected return, covariance, and semi-covariance returns are termed as nominal values. Now those nominal values are given in matrix form as

$$\text{Expected Returns} = \begin{bmatrix} 0.09034 \\ 0.07334 \\ 0.49579 \\ 0.12463 \end{bmatrix},$$

$$\text{Covariance Returns} = \begin{bmatrix} 0.019172 & 0.001483 & 0.060955 & 0.013882 \\ 0.001483 & 0.011865 & -0.042816 & 0.005672 \\ 0.060955 & -0.042816 & 1.787858 & 0.003736 \\ 0.013882 & 0.005672 & 0.003736 & 0.047606 \end{bmatrix},$$

$$\text{Semi-covariance Returns} = \begin{bmatrix} 0.007831 & 0.001610 & 0.009943 & 0.006050 \\ 0.001610 & 0.003719 & -0.003172 & 0.001324 \\ 0.009943 & -0.003172 & 0.127163 & 0.012444 \\ 0.006050 & 0.001324 & 0.012444 & 0.013372 \end{bmatrix}$$

Let our target rate of return be $\tau = 0.25$. Our aim is to find the optimal weight of each asset and the optimal risk associated with the whole portfolio to achieve the given target rate of return.

Now the nominal mean–variance model is formulated as

$$\min \frac{1}{2}\left\{\begin{array}{l} 0.019172x_1^2 + 0.011865x_2^2 + 1.787858x_3^2 + 0.047606x_4^2 \\ +2 \cdot 0.001483x_1x_2 + 2 \cdot 0.060955x_1x_3 + 2 \cdot 0.013882x_1x_4 \\ +2 \cdot (-0.042816)x_2x_3 + 2 \cdot 0.005672x_2x_4 + 2 \cdot 0.003736x_3x_4 \end{array}\right\}$$

$$s.t.: 0.09034x_1 + 0.07334x_2 + 0.49579x_3 + 0.12463x_4 \geq 0.25,$$

$$x_1 + x_2 + x_3 + x_4 = 1, \quad x_1, x_2, x_3, x_4 \geq 0$$

On solving this problem we get the optimal weights of the four assets as $x_1 = 0$, $x_2 = 0$, $x_3 = 0.337779$, $x_4 = 0.662221$. And the optimal risk is obtained as 0.1132666.

Similarly, the nominal mean-semi-variance model is formulated as

$$\min \frac{1}{2}\left\{\begin{array}{l} 0.007831x_1^2 + 0.003719x_2^2 + 0.127163x_3^2 + 0.013372x_4^2 \\ +2 \cdot 0.001610x_1x_2 + 2 \cdot 0.009943x_1x_3 + 2 \cdot 0.006050x_1x_4 \\ +2 \cdot (-0.003172)x_2x_3 + 2 \cdot 0.001324x_2x_4 + 2 \cdot 0.012444x_3x_4 \end{array}\right\}$$

$$s.t.: 0.09034x_1 + 0.07334x_2 + 0.49579x_3 + 0.12463x_4 \geq 0.25,$$

$$x_1 + x_2 + x_3 + x_4 = 1, \quad x_1, x_2, x_3, x_4 \geq 0$$

The optimal weights of the assets for this model are $x_1 = 0$, $x_2 = 0.509621$, $x_3 = 0.408203$, $x_4 = 0.082176$ and the optimal risk is obtained as 0.0109356.

Now taking uncertainty into account, the values of expected returns, covariance, and semi-covariance terms can perturb around the nominal values, so that the optimal weights and optimal risks can be influenced. Here we assume the uncertainty set to be a box and for different orders of perturbation, we form the robust counterparts of both the models as given in Eqs. (8) and (10). For simplicity, we assume each $\delta_i, \delta_{ij}, \delta_{ij}^-$ to be equal. We solve the robust counterpart problems of both the models for different orders of perturbation. The results showing optimal weight vectors and the optimal portfolio risk for both the models are incorporated in Table 1. As in robust

Table 1 Robust solutions under box uncertainty for different orders of perturbation

Perturbation	Optimal attribute	Robust mean–variance model	Robust mean-semi-variance model
10^{-7}	Weight vector	[0.000000 0.000000 0.337779 0.662221]	[0.000000 0.509621 0.408203 0.082176]
	Portfolio risk	0.1132668	0.0109357
10^{-6}	Weight vector	[0.000000 0.000000 0.337782 0.662218]	[0.000000 0.509619 0.408205 0.082176]
	Portfolio risk	0.1132682	0.0109358
10^{-5}	Weight vector	[0.000000 0.000000 0.337806 0.662194]	[0.000000 0.509600 0.408227 0.082173]
	Portfolio risk	0.1132821	0.0109368
10^{-4}	Weight vector	[0.000000 0.000000 0.338048 0.661952]	[0.000000 0.509408 0.408443 0.082150]
	Portfolio risk	0.1134212	0,0109476
10^{-3}	Weight vector	[0.000000 0.000000 0.340473 0.659527]	[0.000000 0.507486 0.410602 0.081912]
	Portfolio risk	0.1148186	0.0110558
10^{-2}	Weight vector	[0.000000 0.000000 0.364721 0.635278]	[0.000000 0.477357 0.430687 0.091956]
	Portfolio risk	0.1293840	0.0138994

approach, we take the worst-case parameters, so we get higher optimal portfolio risk in comparison to the optimal portfolio risk for nominal problem. The percentage difference of robust portfolio risks from the nominal portfolio risk is given in Table 2.

Table 2 Relative increase in robust portfolio risk than nominal portfolio risk for both the models at different orders of perturbation

Perturbation	Relative increase in robust portfolio risk than nominal portfolio risk	
	Mean–variance model (%)	Mean-semi-variance model (%)
10^{-7}	0.0002	0.0009
10^{-6}	0.0014	0.0018
10^{-5}	0.0137	0.0110
10^{-4}	0.1365	0.1097
10^{-3}	1.3702	1.0992
10^{-2}	14.2296	27.1023

5 Concluding Remarks

The results given in Table 1 show that with increase in order of perturbation, the optimal robust portfolio risk increases. That means if the perturbation is higher, we have to take more risk to make sure of getting the target rate of return. From Table 2 we get that at some orders of perturbation the relative increase in robust portfolio risks is higher for mean–variance model and in other cases it is higher for mean-semi-variance model. So we cannot say particularly which model can be more affected by uncertainty.

References

1. H. Markowitz, Portfolio selection. J. Financ. **7**, 77–91 (1952)
2. H. Markowitz, *Portfolio Selection: Efficient Diversification of Investments* (Basil Blackwell, New York, 1959).
3. H. Konno, H. Waki, A. Yuuki, Portfolio optimization under lower partial risk measures. Asia Pacific Finan. Mark. **9**(2), 127–140 (2002)
4. J. Estrada, Systematic risk in emerging markets: the D-CAPM. Emerg.. Mark. Rev. **3**(4), 365–379 (2002)
5. J. Estrada, Mean-semivariance behavior: downside risk and capital asset pricing. Int. Rev. Econ. Financ. **16**(2), 169–185 (2007)
6. A. Ben-Tal, A. Nemirovski, Robust solutions of uncertain linear programs. Oper. Res. Lett. **25**(1), 1–13 (1999)
7. A. Ben-Tal, A. Nemirovski, Robust optimization-methodology and applications. Math. Program, **92**(3), 453–480 (2002)
8. D. Bertsimas, B. Brown, C. Caramanis, Theory and applications of robust optimization problem. SIAM Rev. **53**(3), 464–501 (2011)
9. D. Goldfarb, G. Iyengar, Robust portfolio selection problems. Math. Oper. Res. **28**(1), 1–38 (2003)
10. R.H. Tütüncü, M. Koenig, Robust asset allocation. Ann. Oper. Res. **132**(1–4), 157–187 (2004)
11. M.P. Rajan, N. Rana, A robust portfolio optimization in Indian Stock market, in *2011 World Congress on Information and Communication Technologies* (2011), pp. 645–650

Oscillation Theorems for a Class of Nonlinear Difference Equations with Fractional Order

A. George Maria Selvam, Mary Jacintha, and R. Janagaraj

Abstract With the application of an inequality and generalized Riccati technique, we arrive at certain oscillatory behavior of a discrete fractional order nonlinear equations

$$\Delta\{\rho(\ell)[\Delta(\gamma(\ell)\Psi(\Delta^{\mu}x(\ell)))]^{\eta}\} + q(\ell)F(G(\ell)) = 0, \ell \in N_{\ell_0},$$

where $G(\ell) = \sum_{v=\ell_0}^{\ell-1+\mu} (\ell - v - 1)^{(-\mu)}x(v)$ and $\eta \geq 1$ is a quotient of two odd positive integers, $0 < \mu \leq 1$ is a constant, Δ^{μ} denotes the RL fractional difference operator of order μ and $N_{\ell_0} = \{\ell_0, \ell_0 + 1, \ell_0 + 2, \cdots\}$. Numerical example verify the theoretical result.

Keywords Oscillation · Fractional order · Difference equations · Riccati technique

AMS Mathematics Subject Classification 34A08 · 39A12 · 39A13 · 39A21

1 Introduction

The fundamental concept of fractional calculus is widely believed to originate from a question raised by Marquis to Leibniz in the year 1695. A mathematical discipline dealing with integrals and derivatives of a fractional order is called fractional calculus. Nowadays, various types of methods are applied to estimate fractional integrals and derivative. In the past few decades, many authors have been concerned with existence and uniqueness and stability of solutions of fractional differential equations (FDE).

A. G. M. Selvam (✉) · M. Jacintha
Department of Mathematics, Sacred Heart College (Autonomous), Vellore Dist, Tirupattur 635601, Tamil Nadu, India
e-mail: agmshc@gmail.com

R. Janagaraj
Department of Mathematics, Faculty of Engineering, Karpagam Academy of Higher Education, Coimbatore 641021, Tamil Nadu, India

S. R. Mishra et al. (eds.), *Recent Trends in Applied Mathematics*, Lecture Notes in Mechanical Engineering, https://doi.org/10.1007/978-981-15-9817-3_24

Recently, some papers examined oscillatory theorems for nonlinear FDE, [21, 23] and there are books by Miller and Ross, Podlubny and Kilbas et al. [16, 20, 22].

The qualitative study of the differential equations of fractional order has hit bull's eye among mathematicians and researchers with its applications in varied areas, namely, fluid flow, control theory of dynamical systems, electrical networks, chemical reaction, economics, statistics, etc. FDE are apt to model physical processes which vary with time and space, and its non-local property enables model systems with memory effect. Fractional calculus endorses integration and differentiation to a fractional order whose values could be real and be extended to imaginary.

The discrete version of FDE is the fractional difference equations with fractional order sum and difference operators as basic notions. Ever since, Kuttner [18] mentioned for the first time the fractional order differences in 1956, the theory of difference equations of fractional order has systematically evolved over the past decades [2, 7–9, 15]. The study of the qualitative analysis of these equations has gained momentum in recent years with numerous publications investigate existences of solutions, uniqueness of solutions, monotonicity behavior and asymptotic behavior of these equations [3, 4, 11, 14].

2 Some Significant Previous Works

Of late, the learning of the oscillatory and non-oscillatory solutions of discrete fractional order equations has accelerated with new findings [12, 13, 17, 26]. Hakan Adiguzel in [1] investigated the oscillation theorems for a class of discrete nonlinear fractional order equation

$$\Delta(r(\ell)\Delta(\rho(\ell)\Delta(\gamma(\ell)\Delta^{\mu}x(\ell)))) + q(\ell)G(\ell) = 0.$$

A. Secer and H. Adiguzel in [25] studied and obtained some oscillatory criteria for the given below discrete fractional order equation of $\mu(0 < \mu \leq 1)$

$$\Delta(\gamma(\ell)[\Delta(\rho(\ell)(\Delta^{\mu}x(\ell))^{\eta_1})]^{\eta_2}) + q(\ell)f\left(\sum_{v=\ell_0}^{\ell-1+\mu}(\ell - v - 1)^{(-\mu)}x(v)\right)$$
$$= 0, \ell \in N_{\ell_0+1-\mu}.$$

In [6], Chatzarakis et al. provided conditions for oscillation of discrete fractional order equations of the form

$$\Delta(\Delta^{\mu}y(\ell))^{\lambda} + q(\ell)g(y(\ell)) = 0, \ell \in N_{\ell_0+1-\mu}.$$

Li [19] established the oscillation theorems for a discrete nonlinear fractional order equations

$$(1+\rho(\ell))\Delta(\Delta^{\mu}x(\ell))+\rho(\ell)\Delta^{\mu}x(\ell)+g(\ell,x(\ell))=e(\ell),\ \ell\in\mathbb{N}_0.$$

The asymptotic behavior of solutions for a class of discrete damped nonlinear fractional order equations

$$\Delta\big[r(\ell)[\Delta(\gamma(\ell)\Delta^{\mu}x(\ell))]^{\eta}\big]+\rho(\ell)[\Delta(\gamma(\ell)\Delta^{\mu}x(\ell))]^{\eta} + q(\ell)f\left(\sum_{v=\ell_0}^{\ell-1+\mu}(\ell-v-1)^{(-\mu)}x(v)\right)=0,\ k\in N_{\ell_0}$$

was presented by Bai and Xu [5]. Motivated by the above work, in this present work, we aim at obtaining the criteria for the oscillatory behavior of the solutions for the following nonlinear discrete equation with fractional order,

$$\Delta\{\rho(\ell)[\Delta(\gamma(\ell)\Psi(\Delta^{\mu}x(\ell)))]^{\eta}\}+q(\ell)f(G(\ell))=0,\ \ell\in N_{\ell_0}, \tag{1}$$

where $G(\ell)=\sum\limits_{v=\ell_0}^{\ell-1+\mu}(\ell-v-1)^{(-\mu)}x(v)$ and $\eta\geq 1$ is a quotient of two odd positive integers, $0<\mu\leq 1$ is a constant, Δ^{μ} denotes the RL difference operator of fractional order μ and $N_{\ell_0}=\{\ell_0,\ell_{0+1},\ell_{0+2},\ldots\}$.

The following conditions hold throughout this work:

(H1). $\gamma(\ell)$, $\rho(\ell)$, and $q(\ell)$ are sequences of positive real numbers.

(H2). Ψ is a function which is an odd monotonically increasing and there exists a constant m(positive) such that $\frac{x}{\Psi(x)}\geq m>0$ for $x\Psi(x)\neq 0$, $\Psi^{-1}\in C(R,R)$ is a continuous function with $x\Psi^{-1}(x)>0$ for $x\neq 0$ and there exists a constant θ_1(positive) such that $\Psi^{-1}(uv)\geq\theta_1 u\Psi^{-1}(v)$ for $uv\neq 0$.

(H3). F is a monotone decreasing function satisfying $xF(x)>0$ such that $\frac{F(x)}{x^{\eta}}\geq$ B >0 for $x\neq 0$.

By a solution of Eq. (1), we mean a sequence $x(\ell)\in(R_+;R)$ that satisfies Eq. (1) for $\ell\in\mathbb{N}_0$. If a solution $x(\ell)$ of (1) has arbitrarily large zeros, then it is called oscillatory, otherwise the solution $x(\ell)$ of (1) is called non-oscillatory. Equation (1) is called oscillatory if all its solutions are oscillatory.

3 Basic Lemmas and Preliminaries

In this work, the following definitions, notations, and properties of fractional difference operator will contribute in proving the main results.

Definition 1 For any $\mu\geq 0$, the falling factorial is defined by

$$s^{(\mu)}=\frac{\Gamma(s+1)}{\Gamma(s+1-\mu)}. \tag{2}$$

Definition 2 Reference [2] The μth fractional sum for $\mu > 0$ is defined by

$$\Delta^{-\mu} f(\nu) = \frac{1}{\Gamma(\mu)} \sum_{v=a}^{\nu-\mu} (\nu - v - 1)^{(\mu-1)} f(v). \tag{3}$$

The fractional sum $\Delta^{-\mu}$ is mapped from N_a to $N_{a+\mu}$, where $N_\ell = \{\ell, \ell+1, \ell+2, \ldots\}$.

Definition 3 Reference [2] Let $\mu > 0$ and $n - 1 < \mu < n$, where n denotes a positive integer, $n = \lceil \mu \rceil$. Set $\nu = n - \mu$, then μth fractional difference is defined as

$$\Delta^{\mu} f(\ell) = \Delta^{n-\nu} f(\ell) = \Delta^{n} \Delta^{-\nu} f(\ell). \tag{4}$$

Lemma 1 *Reference* [12] *Let $x(\ell)$ be a solution of (1), $G(\ell) = \sum_{v=\ell_0}^{\ell-1+\mu} (\ell - v - 1)^{(-\mu)} x(v)$, Then*

$$\Delta G(\ell) = \Gamma(1-\mu) \Delta^{\mu} x(\ell). \tag{5}$$

Lemma 2 *Reference* [10] *The product and quotient rules of the difference operator Δ are as follows:*

$$\Delta[x(\ell)y(\ell)] = x(\ell+1)\Delta y(\ell) + \Delta x(\ell) y(\ell) \tag{6}$$

$$\Delta\left[\frac{x(\ell)}{y(\ell)}\right] = \frac{\Delta x(\ell) y(\ell) - x(\ell)\Delta y(\ell)}{y(\ell) y(\ell+1)}, \tag{7}$$

where $\Delta x(\ell) = x(\ell+1) - x(\ell)$.

Lemma 3 *Reference* [5] *If $\eta \geq 1$ is a quotient of two odd positive integers, then the following two inequalities are established:*

if $G(\ell+1) > G(\ell) > 0$, then

$$\Delta G^{\eta}(\ell) \geq (\Delta G(\ell))^{\eta}, \tag{8}$$

if $G(\ell+1) < G(\ell) < 0$, then

$$\Delta G^{\eta}(\ell) \leq (\Delta G(\ell))^{\eta}. \tag{9}$$

4 Main Results

In this section, the oscillation theorems for the solutions of nonlinear discrete fractional order Eqs. (1) are established using the properties of R-L sum, difference operators, and generalized Riccati technique.

Lemma 4. *Assume that (H_1)–(H_3) hold and $x(\ell)$ is eventually positive solution of (1). If*

$$\lim_{\ell\to\infty}\sum_{v=\ell_0}^{\ell-1}\frac{1}{[\rho(v)]^{\frac{1}{\eta}}}=\infty, \tag{10}$$

$$\lim_{\ell\to\infty}\sum_{v=\ell_0}^{\ell-1}\Psi^{-1}\left(\frac{1}{\gamma(v)}\right)=\infty, \tag{11}$$

$$\lim_{\ell\to\infty}\sum_{\xi=\ell_0}^{\ell-1}\Psi^{-1}\left[\frac{1}{\gamma(\xi)}\sum_{\tau=\xi}^{\infty}\left[\frac{1}{\rho(\tau)}\sum_{v=\tau}^{\infty}q(v)\right]^{\frac{1}{\eta}}\right]=\infty, \tag{12}$$

then, there exists a sufficiently large $L\in N_{\ell_0}$ such that $\Delta(\gamma(\ell)\Psi(\Delta^{\mu}x(\ell)))>0$ on $[L,\infty)$ and one of the following two conditions hold: (i) $\Delta^{\mu}x(\ell)>0$ on $[L,\infty)$ and (ii) $\Delta^{\mu}x(\ell)<0$ on $[L,\infty)$ and $\lim_{\ell\to\infty}G(\ell)=0$.

Proof From the assumption, as $x(\ell)$ is an eventually positive solution of (1), there exists a sufficiently large $\ell_1,\ell_1\geq\ell_0$ such that $x(\ell)>0$ on $[\ell_1,\infty)$ so that $G(\ell)>0$ on $[\ell_1,\infty)$ Therefore, from (H_3) and (1) we obtain

$$\Delta\{\rho(\ell)[\Delta(\gamma(\ell)\Psi(\Delta^{\mu}x(\ell)))]^{\eta}\}=-q(\ell)F(G(\ell))\leq-Bq(\ell)G^{\eta}(\ell)<0,\ell\in[\ell_1,\infty). \tag{13}$$

Hence $\{\rho(\ell)[\Delta(\gamma(\ell)\Psi(\Delta^{\mu}x(\ell)))]^{\eta}\}$ is strictly decreasing on $[\ell_1,\infty)$ and $\Delta(\gamma(\ell)\Psi(\Delta^{\mu}x(\ell)))$ is eventually of one sign. For $\nu_2>\nu_1$ is sufficiently large, we claim that $\Delta(\gamma(\ell)\Psi(\Delta^{\mu}x(\ell)))>0$ on $[\ell_2,\infty)$. If not, assume that there exists a sufficiently large $\ell_3>\ell_2$ such that $\Delta(\gamma(\ell_3)\Psi(\Delta^{\mu}x(\ell_3)))<0$ for $\ell\in[\ell_3,\infty)$. Then we arrive at

$$\sum_{v=\ell_3}^{\ell-1}\Delta\{\rho(v)[\Delta(\gamma(v)\Psi(\Delta^{\mu}x(v)))]^{\eta}\}<0$$

$$\{\rho(\ell)[\Delta(\gamma(\ell)\Psi(\Delta^{\mu}x(\ell)))]^{\eta}\}<\{\rho(\ell_3)[\Delta(\gamma(\ell_3)\Psi(\Delta^{\mu}x(\ell_3)))]^{\eta}\}=D<0,\ell\in[\ell_3,\infty), \tag{14}$$

which implies

$$\Delta(\gamma(\ell)\Psi(\Delta^{\mu}x(\ell))) < \left[\frac{D}{\rho(\ell)}\right]^{\frac{1}{\eta}} < 0, \ell \in [\ell_3, \infty). \tag{15}$$

Summing both sides of (15) from ℓ_3 to $\ell - 1$, we get

$$(\gamma(\ell)\Psi(\Delta^{\mu}x(\ell))) - (\gamma(\ell_3)\Psi(\Delta^{\mu}x(\ell))) < D^{\frac{1}{\eta}} \sum_{v=\ell_3}^{\ell-1} \frac{1}{\rho^{\frac{1}{\eta}}(v)}, \tag{16}$$

From (10), we obtain, $\lim_{\ell\to\infty} \gamma(\ell)\Psi(\Delta^{\mu}x(\ell)) = -\infty$ which suggests that for a certain $\ell_4 > \ell_3$, $\gamma(\ell)\Psi(\Delta^{\mu}x(\ell)) < 0$, for $\ell \in [\ell_4, \infty)$. Hence

$$\sum_{v=\ell_4}^{\ell-1} \Delta\big[\gamma(v)\Psi(\Delta^{\mu}x(v)))\big] < 0$$

$$\gamma(\ell)\Psi(\Delta^{\mu}x(\ell)) < \gamma(\ell_4)\Psi(\Delta^{\mu}x(\ell_4)) = c > 0, \ell \in [\ell_4, \infty). \tag{17}$$

From (H_2), we obtain

$$\frac{\Delta G(\ell)}{\Gamma(1-\mu)} = \Delta^{\mu}x(\ell) < \Psi^{-1}\left[\frac{c}{\gamma(\ell)}\right] \leq c\theta_1\Psi^{-1}\left[\frac{1}{\gamma(\ell)}\right], \ell \in [\ell_4, \infty). \tag{18}$$

Now summing the above inequality from ℓ_4 to $\ell - 1$,

$$G(\ell) < G(\ell_4) + \Gamma(1-\mu)c\theta_1 \sum_{v=\ell_4}^{\ell-1} \Psi^{-1}\left[\frac{1}{\gamma(v)}\right]. \tag{19}$$

Then allowing $\ell \to \infty$ and using (11), we arrive at $\lim_{\ell\to\infty} G(\ell) = -\infty$, which contradicts the fact that $G(\ell) > 0$ on $[\ell_1, \infty)$. Accordingly $\Delta\big[\gamma(\ell)\Psi(\Delta^{\mu}x(\ell))\big] > 0$, $\ell \in [\ell_2, \infty)$. Then from (H_2) we get that $\Delta^{\mu}x(\ell)$ is eventually of one sign, with two possibilities: (i) $\Delta^{\mu}x(\ell) > 0$ on $[L, \infty)$, (ii) $\Delta^{\mu}x(\ell) < 0$ where L is sufficiently large.

Now assume that $\Delta^{\mu}x(\ell) < 0$, $\ell \in [\ell_5, \infty)$ where $\ell_5 > \ell_4$ is sufficiently large. Then by Lemma 1, we have $\Delta G(\ell) = \Gamma(1-\mu)\Delta^{\mu}x(\ell) < 0$, $\ell \in [\ell_5, \infty)$. Since $G(\ell) > 0$, $\ell \in [\ell_1, \infty)$, we have $\lim_{\ell\to\infty} G(\ell) = \lambda \geq 0$, we claim $\lambda = 0$. On the contrary assume that $\lambda > 0$, then $G(\ell) \geq \lambda$ on $[\ell_5, \infty)$. By (13), we have

$$\Delta\{\rho(\ell)[\Delta(\gamma(\ell)\Psi(\Delta^{\mu}x(\ell)))]^{\eta}\} \leq -Bq(\ell)G^{\eta}(\ell) \leq -Bq(\ell)\lambda^{\eta}. \tag{20}$$

Summing up (20) with respect to k from ℓ to ∞ leads to

$$\lim_{\ell\to\infty} \rho(v)[\Delta(\gamma(v)\Psi(\Delta^\mu x(v)))]^\eta - \rho(\ell)[\Delta(\gamma(\ell)\Psi(\Delta^\mu x(\ell)))]^\eta \le -B\lambda^\eta \sum_{v=\ell}^{\infty} q(v). \tag{21}$$

Hence,

$$\Delta(\gamma(\ell)\Psi(\Delta^\mu x(\ell))) \ge B^{\frac{1}{\eta}}\lambda \left[\frac{1}{\rho(\ell)}\sum_{v=\ell}^{\infty} q(v)\right]^{\frac{1}{\eta}}. \tag{22}$$

Summing (22) with respect to τ from ℓ to ∞ yields,

$$\sum_{\tau=t}^{\infty} \Delta(\gamma(\ell)\Psi(\Delta^\mu x(\ell))) \ge B^{\frac{1}{\eta}}\lambda \sum_{\tau=\ell}^{\infty}\left[\frac{1}{\rho(\tau)}\sum_{v=\ell}^{\infty} q(v)\right]^{\frac{1}{\eta}}, \tag{23}$$

$$-(\gamma(\ell)\Psi(\Delta^\mu x(\ell))) \ge -\lim_{\tau\to\infty} \gamma(\tau)\Psi(\Delta^\mu x(\tau)) + B^{\frac{1}{\eta}}\lambda \sum_{\tau=\ell}^{\infty}\left[\frac{1}{\rho(\tau)}\sum_{v=\ell}^{\infty} q(v)\right]^{\frac{1}{\eta}}$$

$$-(\gamma(\ell)\Psi(\Delta^\mu x(\ell))) \ge B^{\frac{1}{\eta}}\lambda \sum_{\tau=\ell}^{\infty}\left[\frac{1}{\rho(\tau)}\sum_{v=\ell}^{\infty} q(v)\right]^{\frac{1}{\eta}}, \tag{24}$$

or

$$\Delta^\mu x(\ell) < -\Psi^{-1}\left[\frac{1}{\gamma(\ell)} B^{\frac{1}{\eta}}\lambda \sum_{\tau=\ell}^{\infty}\left[\frac{1}{\rho(\tau)}\sum_{v=\ell}^{\infty} q(v)\right]^{\frac{1}{\eta}}\right], \tag{25}$$

which in turn yields

$$\Delta G(\ell) \le -\Gamma(1-\mu)\theta_1 B^{\frac{1}{\eta}}\lambda\Psi^{-1}\left[\frac{1}{\gamma(\ell)}\sum_{\tau=\ell}^{\infty}\left[\frac{1}{\rho(\tau)}\sum_{v=\tau}^{\infty} q(v)\right]^{\frac{1}{\eta}}\right]. \tag{26}$$

Summing both sides of (26),

$$G(\ell) < G(\ell_5) - \Gamma(1-\mu)\theta_1 B^{\frac{1}{\eta}}\lambda \sum_{\xi=\ell_5}^{\ell} \Psi^{-1}\left[\frac{1}{\gamma(\xi)}\sum_{\tau=\xi}^{\infty}\left[\frac{1}{\rho(\tau)}\sum_{v=\tau}^{\infty} q(v)\right]^{\frac{1}{\eta}}\right]. \tag{27}$$

Therefore, from (12), we arrive at $\lim_{\ell\to\infty} G(\ell) = -\infty$ which contradicts the fact that $G(\ell) > 0$, $\ell \in [\ell_1, \infty)$. Hence, we get $\lambda = 0$, which is $\lim_{\ell\to\infty} G(\ell) = 0$. The proof is complete.

In the case, when $x(\ell)$ is an eventually negative solution of (1), proceeding as in the above discussion we arrive at $\Delta\rho(\ell)[\Delta(\gamma(\ell)\Psi(\Delta^{\mu}x(\ell)))]^{\eta} < 0$ and one of the following two conditions hold: (i) $\Delta^{\mu}x(\ell) > 0$ on $[L, \infty)$ and (ii) $\Delta^{\mu}x(\ell) < 0$ on $[L, \infty)$ and $\lim_{\ell\to\infty} G(\ell) = 0$.

Lemma 5 *Assume that $x(\ell)$ is an eventually positive solution of Eq. (1) such that $\Delta(\gamma(\ell)\Psi(\Delta^{\mu}x(\ell))) > 0$, $\Delta^{\mu}x(\ell) > 0$ on $[\ell_1, \infty)$ where $\ell_1 \geq \ell_0$ is sufficiently large, then we have*

$$\Delta G(\ell) \geq \frac{m\Gamma(1-\mu)\rho^{\frac{1}{\eta}}(\ell)\big[\Delta(\gamma(\ell)\Psi(\Delta^{\mu}x(\ell)))\big]}{\gamma(\ell)} E_1(\ell, \ell_1), \tag{28}$$

where $E_1(\ell, \ell_1) = \sum\limits_{v=\ell_1}^{\ell-1} \frac{1}{\rho^{\frac{1}{\eta}}(v)}$.

Proof From (13) we deduce that $\rho(\ell)[\Delta(\gamma(\ell)\Psi(\Delta^{\mu}x(\ell)))]^{\eta}$ is strictly decreasing on $[\ell_1, \infty)$, so we get

$$\begin{aligned}
\Psi(\Delta^{\mu}x(\ell))\gamma(\ell) &\geq \Psi(\Delta^{\mu}x(\ell))\gamma(\ell) - \Psi(\Delta^{\mu}x(\ell_1))\gamma(\ell_1) \\
&= \sum_{v=\ell_1}^{\ell-1} \frac{\rho^{\frac{1}{\eta}}(v)\big[\Delta(\gamma(v)\Psi(\Delta^{\mu}x(v)))\big]}{\rho^{\frac{1}{\eta}}(v)} \\
&\geq \rho^{\frac{1}{\eta}}(\ell)\big[\Delta(\gamma(\ell)\Psi(\Delta^{\mu}x(\ell)))\big] \sum_{v=\ell_1}^{\ell-1} \frac{1}{\rho^{\frac{1}{\eta}}(v)} \\
&= E_1(\ell, \ell_1)\rho^{\frac{1}{\eta}}(\ell)\big[\Delta(\gamma(\ell)\Psi(\Delta^{\mu}x(\ell)))\big].
\end{aligned}$$

From (H_2) and Lemma (1)

$$\frac{1}{m}\gamma(\ell)\Delta^{\mu}x(\ell) \geq \gamma(\ell)\Psi(\Delta^{\mu}x(\ell)) \geq E_1(\ell, \ell_1)\rho^{\frac{1}{\eta}}(\ell)\big[\Delta(\gamma(\ell)\Psi(\Delta^{\mu}x(\ell)))\big],$$

which leads to

$$\Delta G(\ell) \geq \frac{m\Gamma(1-\mu)\rho^{\frac{1}{\eta}}(\ell)\big[\Delta(\gamma(\ell)\Psi(\Delta^{\mu}x(\ell)))\big]}{\gamma(\ell)} E_1(\ell, \ell_1). \tag{29}$$

Following the proof of Lemma (1), if $x(\ell)$ is eventually negative solution of (1) such that $\Delta(\gamma(\ell)\Psi(\Delta^{\mu}x(\ell))) < 0$, $\Delta^{\mu}x(\ell) < 0$ on $[\ell_1, \infty)$, where ℓ_1 is sufficiently large and $\ell_1 \geq \ell_0$ then

$$\Delta G(\ell) \leq \frac{m\Gamma(1-\mu)\rho^{\frac{1}{\eta}}(\ell)\big[\Delta(\gamma(\ell)\Psi(\Delta^{\mu}x(\ell)))\big]}{\gamma(\ell)} E_1(\ell, \ell_1).$$

Theorem 1 *Assume that (10)–(12) hold. If there exists a positive function $\phi(\ell)$, such that*

$$\lim_{\ell\to\infty} \sup \sum_{v=\ell_2}^{\ell-1}\left(B\phi(v)q(v) - \frac{[\Delta\phi(v)]^2}{4u(v)\phi(v)}\right) = \infty, \tag{30}$$

for all sufficiently large ℓ_2, where $E_1(\ell,\ell_1) = \sum_{v=\ell_1}^{\ell-1}\frac{1}{\rho^{\frac{1}{\eta}}(v)}$ and $u(\ell) = \left[\frac{m\Gamma(1-\mu)}{\gamma(\ell)}E_1(\ell,\ell_1)\right]^{\eta}$ then every solution of Eq. (1) is oscillatory or satisfies $\lim_{\ell\to\infty} G(\ell) = 0$.

Proof Suppose that Eq. (1) has a non-oscillatory solution $x(\ell)$ on $[\ell_0,\infty)$ without loss of generality assume that $x(\ell) > 0$ on $[\ell_1,\infty)$ where $\ell_1 > \ell_0$. By Lemma 4, $[\Delta(\gamma(\ell)\Psi(\Delta^{\mu}x(\ell)))] > 0$, $\ell \in [\ell_2,\infty)$ where $\ell_2 > \ell_1$ is sufficiently large and either $\Delta^{\mu}x(\ell) > 0$ on $[\ell_2,\infty)$ or $\lim_{\ell\to\infty} G(\ell) = 0$. If $\Delta^{\mu}x(\ell) > 0$ on $[\ell_2,\infty)$. Define the generalized Riccati function as

$$w(\ell) = \phi(\ell)\left\{\frac{-\rho(\ell)[\Delta(\gamma(\ell)\Psi(\Delta^{\mu}x(\ell)))]^{\eta}}{G^{\eta}(\ell)}\right\}, \ell \in [\ell_2,\infty). \tag{31}$$

Then it is clear that $w(\ell)$ is well defined. Hence by Lemma 2 on $\ell \in [\ell_2,\infty)$, we get

$$\begin{aligned}\Delta w(\ell) &= -\phi(\ell+1)\\ &\left[\frac{G^{\eta}(\ell)\Delta\rho(\ell)[\Delta(\gamma(\ell)\Psi(\Delta^{\mu}x(\ell)))]^{\eta} - \rho(\ell)[\Delta(\gamma(\ell)\Psi(\Delta^{\mu}x(\ell)))]^{\eta}\Delta G^{\eta}(\ell)}{G^{\eta}(\ell)G^{\eta}(\ell+1)}\right]\\ &- \Delta\phi(\ell)\frac{\rho(\ell)[\Delta(\gamma(\ell)\Psi(\Delta^{\mu}x(\ell)))]^{\eta}}{G^{\eta}(\ell)}.\end{aligned}$$

From (13), (31) and Lemma 4, we obtain

$$\begin{aligned}w(\ell) &= \frac{\phi(\ell+1)q(\ell)F(G(\ell))}{G^{\eta}(\ell+1)} + \phi(\ell+1)\frac{\Delta G^{\eta}(\ell)\rho(\ell)[\Delta(\gamma(\ell)\Psi(\Delta^{\mu}x(\ell)))]^{\eta}}{G^{\eta}(\ell)G^{\eta}(\ell+1)}\\ &+ \Delta\phi(\ell)\left[\frac{w(\ell)}{\phi(\ell)}\right] > \frac{\phi(\ell+1)q(\ell)F(G(\ell))}{G^{\eta}(\ell)}\\ &+ \phi(\ell+1)\frac{(\Delta G(\ell))^{\eta}\rho(\ell)[\Delta(\gamma(\ell)\Psi(\Delta^{\mu}x(\ell)))]^{\eta}}{[G^{\eta}(\ell)]^2} + \Delta\phi(\ell)\left[\frac{w(\ell)}{\phi(\ell)}\right]\\ &> B\phi(\ell+1)q(\ell) + \Delta\phi(\ell)\left[\frac{w(\ell)}{\phi(\ell)}\right] + \phi(\ell+1)\frac{\rho(\ell)[\Delta(\gamma(\ell)\Psi(\Delta^{\mu}x(\ell)))]^{\eta}}{G^{2\eta}(\ell)}\\ &\left[\frac{m\Gamma(1-\mu)E_1(\ell,\ell_1)\rho^{\frac{1}{\eta}}(\ell)[\Delta(\gamma(\ell)\Psi(\Delta^{\mu}x(\ell)))]}{\gamma(\ell)}\right]^{\eta}\end{aligned}$$

$$> B\phi(\ell+1)q(\ell)+\phi(\ell)u(\ell)\left[\frac{w(\ell)}{\phi(\ell)}\right]^2+\Delta\phi(\ell)\left[\frac{w(\ell)}{\phi(\ell)}\right]$$

i.e.,

$$\Delta w(\ell) \geq B\phi(\ell)q(\ell)+u(\ell)\frac{w^2(\ell)}{\phi(\ell)}+\Delta\phi(\ell)\left[\frac{w(\ell)}{\phi(\ell)}\right]. \tag{32}$$

From (31), we arrive at

$$\Delta w(\ell) > B\phi(\ell)q(\ell)+\left[w(\ell)\left\{\frac{u(\ell)}{\phi(\ell)}\right\}^{\frac{1}{2}}+\frac{\Delta\phi(\ell)}{[4\phi(\ell)u(\ell)]^{\frac{1}{2}}}\right]-\frac{[\Delta\phi(\ell)]^2}{4\phi(\ell)u(\ell)}$$

$$\Delta w(\ell) > B\phi(\ell)q(\ell)-\frac{[\Delta\phi(\ell)]^2}{4\phi(\ell)u(\ell)}.$$

Summing the above the inequality from ℓ_2 to $\ell-1$

$$\sum_{v=\ell_2}^{\ell-1}\Delta w(v) \geq \sum_{v=\ell_2}^{\ell-1} z(v)$$

$$\sum_{v=\ell_2}^{\ell-1} z(v) \leq w(\ell)-w(\ell_2) \leq w(\ell_2).$$

Letting $\ell \to \infty$, we get a contradiction to (30), where $u(\ell)=\left[\frac{m\Gamma(1-\mu)E_1(\ell,\ell_1)}{\gamma(\ell)}\right]^{\eta}$ and $z(v)=\left[B\phi(v)q(v)-\frac{(\Delta\phi(v))^2}{4\phi(v)u(v)}\right]$.

The proof is similar if $x(\ell)$ is eventually negative, which concludes the proof.

Notation Through the proofs of the theorems that follow, [24] we use $H(\ell,k)$: $\ell,k \in N, \ell \geq k \geq 0$ to denote the double sequence satisfying

$$\begin{aligned} H(\ell,\ell) &= 0 \text{ for } \ell \geq \ell_0; \quad H(\ell,v)>0 \text{ for } \ell>v\geq \ell_0; \\ \Delta_v H(\ell,v) &= H(\ell,v+1)-H(\ell,v)<0 \text{ for } \ell>v\geq\ell_0. \end{aligned}$$

Theorem 2 *Assume that* (10)–(12) *hold. If*

$$\limsup_{\ell\to\infty}\frac{1}{H(\ell,\ell_0)}\sum_{v=\ell_0}^{\ell-1}\left[B\phi(v)q(v)H(\ell,v)-\frac{h_+^2(\ell,v)\phi(v)}{4u(v)H(\ell,v)}\right]=\infty \tag{33}$$

where $h_+(\ell,v)=\Delta_v H(\ell,v)+\Delta\phi(v)\frac{H(\ell,v)}{\phi(v)}$ and $u(\ell)$ is same as in Theorem 1, then (1) is oscillatory or satisfies $\lim_{\ell\to\infty} G(\ell)=0$.

Proof Suppose on the contrary that $x(\ell)$ is a non-oscillatory solution of (1). Then $x(\ell)$ is either eventually positive or eventually negative. In the case when $x(\ell)$ is eventually positive, we assume that $x(\ell) > 0$ on $[\ell_1, \infty)$ where $\ell_1 \in N_{\ell_0}$ is sufficiently large. Then from the proof of Theorem 1 if $\Delta^{\mu} x(\ell) > 0$ on $[\ell_2, \infty)$, then (32) holds. Multiplying both sides of (32) by $H(\ell, v)$ and summing from ℓ_2 to $\ell - 1$ yields,

$$\sum_{v=\ell_2}^{\ell-1} BH(\ell, v)\phi(v)q(v) < \sum_{v=\ell_2}^{\ell-1} H(\ell, v)\Delta w(v) - \sum_{v=\ell_2}^{\ell-1} H(\ell, v)u(v)\frac{w^2(v)}{\phi(v)} - \sum_{v=\ell_2}^{\ell-1} H(\ell, v)\Delta\phi(v)\left[\frac{w(v)}{\phi(v)}\right] \tag{34}$$

Now applying summation by parts formula, we obtain

$$\sum_{v=\ell_2}^{\ell-1} \Delta w(v)H(\ell, v) = -w(\ell_2)H(\ell, \ell_2) - \sum_{v=\ell_2}^{\ell-1} \Delta_v H(\ell, v)w(v+1). \tag{35}$$

Now substitute (35) in (34)

$$\sum_{v=\ell_2}^{\ell-1} B\phi(v)q(v)H(\ell, v) < -w(\ell_2)H(\ell, \ell_2) - \sum_{v=\ell_2}^{\ell-1} \Delta_v H(\ell, v)w(v+1) - \sum_{v=\ell_2}^{\ell-1} u(v)\frac{H(\ell, v)}{\phi(v)}w^2(v) - \sum_{v=\ell_2}^{\ell-1} \Delta\phi(v)\left[\frac{H(\ell, v)}{\phi(v)}\right]w(v)$$

$$- \sum_{v=\ell_2}^{\ell-1} B\phi(v)q(v)H(\ell, v) > w(\ell_2)H(\ell, \ell_2) + \sum_{v=\ell_2}^{\ell-1}\left[w(v+1)h_+(\ell, v) + u(v)\frac{H(\ell, v)}{\phi(v)}w^2(v+1)\right]$$

$$-\sum_{v=\ell_2}^{\ell-1} B\phi(v)q(v)H(\ell, v) > w(\ell_2)H(\ell, \ell_2) - \sum_{v=\ell_2}^{\ell-1} \frac{h_+^2(\ell, v)\phi(v)}{4u(v)H(\ell, v)} + \sum_{v=\ell_2}^{\ell-1}\left[\left[\frac{u(v)H(\ell, v)}{\phi(v)}\right]^{\frac{1}{2}} w(v+1) + \frac{h_+(\ell, v)}{2\left[\frac{u(v)H(\ell,v)}{\phi(v)}\right]^{\frac{1}{2}}}\right]^2$$

$$\sum_{v=\ell_2}^{\ell-1} B\phi(v)q(v)H(\ell, v) < -w(\ell_2)H(\ell, \ell_2) + \sum_{v=\ell_2}^{\ell-1} \frac{h_+^2(\ell, v)\phi(v)}{4u(v)H(\ell, v)}$$

$$\sum_{v=\ell_2}^{\ell-1}\left[B\phi(v)q(v)H(\ell, v) - \frac{h_+^2(\ell, v)\phi(v)}{4u(v)H(\ell, v)}\right] < -w(\ell_2)H(\ell, \ell_2)$$

$$\sum_{v=\ell_2}^{\ell-1}\left[B\phi(v)q(v)H(\ell, v) - \frac{h_+^2(\ell, v)\phi(v)}{4u(v)H(\ell, v)}\right] < -w(\ell_2)H(\ell, \ell_0), \tag{36}$$

for $\ell > \ell_2 > \ell_1 > \ell_0$,

$$\sum_{v=\ell_0}^{\ell-1}\left[B\Psi(v)q(v)H(\ell,v)-\frac{h_+^2(\ell,v)\Psi(v)}{4u(v)H(\ell,v)}\right]=\sum_{v=\ell_0}^{\ell_2-1}\left[B\phi(v)q(v)H(\ell,v)-\frac{h_+^2(\ell,v)\phi(v)}{4u(v)H(\ell,v)}\right]$$
$$+\sum_{v=\ell_2}^{\ell-1}\left[B\phi(v)q(v)H(\ell,v)-\frac{h_+^2(\ell,v)\phi(v)}{4u(v)H(\ell,v)}\right]$$

From (36)

$$\sum_{v=\ell_0}^{\ell-1}\left[B\phi(v)q(v)H(\ell,v)-\frac{h_+^2(\ell,v)\phi(v)}{4u(v)H(\ell,v)}\right]<\sum_{v=\ell_0}^{\ell_2-1}\left[B\phi(v)q(v)H(\ell,v)-\frac{h_+^2(\ell,v)\phi(v)}{4u(v)H(\ell,v)}\right]$$
$$-H(\ell,\ell_0)w(\ell_2)<H(\ell,\ell_0)\sum_{v=\ell_0}^{\ell_2-1}B\phi(v)q(v)$$
$$-H(\ell,\ell_0)w(\ell_2),$$

which means

$$\frac{1}{H(\ell,\ell_0)}\sum_{v=\ell_0}^{\ell-1}\left[B\phi(v)q(v)H(\ell,v)-\frac{h_+^2(\ell,v)\phi(v)}{4u(v)H(\ell,v)}\right]<\sum_{v=\ell_0}^{\ell_2-1}B\phi(v)q(v)-w(\ell_2).$$

Taking limit sup as $\ell \to \infty$, we get,

$$\limsup_{\ell\to\infty}\frac{1}{H(\ell,\ell_0)}\sum_{v=\ell_0}^{\ell-1}\left[B\phi(v)q(v)H(\ell,v)-\frac{h_+^2(\ell,v)\phi(v)}{4u(v)H(\ell,v)}\right]$$
$$<\sum_{v=\ell_0}^{\ell-1}B\phi(v)q(v)-w(\ell_2)<\infty,$$

which is a contradiction to (33).

In the case when $x(\ell)$ is eventually negative, it can be proved in the similar manner. Hence the proof is complete.

5 Application

In this section, we present an example to validate the theoretical results. Consider the following equation

$$\Delta\left\{\frac{1}{\ell^3}\left[\Delta\{\ell^2\Delta^\mu x(\ell)\}\right]^5\right\}+\frac{1}{\ell}\left(\sum_{v=1}^{\ell-1+\mu}(\ell-v-1)^{(-\mu)}x(v)\right)^{-3}=0,\ \ell\geq 1, \quad (37)$$

where $\mu \in (0,1]$. Comparing (37) with Eq. (1), we have $\eta = 5$, $\ell_0 = 1$, $\rho(\ell) = \ell^{-3}$, $\gamma(\ell) = \ell^2$, $q(\ell) = \ell^{-1}$, $F(x) = x^{-3}$, $\Psi(x) = x$, $\frac{F(x)}{x^\eta} = \frac{1}{x^8} > \delta = B > 0$, $\frac{x}{\Psi(x)} \geq m = 1$, $\phi(\ell) = \ell^2$ where $\delta \in (0,\infty)$. Hence we observe that assumptions $(H_1) - (H_3)$ hold and also

$$\sum_{v=\ell_0}^{\infty}\frac{1}{\rho^{\frac{1}{\eta}}(v)}=\sum_{v=1}^{\infty}\frac{1}{\left[v^{-3}\right]^{\frac{1}{5}}}=\sum_{v=1}^{\infty}v^{\frac{3}{5}}=\infty,$$

$$\sum_{v=\ell_0}^{\infty}\Psi^{-1}\left(\frac{1}{\gamma(v)}\right)=\sum_{v=1}^{\infty}\frac{1}{\gamma(v)}=\sum_{v=1}^{\infty}\frac{1}{v^2}=\infty,$$

$$\sum_{\xi=t_0}^{\infty}\Psi^{-1}\left[\frac{1}{\gamma(\xi)}\sum_{\tau=\xi}^{\infty}\left[\frac{1}{\rho(\tau)}\sum_{v=\tau}^{\infty}q(v)\right]^{\frac{1}{\eta}}\right]=\sum_{\xi=1}^{\infty}\frac{1}{\xi^2}\sum_{\tau=\xi}^{\infty}\left[\tau^3\sum_{v=\tau}^{\infty}\frac{1}{v}\right]^{\frac{1}{5}}=\infty,$$

which implies that (10), (11), (12) hold. For sufficiently large ℓ_1,

$$E_1(\ell,\ell_1)=\sum_{v=\ell_1}^{\ell-1}\frac{1}{[\rho(v)]^{\frac{1}{\eta}}}=\sum_{v=\ell_1}^{\ell-1}v^{\frac{3}{5}},$$

and

$$u(\ell)=\left[\frac{m\Gamma(1-\mu)}{\gamma(\ell)}E_1(\ell,\ell_1)\right]^\eta=\left[\frac{m\Gamma(1-\mu)}{\ell^2}\sum_{v=\ell_1}^{\ell-1}v^{\frac{3}{5}}\right]^5,$$

also for $\ell_2 > \ell_1$,

$$\sum_{v=\ell_2}^{\ell-1}\left[B\phi(v)q(v)-\frac{[\Delta\phi(v)]^2}{4u(v)\phi(v)}\right]=\sum_{v=\ell_2}^{\ell-1}\left[\delta v^2\frac{1}{v}-\frac{(2v+1)^2}{4v^2\left[\frac{m\Gamma(1-\mu)}{v^2}\sum_{\xi=\ell_1}^{v-1}\xi^{\frac{3}{5}}\right]^5}\right]$$

$$>\sum_{v=\ell_2}^{\ell-1}\left[\delta v-\frac{(2v+1)^2v^8}{4\ell_1^3[m\Gamma(1-\mu)]^5}\right].$$

Therefore,

$$\lim_{\ell\to\infty}\sup\sum_{v=\ell_2}^{\ell-1}\left[B\phi(v)q(v)-\frac{[\Delta\phi(v)]^2}{4u(v)\phi(v)}\right]\geq\lim_{\ell\to\infty}\sup\sum_{v=\ell_2}^{\ell-1}\left[\delta v-\frac{(2v+1)^2v^8}{4\ell_1^3[m\Gamma(1-\mu)]^5}\right]=\infty,$$

which indicates that (30) is satisfied. Hence in accordance with Theorem 1, we conclude that every solution of (37) is oscillatory or the condition $\lim_{\ell\to\infty} G(\ell)=0$ is satisfied.

References

1. H. Adiguzel, Oscillatory Behavior of solutions of certain fractional difference equations. Adv. Differ. Equ. **2018**(445), 1–13 (2018)
2. F.M. Atici, P.W. Eloe, A transform method in discrete fractional calculus. Int. J. Differ. Equ.**2**(2), 165–176 (2007). https://ecommons.udayton.edu/mth_fac_pub/110
3. F.M. Atici, P.W. Eloe, Initial value problems in discrete fractional calculus. Proc. Am. Math. Soc. **137**(3), 981–989 (2008). https://doi.org/10.1090/S0002-9939-08-09626-3
4. F.M. Atici, P.W. Eloe, Linear systems of fractional nabla difference equations. Rocky Mt. J. Math. **41**, 353–370 (2011)
5. Z. Bai, R. Xu, The asymptotic behavior of solutions for a class of nonlinear fractional difference equations with damping term. Discret. Dyn. Nat. Soc. **2018**, 1–11 (2018)
6. G.E. Chatzarakis, P. Gokulraj, T. Kalaimani, Oscillation test for fractional difference equations. Tatra Mt. Math. Publ.**71**(1), 53–84 (2018). https://doi.org/10.2478/tmmp-2018-0005
7. J.F. Cheng, Y.M. Chu, On the fractional difference equations of order (2, q). Abstr. Appl. Anal. **2011**, 497259 (2011)
8. J.F. Cheng, Y.M. Chu, Fractional difference equations with real variable. Abstr. Appl. Anal. **2012**, 918529 (2012)
9. J.B. Diaz, T.J. Osler, Differences of fractional order. Math. Comp. **28**(125), 185–202 (1974). https://doi.org/10.2307/2005825
10. S. Elaydi, An Introduction to Difference Equations, 3-e. Springer International Edition (1996). https://doi.org/10.1007/978-1-4757-9168-6
11. L. Erbe, C.S. Goodrich, B. Jia, A. Peterson, Survey of the qualitative properties off rational difference operators: monotonicity, convexity, and asymptotic behavior of solutions. Adv. Differ. Equ. **2016**, 43 (2016)
12. A. George Maria Selvam, R. Janagaraj, Oscillation theorems for damped fractional order difference equations. AIP Conf. Proc.**2095**(030007), 1–7 (2019). https://doi.org/10.1063/1.5097518
13. A. George Maria Selvam, R. Janagaraj, Oscillation criteria of a class of fractional order damped difference equations. Int. J. Appl. Math.**32**(3), 433–441 (2019). https://doi.org/10.12732/ijam.v32i3.5
14. C.S. Goodrich, On a discrete fractional three-point boundary value problem. J. Differ. Equ. Appl. **18**, 397–415 (2012)
15. C.S. Goodrich, A.C. Peterson, *Discrete Fractional Calculus* (Springer International Publishing, Switzerland, 2015).https://doi.org/10.1007/978-3-319-25562-0
16. A.A. Kilbas, H.M. Srivastava, J.J. Trujillo, Theory and Applications of Fractional Differential Equations, 1-e, North-Holland Mathematics Studies, 204 (2006)
17. S. Kisalar, M.K. Yildiz, E. Aktoprak, Oscillation of higher order fractional nonlinear difference equations. Int. J. Differ. Equ **10**(2), 201–212 (2015)

18. B. Kuttner, On differences of fractional order. Proc. Lond. Math. Soc. **3–7**(1), 453–466 (1957). https://doi.org/10.1112/plms/s3-7.1.453
19. W.N. Li, Oscillation results for certain forced fractional difference equations with damping term. Adv. Differ. Equ.**2016**(1) (2016), article no. 70, 1–9
20. K.S. Miller, B. Ross, *An Introduction to the Fractional Calculus and Fractional Differential Equations* (Wiley, New York, USA, 1993).
21. V. Muthulakshmi, S. Pavithra, Oscillatory behavior of fractional differential equations with damping. Int. J. Math. Its Appl. **5**(4-c), 383–388 (2017)
22. I. Podlubny, *Fractional Differential Equations*, Mathematics in Science and Engineering, vol. 198. (Academic Press, San Diego, Calif, USA, 1999)
23. C. Qi, J. Cheng, Interval oscillation criteria for a class of fractional differential equations with damping term . Math. Probl. Eng. **301085**, 1–8 (2013)
24. S. Saker, *Oscillation Theory of Delay Differential and Difference Equations* (VDm Verlag, Second and Third Orders, 2010).
25. A. Secer, H. Adiguzel, Oscillation of solutions for a class of nonlinear fractional difference equations. J. Nonlinear Sci. Appl. **9**, 5862–5869 (2016)
26. J. Yang, A. Liu, T. Liu, Forced oscillation of nonlinear fractional difference equations with damping term. Adv. Differ. Equ. **2015**(1), 1–7 (2015). https://doi.org/10.1186/s13662-014-0331-4

Numerical Treatment on the Analysis of Heat Transfer of a Magneto-micropolar Fluid over a Continuously Moving Surface with Heat Source/Sink

R. S. Tripathy and B. Nayak

Abstract We analyze the heat transfer effect of an electrically conducting micropolar fluid on a boundary layer flow past, a moving plate depending on a transverse magnetic field in the incidence of uniform heat source/sink. The thermal buoyancy effect is also discussed in this study. The involved partial differential equations are altered into ordinary nonlinear differential equations using a similarity variable. Fourth-fifth ranked Runge–Kutta method associated by means of shooting technique is used to obtain a numerical solution. Graphical representation of the solutions is aimed at dimensionless velocity and temperature profiles, whereas numerical characteristics of the Nusselt number and skin friction are obtainable in the form of a table by considering numerous values of pertinent parameters affecting the movement and heat transfer phenomena.

Keywords Buoyancy induced flow · Magnetofluid · Micropolar · Uniform heat source/sink · Runge–Kutta · Shooting technique

Nomenclature

u	x-components vector
υ	Kinematic viscosity
v	y-components vector
S	Heat generation/absorption parameter
T	Temperature of the fluid
T_∞	Plate temperature at the boundary
T_w	Plate temperature at the result boundary
Gr	Thermal Grashof number
R	Radiation parameter
Pr	Prandtl number

R. S. Tripathy (✉) · B. Nayak
Department of Mathematics, Siksha 'O' Anusandhan Deemed to be University, Khandagiri, Bhubaneswar 751030, Odisha, India
e-mail: rstspr@gmail.com

S. R. Mishra et al. (eds.), *Recent Trends in Applied Mathematics*, Lecture Notes in Mechanical Engineering, https://doi.org/10.1007/978-981-15-9817-3_25

b Rate constant
α Thermal diffusivity
Ec Eckert number,
f_w Suction parameer,
μ Dynamic viscosity,
ρ Fluid density
g Gravitational acceleration
σ Fluid electrical conductivity
B_0 Magnetic induction
c_p Specific heat at constant pressure
Q Dimensional heat source coefficient
K_1 Coupling parameter

1 Review and Introduction

In the current years, the concept of micropolar fluid is of extensive consideration to the investigators as the old-style Newtonian fluids are not capable to exactly describe the property of fluid consisting of adjourned particles. These micro fluids are the specific form of the non-Newtonian fluid comprising of stiff and small fundamentals of cylinder shape called dumb-bell molecules. For example, polymers, animal body fluid, and fluids deferments. These fluids are used to model the existence of smoke or dust, mainly in a gas. Moreover, Micropolar fluids are of substantial consideration to study the thermal boundary layer movements through affecting boundaries [1]. Boundary layer regulator and thermal shield in a flow produced by high energy through wall velocity and mass transfer are some practical examples of such phenomena. Micropolar fluid possesses a microstructure and is an element of a specific type of fluids that have a nonsymmetrical stress tensor. The involved particles possess random orientation adjourned in a sticky medium and classified as Micropolar. The convection produced by revolving a cone in micropolar fluids with a random change in surface temperature is measured by Gorla and Nakamura [2]. Takhar et al. [3] observed the springy effects by considering an incompressible micropolar fluid forced flow of three-dimensional nonsteady motion in the neighborhood of the onward sluggishness point of a rounded nosed body.

The analysis of convective transference in a porous medium is of notable concern in the literature because of its massive usages in thermal engineering associated with insulation, packed bed catalytic reactors, and geothermal pools. When the fluid transfers through twisting paths in porous media, then it produces recirculation and mixing of already present fluid streams. This hydrodynamic mingling of fluid on the pore level becomes a source of thermal, as well as solutal diffusion in an absorbent medium. The model of varying mixed convection movement with constant surface suction or injection in an isothermal vertical plate in porous media is discussed by Hopper et al. [4]. In view of their model, the complete picture of the domain of free-forced-mixed convection is obtainable by using one parameter. The outcomes of three-dimensional electroosmotic movements inside the charged micro and nanoscale arbitrary porous media of homogeneous nature are modeled and analyzed numerically by Wang and Chen [5]. The electroosmotic flows over electric anisotropic porous media in view of the lattice Poisson–Boltzmann technique is studied by Wang et al. [6]. These types of problems can be investigated more accurately with the manifestation of magnetic field and thermal energy for micropolar fluid. The movement field in diverse convection boundary layer through an upright surface fixed in a porous medium is strongly inclined with Soret and Dufour effects. These effects are measured with the commencement of convection in an upright porous layer in view of unvarying heat flux by Bourich et al. [7]. They computed the solutions by using both the analytical and numerical approaches. Mishra et al. [8] investigated unrestricted convective movement in the viscoelastic fluid using an upright channel considering the occurrence of the Dufour effect. Postelnicu [9] have investigated the results by considering the flow in unrestricted convection outer sheet of an upright surface fixed in a porous medium. They also assumed the existence of Soret and Dufour.

The usual convection movement of micropolar fluid around a sphere by means of puffing and pressure is investigated by Lien et al. [10]. Mohanty and coauthors [11] considered the mathematical exploration for temperature and mass transmission outcome of micropolar fluid beyond an extending sheet. Lien and his collaborators [12] surveyed the unrestricted convection movement of micropolar fluid in view of a flat holey cylinder with a varying thermal state. Takhar and coauthors [13] calculated the diverse convection movement of a micropolar fluid through an enlarging layer. Mishra and his collaborators [14] observed the stream of temperature and mass transmission in a micropolar fluid due to heat medium. Dash and his collaborators [15] discussed the numerical method to boundary layer delay-point stream passing through a reduction sheet. Further, Tripathy and coauthors [16] studied hydromagnetic micropolar fluid beside an extending sheet using chemical response and nonuniform heat medium. Cheng [17] observed the unrestricted convection of mass and heat transmission adjacent to elliptical cylinders inside the micropolar fluid. Lately, some researchers have considered several physical characteristics of the combination of mass and heat transmission problem. Tripathy and collaborators [18] considered the impact of the chemical response for unrestricted MHD convective boundary beyond a moving upright plane inside the porous medium. Muthukumaraswamy and coauthors [19] have considered perpendicular plate by heat and mass transmission and observed the short-term free-convection movement. The results of mutual buoyancy forces in the presence of natural convection movement from a vertical wavy surface are investigated by Hossain and Rees [20]. They got results by using numerical calculations and illustrated the development of the shear stress, amount of heat transfer, and absorption gradient for the involved surface. Chamkha and Khaled [21] explored the formulation of joined mass and heat transmission in view of unrestricted magnetohydrodynamic convection using an inclined plate by considering the existence of internal heat engagement. Furthermore, Cortell [22], Bataller [23], Ishak [24], and

Aliakbar et al. [25] have done further analysis to investigate the consequences of thermal radiation.

Lately, Seddeek et al. [26] explored an exact solution involving the significance of energy for a magneto-micropolar fluid passing through a moving vertical plate by means of blowing and suction. In their study, they have not discussed the role of buoyancy on the magneto-micropolar fluid. Hence our aim is to investigate the significance of thermal buoyancy present in the momentum equation and the result of unchanging heat cause in the energy equation, respectively. The nonlinear nonhomogenous ordinary differential equalities are converted to the set of first order differential equations. We solve these equations by using the efficient Ruuge–Kutta fourth-fifth order scheme tracked by shooting procedure. The current results are compared with the previously established results of Seddeek et al. [26] as a specific situation by withdrawing the thermal buoyancy and heat cause parameter. For further related studies, we refer the interested reader to [27–33].

2 Formulation of the Problem

Consider a stable, laminar edge layer movement and heat transmission of viscous incompressible and electrically conducting fluid through a continuously moving plate. As usual, the horizontal path is directed along the x-axis and the vertical path along the y-axis. It is presumed that a creek of cold fluid is at temperature T_∞. An unchanging transverse magnetic field with strength B_0 is operational along y-axis perpendicular to the main direction of the flow. We neglect the Hall effect and Joule heating in the nonappearance of the applied electric field. The converted magnetic field and the electric field are presumed to be negligible because of the polarization of charges. Under Boussinesq estimate, the key outer layer mathematical expressions for this formulation are given as follows [26] (Fig. 1):

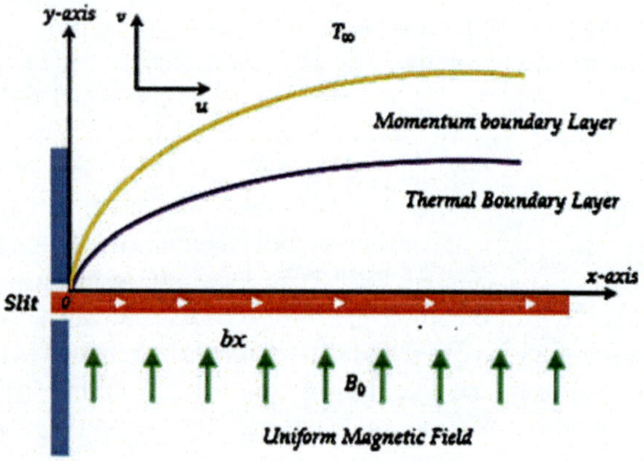

Fig. 1 Flow geometry

$$\frac{\partial u}{\partial x} + \frac{\partial v}{\partial y} = 0 \tag{1}$$

$$\left(u\frac{\partial u}{\partial x} + v\frac{\partial u}{\partial y}\right)\rho = \frac{\partial^2 u}{\partial y^2}\mu + k_1\rho\frac{\partial N}{\partial y} - \sigma B_0^2 u + g\rho\beta(T - T_\infty) \tag{2}$$

$$G_1\frac{\partial^2 N}{\partial y^2} - 2N - \frac{\partial u}{\partial y} = 0 \tag{3}$$

$$\rho c_p\left(u\frac{\partial T}{\partial x} + v\frac{\partial T}{\partial y}\right) = \kappa\frac{\partial^2 T}{\partial y^2} + \frac{\partial q_r}{\partial y} + \mu\left(\frac{\partial u}{\partial y}\right)^2 + Q(T - T_\infty) \tag{4}$$

The settings for the outer layer (boundary) are

$$\begin{aligned} &u = bx \;\; v = v_w \quad N = 0 \quad T = T_w \quad \text{at y} = 0 \\ &u \to 0 \;\; N \to 0 \;\; T \to T_\infty \text{ as } y \to \infty \end{aligned} \tag{5}$$

where $k_1 = a/\rho$, the coupling constant ($k_1 > 0$), a, the constant characteristics of the fluid, G_1, the microrotation constant.

The radiative heat fluctuation is denoted by q_r

$$q_r = \frac{-4\sigma^*}{3k^*}\frac{\partial T^4}{\partial y}, \tag{6}$$

and measured in view of Rosseland diffusion approximation (Hossain and Rees [20] and Raptis [27]), where

$k^* \to$ Rosseland mean absorption coefficient; $\sigma^* \to$ Stefan – Boltzman constant.

By considering the appropriately slight changes in the temperature of inside flow wherever the terms involving the higher powers of T can be neglected and the factor T^4 is stated as the function of temperature

$$T^4 \cong 4T_\infty^3 T - 3T_\infty^4. \tag{7}$$

Introducing flow function, $\psi(x, y)$ the statement numbered as (1) is satisfied by using the following functions

$$u = \frac{\partial \psi}{\partial y},\; v = -\frac{\partial \psi}{\partial x} \tag{8}$$

With the purpose of conversion of the Eqs. (2–5) into a system of equations involving the ordinary derivative, the ensuing similarity conversions and nondimensional variables are declared here as follows:

$$\left.\begin{aligned} &\eta = \sqrt{\frac{b}{\nu}}y, \quad \psi = \sqrt{b\nu x} f(\eta), \\ &N = \left(\frac{b^3}{\nu}\right)^{\frac{1}{2}} x g(\eta),\ \theta(\eta) = \frac{T - T_\infty}{T_w - T_\infty}, \end{aligned}\right\} \tag{9}$$

Substituting the value of Eq. (9) in Eqs. (2–4), and the boundary condition (5), we get

$$f''' + ff'' - (f')^2 - M f' + K g' + Gr\,\theta = 0 \tag{10}$$

$$G g'' - f'' - 2g = 0 \tag{11}$$

$$\left(\frac{1+R}{R\,\Pr}\right)\theta'' + f\theta' - \gamma f'\theta + Ecf''^2 + S\theta = 0 \tag{12}$$

$$\left.\begin{aligned} &f(0) = f_w, \quad f'(0) = 1, \quad g(0) = 0, \quad \theta(0) = 1 \\ &f'(\infty) = 0, \quad g(\infty) = 0, \quad \theta(\infty) = 0 \end{aligned}\right\} \tag{13}$$

Here, $M = \frac{\sigma B_0^2}{b\rho}$, the magnetic parameter, $\Pr = \frac{\rho c_p \nu}{\kappa}$, the Prandtl number, $R = \frac{3\kappa k^*}{16\sigma^* T_\infty^3}$, the radiation parameter, $K = \frac{k_1}{\nu}$, the coupling constant parameter, $Ec = \frac{b^2 x^2}{c_p (T_w - T_\infty)}$, the Eckert number, and $G = \frac{G_1 b}{\nu}$, the microrotation parameter.

3 Physical Quantities of Interest

The wall shear stress, τ_w, the skin friction coefficient c_f, Reynolds number, Re, heat flux, q_w at the wall, Nusselt number, Nu are defined as

$$\tau_w = -\mu_w \left(\frac{\partial u}{\partial y}\right)_{y=0} = -\mu_w u_w \sqrt{\frac{u_w}{\nu x}} f''(0) \tag{14}$$

$$c_f = \frac{\tau_w}{\rho u_w^2} = -(\mathrm{Re})^{-\frac{1}{2}} f''(0) \tag{15}$$

where, the Reynolds number, $\mathrm{Re} = \frac{u_w x}{\upsilon}$.

$$q_w = -\kappa \left(\frac{\partial T}{\partial y}\right)_{y=0} = -\kappa (T_w - T_\infty) \sqrt{\frac{u_w}{\nu x}} \theta'(0) \tag{16}$$

$$Nu = \frac{x q_w}{\kappa (T_w - T_\infty)} = -(\mathrm{Re})^{\frac{1}{2}} \theta'(0) \tag{17}$$

4 Numerical Method

The solution of leading boundary layer Eqs. (10), (11), and (12) with respect to boundary settings (13) are determined by using the numerical shooting method. In the first step, we convert the differential Eqs. (10)–(12) into linear differential equations of the first order. In the second step, we transform these equations in the form of initial value problem by using the shooting numerical method. The temperature of the plate surface, coefficient of the local skin friction, and the local Nusselt numeral are investigated and their numerical values are listed in the form of a table by using the involved numerical computational method. Newton shooting approach associated with Runge–Kutta fourth-order methods are described as follows:

$$f = y_1, f' = y_2, f'' = y_3, g = y_4, g' = y_5, \theta = y_6, \theta' = y_7$$

$$f''' = -\left(ff'' - (f')^2 - M\,f' + K\,g' + Gr\,\theta\right)$$

$$\Rightarrow y_3' = -y_1 y_3 + y_2^2 + M y_2 - K y_5 - Gr y_6$$

$$g'' = \frac{1}{G}\left(f'' + 2g\right)$$

$$\Rightarrow y_5' = \frac{1}{G}(y_3 + 2y_4)$$

$$\theta'' = \frac{R\,\mathrm{Pr}}{1+R}\left(-f\theta' + \gamma f'\theta - Ecf''^2 - S\theta\right)$$

$$\Rightarrow y_7' = \frac{R\,\mathrm{Pr}}{1+R} - y_1 y_7 + \gamma y_2 y_6 - Ecy_3^2 - Sy_6$$

$$y_1(0) = f_w, y_2(0) = 1, y_3(0) = s_1, y_4(0) = 0, y_5(0) = s_2, y_6(0) = 1, y_7(0) = s_3$$

where $s_1, s_2,$ *and* s_3 are initial guesses. The inner iteration is then counted until the nonlinear solution converges with a convergence criterion of 10^{-6} in all cases.

5 Consequences and Discussion

Considered the natural convection movement for a micropolar fluid passing through a porous plate with unvarying heat source/sink. We solved the problem numerically in the presence of electrical conduction. Also, the heat and mass transmission impact on an edge layer radiative MHD flow subject to a transverse magnetic field has been analyzed. Computations are performed for an extensive range of physical parameters of the problem and discussed through Figs. 2, 3, 4, 5, 6, 7, 8, 9, 10 and 11 and the numerical results are presented through Tables 1 and 2 to illustrate the consequences of various controlling parameters for the shear stress, couple stress, and proportion of heat transmission. The high value Prandtl number (Pr = 10) is considered for

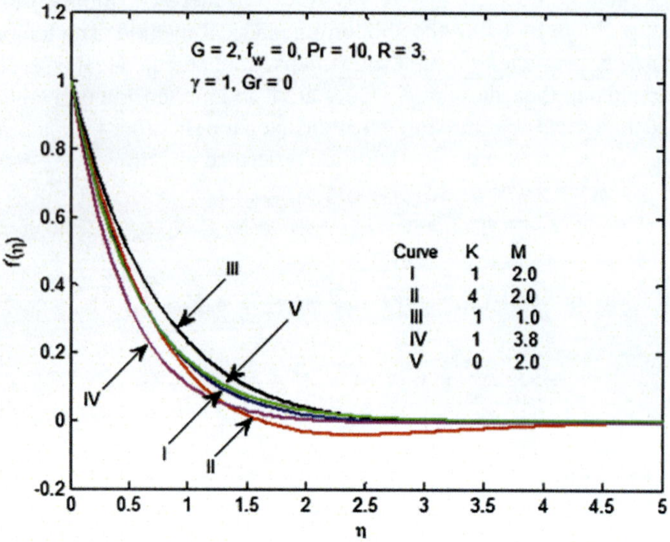

Fig. 2 Comparison plot of velocity for the influence of K, M and $Gr = 0$

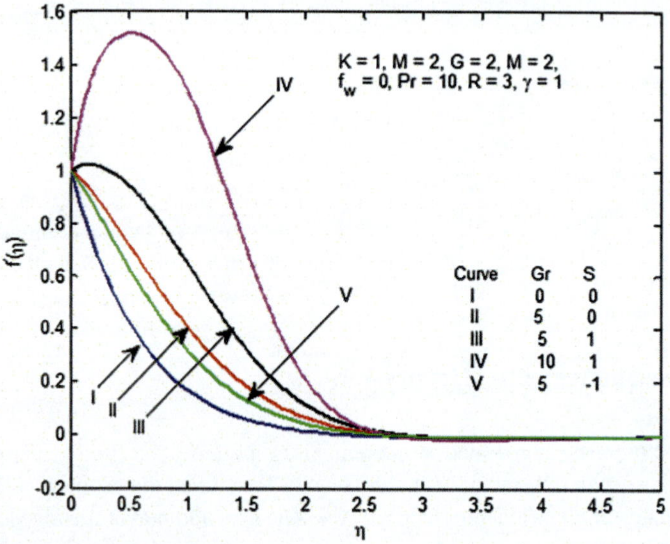

Fig. 3 Influence of Gr and S on velocity profiles

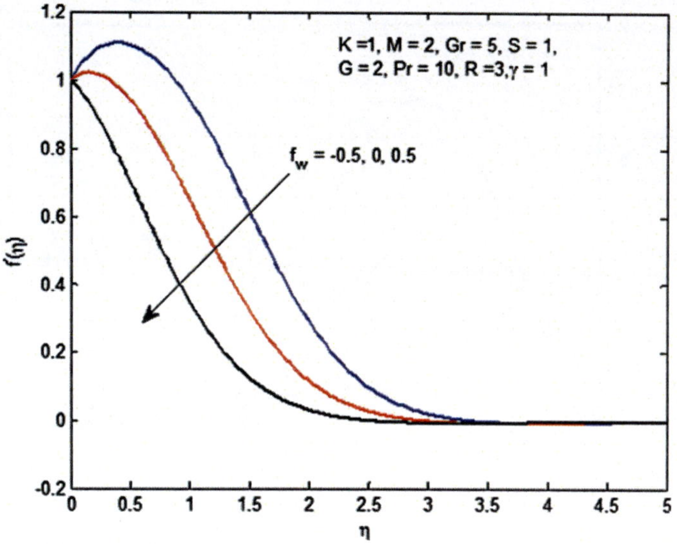

Fig. 4 Influence of f_w on the velocity profiles

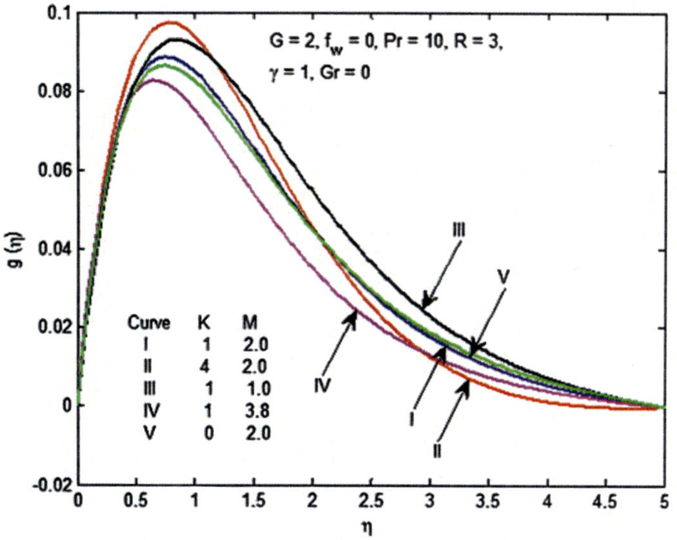

Fig. 5 Comparison plot of microrotation for the influence of K, M and $Gr = 0$

water at low temperature. Consideration is concentrated on the positive value of the buoyancy parameter, i.e., thermal Grashof number, $Gr > 0$. From Eq. (12) it is clear that $S > 0$ indicates the heat source and $S < 0$ indicates the contribution of heat sink in the heat transfer phenomena. Validation with the earlier published work is given

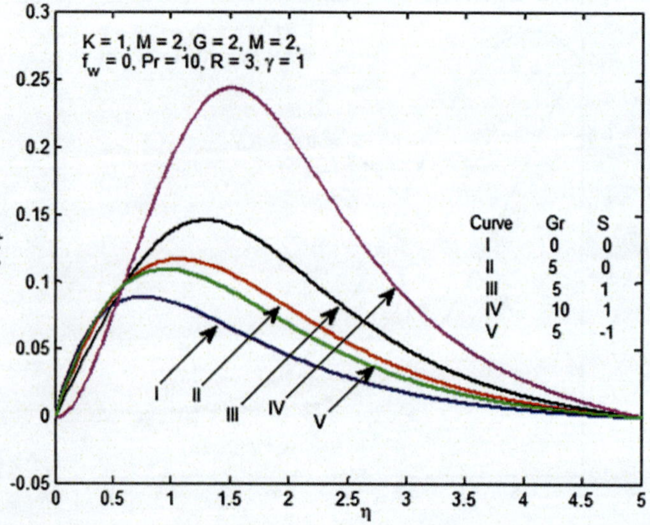

Fig. 6 Influence of Gr and S on microrotation velocity profiles

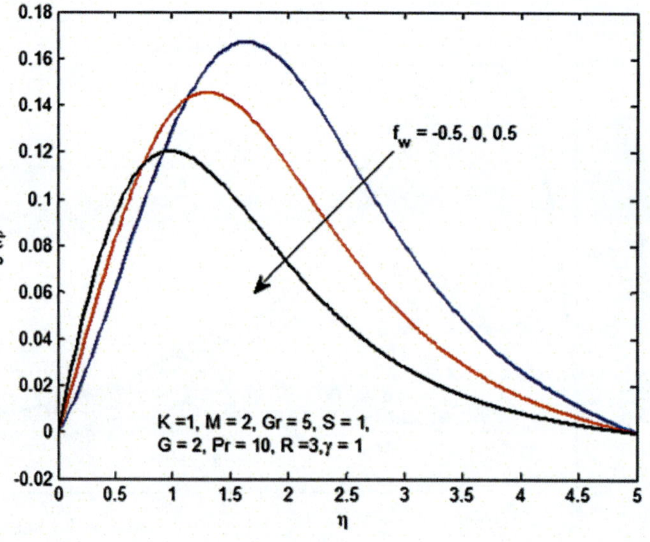

Fig. 7 Influence of f_w on the microrotation profiles

by discussing the particular cases $Gr = 0, S = 0$ and other parameters are taken to be fixed.

Figure 2 Shows the variation of K and M for the velocity function in the nonappearance of thermal buoyancy parameter Gr ($Gr = 0$) and suction parameter $f_w = 0$. For $K = 1$, $M = 2$ (Curve-I), the current outcomes are consistent with the results of

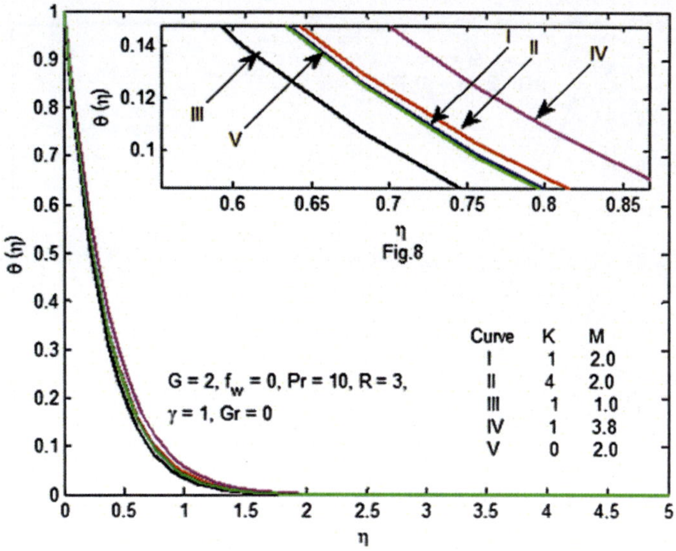

Fig. 8 Influence of K and M on the temperature profiles

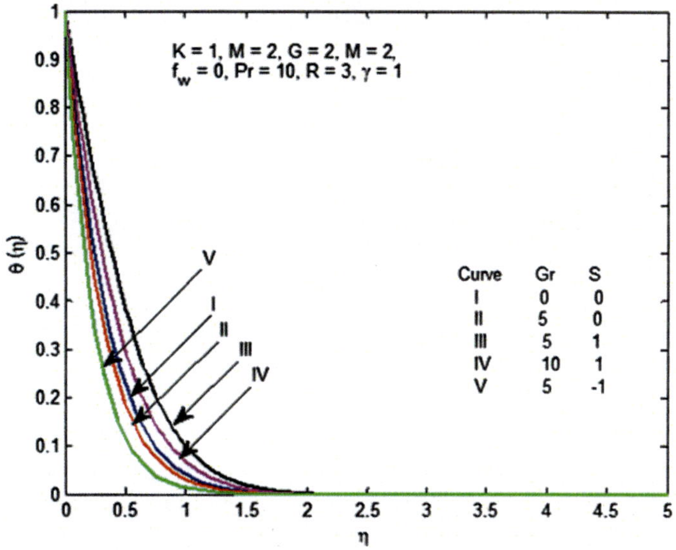

Fig. 9 Influence of Gr and S on the temperature profiles

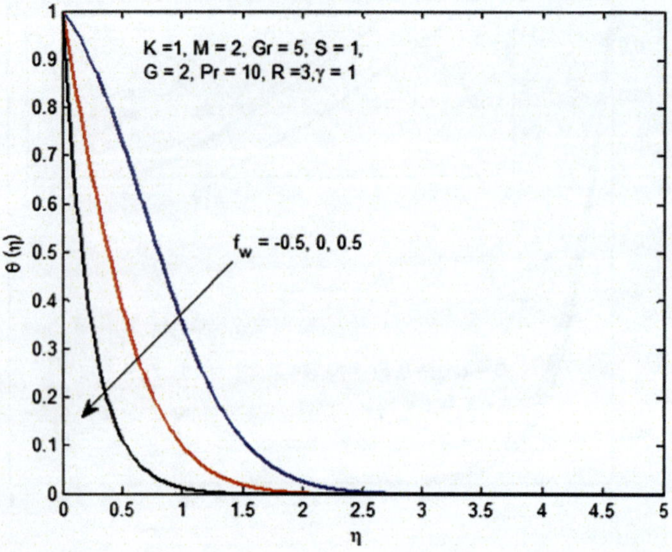

Fig. 10 Influence of f_w on the temperature profiles

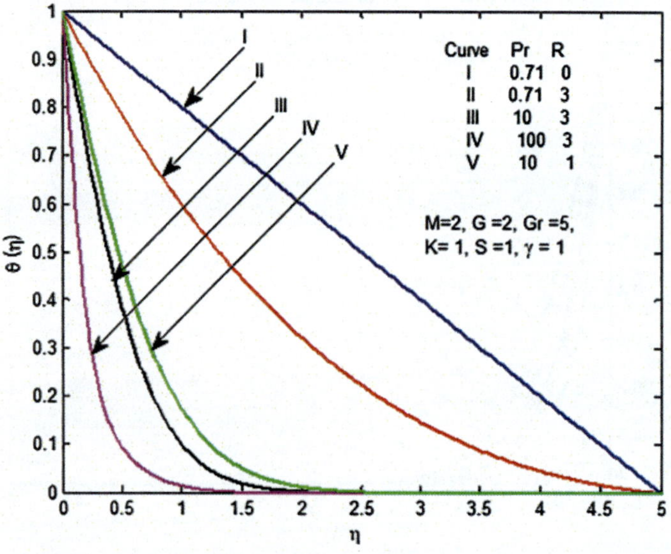

Fig. 11 Influence of Pr and S on the temperature profiles

[26]. It is observed that the variation is insignificant for the increasing value of K up to a certain region, i.e., $\eta < 1$. Soon after this, the velocity profile reduces asymptotically to meet the boundary conditions (Curve-I and II). Further, the velocity profile slows down as the value of the magnetic parameter increases. It is because a resistive force namely the Lorentz force gives rise to an electromagnetic origin, in the form of magnetic parameter, cause retardation in the velocity distribution. However, maximum retardation is found in the nonappearance of K that means $K = 0$.

Figures 3 and 6 exhibit the deduction of thermal buoyancy parameter in the absence/presence of heat source/sink for the velocity and microrotation profiles with the fixed values of other physical parameters depicting the flow phenomena. Note that a rise in buoyancy enhances the velocity and microrotation boundary layer significantly. Further, high value of Gr ($Gr = 10$), enhancement is maximum near the boundary layer and then decreases smoothly. This is because of the association of magnetic parameters. Another interesting result is observed in the incidence of heat source/sink. By increasing the heat source, both profiles increase but the opposite effect is encountered in case of sink, i.e., sink gives rise to a retardation in the profiles on every point in the velocity outer layer.

Figures 4 and 10 illustrate the corporeal consequence of the suction/injection for velocity and temperature profiles. In comparison, together the injection ($f_w < 0$)/suction ($f_w > 0$) are related to the resistant situation ($f_w = 0$). One can observe that suction retards together the velocity and temperature profiles, whereas injection enhances the profiles meaningfully.

The effects of K and M on the microrotation profile in the nonappearance of thermal buoyancy and suction are obvious in Fig. 5. The current findings are consistent with the earlier results of [26] for $K = 1, M = 2$ (Curve-I). The variation is insignificant near the boundary layer, i.e., $\eta < 0.5$. It is seen that two-layer characteristics are exhibited for the variation of K in the microrotation profiles. Also, the effect of K is reverse as compared to Fig. 2 for velocity profile, i.e., with the growth in K, the microrotation profile rises within the middle of the region and then it reduces. Further, the microrotation profile reduces with the intensification in the magnetic parameter as revealed in the velocity profile, Fig. 2.

Figure 7 shows the variation of suction/injection for the microrotation profile having static values of the related parameters. Further, it is exciting to notice that near the boundary layer, i.e., for $\eta < 1$ microrotation profile develops with the increasing value of suction, whereas injection decelerates it, but the reverse effect is reported for $\eta > 1$. As compared to that of the impermeable region, $f_w = 0$ suction retards the profile, whereas injection is favorable to enhance the microrotation profile significantly.

Figure 8 demonstrates the result of K and M for the temperature profile. The enhancement in the fluid temperature intensifies the thermal boundary layer as both K and M grow further. The non-Newtonian parameter $K > 0$ in the manifestation of magnetic parameter improves the temperature for every point in the corresponding outer sheet. Variation of the temperature profile is insignificant due to high Prandtl fluid. Thus, it is to conclude that interaction of high Prandtl number with magnetic parameter the temperature profile is thinner. It is also interesting to remark that the current outcome is consistent with the previously computed results of [26] in the nonexistence of thermal cause and buoyancy.

The temperature profile variation for some values of thermal buoyancy parameter and nondimensional heat source/sink is exhibited in Fig. 9. The comparison is displayed for the absence of buoyancy parameter Gr ($Gr = 0$) and source parameters S ($S = 0$) (Curve-I). Further to this, we make an observation that an increase in Gr reduces the temperature in the thermal boundary layer with the nonappearance of basic parameter (Curve-II), whereas the presence of source enhances it (Curve-IV). Further, the temperature of the fluid grows with the occurrence of heat source, whereas sink reduces it significantly.

Figure 11 reveals the effect of Prandtl number and thermal radiation on temperature profiles. In the present discussion, different values of the involved Prandtl number are considered such as Pr = 0.71 (air) and Pr = 10 (water). It is clear to note that the temperature is linear in case of low Prandtl fluid, Pr = 0.71 and in the absence of radiation. Also, in the presence of a radiation parameter R ($R = 3$), the fluid temperature reduces. An increase in Prandtl number also reduces the temperature for each point in the thermal boundary sheet.

Table 1 presents the coefficient of couple stress ($g'(0)$), skin friction ($f''(0)$), and proportion of heat transmission ($\theta'(0)$) for the various values of K and M in the absence/presence of thermal buoyancy and heat source parameters. It is observed that growth in K reduces the skin friction, whereas couple stress and percentage of heat transmission continue to increase in the absence of heat source and presence/absence of thermal buoyancy. It is interesting to note that enhancement in heat source with an increase in K decreases for all the coefficients. Also, a decrease in thermal buoyancy decreases the couple stress and skin friction, but the rate of heat transmission increases. Further, an increase in magnetic parameter increases the skin friction and couple stress, nevertheless, the degree of heat transmission decreases in the presence/absence of Gr and S.

Table 2 illustrates the influence of Prandtl number and suction on couple stress, skin friction, and degree of heat transfer. Degree of heat transmission reduces by the increase in Prandtl number Gr and S are vague but suction increases it. Further, in the presence of Gr and S, as Pr increases skin friction, couple stress and proportion of heat transmission upsurges.

Table 1 Coefficient of skin friction ($f''(0)$), couple stress ($g'(0)$), Nusselt Number ($-\theta'(0)$)

K	Gr	S	$f''(0)$	$g'(0)$	$\theta'(0)$	M	Gr	S	$f''(0)$	$g'(0)$	$\theta'(0)$
1	0	0	**1.676335**	**0.317929**	**2.93282**	1	0	0	**1.358266**	**0.294829**	**3.034388**
1	1	0	1.478569	0.310327	2.981009	1	1	0	1.157684	0.286946	3.078478
1	1	1	1.268082	0.284662	0.263141	1	1	1	1.041472	0.275089	1.176399
4	0	0	**1.494735**	**0.321982**	**2.965145**	2	0	0	**1.676335**	**0.317929**	**2.93282**
4	1	0	1.294381	0.314305	3.013458	2	1	0	1.478569	0.310327	2.981009
4	1	1	1.014493	0.273598	0.127768	2	1	1	1.268082	0.284662	0.263141

Table 2 Coefficient of skin friction ($f''(0)$), couple stress ($g'(0)$), Nusselt Number ($-\theta'(0)$)

Pr	Gr	S	$f''(0)$	$g'(0)$	$\theta'(0)$	f_w	Gr	S	$f''(0)$	$g'(0)$	$\theta'(0)$
0.71	0	0	**1.676335**	**0.317929**	**0.541513**	0.5	0	0	**1.946083**	**0.334816**	**5.527683**
0.71	1	0	1.244556	0.266479	0.645027	0.5	1	0	1.811721	0.331856	5.547313
0.71	1	1	1.064772	0.212745	0.156912	0.5	1	1	1.786446	0.330616	4.324762
10	0	0	**1.676335**	**0.317929**	**2.93282**	0	0	0	**1.676335**	**0.317929**	**2.93282**
10	1	0	1.478569	0.310327	2.981009	0	1	0	1.478569	0.310327	2.981009
10	1	1	1.26808	0.284662	0.263141	0	1	1	1.268082	0.284662	0.263141

6 Conclusions

The numerical computation has been used to study the consequence of uniform heat source/sink on unrestricted convective outer layer movement of an electrically conducting micropolar fluid depending on the transverse magnetic field through porous medium passing through a moving plate. The main partial differential equations are changed in the form of ordinary differential equations by using similarity variables and transformation. After that, we found a numerical solution by making use of Runge–Kutta technique along with the shooting procedure. Some motivating and useful observations are obtained graphically for these pertinent parameters. The numerical computations for the amount of shear stress, couple stress, and proportion of heat transmission are accessed by means of tables, which are summarized below:

1. Retardation in the velocity distribution occurs due to the interaction of Lorentz force.
2. Increase in thermal buoyancy enhances the velocity and microrotation boundary layer significantly.
3. For $\eta < 1$ microrotation profile enhances with the increasing value of suction, whereas injection decelerates it.
4. Temperature profile is thinner due to high Prandtl number and radiation parameter inclusion with the magnetic parameter.
5. Rise in material parameter decays the skin friction, whereas couple stress and proportion of heat transmission continues to increase in the presence/absence of thermal buoyancy and the absence of heat source.

References

1. R. Bhargava, H.S. Takhar, Numerical study of heat transfer characteristics of the micropolar boundary layer near a stagnation point on a moving wall. Int. J. Eng. Sci. **38**, 383–394 (2000)
2. R.S.R. Gorla, S. Nakamura, Mixed convection from a rotating cone to micropolar fluids. Int. J. Heat Fluid Flow **16**, 69–73 (1975)
3. H.S. Takhar, R.S. Agarwal, R. Bhargava, S. Jain, Mixed convective non-steady three dimensional micropolarfluid at a stagnation point. Heat Mass Transf. **33**, 443–448 (1998)

4. W.B. Hooper, T.S. Chen, B.F. Armaly, Mixed convection from a vertical plate in porous media with surface injection or suction. Numer. Heat Transf. **25**, 317–329 (1993)
5. M. Wang, S. Chen, Electroosmosis in homogeneously charged micro- and nanoscale random porous media. J. Colloids Interf. Sci. **314**(1), 264–273 (2007)
6. M. Wang, J. Wang, N. Pan, S. Chen, Lattice Poisson–Boltzmann simulations of electroosmotic flows in charged anisotropic porous media. Commun. Comput. Phys. **2**(6), 1055–1070 (2007)
7. M. Bourich, M. Hasanaoui, A. Mahmind, Soret convection is a shallow porous cavity submitted to uniform fluxes of heat and mass. Int. Commun. Heat Mass Transf. **31**, 773–782 (2004)
8. S.R. Mishra, G.C. Dash, M. Acharya, Free convective flow of viscoelastic fluid in a vertical channel with Dufour effect. World Appl. Sci. J. **28**(9), 1275–1280 (2013)
9. A. Postenicu, Influence of magnetic field on heat and mass transfer from a vertical surfaces in porous media considering Soret and Dufour effects. Int. J. Heat Mass Transf. **47**, 1467–1472 (2004)
10. F.S. Lien, C.K. Chen, J.W. Cleaver, Analysis of natural convection flow of micropolar fluid about a sphere with blowing and suction. ASME J. Heat Transf. 967–970 (1986)
11. B. Mohanty, S.R. Mishra, H.B. Pattnaik, Numerical investigation on heat and mass transfer effect of micropolar fluid over a stretching sheet. Alex. Eng. J. **54**(2), 223–232 (Elsevier) (2015)
12. F.S. Lien, T.M. Chen, C.K. Chen, Analysis of a free-convection micropolar boundary layer about a horizontal permeable cylinder at a non-uniform thermal conditions. ASME J. Heat Transf. **112**, 504–506 (1990)
13. H.S. Takhar, R.S. Agarwal, B. Bhargava, S. Jain, Mixed convection flow of a micropolarfluid over a stretching sheet. Heat Mass Transf. **34**, 213–219 (1998)
14. S.R. Mishra, G.C. Dash, P.K. Pattnaik, Flow of heat and mass transfer on MHD free convection in a micropolar fluid with heat source. Alex. Eng. J. **54**(3), 681–689 (Elsevier) (2015)
15. G.C. Dash, R.S. Tripathy, M.M. Rashidi, S.R. Mishra, Numerical approach to boundary layer stagnation-point flow past a stretching/shrinking sheet. J. Mol. Liq. **221**, 860–866 (2016)
16. R.S. Tripathy, S.R. Mishra, G.C. Dash, M.M. Hoque, Numerical analysis of hydromagnetic micropolar fluid along a stretching sheet with non-uniform heat source and chemical reaction. Int. J. Eng. Sci. Technol. **19**, 1573–1581 (2016)
17. C.Y. Cheng, Free convection heat and mass transfer from a horizontal cylinder of elliptic cross section in micropolar fluids. Int. Commun. Heat Mass Transf. **33**, 311–318 (2006)
18. R.S. Tripathy, G.C. Dash, S.R. Mishra, S. Baag, Chemical reaction effect on MHD free convective surface over a moving vertical plane through porous medium. Alex. Eng. J. **54**(3), 673–679 (Elsevier) (2015)
19. R. Muthukumaraswamy, P. Ganesan, Unsteady flow past an impulsively started vertical plate with heat and mass transfer. Heat Mass Transf. **34**, 187–193 (1998)
20. M.A. Hossain, D.A.S. Rees, Combined heat band mass transfer in natural convection flow from a vertical wavy surface. Acta Mech. **136**, 33–41 (1999)
21. A.J. Chamkha, A.R.A. Khaled, Similarity solutions for hydromagnetic simultaneous heat and mass transfer by natural convection from an inclined plate with internal heat generation or absorption. Heat Mass Transf. **37**, 117–123 (2001)
22. R. Cortell, Effects of viscous dissipation and radiation on the thermal boundary layer over a nonlinearly stretching shee. Phys. Lett. A **372**, 631 (2008)
23. R.C. Bataller, Radiation effects for the Blasius and Sakiadisflows with a convective surface boundary condition. Appl. Math. Comput. **206**, 832 (2008)
24. A. Ishak, Radiation effects on the flow and heat transfer over a moving plate in a parallel stream. Chin. Phys. Lett. **26**, 034701 (2009)
25. V. Aliakbar, A. Alizadeh-Pahlavan, K. Sadeghy, The influence of thermal radiation on MHD flow of Maxwellian fluids above stretching sheets. Commun. Nonlinear Sci. Numer. Simul. **14**(3), 779 (2009)
26. M.A. Seddeek, S.N. Odda, M.Y. Akl, M.S. Abdelmeguid, Analytical solution for the effect of radiation on flow of a magneto-micropolar fluid past a continuously moving plate with suction and blowing. Comput. Mater. Sci. **45**, 423–428 (2009)

27. A. Raptis, Flow of a micropolar fluid past a continuously moving plate by the presence of radiation. Int. J. Heat Mass Transf. **41**, 2865–2866 (1998)
28. R. Ellahi, A. Zeeshan, N. Shehzad, S.Z. Alamri, Structural impact of Kerosene-Al_2O_3 nanoliquid on MHD Poiseuille flow with variable thermal conductivity: application of cooling process. J. Mol. Liq. **264**, 607–615 (2018)
29. R. Ellahi, C. Fetecau, M. Sheikholeslami, Recent advances in the application of differential equations in mechanical engineering problems. Math. Probl. Eng. 2018, Article ID 1584920, 3 pp (2018)
30. R. Ellahi, A. Zeeshan, F. Hussain, T. Abbas, Study of shiny film coating on multi-fluid flows of a rotating disk suspended with nano-sized silver and gold particles: a comparative analysis. Coatings **8**, 422 (2018)
31. R. Ellahi, S.Z. Alamri, A. Basit, A. Majeed, Effects of MHD and slip on heat transfer boundary layer flow over a moving plate based on specific entropy generation. J. Taibah Univ. Sci. **12**(4), 476–482 (2018)
32. R. Ellahi, M. Hassan, A. Zeeshan, Shape effects of spherical and nonspherical nanoparticles in mixed convection flow over a vertical stretching permeable sheet. J. Mech. Adv. Mater. Struct. **24**(15), 1231–1238 (2017)
33. R. Ellahi, M.H. Tariq, M. Hassan, K. Vafai, On boundary layer magnetic flow of nano-Ferroliquid under the influence of low oscillating over stretchable rotating disk. J. Mol. Liq. **229**, 339–345 (2017)

A Two Level Supply Chain Model Where Demand Is Stochastic Additive Under Buyback Policy

Rubi Das and P. K. De

Abstract In this paper, attention has been paid to the study of a supply chain model consisting of one manufacturer and two retailers where buyback policy has been incorporated. We have modeled the manufacturer and retailers optimality system as a profit maximization problem to determine the stochastic order quantity and selling price of the retailers. The demand focused by the manufacturer and retailers is stochastic additive in nature which is also price sensitive and also it depends on the preference of the number of customers to the retailers. Both centralized and decentralized cases of the supply chain have been analyzed by using the normal distribution. One numerical example is provided to illustrate and demonstrate the proposed model. A comprehensive sensitivity analysis with respect to major parameters is carried out and some managerial inferences are obtained.

1 Introduction

Profit of a supply chain depends on various parameters like market demand, stock of inventory, holding cost, cost of transportation, and many others. Further market demand depends on some other vital factors selling price of the product, market availability of the product, quality of the product, and lead time. This paper presents a supply chain model with one manufacturer and two retailers where demand is stochastic additive in nature. Due to the competitive environment created in the multichannel supply chain model, a huge number of customers are benefitted by it.

In this paper, we have incorporated the newsboy problem, which is referred to Porteus and Cachon and Netessine [2] for a detailed survey and extension. The problem is designed in the form of a mathematical model to optimize the expected profit functions to determine the optimal order quantity. We have incorporated

R. Das (✉) · P. K. De
National Institute of Technology Silchar, Assam 788010, India
e-mail: rubidas1118@gmail.com

P. K. De
e-mail: pijusde@gmail.com

S. R. Mishra et al. (eds.), *Recent Trends in Applied Mathematics*, Lecture Notes in Mechanical Engineering, https://doi.org/10.1007/978-981-15-9817-3_26

buyback policy in our model and buyback policy is extensively applied in a brief lifecycle product of industries like books, fashion apparel, computers, toys, etc. According to buyback policy, unsold products are taken back from the retailers, and instead of that manufacturer provide some credit to the retailers for those return products. The main objective of the buyback policy in this supply chain is to minimize the risk of overstocking of the retailers. Generally in a manufacturer-Stackelberg system, a manufacturer is considered as the Stackelberg leader and the retailers are considered as the followers. Ertek and Griffin [3] developed a model by incorporating pricing scheme to the concept of manufacturer-Stackelberg and retailer-Stackelberg system.

2 Literature Review

In day to day, many concepts have been used in the newsboy problem. Minimization of the value-at-risk (CVaR) in a single period newsboy problem has been developed by Gotoh and Takano [4]. Arcelus et al. [1] solved the problem by the development of reordering points distinguished by the random length of time interval and of uncertain period within the temporary price discount problem. He et al. [5] studied a supply chain model with a random demand depending on the price and sales effort of the retailer by assembling the chain with supplier and retailer. They have paid attention to left over inventory and buyback policy. Game theory has an important role in the supply chain. Wang [15] studied a multiple newsvendor problem. Huang et al. [6] analyze the disordering of demand for dual-channel supply chain. Panda et al. [12] analysis displayed the outcome of product compatibility on the noteworthy work of the dual-channel supply chain. We considered the influence of product compatibility (transfer rate of customers from one retailer to another). In product compatibility, noteworthy works of Xiao and Shi, Yan et al. [16] should be mentioned.

Roy et al. [13] studied one supply chain model where retailers are considered in a competitive environment and buyback policy has been incorporated. Also they have considered demand in an uncertain environment. In the present paper the concept of Roy et al. has been extended. Authors have also not incorporated holding costs in their model and cross-price effect in their demand function. These short comings have been addressed in our model and the demand functions have been formulated accordingly. In our paper, we also consider the gaming structure retailer-Stackelberg approach. Modak and Kelle [8] analyzed a supply chain model where they have considered an online channel or offline channel simultaneously. They have considered random demand which is sensitive to both selling price and lead time. Further, they have considered their model with both distribution-free and known distribution approaches. The authors have not included the buyback policy. We have extended the work of Modak and Kelle [8] by using the Cournat–Bertrand approach for normal distribution case and included the buyback policy. We developed a mathematical model by combining some extra issues in existing literature for the result of profit maximization. A numerical example is also discussed for theoretical illustration.

The rest part of the work is categorized as Sect. 3 introduces the assumptions and notations considering for introducing the model. In Sect. 4 the formulation and analysis of the model have been evaluated. In Sect. 5, we have demonstrated the model with one numerical example and the results have been analyzed. A sensitivity analysis has been shown in this section. Section 6 provides some conclusions and future research directions.

3 Assumptions and Notations

The following assumptions and notations are adopted to develop the model.

3.1 Assumptions

(a) The supply chain consists of a manufacturer and two retailers.
(b) In this chain, a single product is produced by the manufacturer.
(c) Uncertain demand depends on selling prices of the items fixed by the retailers.
(d) Our model considered the return policy.
(e) The lead time is not considered.
(f) Shortages at the manufacturer are not allowed but at the retailers are permitted.
(g) Replenishment rate is instaneously infinite in our model but its size is finite.

3.2 Notations

For $(i = 1, 2)$ the parameters and decision variables that are required for mathematical formulation are given below

Parameters

c = Manufacturer production cost
w = Manufacturer wholesale price to the retailer (per unit)
h = Retailers holding cost per unit
s_i = Salvage value of unsold items of retailers
r_i = Retailer's shortage cost
d_i = Deterministic demand rate
b_i = Number of customers prefer the retailers
γ_i = Retailer's price sensitivity
δ = Rate of transfer of customers (number of customers switching from retailer 1 to retailer 2 per unit increase in price difference between p_1 and p_2)
x_i = A part of demand quantity during a period, which is a random variable following probability distribution (units/month)

$f_i(x)$ = Probability density distribution function of x_i
$F_i(x)$ = Cumulative distribution function of x_i
$F_i^{-1}(x)$ = Inverse function of F_i.

Decision variables

q_i = Order quantity of retailers to satisfy the stochastic portion of demand
p_i = Selling price of retailers
ψ_m = Profit function of manufacturer
ψ_{r_i} = Profit function of retailers
ψ_{eip} = Integrated profit function.

4 Model Framework

According to the references [7], Modak and Kelle [8] both the retailers are facing the following demand functions in the model:

$$d_1 = b_1 - \gamma_1 p_1 - \delta(p_1 - p_2)$$

$$d_2 = b_2 - \gamma_2 p_2 + \delta(p_1 - p_2)$$

where d_1 and d_2 are, respectively, the demand functions of retailer 1 and retailer 2.

Changeability in demand can be attained in either by multiplicative form or additive form [9, 11]. As the additive case is more submissive than multiplicative, we add a different random variable for both the retailers demand. The form of stochastic demand is given as follows:

$$d_{1s} + \xi_1 = d_1 + \xi_1 = b_1 - \gamma_1 p_1 - \delta(p_1 - p_2) + \xi_1$$

$$d_{2s} + \xi_2 = d_2 + \xi_2 = b_2 - \gamma_2 p_2 + \delta(p_1 - p_2) + \xi_2$$

where ξ_1 is a random variable with range $[a_1, b_1]$ which has mean μ_1 and standard deviation σ_1 and ξ_2 is a random variable with range $[a_2, b_2]$ which has mean μ_2 and standard deviation σ_2.

Profit functions of manufacturer and retailers are

$$\psi_m = (w - c)(d_1 + q_1 + d_2 + q_2) - s_1 E(q_1 - x_1)^+ - s_2 E(q_2 - x_2)^+ \quad (1)$$

$$\psi_{r_1} = (p_1 - w)(d_1 + \mu_1) - h\Lambda_1 + s_1 E(q_1 - x_1)^+ - r_1\Theta_1 \quad (2)$$

$$\psi_{r_2} = (p_2 - w)(d_2 + \mu_2) - h\Lambda_2 + s_2 E(q_2 - x_2)^+ - r_2\Theta_2 \quad (3)$$

where

$\Lambda_1 = \int_{a_1}^{q_1} (q_1 - u) f(u) d(u)$, $\Lambda_2 = \int_{a_2}^{q_2} (q_2 - v) f(v) dv$, expected values of random demand for both the retailers

$\Theta_1 = \int_{q_1}^{b_1} (u - q_1) f(u) d(u)$, $\Theta_2 = \int_{q_2}^{b_2} (v - q_2) f(v) dv$, expected values of random demand lost in both the retailers

$$E(q_i - x_i)^+ = \int_{-\infty}^{q_i} (q_i - x_i) f_i(x_i) dx_i, \int_{-\infty}^{\infty} f_i(x_i) dx_i = 1, F_i(q_i) = \int_{-\infty}^{q_i} f_i(x_i) dx_i$$

4.1 Centralized Decision Model Through Normal Distribution

Under the centralized situation, all decisions are controlled by a central department. The decision is maximized by the total profit of the whole supply chain. The relevant function is as follows:

$$\begin{aligned} \text{Max}[\psi_{eip}] &= \text{Max}[\psi_m + \psi_{r_1} + \psi_{r_2}] \\ &= (w - c)(d_1 + q_1 + d_2 + q_2) + (p_1 - w)(d_1 + \mu_1) \\ &\quad + (p_2 - w)(d_2 + \mu_2) - h\Lambda_1 - r_1\Theta_1 - h\Lambda_2 - r_2\Theta_2 \end{aligned} \tag{4}$$

Now differentiating ψ_{eip} with respect to p_1, p_2, q_1, q_2 we get

$$\frac{\partial \psi_{eip}}{\partial p_1} = b_1 + \gamma_1 c - 2\gamma_1 p_1 - 2\delta p_1 + 2\delta p_2 + \mu_1 \tag{5}$$

$$\frac{\partial \psi_{eip}}{\partial p_2} = b_2 + \gamma_2 c - 2\gamma_2 p_2 + 2\delta p_1 - 2\delta p_2 + \mu_2 \tag{6}$$

$$\frac{\partial \psi_{eip}}{\partial q_1} = (w - c) - hF_1(q_1) + r_1[1 - F_1(q_1)] \tag{7}$$

$$\frac{\partial \psi_{eip}}{\partial q_2} = (w - c) - hF_2(q_2) + r_2[1 - F_2(q_2)] \tag{8}$$

$$\frac{\partial^2 \psi_{eip}}{\partial p_1^2} = -2\gamma_1 - 2\delta < 0 \tag{9}$$

$$\frac{\partial^2 \psi_{eip}}{\partial p_2^2} = -2\gamma_2 - 2\delta < 0 \tag{10}$$

$$\frac{\partial^2 \psi_{eip}}{\partial q_1^2} = -(h + r_1) f_1(q_1) < 0 \tag{11}$$

$$\frac{\partial^2 \psi_{eip}}{\partial q_2^2} = -(h + r_2) f_2(q_2) < 0 \tag{12}$$

$$\frac{\partial^2 \psi_{eip}}{\partial p_1 \partial p_2} = 2\delta \tag{13}$$

$$\frac{\partial^2 \psi_{eip}}{\partial q_1 \partial q_2} = 0 \tag{14}$$

$$\frac{\partial^2 \psi_{eip}}{\partial p_1 \partial q_1} = 0 \tag{15}$$

$$\frac{\partial^2 \psi_{eip}}{\partial p_2 \partial q_1} = 0 \tag{16}$$

$$\frac{\partial^2 \psi_{eip}}{\partial p_1 \partial q_2} = 0 \tag{17}$$

$$\frac{\partial^2 \psi_{eip}}{\partial p_2 \partial q_2} = 0 \tag{18}$$

By solving Eqs. (5)–(8) we obtained the optimum values. Equations (5), (6) provide

$$p_1^{C*} = \frac{\delta b_2 + \delta\gamma_1 c + \delta\mu_2 + \gamma_2 b_1 + \gamma_1\gamma_2 c + \gamma_2\mu_1 + \delta b_1 + \delta\gamma_1 c + \delta\mu_1}{2(\gamma_1\gamma_2 + \gamma_1\delta + \gamma_2\delta)} \tag{19}$$

$$p_2^{C*} = \frac{\gamma_1 b_2 + \delta b_2 + \delta\mu_1 + \delta\gamma_2 c + \gamma_1\gamma_2 c + \gamma_1\mu_2 + \delta\mu_2 + \delta\gamma_1 c + \delta b_1}{2(\gamma_1\gamma_2 + \gamma_1\delta + \gamma_2\delta)} \tag{20}$$

Equations (7), (8) provide

$$q_1^{C*} = F_1^{-1}\left(\frac{r_1 + w - c}{h + r_1}\right) \tag{21}$$

$$q_2^{C*} = F_2^{-1}\left(\frac{r_2 + w - c}{h + r_2}\right) \tag{22}$$

From Eq. (4) Hessian matrix for p_1, p_2, q_1, q_2 is established

$$H_{\psi_{eip}} = \begin{pmatrix} -(h+r_1)f_1(q_1) & 0 & 0 & 0 \\ 0 & -(h+r_2)f_2(q_2) & 0 & 0 \\ 0 & 0 & -2\gamma_1 - 2\delta & 2\delta \\ 0 & 0 & 2\delta & -2\gamma_2 - 2\delta \end{pmatrix}$$

If Δ_m (mth order principal minor) satisfies the sign $(-1)^m$ then ψ_{eip} is a concave function, i.e., it becomes optimum at $(p_1^{C*}, p_2^{C*}, q_1^{C*}, q_2^{C*})$

$$\Delta_1 = -(h+r_1)f_1(q_1) < 0$$

$$\Delta_2 = (h+r_1)(h+r_2)f_1 f_2 > 0$$

$$\Delta_3 = -2(h+r_1)(h+r_2)(\gamma+\delta)f_1 f_2 < 0$$

$$\Delta_4 = 4(h+r_1)(h+r_2)(\gamma_1\gamma_2 + \gamma_1\delta + \gamma_2\delta)f_1 f_2 > 0$$

4.2 *Decentralized Decision Model Through Normal Distribution (DCS)*

In this system, manufacturer and both the retailers purchase materials individually and separately and they mainly focused on maximizing their own expected profit without considering the profit of the whole chain.

4.2.1 DCS Through Cournot–Bertrand Approach

In this case, manufacturer is new to the market, and he does not have much more idea about the market condition. Retailers face directly to the customers so they have knowledge about the demand trend of market. For maximizing their own profits, retailers decided their own optimal order quantities and sales price. Therefore partial derivatives of ψ_{r_1} and ψ_{r_2} are

$$\frac{\partial \psi_{r_1}}{\partial p_1} = b_1 - 2\gamma_1 p_1 - 2\delta p_1 + \delta p_2 + \mu_1 + \gamma_1 w + \delta w \tag{23}$$

$$\frac{\partial \psi_{r_2}}{\partial p_2} = b_2 - 2\gamma_2 p_2 - 2\delta p_2 + \delta p_1 + \mu_2 + \gamma_2 w + \delta w \tag{24}$$

$$\frac{\partial \psi_{r_1}}{\partial q_1} = -hF_1(q_1) + s_1 F_1(q_1) + r_1[1 - F_1(q_1)] \tag{25}$$

$$\frac{\partial \psi_{r_2}}{\partial q_2} = -hF_2(q_2) + s_2F_2(q_2) + r_2[1 - F_2(q_2)] \tag{26}$$

$$\frac{\partial^2 \psi_{r_1}}{\partial p_1^2} = -2\gamma_1 - 2\delta < 0 \tag{27}$$

$$\frac{\partial^2 \psi_{r_2}}{\partial p_2^2} = -2\gamma_2 - 2\delta < 0 \tag{28}$$

$$\frac{\partial^2 \psi_{r_1}}{\partial q_1^2} = -(h + r_1)f_1(q_1) + s_1 f_1(q_1) < 0 \text{ as} (h + r_1) > s_1 \tag{29}$$

$$\frac{\partial^2 \psi_{r_2}}{\partial q_2^2} = -(h + r_2)f_2(q_2) + s_2 f_2(q_2) < 0 \text{ as} (h + r_2) > s_2 \tag{30}$$

$$\frac{\partial^2 \psi_{r_1}}{\partial p_1 \partial q_1} = 0 \tag{31}$$

$$\frac{\partial^2 \psi_{r_1}}{\partial p_1 \partial q_1} = 0 \tag{32}$$

$$\frac{\partial^2 \psi_{r_2}}{\partial p_2 \partial q_2} = 0 \tag{33}$$

By solving Eqs. (23)–(26) we obtained the optimum values. Equations (23), (24) provide

$$p_1^{DC*} = \frac{2\gamma_2 b_1 + 2\delta b_1 + 2\gamma_2\mu_1 + 2\delta\mu_1 + 2\gamma_1\gamma_2 w + 2\delta\gamma_1 w + 3\delta\gamma_2 w + 3\delta^2 w + \delta b_2 + \delta\mu_2}{4\gamma_1\gamma_2 + 4\gamma_1\delta + 4\gamma_2\delta + 3\delta^2} \tag{34}$$

$$p_2^{DC*} = \frac{2\gamma_1 b_2 + 2\delta b_2 + 2\gamma_1\gamma_2 w + 2\gamma_2\delta w + 3\gamma_1\delta w + 2\gamma_1\mu_2 + 2\delta\mu_2 + 3\delta^2 w + \delta b_1 + \delta\mu_1}{4\gamma_1\gamma_2 + 4\gamma_1\delta + 4\gamma_2\delta + 3\delta^2} \tag{35}$$

Equations (25) and (26) provide

$$q_1^{DC*} = F_1^{-1}\left(\frac{r_1}{h + r_1 - s_1}\right) \tag{36}$$

$$q_2^{DC*} = F_2^{-1}\left(\frac{r_2}{h + r_2 - s_2}\right) \tag{37}$$

$$\left(\frac{\partial^2 \psi_{r_1}}{\partial p_1^2}\right)\left(\frac{\partial^2 \psi_{r_1}}{\partial q_1^2}\right) - \left(\frac{\partial^2 \psi_{r_1}}{\partial p_1 \partial q_1}\right) = (2\gamma + 2\delta) f_1(q_1)(h + r_1 - s_1) > 0 \text{ as} (h + r_1) > s_1 \tag{38}$$

Table 1 Optimal result of the proposed model

Optimal values	p_1^*	p_2^*	q_1^*	q_2^*	ψ_m^*	ψ_{r1}^*	ψ_{r2}^*	ψ_{eip}^*
Centralized	190	177.5	35.97	29.66	1484.49	8481.66	6583.54	16549.69
Decentralized	184.04	170.9	39.09	30.86	1556.13	8406.32	6545.13	16507.58

$$\left(\frac{\partial^2\psi_{r_1}}{\partial p_2^2}\right)\left(\frac{\partial^2\psi_{r_2}}{\partial q_2^2}\right)-\left(\frac{\partial^2\psi_{r_2}}{\partial p_2\partial q_2}\right)$$
$$=(2\gamma+2\delta)f_2(q_2)(h+r_2-s_2)>0 \text{ as}(h+r_2)>s_2 \tag{39}$$

So, ψ_{r_1} and ψ_{r_2} are strictly concave functions.

5 Analysis of an Example

We consider the following values of the key parameters in appropriate units in both centralized and decentralized cases and optimal results are shown in Table 1.

$$\left\{f(x)=\frac{1}{\sigma\sqrt{2\pi}}e^{-\frac{1}{2}\left(\frac{x-\mu_i}{\sigma_i}\right)^2},\forall-\infty\le x\le\infty\right\}$$

$w=\$90$, $b_1=200$ units, $b_2=190$ units, $\gamma_1=.75$, $\gamma_2=.8$, $\delta=.2$, $h=\$15$, $r_1=\$28$,

$r_2=\$26$, $\mu_1=30$ units, $\mu_2=25$ units, $\sigma_1=5$ units, $\sigma_2=4$ units, $c=\$80$, $s_1=\$14$, $s_2=\$13$.

6 Sensitivity Analysis and Discussion

To analyze the effect of different parameters on the optimal values (p_1, p_2, q_1, q_2), sensitivity analysis is conducted for the above numerical example. The results of the sensitivity analysis are presented in Tables 2, 3.

By increasing and decreasing the values of the key parameters $(\gamma_1, \gamma_2, \delta)$ the variations in the optimal solutions of $p_1, p_2, \psi_m, \psi_{r_1}, \psi_{r_2}$ within the range -10% to $+10\%$, are listed in the following tables. In both the cases, since the formulation of q_1, q_2 are independent of $(\gamma_1, \gamma_2, \delta)$ so the optimal solutions q_1, q_2 are insensitive to the changes in those parameters. On the other hand, the expected profit function of manufacturer and both the retailers changes remarkably with the changes of (p_1, p_2).

For centralized case

Table 2 Sensitivity analysis for Centralized case

Percentage change of parameters	p_1	p_2	q_1	q_2	ψ_m	ψ_{r_1}	ψ_{r_2}	ψ_{eip}
−10%	202.99	180.19	35.97	29.66	1517.73	10049.6	6778.34	18345.67
−5%	195.99	178.79	35.97	29.66	1502.76	9271.11	6672.75	17446.62
$\{\gamma_1 = 0.75\}$	190	177.5	35.97	29.66	1484.49	8481.66	6583.54	16549.69
+5%	183.64	176.31	35.97	29.66	1472.82	7916.85	6486.59	15876.26
+10%	178.17	175.22	35.97	29.66	1457.84	7325.61	6404.16	15187.61
−10%	185.44	190.03	35.97	29.66	1570.45	8823.79	7705.72	18099.96
−5%	184.38	183.49	35.97	29.66	1552.09	8698.07	7056.86	17307.03
$\{\gamma_2 = 0.8\}$	190	177.5	35.97	29.66	1484.49	8481.66	6583.54	16549.69
+5%	182.51	171.99	35.97	29.66	1515.96	8478.39	5932.45	15926.8
+10%	181.68	166.9	35.97	29.66	1498.13	8381.88	5443.69	15323.7
−10%	190.23	177.29	35.97	29.66	1484.49	8578.11	6567.43	16630.03
−5%	190.11	177.39	35.97	29.66	1484.49	8568.26	6575.63	16628.39
$\{\delta = 0.2\}$	190	177.5	35.97	29.66	1484.49	8481.66	6583.54	16549.69
+5%	189.89	177.6	35.97	29.66	1484.49	8549.58	6591.19	16625.27
+10%	189.79	177.69	35.97	29.66	1484.49	8540.70	6598.59	16623.78

Table 3 Sensitivity analysis for decentralized case

Percentage change of parameters	p_1	p_2	q_1	q_2	ψ_m	ψ_{r_1}	ψ_{r_2}
−10%	196.09	172.11	39.09	30.86	1603.19	9854.51	6730.02
−5%	182.34	171.48	39.09	30.86	1632.69	9045.92	6505.42
$\{\gamma_1 = 0.75\}$	184.04	170.9	39.09	30.86	1556.13	8406.32	6545.13
+5%	178.71	171.14	39.09	30.86	1557.58	8057.84	6446.88
+10%	173.77.	169.88	39.09	30.86	1511.13	7197.37	6368.31
−10%	185.21	181.98	39.09	30.86	1604.42	8616.18	7771.35
−5%	185.01	176.6	39.09	30.86	1573.99	8513.93	7129.57
$\{\gamma_2 = 0.8\}$	184.04	170.9	39.09	30.86	1556.13	8406.32	6545.13
+5%	183.53	166.01	39.09	30.86	1532.77	8314.82	5996.98
+10%	183.05	161.49	39.09	30.86	1509.77	8230.38	5507.09
−10%	185.27	171.71	39.09	30.86	1540.51	8446.31	6530.79
−5%	184.65	171.31	39.09	30.86	1548.39	8426.46	6532.41
$\{\delta = 0.2\}$	184.04	170.9	39.09	30.86	1556.13	8406.32	6545.13
+5%	183.44	171.29	39.09	30.86	1557.61	8402.06	6533.44
+10%	182.85	170.11	39.09	30.86	1571.48	8366.84	6534.13

- When price sensitivity of retailer 1, i.e., γ_1 and price sensitivity of retailer 2, i.e., γ_2 increases then prices of both the retailers decreases.
- When γ_1 decreases then prices of both the retailers increases.
- But when γ_2 decreases then price of retailer 1 first decreases and then increases whereas opposite phenomenon is observed in case of retailer 2 where price increases constantly.
- It is observed that the expected profits of the manufacturer, both the retailers and integrated profit increases simultaneously as γ_1 decreases. Opposite phenomenon is noticed, i.e., all the above profits decreases simultaneously with the increasing values of γ_1.
- Similarly the above four profits increases with the decreasing values of γ_2. And further decreases with the increasing values of γ_2 except profit of the manufacturer. The profit of the manufacturer first increases and then decreases for increasing value of γ_2.
- A different phenomenon is observed in case of parameter δ. Price of retailer 1 decreases and price of retailer 2 increases for increasing value of δ. Opposite phenomenon is observed for decreasing value of δ, i.e., price of retailer 1 increases and price of retailer 2 decreases.
- It is noticed that the expected profit of manufacturer is independent with the changes of parameter δ.
- Expected profit of retailer 1 increases with the decrease and increases of the parameter δ. Whereas the expected profit of retailer 2 decreases with the decreasing values of the parameter δ. And expected profit of retailer 2 increases with the increasing values of δ. From this phenomenon, we can infer that expected profit of retailer 2 changes proportionally with δ.
- Like expected profit of retailer 1, a similar phenomenon is observed for the integrated profit. That means integrated profit changes proportionally with parameter δ.

For decentralized case

- For increasing values of γ_1, prices of retailer 1 decreasing and prices of retailer 2 first increases and then decreases.
- It is noticed that expected profits of both the retailers decreases simultaneously for increasing value of γ_1. Expected profit of the manufacturer first increases and then decreases for the increasing values of γ_1.
- For decreasing value of γ_1 expected profit of the manufacturer and retailer 1 increases simultaneously. But expected profit of retailer 2 first decreases and then increases.
- For decreasing values of γ_2, prices of both the retailers, expected profit of the manufacturer and both the retailers are increasing. And for increasing values of γ_2, prices of both the retailers and all these profits are decreasing simultaneously.
- For decreasing value of δ prices of both the retailers increases. And for increasing value of δ prices of retailer 1 decreases and prices of retailer 2 first increases and then decreases.

- It is observed that expected profit of both the retailers decreasing simultaneously and expected profit of the manufacturer increases for increasing value of δ. An opposite phenomenon is observed for decreasing value of δ.

7 Conclusion

This paper has introduced a two-layered supply chain model with buyback policy under demand uncertainty by considering a game-theoretic approach. Our two-echelon supply chain model is developed under a normal distribution approach where demand is stochastic additive. We have proposed this model with price and preference of the number of customers-dependent demand. The objective of this paper is to find the optimal order quantity for a single product for a stochastic additive demand pattern which maximizes the expected profit of all the channels. For uncertainty of demand, retailers have to pay holding costs and shortage cost for unsold products. Our model is examined under Cournot–Bertrand and centralized approaches for normal distribution case. According to the results, the centralized system has a better performance solution compared to a decentralized system.

Acknowledgements First author is thankful to MHRD to giving financial support for carrying out research at NIT Silchar and also she is thankful to TEQIP III for providing financial support to present this paper in AMSE 2019 at Bhubaneshwar.

References

1. F.J. Arcelus, T.P.M. Pakkala, G. Srinivasan, On the interaction between retailers inventory policies and manufacturer trade deals in response to supply-uncertainty occurrences. Ann. Oper. Res. **143**(1), 45–58 (2006)
2. G.P. Cachon, S. Netessine, Supply Chain Analysis in the eBusiness Era. Game theory in supply chain analysis (2004)
3. G. Ertek, P.M. Griffin, Supplier-and buyer-driven channels in a two-stage supply chain. IIE Trans. **34**(8), 691–700 (2002)
4. J.Y. Gotoh, Y. Takano, Newsvendor solutions via conditional value-at-risk minimization. Eur. J. Oper. Res. **179**(1), 80–96 (2007)
5. Y. He, X. Zhao, L. Zhao, J. He, Coordinating a supply chain with effort and price dependent stochastic demand. Appl. Math. Model. **33**(6), 2777–2790 (2009)
6. S. Huang, C. Yang, X. Zhang, Pricing and production decisions in dual-channel supply chains with demand disruptions. Comput. Ind. Eng. **62**(1), 70–83 (2012)
7. G. Hua, S. Wang, T.E. Cheng, Price and lead time decisions in dual-channel supply chains. Eur. J. Oper. Res. **205**(1), 113–126 (2010)
8. N.M. Modak, P. Kelle, Managing a dual-channel supply chain under price and delivery-time dependent stochastic demand. Eur. J. Oper. Res. **272**(1), 147–161 (2019)
9. E.S. Mills, Uncertainty and price theory. Q. J. Econ. **73**(1), 116–130 (1959)
10. J. Mostard, R. De Koster, R. Teunter, The distribution-free newsboy problem with resalable returns. Int. J. Prod. Econ. **97**(3), 329–342 (2005)

11. N.C. Petruzzi, M. Dada, Pricing and the newsvendor problem: a review with extensions. Oper. Res. **47**(2), 183–194 (1999)
12. S. Panda, N.M. Modak, S.S. Sana, M. Basu, Pricing and replenishment policies in dual-channel supply chain under continuous unit cost decrease. Appl. Math. Comput. **256**, 913–929 (2015)
13. A. Roy, S.S. Sana, K. Chaudhuri, Optimal Pricing of competing retailers under uncertain demand-a two layer supply chain model. Ann. Oper. Res. **260**(1–2), 481–500 (2018)
14. D. Simchi-Levi, S.D. Wu, Z.J.M. Shen, (Eds.), Handbook of Quantitative Supply Chain Analysis: Modeling in the E-Business Era , vol 74 (Springer Science & Business Media, 2004)
15. C.X. Wang, The loss-averse newsvendor game. Int. J. Prod. Econ. **124**(2), 448–452 (2010)
16. W. Yan, Y. Xiong, J. Chu, G. Li, Z. Xiong, Clicks versus bricks: the role of durability in marketing channel strategy of durable goods manufacturers. Eur. J. Oper. Res. **265**(3), 909–918 (2018)

Analytical Study of MHD Free Convective Flow in a Composite Medium Between Coaxial Vertical Cylinders Partially Filled with Porous Material

M. Senapati, S. K. Parida, and G. C. Dash

Abstract The natural convective flow between two coaxial vertical cylinders partially filled with a porous material has been studied. The surface of the inner cylinder is subject to the constant heat flux and the outer cylinder is maintained at a constant temperature. The Brinkman extended Darcy model has been applied to porous media flow. The interface of the two regions is subjected to shear stress jump and continuity of the velocity. The analytical solutions of the physical model are carried out with the help of the modified Bessel function. The important findings are: the permeability of the medium and interface condition plays a vital role for the output of the desired flow rate and consistency of flow, the heat transfer enhances with another gap of the cylinders, i.e. the squeezing of the annular gap produces a cooling effect on/in cylinder surfaces. Some interesting outcomes of the analysis are: the noticeable momentum transport occurs in the region close to the interface of fluid and porous region. The consistency of flow pattern and fluidity of the fluid model depends on the permeability of the porous region. The adjustable magnetic field (force-act-at-a distance) and stress jump condition (act-at-the contact) are to be simulated for obtaining the desired smooth flow pattern.

Keywords Brinkman extended Darcy model · Free convection · Heat flux · Stress jump · Magnetic field

Nomenclature

Da Darcy number
T_0' Temperature of the outer cylinder
d' Radial distance of interface
u' Velocity along the axis of cylinder
D Radial distance of interface in non-dimensional form

M. Senapati · S. K. Parida (✉) · G. C. Dash
Department of Mathematics, Faculty of Engineering, Siksha 'O' Anusandhan Deemed to be University, Bhubaneswar 751030, Odisha, India
e-mail: sampadaparida@soa.ac.in

S. R. Mishra et al. (eds.), *Recent Trends in Applied Mathematics*, Lecture Notes in Mechanical Engineering, https://doi.org/10.1007/978-981-15-9817-3_27

U	Velocity along the axis of cylinder in non-dimensional form
g	Acceleration due to gravity
Q	Rate of heat transfer
R^*	Radius of the outer cylinder
β	Coefficient of thermal expansion
γ	Ratio of dynamic viscosity
T'_f	Temperature of the fluid layer
μ	Dynamic viscosity of the fluid
T'_p	Temperature of the porous layer
θ	Temperature in non-dimensional form
K'	Permeability of the porous medium
R	Radius of the inner cylinder
α	Adjustable coefficient in the stress jump condition

1 Introduction

The fluid flow and heat transfer in a composite system, partially filled with a porous material, find numerous applications in thermal engineering, pertaining to heat and mass transfer processes, oil extraction and heat exchangers, etc. Many researchers Singh et al. [1], Ramanaiah et al. [2], Pop et al. [3], Lesinigo et al. [4] contributed significantly using different fluid models of the flow through porous media using Darcy and Brinkman models for different thermal conditions and geometrical configurations. Further, Paul and Singh [5] studied fully developed free convection between two coaxial vertical cylinders. Free convective flow for low-Prandtl-number fluid ($0.2 < \mathrm{Pr} < 1$) in a horizontal annular region has been studied by Yoo [6]. Their study leads to dual steady solutions when the Rayleigh number exceeds a critical value.

Though the above investigations on free convective flow are theoretical in nature, still then some experimental studies support some of the theoretical predictions. For example, a study on flow through vertical and inclined elliptical tubes with constant heat flux studied by Elshazly [7]. Further, experimental studies carried out by Seghir-Quali et al. [8] on an axial air flow pertaining to convective heat transfer inside a rotating cylinder and convective flow in a vertical circular cylinder with constant heat flux by Mohammed and Salman [9] are of great interest.

A numerical study of buoyancy-driven unsteady natural convection boundary layer flow past a vertical cone embedded in a non-Darcian isotropic porous regime with a transverse magnetic field is considered by Prakash et al. [10]. The transient fully developed free convective flow of viscous incompressible fluid between two concentric vertical cylinders filled with a porous material and saturated with the same fluid has been analyzed when the outer surface of the inner cylinder is subject to isothermal or isoflux heating by Jha et al. [11]. Steady two-dimensional MHD laminar free convective boundary layer flows of an electrically conducting Newtonian nanofluid over a solid stationary vertical plate in a quiescent fluid is investigated numerically by Uddin et al. [12] when the bounding surface is subject to Newtonian heating.

The captivating attributes of carbon nanotubes (CNT) comprising chemical and mechanical steadiness, outstanding electrical and thermal conductivities, featherweight, and physiochemical consistency make them useful in the manufacturing of electrochemical devices. Keeping in view such exciting features of carbon nanotubes, our objective in the present study is to examine the flow of aqueous-based nanofluid comprising single and multi-wall carbon nanotubes (CNTs) past a vertical cone encapsulated in a permeable medium with convective heat and solutal stratification by Ramzan et al. [13]. The MHD flow of an incompressible viscous electrically conducting fluid past a porous plate through porous medium rotating with uniform angular velocity about an axis normal to the plate and the fluid at infinity rotates with the same angular velocity about a non coincident parallel axis. The governing equations of motion and heat transfer are investigated by Parida et al. [14] using a special technique. A finite element study of combined heat and mass transfer flow through a porous medium in a circular cylindrical annulus with Soret and Dufour effects in the presence of heat sources has been analyzed by Gnaneswar [15]. In this paper, the heat and mass transfer characteristics of mixed convection about a circular cylindrical annulus in a porous medium, by taking into account the Thermo-Diffusion (Soret) and Diffusion–Thermo (Dufour) effects have been analyzed by Reddy et al. [16].

The novelty of the present study rests upon the flow of electrically conducting fluid subject to an externally applied magnetic field producing a body force act at a distance and a stress jump condition act at the contact of the composite medium along with constant heat flux at the surface of the inner cylinder. Further, the flow through a saturated porous medium with uniform permeability has also been analyzed by applying Brinkman extended Darcy model. Many industrial fluids and biological fluids are electrically conducting in nature. Therefore, an electrically conducting fluid subject to a transverse magnetic field has been considered in the present study. Consequently, additional electromagnetic force in the form of body force (Lorentz force) comes into play and modifies the momentum equation.

The analytical solution with the help of modified Bessel function provides an edge over the error bound numerical solution and ensures the reliability. The present analysis is more realistic and compatible with real-world problems than the constant surface heat flux assumptions.

2 Formulation of the Problem

Consider the problem of free convective flow in an annular region between two coaxial vertical cylinders in a composite medium, i.e. clear fluid region and a saturated porous region. The flow is generated due to the difference in the temperature of the outer cylinder maintained at constant temperature and the inner cylinder subjected to heat flux. The direction of the flow is along the axis of the cylinder, while another

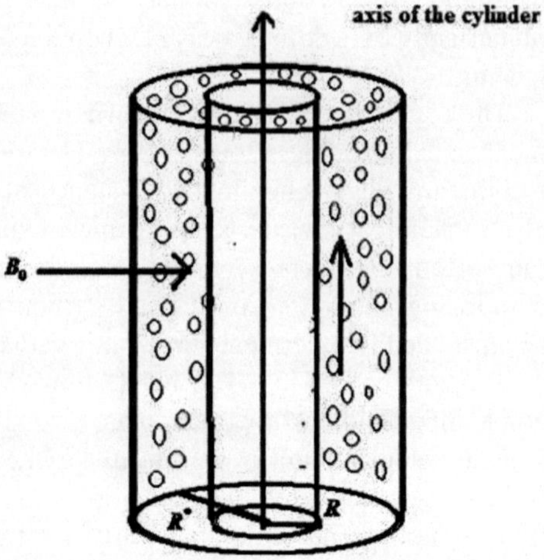

Fig. 1 Flow geometry of the problem

axis is along the radius of the cylinder. At the interface of the porous medium and clear fluid zone, the continuity of the flow and heat transfer is assured (Fig. 1).

Under the usual Boussinesq's approximation, the governing equations related to the problem are as follows:

Fluid region:

$$\mu_f \frac{d^2 u'_f}{dr'^2} + \frac{\mu'_f}{r'}\frac{du'_f}{dr'} - \sigma_f B_0^2 u'_f + \rho g \beta \left(T'_f - T'_c\right) = 0 \tag{1}$$

$$\frac{d^2 T'_f}{dr'^2} + \frac{1}{r'}\frac{dT'_f}{dr'} = 0 \tag{2}$$

Porous region:

$$\mu_P \frac{d^2 u'_P}{dr'^2} + \frac{\mu'_P}{r'}\frac{du'_P}{dr'} - \sigma_p B_0^2 u_p - \frac{\mu_f}{K'_p} u'_P + \rho g \beta \left(T'_p - T'_c\right) = 0 \tag{3}$$

$$\frac{d^2 T'_P}{dr'^2} + \frac{1}{r'}\frac{dT'_P}{dr'} = 0 \tag{4}$$

The corresponding boundary conditions are

$$\left.\begin{array}{r}u'_f = 0,\ \ q' = -k\dfrac{dT'_f}{dr'}\ \text{ at } r' = R \\ u'_P = 0,\ \ T'_P = T'_c\ \ \text{ at } r' = R^* \\ \left.\begin{array}{r} u'_f = u'_P,\ \mu_P\dfrac{du'_P}{dr'} - \mu'_f\dfrac{du'_f}{dr'} = \gamma\dfrac{\mu_f}{\sqrt{K'}}u'_P \\ T'_f = T'_P,\ \dfrac{dT'_f}{dr'} = \dfrac{dT'_P}{dr'}\end{array}\right)\text{ at } r' = d'\end{array}\right\} \tag{5}$$

Using the following non-dimensional quantities

$$\begin{aligned} &Da = \frac{K'_p}{R^2}, r = \frac{r'}{R}, d = \frac{d'}{R}, \lambda = \frac{R^*}{R}, u_f = \frac{u'_f \upsilon_f K}{g\beta q R^3}, \\ &u_p = \frac{u'_p \upsilon_p K}{g\beta q R^3}, \theta_f = \frac{\left(T'_f - T'_c\right)K}{Rq}, \theta_p = \frac{\left(T'_p - T'_c\right)K}{Rq} \end{aligned} \tag{6}$$

the Eqs. (1)–(4) become

$$\frac{d^2u_f}{dr^2} + \frac{1}{r}\frac{du_f}{dr} - M_f^2 u_f + \theta_f = 0 \tag{7}$$

and

$$\frac{d^2\theta_f}{dr^2} + \frac{1}{r}\frac{d\theta_f}{dr} = 0 \quad \text{(For fluid region)} \tag{8}$$

$$Rv\frac{d^2u_p}{dr^2} + \frac{Rv}{r}\frac{du_p}{dr} - \frac{1}{Da}u_p - M_p^2 u_p + \theta_p = 0 \tag{9}$$

and

$$\frac{d^2\theta_p}{dr^2} + \frac{1}{r}\frac{d\theta_p}{dr} = 0 \quad \text{(For porous region)} \tag{10}$$

The boundary and matching conditions in dimensionless form are

$$\left.\begin{array}{r}u_f = 0,\ \dfrac{d\theta_f}{dr} = -1 \text{ at } r = 1 \\ u_p = 0,\ \theta_p = 0 \text{ at } r = \lambda \\ \left.\begin{array}{r} u_f = u_p,\ Rv\dfrac{du_p}{dr} - \dfrac{du_f}{dr} = \dfrac{\gamma}{\sqrt{Da}}u_p \\ \theta_f = \theta_p,\ \dfrac{d\theta_f}{dr} = \dfrac{d\theta_p}{dr}\end{array}\right\}\text{ at } r = d\end{array}\right\} \tag{11}$$

The solutions of Eqs. (7)–(10) with boundary conditions (11) are given by

$$u_f = C_1 + C_2 \log r + \frac{r^2}{4}(\log r - \log \lambda - 1) \tag{12}$$

$$u_P = C_3\, I_0\left(\frac{r}{\sqrt{Rv\, Da}}\right) + C_4\, K_0\left(\frac{r}{\sqrt{Rv\, Da}}\right) + Da\,(\log \lambda - \log r) \tag{13}$$

$$\theta_f = \theta_P = \log \lambda - \log r \tag{14}$$

where I_0, K_0, I_1 *and* K_1 are the modified Bessel functions of the first kind and second kind of order zero and one, respectively.

Using Eqs. (12) and (13), the skin friction on the walls at $r = 1$ and λ is calculated as

$$\tau_1 = -\left(\frac{du_f}{dr}\right)_{r=1} = -C_2 + \frac{1}{2}(\log \lambda + 1) - \frac{1}{4} \tag{15}$$

$$\tau_2 = -\left(\frac{du_f}{dr}\right)_{r=\lambda} = -\frac{1}{\sqrt{RvDa}} C_3 I_1\left(\frac{\lambda}{\sqrt{RvDa}}\right) + \frac{1}{\sqrt{RvDa}} C_4 K_1\left(\frac{\lambda}{\sqrt{RvDa}}\right) + \frac{Da}{\lambda} \tag{16}$$

3 Results and Discussion

The exact analytical solutions are obtained for the flow through a composite medium under the boundary and interface conditions and are presented through graphs. The effect of an applied transverse magnetic field producing a body force acting at a distance on the flow through an annular region is of special interest which has not been taken care of in the earlier studies.

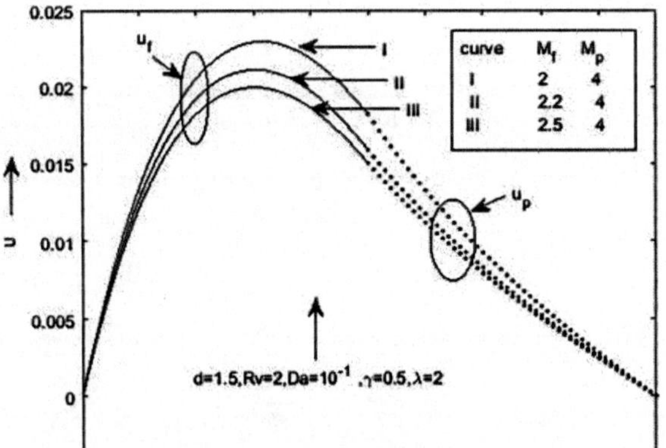

Fig. 2 Velocity profile for various values of M_f and M_p

Figure 2 shows the velocity profiles for different values of the magnetic parameter, a representative of electromagnetic force generated due to interaction of the electric current and magnetic field assuming the magnetic Reynolds number which determines the diffusion of the magnetic field along the streamlines analogous to the ordinary Reynolds number for the diffusion of vorticity along the streamline, is small compared to the unity. The magnetic nature is very small compared to the unity, the magnetic field is not distorted by the flow. When it is very large, on the other hand, the magnetic field moves with the flow and is called frozen-in. In engineering problems, it is rare to obtain a magnetic Reynolds number greater than unity because of the low electrical conductivity of the useful fluids. From Fig. 2, it is seen that due to resistive force, the fluid velocity reduces in both fluid and porous regions. In this figure, the position of the interface is at $d = 1.5$. In the porous region ($d > 1.5$), the velocity falls rapidly to attend the prescribed value. The maximum value of the profile occurs in the fluid region just before the transition zone.

Figure 3 depicts the velocity profile for the various values of the viscosity ratio Rv. It is observed that an increase in Rv decreases the velocity throughout the flow domain. Thus, it is concluded that under the influence of the dominating effect of viscosity in the fluid layer, momentum transport decreases and the decrease of magnitude is significant in the adjacent layers of the interface ($d = 1.5$). The velocity becomes maximum when $Rv = 1$, i.e. both the viscosities are of the same order magnitude.

Figure 4 shows the velocity profiles for different values of the Darcy number, characterizing the permeability of the porous medium. It is seen that as Da increases the velocity of the fluid flow increases under the influence of the dominating magnetic field parameter of the porous region over the fluid region ($M_p = 4$ and $M_f = 2$). On careful observation of curve II $\left(Da = 10^{-2}\right)$, it is seen that at the interface, the fluidity of flow is detracted. Such an effect is also marked in small measure when $Da = 0.2$ and $Da = 0.5$. Therefore, it is concluded that for maintaining smooth

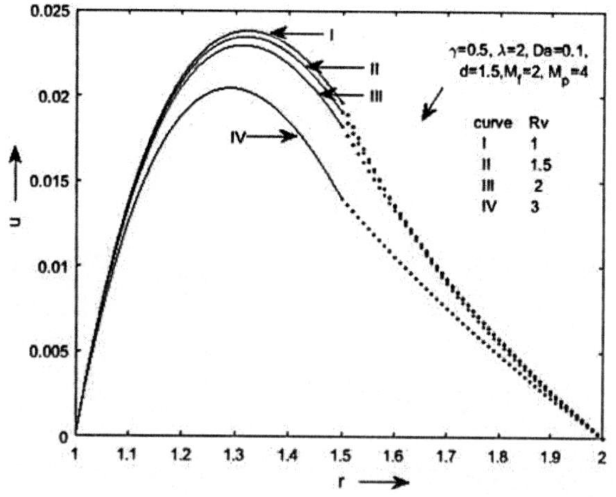

Fig. 3 Velocity profile for various values of Rv

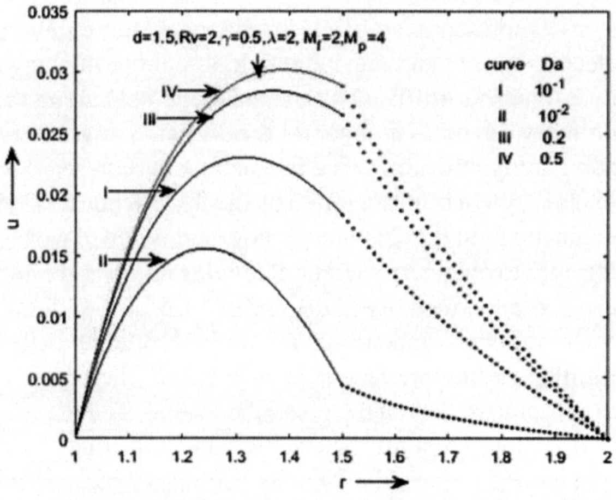

Fig. 4 Velocity profile for various values of Da

flow, the permeability of the medium plays a vital role in the output of the desired flow rate.

From Fig. 5, it is seen that as γ increases the velocity increases, whereas the effect of Rv is to decrease the velocity. The parameter γ represents the adjustable coefficient of stress jump. The higher stress jump leads to a greater deformation, consequently the velocity increases.

Figure 6 shows the velocity distribution for different positions of the interface of fluid and porous regions. This shows that the onset of flow behaviour in the respective region recedes as per the position of the interface. Further, it is seen that an increase in spatial distance of the interface enhances the velocity distribution in both the regions.

The temperature distribution is almost linear and increases with the annular gap λ which is evident from the Eq. (14) also.

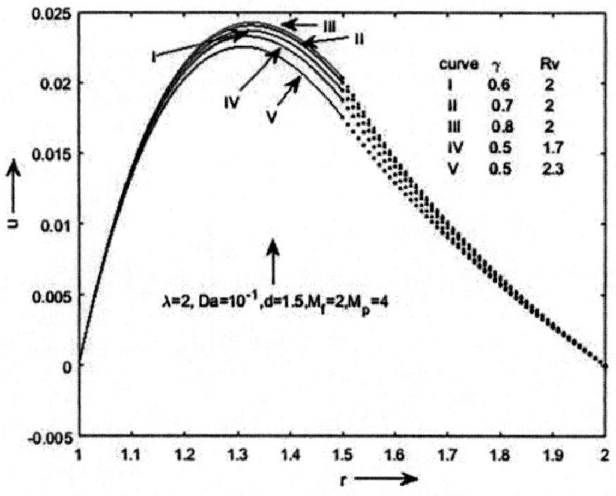

Fig. 5 Velocity profile for various values of γ and Rv

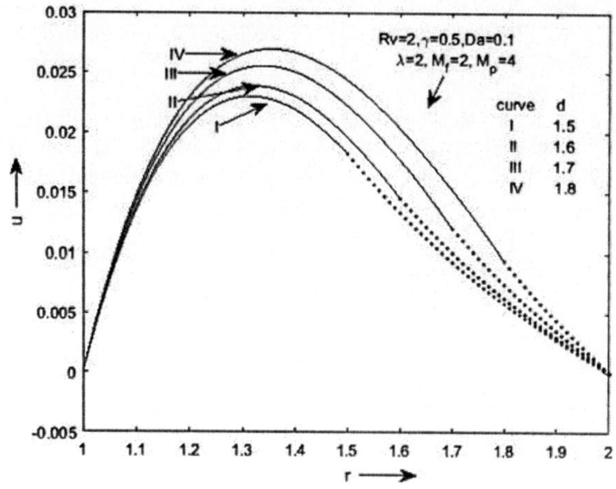

Fig. 6 Velocity profile for various values of *d*

4 Conclusion

- The significant momentum transport occurs in the region close to the interface of two zones.
- For maintaining smooth flow the permeability of the porous region plays a vital role.
- The strength of the external applied magnetic field is to be regulated to obtain the desired flow rate.
- Higher the stress jump leads to greater deformation so the velocity distribution increases.
- The temperature distribution is almost linear across the flow domain and it increases with the spatial distance of the annular region. Also the velocity distribution both the region increases as increase in spatial distance.

References

1. A.K. Singh, Natural convection in unsteady Couette motion. Def. Sci. J. **38**, 35–41 (1988)
2. G. Ramanaiah, G. Malarvizhi, Unified treatment of free and mixed convection on a permeable vertical cylinder in a saturated porous medium. Indian J. Technol. **28**, 604–608 (1990)
3. I. Pop, P. Cheng, Flow past a circular cylinder embedded in a porous medium based on the Brinkman model. Int. J. Eng. Sci. **30**, 257–262 (1992)
4. M. Lesinigo, C.D. Angelo, A. Quarteroni, A multiscale Darcy–Brinkman model for fluid flow in fractured porous media. Numerische Mathematik **117**(4), 717–752 (2011)
5. T. Paul, A.K. Singh, Natural convection between coaxial vertical cylinders partially filled with a porous material. Forsch. Ingenieurwes **64**, 157–162 (1998)
6. Joo-Sik Yoo, Dual free-convective flows in a horizontal annulus with a constant heat flux wall. Int. J. Heat Mass Transf. **46**, 2499–2503 (2003)

7. K. Elshazly, M. Moawed, E. Ibrahim, M. Emara, Heat transfer by free convection from the inside surface of the vertical and inclined elliptic tube. Energy Convers. Manag. **46**, 1443–1463 (2005)
8. S. Seghir-Quali, D. Saury, S. Harmand, O. Phillipart, D. Laloy, Convective heat transfer inside a rotating cylinder with an axial air flow. Int. J. Therm. Sci. **45**, 1166–1178 (2006)
9. H.A. Mohammed, Y.K. Salman, Combined natural and forced convection heat transfer for assisting thermally developing flow in a uniformly heated vertical circular cylinder. Int. Commun. Heat Mass Transf. **34**, 474–491 (2007)
10. J. Prakash, S.G. Mohiddin, S.V.K. Varma, Free convective MHD flowpast a vertical cone with variable heat and mass flux. J. Fluids **2013**, 1–9 (2013)
11. B.K. Jha, T.S. Yusuf, Transient free convective flow in an annular porous medium: a semi-analytical approach. Eng. Sci. Technol. Int. J. **30**, 1–13 (2016)
12. M.J. Uddin, W.A. Khan, A.I. Ismail, MHD free convective boundary layer flow of a nanofluid past a flat vertical plate with newtonian heating boundary condition. PLoS ONE 1–12 (2012)
13. M. Ramzan, M. Mohammad, F. Howari, Magnetized suspended carbon nanotubes based nanofluid flow with bio-convection and entropy generation past a vertical cone. Sci. Rep. Nat. Res. 1–15 (2019)
14. S.K. Parida, M.R. Acharya, G.C. Dash, S. Panda, MHD flow past a rotating porous plate through porous medium rotating about a non-coincident parallel axis. Model. Meas. Control B **80**(1–2), 88–103 (2011)
15. K. Gnaneswar, Effect of thermo-diffusion and chemical reaction on mixed convective heat and mass transfer through a porous medium in cylindrical annulus with heat source. Int. J. Eng. Res. Appl. **4**, 01–16 (2014)
16. P.S. Reddy, V.P. Rao, Thermo-diffusion and diffusion-thermo effects on convective heat and mass transfer through a porous medium in a circular cylindrical annulus with quadratic density temperature variation—finite element study. J. Appl. Fluid Mech. **5**(4), 139–144 (2012)

Effects of Dissipative Heat Energy and Chemical Reaction on MHD Nanofluid Flow Over a Nonlinearly Stretching Sheet

S. Baag, B. Nayak, and S. R. Mishra

Abstract The present investigation is intended to study the effects of heat source and chemical reaction of viscous MHD nanofluid flow over a nonlinearly stretching sheet with viscous dissipation. Similarity transformation is used to transform the governing equations of the problem into nonlinear ordinary differential equations. The transformed ODEs are solved numerically by employing the fourth order Runge–Kutta method associated with the shooting technique. The effect of various physical parameters on the velocity, temperature, and concentration are presented graphically and numerical results for local skin friction, local Nusselt number, and Sherwood number are tabulated for various physical parameters. For validation, the present result is compared with the earlier published result in a particular case. The boundary layer thickness decreases with an increase in magnetic parameters for both the absence/presence of a porous matrix. The temperature of the nanofluid and dimensionless concentration increases with increasing values of thermophoretic parameter, whereas the Brownian motion parameter retards the nanoparticle volume fraction. Increase in Lewis number both the temperature and nanoparticle concentration decreases. Brownian motion parameter and thermophoresis parameter have an opposite effect on Nusselt and Sherwood number.

Keywords MHD · Nonlinear stretching sheet · Heat source · Chemical reaction · Runge–kutta method · Shooting technique

---

S. Baag
Department of Physics, College of Basics Sciences and Humanities, OUAT, Bhubaneswar, India
e-mail: sbaag22@gmail.com

B. Nayak · S. R. Mishra (✉)
Department of Mathematics, Siksha 'O' Anusandhan Deemed to be University, Bhubaneswar 751030, Odisha, India
e-mail: satyaranjan_mshr@yahoo.co.in

S. R. Mishra et al. (eds.), *Recent Trends in Applied Mathematics*, Lecture Notes in Mechanical Engineering, https://doi.org/10.1007/978-981-15-9817-3_28

1 Introduction

Due to the various practical and real-life applications, the investigation of influences of the thermal radiation and external heat source for the magnetohydrodynamic flow of viscous fluid over the nonlinear stretching sheet creates a significant challenge among the young researchers and scientists. In particular, the polymer extrusion in a dying process, metallic plate condensation in a cooling bath, etc., are the basic examples from industrial applications. Sakiadis [1] performed his pioneering work on the study of stretching surfaces. However, for the production of several materials in industrial manufacturing processes, the inclusion of additional effects and contribution of these effects on the stretching problems have many physical aspects. Metallurgical processing is a new application in the research area of magnetohydrodynamics (MHD) flow phenomena. Exchange of the knowledge on the study of various flow over an exponentially stretching sheet is also limited (Magyari and Kellar [2], Elbashbeshy [3], Vajravelu [4], Vajravelu and Cannon [5], and Cortell [6]). In space technology applications, the role of thermal radiation is important. In some industrial applications (propulsion systems, aerodynamics rocket, etc.) due to the requirement of high temperatures at the time of operating the system, the use of thermal radiation is important. Cortell [7], Hady et al. [8], and Vajravelu and Rollins [9] have investigated the flow phenomena in an electrically conducting fluid over a stretching surface.

Ellahi and his co-workers [10–14] have proposed their study for the behavior of dissipative heat energy in various types of viscous as well as nanofluids. They have employed both analytical and numerical techniques along with approximate methods for the solution of complex nonlinear problems with several slip boundary conditions. Bhukta et al. [15] made an attempt to understand the heat and mass transfer effects in the boundary layer flow using a porous medium and used Kummer's functions to solve the transfer functions. Similar attempt was made by Mohanty et al. using micropolar fluid over a stretching sheet in a uniform magnetic field [16]. However, Dash et al. [17] studied the magnetohydrodynamics flow for heat and mass diffusions in the case of electrically conducting stagnation.

In sighting the aforesaid studies, the aim of our work is to investigate the effects of heat source/sink and chemical reaction of MHD nanofluid flow over a nonlinearly stretching sheet through porous medium with viscous dissipation. The transformed nonlinear ODEs are solved by using the numerical method. For the conformity, the numerical validation is obtained with the earlier published results of Mahbood et al. [18].

2 Mathematical Formulation

Two-dimensional steady, incompressible viscous flow of a water-based electrically conducting nanofluid past over a nonlinear stretching sheet through porous medium is considered in this present investigation. The flow is along the x-direction and

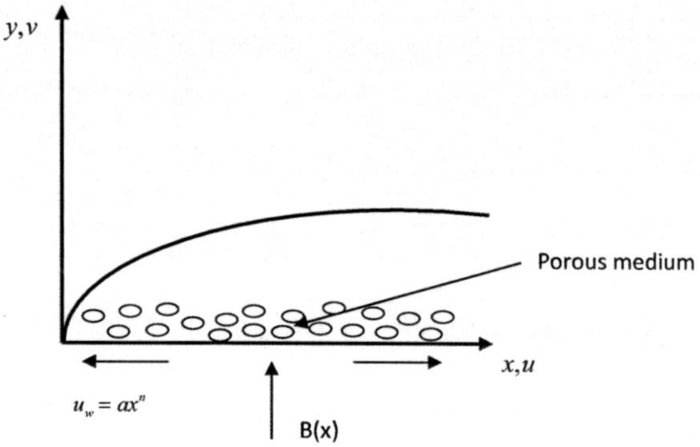

Fig. 1 Flow configuration

the velocity of the sheet stretched with $u_w = ax^n$, where n is the nonlinear stretching parameter and a is constant. y-coordinate is normal to the flow direction. The conducting fluid imposed with the magnetic field $B(x) = B_0 x^{(n-1)/2}$ applied normal to the sheet (Fig. 1). In addition, heat source/sink and the reactant species are also considered. T_w and C_w are assumed as the wall temperature and nanoparticle concentration, respectively, whereas T_∞ and C_∞ are ambient temperature and nanoparticle concentration. It is important to note that $T_w > T_\infty$ and $C_w > C_\infty$. The governing equations of continuity equations, momentum equation, energy equation, and concentration equations with the boundary conditions are given below.

$$\frac{\partial u}{\partial x} + \frac{\partial v}{\partial y} = 0 \tag{1}$$

$$u\frac{\partial u}{\partial x} + v\frac{\partial u}{\partial y} = \frac{\partial^2 u}{\partial y^2} - \frac{\sigma B^2(x)}{\rho_f}u - \frac{\upsilon}{Kp}u \tag{2}$$

$$u\frac{\partial T}{\partial x} + v\frac{\partial T}{\partial y} = \frac{\kappa}{\rho_f c_p}\frac{\partial^2 T}{\partial y^2} + \frac{\mu}{\rho_f c_p}\left(\frac{\partial u}{\partial x}\right)^2 + \tau\left(D_m\frac{\partial C}{\partial y}\frac{\partial T}{\partial y} + \frac{D_T}{T_\infty}\left(\frac{\partial T}{\partial y}\right)^2\right) + \frac{Q}{\rho_f c_p}(T - T_\infty) \tag{3}$$

$$u\frac{\partial C}{\partial x} + v\frac{\partial C}{\partial y} = D_m\frac{\partial^2 C}{\partial y^2} + \frac{D_T}{T_\infty}\frac{\partial^2 T}{\partial y^2} - Kc^*(C - C_\infty) \tag{4}$$

$$\begin{aligned} &y = 0, U_w = ax^n, \quad V = 0, \quad T = T_w, \quad C = C_w \\ &Y = \infty, U = 0, \quad V = 0, \quad T = T_\infty, \quad C = C_\infty \end{aligned} \tag{5}$$

Following similarity variables and transformations are used to convert the PDEs to ODEs.

$\eta = y\sqrt{\frac{a(n+1)}{2\upsilon}}x^{\frac{(n-1)}{2}}$, $u = ax^n f'(\eta)$, $v = -\sqrt{\frac{a(n+1)}{2}}x^{\frac{(n-1)}{2}}\left(f(\eta) + \frac{n-1}{n+1}\eta f'(\eta)\right)$, $\theta(\eta) = (T - T_\infty)/(T_W - T_\infty)$, $\phi(\eta) = (C - C_\infty)/(C_W - C_\infty)$

where ψ represents stream functions and is defined as $u = \frac{\partial\psi}{\partial y}$, $v = \frac{-\partial\psi}{\partial y}$ so that Eq. (1) is satisfied identical. The governing Eqs. (2)–(4) and the boundary condition (4) are reduced to

$$f''' + ff'' - \left(\frac{2n}{n+1}\right)f'^2 - \left(M + \frac{1}{Kp}\right)f' = 0 \tag{6}$$

$$\frac{1}{Pr}\theta'' + f\theta' + Nb\phi'\theta' + Nt\theta'^2 + Ecf''^2 + S\theta = 0 \tag{7}$$

$$\phi'' + Lef\phi' + \frac{Nt}{Nb}\theta'' - Le\gamma\phi = 0 \tag{8}$$

$$f(0) = 0,\ f'(0) = 1,\ \theta(0) = 1,\ \phi(0) = 1,$$

$$f'(\infty) = 0,\ \theta(\infty) = 0,\ \phi(\infty) = 0 \tag{9}$$

Here, $\Pr = \frac{v}{a}$, the Pradtl number, $M = \frac{2\sigma B_0^2}{a\rho f(n+1)}$, the magnetic parameter, $Ec = \frac{u_w^2}{c_p(T_w - T_\infty)}$, the Eckert number and the other parameters are described in the results and discussion section.

3 Physical Quantities of Interest

The practical interest of the quantities like the local skin friction C_{fx}, Nusselt number Nu_x which is defined as

$$C_{fx} = \frac{\mu_f}{\rho u_w^2}\left(\frac{\partial u}{\partial y}\right)_{y=0},\ Nu_x = \frac{xq_w}{k(T_w - T_\infty)} \tag{11}$$

where k, the nanofluid thermal conductivity and the heat flux q_w near the surface is given by

$$q_w = -\left[\frac{\partial T}{\partial y}\right]_{y=0}, \tag{12}$$

Substituting the similarity forms into Eqs. (11)–(12), we get
$\mathrm{Re}_x^{1/2} C_{fx} = \sqrt{\frac{n+1}{2}} f''(0)$, $\mathrm{Re}_x^{-1/2} Nu_x = -\sqrt{\frac{n+1}{2}}\theta'(0)$,

where $\mathrm{Re}_x = u_w \frac{x}{v}$ is the local Reynolds number.

4 Results and Discussion

Magnetohydrodynamics flow of steady, incompressible, viscous, and electrically conducting water-based nanofluid past over a nonlinear stretching surface through a porous medium is considered in the present paper. Internal heat generation/absorption and first order chemical reaction are also taken into account in the energy and mass transfer equation, respectively. The transformed governing nonlinear ordinary equations are solved numerically employing Runge–Kutta fourth order along with shooting technique. The physical significance of pertinent parameters that characterizes the flow phenomena are presented in Figs. 2, 3, 4, 5, 6, 7, 8 and 9 and discussed. Also, the numerical computation for the rate of shear stress, Nusset number, and Sherwood number are presented in Table 1.

Figure 2 presents the comparison plot visa-a-vis the variation of magnetic parameter (M) and nonlinear stretching parameter (n) in both the presence/absence of porous matrix and fixed values of other physical parameters shown in the figure caption. To validate the present solution, comparisons have been made with the earlier published result of Mahmood et al. [18] in a particular case by taking $M = 0$ (dotted line) and $M = 1$ (bold line) and are found to be an excellent agreement. Also, it is observed that the velocity of nanofluid reduces with an increase in the magnetic parameter and nonlinear stretching parameter for both the absence of porous matrix, Kp ($Kp = 100$)/the presence of porous matrix, Kp ($Kp = 0.5$). It is due to the fact that the body force, magnetic field acts transversely to the main direction of the flow, has retards the velocity profile for which the boundary layer thickness is

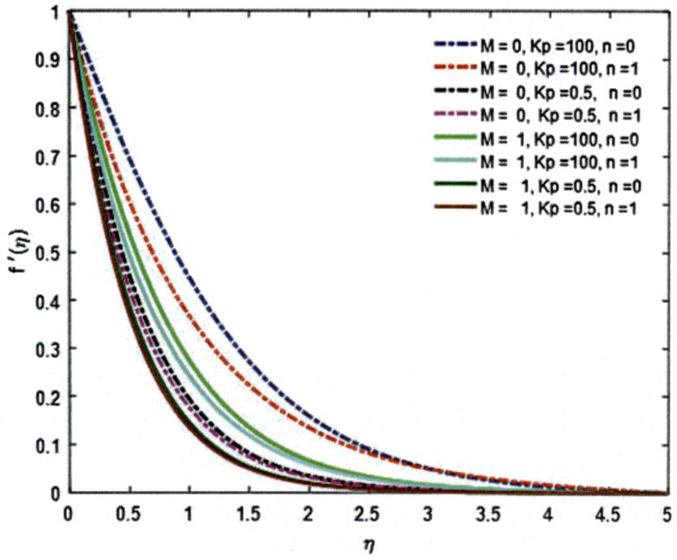

Fig. 2 Effects of M, Kp and n on velocity profiles for $\Pr = 6.2$, $Le = 5$, $Nt = Nb = Ec = 0.1$

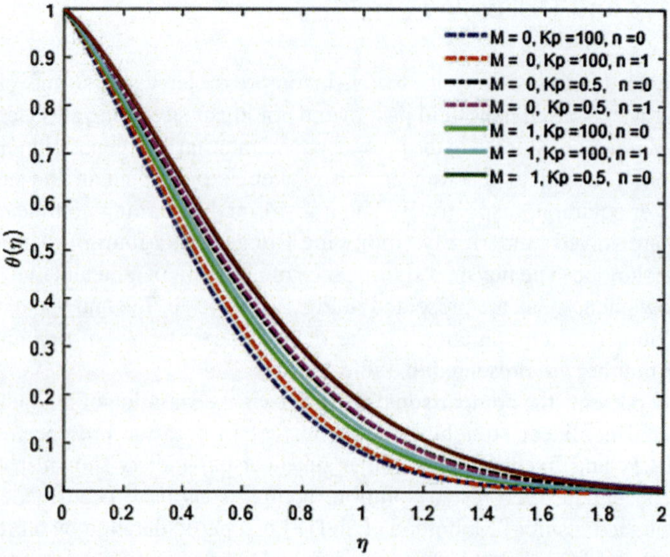

Fig. 3 Effects of M, Kp and n on temperature profiles for $\Pr = 6.2$, $Le = 5$, $Nt = Nb = Ec = 0.1$

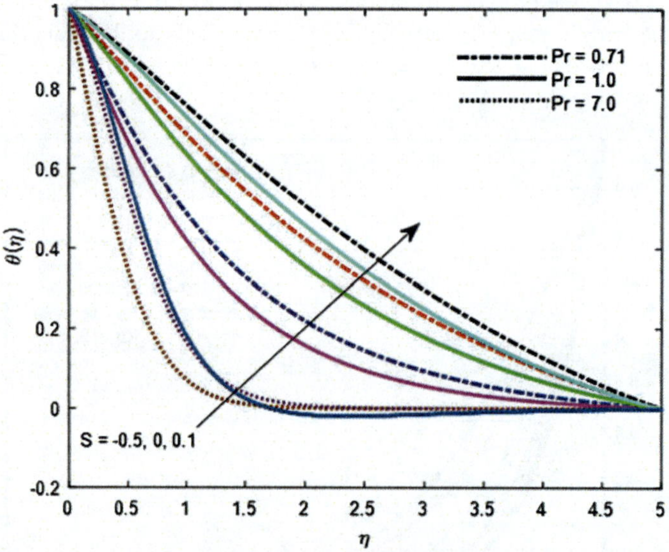

Fig. 4 Effects of Pr and S on temperature profiles for $M = 1$, $Kp = 0.5$, $Le = 5$, $Nt = Nb = Ec = 0.1$

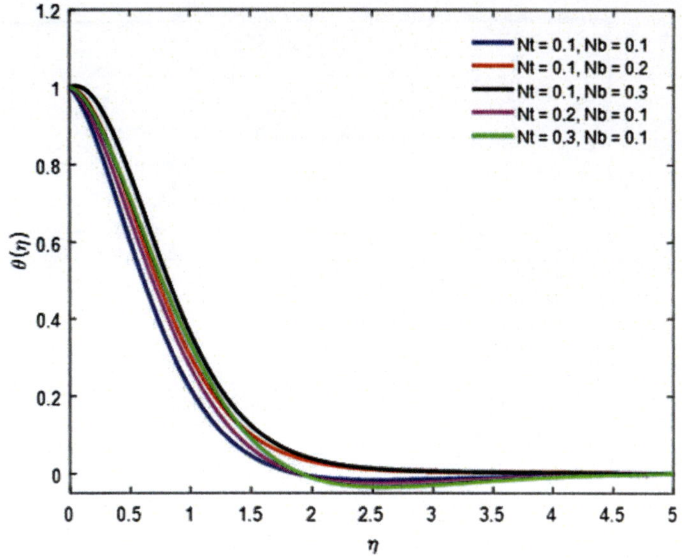

Fig. 5 Effects of Nt and Nb on temperature profiles for $M = 1, Kp = 0.5, Le = 5, Ec = 0.1, n = 2$

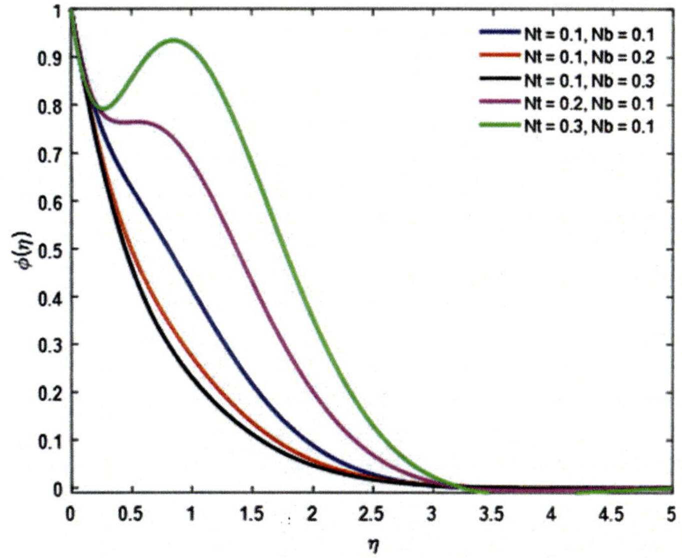

Fig. 6 Effects of Nt and Nb on concentration profiles for $M = 1, Kp = 0.5, Le = 5, Ec = 0.1, n = 2$

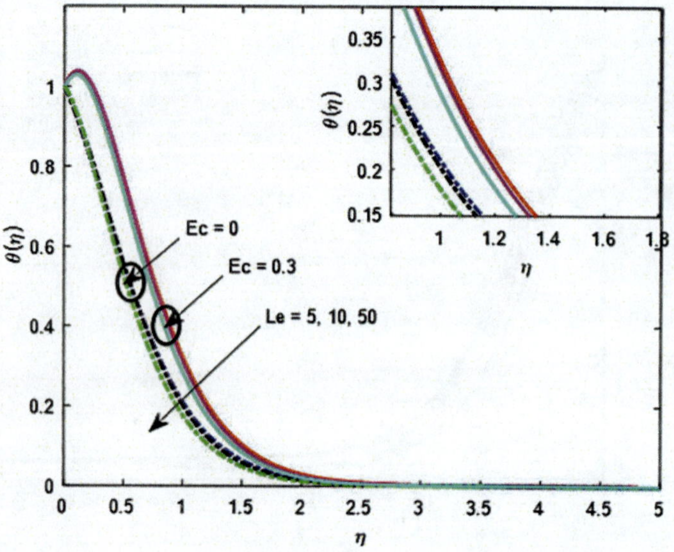

Fig. 7 Effects of Ec and Le on temperature profiles for $M = 1$, $Kp = 0.5$, $Nt = Nb = 0.1$, $n = 2$

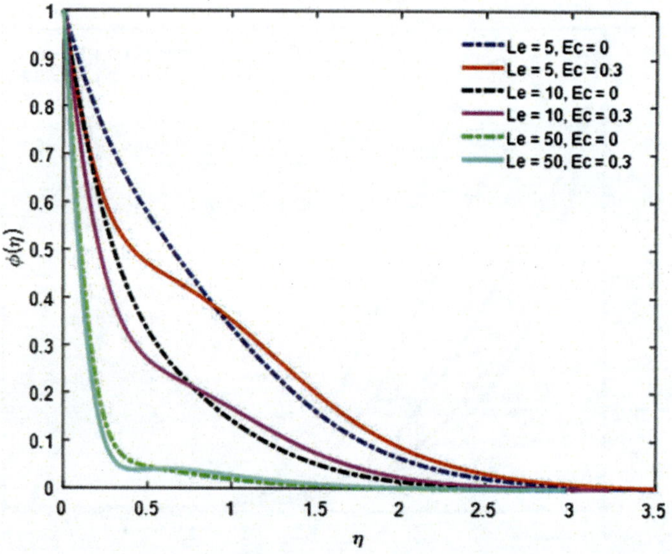

Fig. 8 Effects of Ec and Le on concentration profiles for $M = 1$, $Kp = 0.5$, $Nt = Nb = 0.1$, $n = 2$

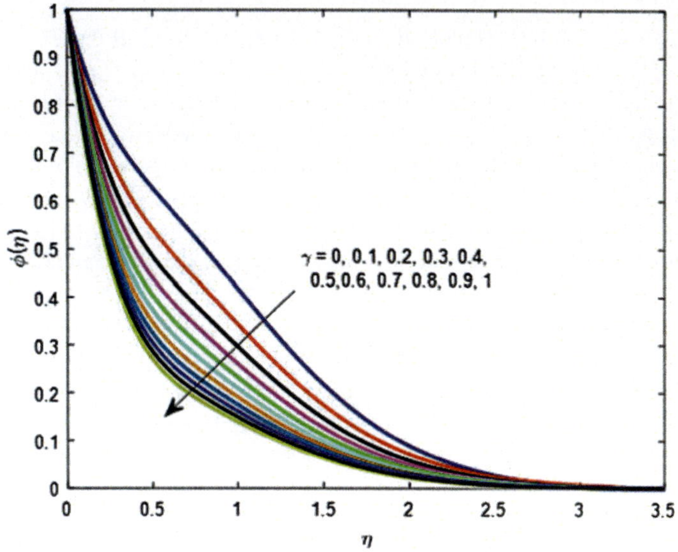

Fig. 9 Effect of γ on concentration profiles for $M = 1, Kp = 0.5, Nt = Nb = 0.1, Ec = 0.1, n = 2$

Table 1 Skin friction coefficient ($f''(0)$), Nusselt number ($-\theta'(0)$) and Sherwood number($-\phi'(0)$)

n	M	Kp	Nb	Nt	S	γ	$f''(0)$	$-\theta'(0)$	$-\phi'(0)$
0	0	100	0.5	0.1	0	0	−0.63698	0.524741	0.920767
0.2	0	100					−0.77492	0.513748	0.899625
	1						−1.25605	0.461033	0.79604
	2						−1.60402	0.424222	0.72809
	1	0.5					−1.88701	0.396327	0.679353
			0.3				−1.88701	0.498033	0.619042
			0.1				–	0.617811	0.300959
				0.3			–	0.548483	−0.36844
				0.5			–	0.486782	−0.82737
				0.1	−0.5		–	1.17046	−0.18712
					0.5		–	0.50351	0.343971
						−0.5	–	0.642853	1.782863
						0.5	–	0.474407	1.340238

also decreases. Similarly, Fig. 3 also exhibits the comparison plot of the temperature profile in a particular case for the absence of heat generation parameter $S(S = 0)$ along with the effects of M and n in both the absence/presence of porous matrix on the dimensionless temperature of the nanofluid. In the absence of S, the present result is in good agreement with the result of Mahmood et al. [18] when $M = 0$,

$n = 0$ and $Kp = 100$. It is clear to note that, the effect of all these parameters is reversed as that of Fig. 2 discussed earlier. That is, an increase in M and n in both the absence/presence of porous matrix enhances the temperature profile at all points in the thermal boundary layer. Since, Lorentz force, a resistive force, opposes the fluid motion, and then heat is produced and as a result temperature of the nanofluid increases.

Figure 4 exhibits the effects of Prandtl number and heat absorption/generation parameter on the temperature profile in the presence of magnetic and porous matrix and other fixed parameters presented in the figure caption. It is clear to note that the higher value of Prandtl number decelerates the temperature of nanofluid as a result thermal boundary layer thickness also decreases, whereas heat generation ($S = 1$) is favorable to enhance the profile. Further, the impact is opposite in the case of heat absorption ($S = 0.5$), i.e., heat absorption parameter retards the profile asymptotically.

Figures 5 and 6 illustrate the effects of thermophoresis parameter Nt and Brownian motion parameter Nb on the temperature and concentration profile, respectively, for fixed values of the other physical parameters. It is seen that both temperatures of the nanofluid and dimensionless concentration increases with the increasing values of the thermophoretic parameter significantly. Thermophoretic parameter generated by the temperature gradient, intended to flow away from the stretching sheet leading to an increase in temperature and concentration of the nanofluid. Further, the increase in Brownian motion parameter retards the nanoparticle volume fraction leads to thicken the concentration boundary layer whereas temperature profile enhances.

Figures 7 and 8 present the effects of Eckert number, Ec and Lewis number, Le on the temperature and concentration profiles, respectively. It is observed that with an increase in Lewis number, both the temperature and nanoparticle concentration decreases and this decrease is insignificant. Reduction in thermal boundary layer thickness is marked with an increase in Lewis number resulted in decrease in temperature and this also clear from the definition of Lewis number. Larger Le will also suppress concentration values. Further, an increase in Eckert number, the temperature of the nanofluid increases. Concentration fluctuation is remarked in the concentration boundary layer with a decrease in nanoparticle concentration. Absence of Ec corresponds to no viscous dissipation.

Figure 9 illustrates the effect of chemical reaction parameter on the concentration profile. The absence of chemical reaction parameter, $\gamma(\gamma = 0)$ corresponds to no chemical reaction and it is found that the present result coincides with the works of Mahbood et al. [18]. Further, with an increase in chemical reaction parameter, the nanoparticle concentration decreases significantly, as well as the concentration boundary layer thickness also decreases. It is clear to mark that, higher value of $\gamma(\gamma = 1)$ the minimum value of concentration is rendered near the plate.

Finally, Table 1 presents the rate of shear stress, rate of heat, and mass transfer for different physical parameters. To validate the present result, comparisons have been made with earlier published result of Rana and Bhargava [19] and Mahbood et al. [18] in a particular case for $n = 0$, $M = Ec = 0$, $\Pr = Le = 2$ and absence of heat source ($S = 0$) and chemical reaction parameter ($\gamma = 0$). It is seen that the present result is in good agreement. From Table 1, it is clear to observe that with an increase in surface parameter n and magnetic parameter (M), the rate of shear stress increases but both the rate of heat and mass transfer decreases for both absence or presence of porous matrix. It is interesting to note that the Brownian motion parameter and thermophoresis parameter have an opposite effect on Nusselt and Sherwood number. An increase in Nb Nusselt number decreases and Sherwood number increases whereas reverse effect is encountered as Nt increases. When we increase the values of S (sink to source), the chemical reaction (constructive to destructive) rate of heat transfer decreases, whereas the rate of mass transfer increases. Hence, it is to conclude that heat sink favors into generate heat near the plate and destructive chemical reaction is account for to increase the rate of mass transfer in a nanofluid.

5 Conclusive Remarks

The study concludes with the following results:

- The boundary layer thickness decreases with an increase in magnetic parameter for both the absence/presence of the porous matrix.
- Temperature of the nanofluid and dimensionless concentration increases with increasing values of the thermophoretic parameter, whereas the Brownian motion parameter retards the nanoparticle volume fraction.
- Increase in Lewis number both the temperature and nanoparticle concentration decreases.
- Brownian motion parameter and thermophoresis parameter have an opposite effect on Nusselt and Sherwood number.

References

1. B.C. Sakiadis, Boundary layer behaviour on continuous solid surfaces. AlChe. J. **7**, 26–28 (1961)
2. B. Magyari, Keller, heat mass transfer in the boundary layer on an exponentially stretching sheet. J. Phys. D Appl. Phys. **32**, 577–585 (1999)
3. E.M.A. Elbashbeshy, Heat transfer over an exponentially stretching continuous surface with suction. Arch. Mech. **53**(6), 643–651 (2001)
4. K. Vajravelu, Viscous flow over a nonlinear stretching sheet. Appl. Math. Comput. **124**, 281–288 (2001)
5. K. Vajravelu, J.R. Cannon, Fluid flow over a nonlinear stretching sheet. Appl. Math. Comput. **181**, 609–628 (2006)
6. R. Cortell, Viscous flow and heat transfer over a nonlinearly stretching sheet. Appl. Math. Comput. **184**, 864–873 (2007)
7. R. Cortell, Effect of viscous dissipation and radiation on the thermal boundary layer over a nonlinearly stretching sheet. Phys. Lett., Sect. A. **372**(5), 631–636 (2008)

8. F.M. Hady, F.S. Ibrahim, S.M. Abdel-Gaied, Md. R. Eid, Radiation effect on viscous flow of a nanofluid and heat transfer over a nonlinearly stretching sheet. Nanoscale Res. Lett. **7**(229), 13 pp. (2012)
9. K. Vajravelu, D. Rollins, Heat transfer in electrically conducting fluid over a stretching sheet. Int. J. Non-Linear Mech. **27**(2), 265–277 (1992)
10. Ellahi, R., Shivanian, E., Abbasbandy, Rahman, S.U., Hayat, T, Analysis of steady flow in viscous fluid with heat\mass transfers and slips effects. Int. J. Heat Mass Transfer. **55**, 6384–6390 (2012)
11. R. Ellahi, M. Hammed, Numerical analysis of steady flows with heat transfer analysis, MHD and nonlinear slip effects. Int. J. Numer. Methods Heat Fluid Flow. **22**(1), 24–38 (2012)
12. R. Ellahi, The effects of MHD and temperature dependent viscosity on the flow of non-Newtonian nanofluid in a pipe: analytic solutions. Appl. Math. Model. **37**(3), 1451–1467 (2013)
13. R. Ellahi, S. Aziz, A. Zeeshan, Non-Newtonian nanofluid flow through a porous medium between two coaxial cylinders with heat transfer and variable viscosity. J. Porous Media. **16**(3), 205–216 (2013)
14. A. Zeeshan, R. Ellahi, Series solutions for nonlinear partial differential equations with slip boundary conditions for non-Newtonian MHD fluid in porous space. J. Appl. Math. Inf. Sci. **7**(1), 253–261 (2013)
15. D. Bhukta, G.C. Dash, S.R. Mishra, heat and mass transfer on MHD flow of a viscoelastic fluid through porous media over a shrinking sheet. Int. Schol. Res. Not. 11 pp. Article ID 572162 (2014)
16. B. Mohanty, S.R. Mishra, H.B. Pattnaik, Numerical investigation on heat and mass transfer effect of micropolar fluid over a stretching sheet. Alexandria Eng. J. **54**(2), 223–232 (2015)
17. G.C. Dash, R.S. Tripathy, M.M. Rashidi, S.R. Mishra, Numerical approach to boundary layer stagnation-point flow past a stretching/shrinking sheet. J. Mol. Liq. **221**, 860–866 (2016)
18. F. Mabood, W.A. Khan, A.I.M. Ismail, MHD boundary layer flow and heat transfer of nano fluids over a nonlinear stretching sheet: a numerical study. J. Magn. Magn. Mater. **374**, 569–576 (2015)
19. P. Rana, R. Bhargava, Flow and heat transfer of a nano fluid over a nonlinearly stretching sheet: a numerical study. Commun. Nonlinear Sci. Numer. Simul. **17**, 212–226 (2012)

An Overview of Transverse Vibration of Axially Travelling String

Shashendra Kumar Sahoo, H. C. Das, and L. N. Panda

Abstract The paper presents a comprehensive review of transverse vibration of axially travelling string. Axially travelling string includes many mechanical systems such as conveyor belt, aerial tramways, magnetic tapes, textile fibres, band-saw blades, thread lines and pipe conveying fluid. The analysis of transverse vibration of axially travelling string has theoretical as well as industrial significance because it is the simplest representation of the gyroscopic system. In the paper, the major emphasis is given on different types of modelling, the governing equations and the method of analysis of axially travelling string. In the discussion of linear model of axially travelling string, the paper describes about governing equation, modal analysis and response solution. In the discussion of non-linear model of axially travelling string, the paper discusses about governing equation, the method of analysis using direct perturbation method and the discretised perturbation method and numerical methods based on Galerkin discretization. A discussion is also made on the modelling of the dissipative mechanism by considering string as viscoelastic material based on Kelvin viscoelastic model. The paper also describes about linear and non-linear parametric excitation of axially travelling string which occurs due to tension and velocity fluctuation.

1 Introduction

Axially travelling strings include many mechanical systems such as conveyor belt, aerial tramways, magnetic tapes, textile fibres, band-saw blades, thread lines and pipe conveying fluid The study of transverse vibration of axially travelling string has

S. K. Sahoo (✉)
Department of Mechanical Engineering, I.T.E.R, Bhubaneswar, India
e-mail: sashendrasahoo@soa.ac.in

H. C. Das
Department of Mechanical Engineering, N.I.T, Shillong, Meghalaya, India

L. N. Panda
Department of Mechanical Engineering, C.E.T, BPUT, Bhubaneswar, India

S. R. Mishra et al. (eds.), *Recent Trends in Applied Mathematics*, Lecture Notes in Mechanical Engineering, https://doi.org/10.1007/978-981-15-9817-3_29

theoretical as well as industrial significance because it is the simplest representation of the gyroscopic system. The gyroscopic system includes both translation and rotation. Mote et al. [1–5] and Wickert et al. [6–8] made extensive research on linear and non-linear analysis of axially translating string. Pakdemirli et al. [9, 10] studied the vibration analysis of axially travelling string using both Galerkin's discretization and perturbation method and made the stability analysis using Floquet theory.

Chen et al. [11–15] analysed the stability analysis of axially accelerating viscoelastic string with two-to-one parametric resonance. Zhang et al. [16–19] investigated the non-linear analysis of axially translating string and studied the stability analysis using the method of multiple scales, a direct perturbation method. Oz et al. [20–22] studied both linear and non-linear vibration analysis of travelling tensioned beam. Panda and Kar [23, 24] studied the combination and principal parametric resonance with 3:1 internal resonance of pipe transporting pulsatile fluid. Zhou et al. [25] studied the dynamics of pipes transporting fluid considering super-harmonic resonance of the second mode with 1:2 internal resonance. Javadi et al. [26] studied the divergence and flutter instabilities of pipes transporting fluid assuming Euler–Bernoulli beam theory and viscoelastic damping.

In the paper, a comprehensive review on the vibration analysis of axially travelling string is presented. The emphasis is given on the recent developments on the vibration analysis of axially travelling string with earlier studies taken for reference. The organisation of the paper is as such: Sect. 2 discusses about the governing equation and the dynamic analysis of linear vibration of axially travelling string. Section 3 discusses about the governing equation and the dynamic analysis of non-linear vibration of axially travelling string. Section 4 represents the governing equation and the dynamic analysis of linear parametric vibration of axially travelling string. Section 5 discusses about the governing equation and the dynamic analysis of non-linear parametric vibration of travelling string. Section 6 discusses about the conclusion and future research to be carried out.

2 Linear Vibration

2.1 *Axially Travelling String with Constant Transport Velocity*

2.1.1 Governing Equation

Swope and Ames [27] derived the equation of motion of axially travelling string using Newton's second law. A continuous string of length L, having mass density ρ, tension P and moving with axial transport velocity v between two fixed eyelets is considered as shown in Fig. 1.

The governing equation of motion for transverse vibration of travelling string with boundary condition is given as

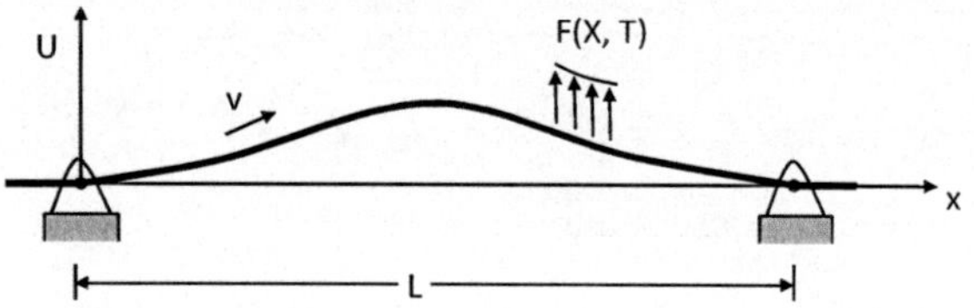

Fig. 1 An axially travelling string subjected to distributed loading

$$\rho(U_{TT} + 2VU_{XT} + V^2U_{XX}) - PU_{XX} = F \tag{1}$$

$$U(0, T) = 0, U(L, T) = 0 \tag{2}$$

where $U(X, T)$ represents the displacement of the string in transverse direction and $F(X, T)$ represents distributed loading.

U_{TT}, $2VU_{XT}$ and V^2U_{XX} represent local acceleration, coriolis acceleration and centripetal acceleration, respectively.

Using the dimensionless variables

$$x = \frac{X}{L}\ u = \frac{U}{L}\ t = T(P/\rho L^2)^{1/2}\ f = FL/P\ v = V(\rho/P)^{1/2} \tag{3}$$

The dimensionless form of the equation becomes

$$Mu_{tt} + Gu_t + ku = f \tag{4}$$

where the mass, gyral and stiffness operators are as follows:

$$M = I\ G = 2v\frac{\partial}{\partial x}\ k = -(1 - v^2)\frac{\partial^2}{\partial x^2} \tag{5}$$

I is the identity operator.

2.1.2 Method of Analysis

Wickert and Mote carried out the dynamic analysis of the transverse vibration of travelling string using modal analysis [28, 29]. The natural frequencies for nth mode for fixed end boundary conditions were calculated as follows:

$$\omega_n = n\pi(1 - v^2) \tag{6}$$

The real and imaginary components of eigenfunctions were obtained as

$$\psi_n^R(\chi) = \frac{1}{n\pi}\sqrt{\frac{2}{1 - v^2}}\sin(n\pi x)\cos(n\pi vx) \tag{7}$$

$$\psi_n^I(\chi) = \frac{1}{n\pi}\sqrt{\frac{2}{1-v^2}}\sin(n\pi x) sin(n\pi v x) \tag{8}$$

The response solution is given as

$$u(x,t) = \sum_{n=1}^{\infty}\frac{1}{n\pi}\sqrt{\frac{2}{1-v^2}}(\xi_n^R(t)\sin(n\pi x)\cos(n\pi v x)$$
$$+ \xi_n^I(t)\sin(n\pi x)\sin(n\pi v x) \tag{9}$$

where $\xi_n^R(t)$ and $\xi_n^I(t)$ are the components of the generalised coordinates.

2.2 *Axially Travelling String Over Elastic Foundation*

2.2.1 Governing Equation

Wickert [8] considered an elastic supported axially travelling string with elastic foundation modulus k and with transport velocity v between two fixed supports as shown in Fig. 2.

The governing equation of motion in non-dimensional form with boundary conditions is given as

$$u_{tt} + 2vu_{,xt} + (v^2 - 1)u_{,xx} + ku = f(x,t) \tag{10}$$

$$u(0,t) = u(1,t) = 0 \tag{11}$$

where $u(x,t)$ is the non-dimensional displacement in transverse direction for $x \in (0, 1)$ and $f(x,t)$ is the exciting force.

2.2.2 Method of Analysis

Using modal analysis [28, 29], the natural frequencies for nth mode were obtained as

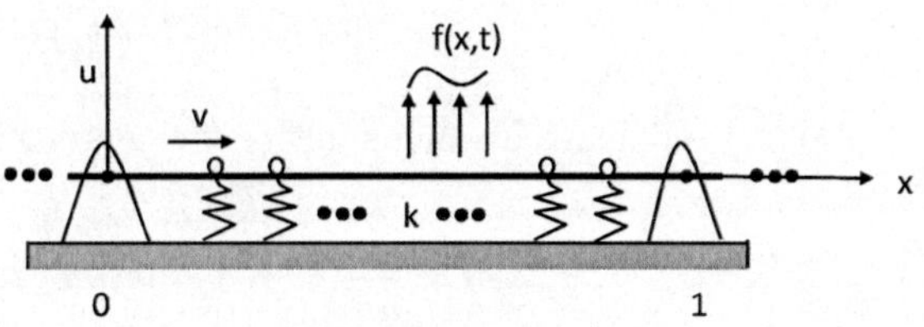

Fig. 2 An axially travelling string over an elastic foundation

$$\omega_n = n\pi(1 - v^2)\sqrt{1 + \frac{k}{(n\pi)^2(1 - v^2)}} \qquad for\, n = 1, 2, \ldots \tag{12}$$

The real and imaginary components of eigenfunctions are

$$\psi_n^R = \left(\frac{1}{n\pi}\sqrt{\frac{2}{1 - v^2}}\right)\left(\frac{1}{\sqrt{1 + \frac{k}{(n\pi^2)(1 - v^2)}}}\right)\sin(n\pi x) \times \cos\left(\sqrt{1 + \frac{k}{(n\pi)^2(1 - v^2)}}n\pi v x\right)$$

$$\psi_n^I = \left(\frac{1}{n\pi}\sqrt{\frac{2}{1 - v^2}}\right)\left(\frac{1}{\sqrt{1 + \frac{k}{(n\pi^2)(1 - v^2)}}}\right)\sin(n\pi x) \times \sin\left(\sqrt{1 + \frac{k}{(n\pi)^2(1 - v^2)}}n\pi v x\right) \tag{13}$$

The general solution for nth mode (normalised) considering free vibration is

$$u(x, t) = \left(\frac{1}{n\pi}\sqrt{\frac{2}{1 - v^2}}\right)\left(\frac{1}{\sqrt{1 + \frac{k}{(n\pi^2)(1 - v^2)}}}\right)\sin(n\pi x) \times \cos\left(\sqrt{1 + \frac{k}{(n\pi)^2(1 - v^2)}}(n\pi(1 - v^2)t + n\pi v x)\right) \tag{14}$$

Perkins [30] examined the dynamic response of an axially travelling string over an elastic foundation. Exact solutions were obtained for travelling string across a discrete foundation and a uniform step foundation. He concluded that the critical speed is not affected by the elastic foundations. The presence of a continuous elastic foundation makes the string dispersive and creates three distinct modes of vibration.

Parker [31] investigated the stability analysis of an axially travelling string over discrete elastic foundation and distributed elastic foundation. He concluded that any elastic supported string gives rise to multiple critical speeds, whereas an unsupported string has a single critical speed and stable at all supercritical speeds. A single region of divergence instability occurs above the first critical speed in case of an elastically supported string.

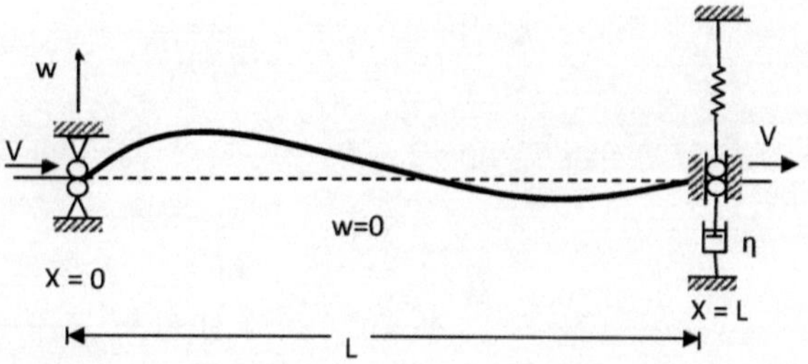

Fig. 3 An axially travelling string under boundary damping

2.3 *Axially Travelling String Under Boundary Damping*

2.3.1 Governing Equation

Malookani et al. [32] analysed an axially travelling string with one end fixed and the other end tied to the spring-dashpot system as shown in Fig. 3.

The transverse equation of motion for the travelling string with boundary conditions is

$$\rho\left(w_{tt} + 2Vw_{xt} + V^2 w_{xx}\right) - T_0 w_{xx} = 0, 0 < x < L, t \geq 0 \tag{15}$$

$$w(0, t) = 0, t \geq 0 \tag{16}$$

$$\eta w_t(L, t) + T_0 w_x(L, t) - \rho V w_t(L, t) - \rho V^2 w_x(L, t) = 0, t \geq 0 \tag{17}$$

where T_0 the tension of the string, η the boundary damping parameter, ρ the mass density and V is the transport velocity of the string.

2.3.2 Method of Analysis

Approximate solutions of the homogenous linear partial differential equations were obtained by two time scales perturbation [33] and characteristic coordinate method [26]. The vertical displacement of the system was obtained under two particular initial conditions. It was concluded that the motion of the travelling string in terms of vertical displacement is damped out by increasing the damping in the system.

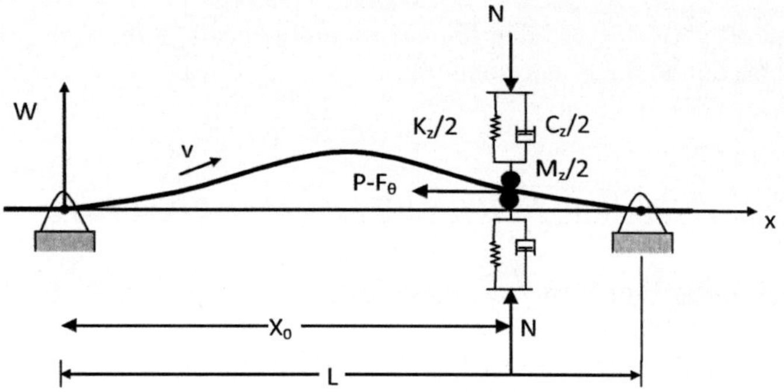

Fig. 4 An axially travelling string with a stationary load system

2.4 Axially Travelling String in Contact with a Stationary Load System

2.4.1 Governing Equation

Chen [34] considered a flexible string that slides between guides that are elastically supported as shown in Fig. 4. The upper and lower parts of the guide are identical and are modelled as spring-mass-dashpot system. A constant tension P is applied to the right of the string at $X = X_0$ and $P{-}F_\theta$ to the left of the string.

The governing equation of motion considering transverse vibration is given as

$$\rho(W_{TT} + 2VW_{XT} + V^2 W_{XX}) - FW_{XX}$$
$$= -(K_Z W + C_Z W_T + M_Z W_{TT} - F_\theta W_X)\delta^*(X - X_0) \quad (18)$$

where

$$F = \begin{cases} P, & X_0 < X < L \\ P - F_\theta, & 0 < X < X_0 \end{cases} \quad (19)$$

$\delta^*(:)$ is the Dirac delta function.

2.4.2 Method of Analysis

Using eigenfunction expansion method [28, 29], the eigenvalues are calculated. It was found that the natural frequencies of the travelling string increases with the stiffness of the load system and decreases with the inertia element of the load system. The damping of the load system brings about the stability in the system in the subcritical speed range. It was observed that flutter instability occurs in the high speed range due to dry friction in the load system.

2.5 *Axially Travelling String with a Viscous Fluid Layer*

2.5.1 Governing Equation

Huang and Mote [35] considered a thin viscous fluid that acts as a damping between translating string and a translating rigid surface.

The governing equation of motion for transverse vibration is

$$L\left(M\frac{\partial^2 v}{\partial t^2}+G\frac{\partial v}{\partial t}+Kv\right)+C\left(v\frac{\partial v}{\partial x}+\frac{\partial v}{\partial t}\right)=0 \tag{20}$$

where L is a symmetric operator describing fluid coupling. The second term in the equation represents viscous damping translating with speed v. M, G, K in the equation represent mass, gyral and stiffness operators, respectively, in the translating elastic medium.

2.5.2 Method of Analysis

The method of the slowly varying parameter was used to calculate the wave modes near the boundary separating stability and instability zones. It was noticed that the unstable modes propagate at speeds close to mean flow speed near the stability boundary when the damping force is predominant.

3 Non-linear Vibration

The linear vibration analysis is applicable to small-amplitude transverse vibration. The string is considered to be of linear material. But in the non-linear vibration analysis, the geometric non-linearity, non-linearity due to damping, material non-linearity are taken into account. The non-linear dynamics of the axially travelling string is being studied extensively in recent years because of the complex dynamical behaviours exhibited by the system.

3.1 Axial Travelling String with Geometric Non-linearity

3.1.1 Governing Equation

Mote [2] applied the Hamilton's principle to derive the governing equation of axially travelling string considering geometric non-linearity. The governing equation considering transverse vibration is given by

$$\rho A v_{tt} + 2\rho A c v_{xt} + (\rho A c^2 - P) v_{xx} + \frac{3}{2} v_{xx} v_x^2 (p - AE) = 0 \tag{21}$$

where v is the transverse displacement, A is the cross-sectional area of the string, c is the constant axial velocity, E is the elastic modulus, P is the initial tension in the string.

Thurman and Mote [4] derived the transverse and longitudinal equation of motion of axially moving strip using Hamilton's principle.

$$\begin{aligned} &\rho A v_{tt} + 2\rho A c v_{xt} + (\rho A c^2 - EA) v_{xx} + EI v_{xxxx} \\ &+ (EA - R_0) \frac{(1 + u_x)^2 v_{xx} - (1 + u_x) v_x u_{xx}}{[(1 + u_x)^2 + v_x^2]^{3/2}} = 0 \end{aligned} \tag{22}$$

$$\begin{aligned} &\rho A u_{tt} + 2\rho A c u_{xt} + (\rho A c^2 - EA) u_{xx} \\ &+ (EA - R_0) \frac{(1 + u_x) v_x v_{xx} - v_x^2 u_{xx}}{[(1 + u_x)^2 + v_x^2]^{3/2}} = 0 \end{aligned} \tag{23}$$

where $v(x, t)$ is the transverse displacement and $u(x, t)$ is the longitudinal displacement and all other notations have the usual meaning.

3.1.2 Method of Analysis

Mote used the method of characteristics [26] to calculate the fundamental period of oscillation of axially travelling string. Thurman and Mote used a perturbation technique combining Lindstedt and Krylov-Bogoliubov methods [25] to calculate the non-linear fundamental period of transverse oscillation. The results indicate that axial transport velocity decreases the fundamental period of oscillation and increases the effect of non-linear terms in the equation of motion.

3.2 *Axially Travelling String with Viscoelastic Damping*

3.2.1 Governing Equation

Zhang and Zu [15] derived the equation of motion for non-linear free vibration of viscoelastic moving belt with geometric non-linearities. The kelvin viscoelastic model composed of linear spring and linear dashpot in parallel is considered to describe the material property of the belt material. The Kelvin viscoelastic model is expressed by the linear differential operator E^*

$$E^* = E^0 + \eta \frac{\partial}{\partial t} \tag{24}$$

where E^0 is the spring stiffness constant and η is the dashpot dynamic viscosity.

The non-linear equation of motion governing transverse vibration is given as

$$\rho \frac{\partial^2 V}{\partial t^2} + 2\rho c \frac{\partial^2 V}{\partial t \partial x} + \left(\rho c^2 - \frac{T}{A}\right) \frac{\partial V}{\partial x^2} = E^*(\frac{1}{2} V_x^2) V_{xx} + V_x \left\{ E^* \left(\frac{1}{2} V_x^2 \right) \right\}_x \tag{25}$$

where V is the transverse displacement of the belt, c is the axial velocity of the belt, ρ is the mass density and A is the cross-sectional area of the belt and T is the initial Tension.

3.2.2 Method of Analysis

Using the method of multiple scales [33], the non-linear frequencies and the amplitude of response were calculated. It was concluded that the amplitude of the response is affected by the damping of the system, whereas the non-linear natural frequencies are not affected. The non-linear natural frequencies decrease with an increase in the axial velocity of the string.

4 Linear Parametric Vibration

Large transverse vibration occurs in the axially travelling string due to periodical fluctuation of their parameters. A parametric excitation gives rise to large instability even if the frequency of the fluctuation is away from linear natural frequencies of the system. The major two factors contributing towards parametric vibration of axially travelling string are the tension fluctuation and the velocity fluctuation.

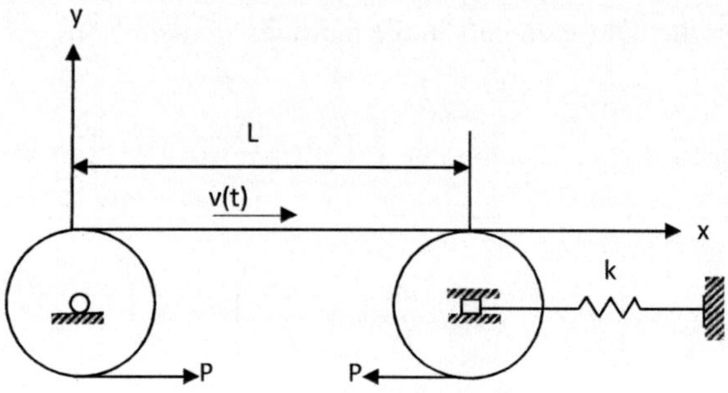

Fig. 5 Schematic diagram of an axially travelling string on pulley mounting system

4.1 Parametric Vibration of Travelling String Due to Velocity Fluctuation

4.1.1 Governing Equation

Pakdermili et al. [9] studied the parametric excitation of an axially travelling string passing over two pulleys at a transport velocity $v(t)$ which is a function of time as shown in Fig. 5.

The governing equation of motion was derived using Hamilton's principle as

$$\rho A(\ddot{y} + \dot{v}y' + 2v\dot{y}') + (\rho A v^2 - P)y'' = 0 \tag{26}$$

where y is the transverse displacement of the string, P is the tension force in the string given as $P = P_0 + \eta\rho A v^2$ where $0 \leq \eta \leq 1$. The pulley support parameter is given as $\kappa = 1 - \eta$. The transport velocity v is a periodical function of time expressed as $v(t) = v_0 \sin \omega_0 t$ where v_0 is the axial velocity amplitude and ω_0 is the frequency of axial velocity variation.

4.1.2 Method of Analysis

Galerkin's method [36] was used to solve the partial differential equation.

The trial function taken is

$$y(x, t) = \sum_{i=1}^{n} q_i(t) \sin(i\pi x/L) \tag{27}$$

where $\sin(i\pi x/L)$ is the ith eigenfunction of the simply supported stationary string and $q_i(t)$ is the generalised displacement.

Applying Galerkin's method, the discretised governing equation obtained is

$$M\ddot{q} + C\dot{q} + K\dot{q} = 0 \tag{28}$$

where the elements in the matrices are defined as

$$m_{ij} = \int_0^L \left(\sin\frac{i\pi x}{L} \sin\frac{j\pi x}{L} \right) dx = \begin{cases} L/2, i = j \\ 0, i \neq j \end{cases} \tag{29}$$

$$c_{ij} = \left(\int_0^L \frac{2\pi i v}{L} \cos\frac{i\pi x}{L} \sin\frac{j\pi x}{L} \right) dx = \begin{cases} 0, i = j \\ 0, i \neq j, i + j = 2n \\ 4ijv/(j^2 - i^2), i \neq j, i + j = 2n + 1 \end{cases} \tag{30}$$

$$k_{ij} = \int_0^L \left\{ \left(\frac{p}{\rho A} - v^2 \right) \left(\frac{i\pi}{L} \right)^2 \sin\frac{i\pi x}{L} \sin\frac{j\pi x}{L} + \frac{\dot{v}\pi i}{L} \cos\frac{i\pi x}{L} \sin\frac{j\pi x}{L} \right\} dx \tag{31}$$

Taking the one term approximation in the series solution yields the Mathieu equation whose solution gives the transition curves separating stability and instability zones. Taking two term approximation in the series solution leads to gyroscopically coupled equation. The stability analysis was obtained using Floquet theory. The two term approximation represents the system behaviour better.

Pakdermili and Ulsoy [10] found an approximate analytical solution using the method of multiple scales [33] applied directly to the partial differential equation. Principal parametric resonance is investigated and transition curves showing stable and unstable zones are calculated for a band-saw vibration. Taking numerical values $P_0 = 76.22$ N, $\rho = 7754$ kg/m^3, $A = 0.5201 \times 10^{-5}$m^2, $k = 0.22$ and $L = 0.3681$ m, the stability boundaries for first five natural frequencies were plotted as shown in Fig. 6. It was observed that the instability zone for the fifth natural frequency appears as a line. They concluded that instability occurs when the fluctuation frequency of the velocity is nearly equal to two times the natural frequency and no instability occurs when changing frequency is close to zero.

4.2 *Parametric Vibration of Travelling String Over Elastic Foundation*

4.2.1 Governing Equation

Ghayesh [37] studied the parametric vibration analysis of an axially travelling string over a partial elastic foundation. The string is divided into three segments. The segment in the span $a < x^* < (a + b)$ is supported by elastic foundation and other segments $0 < x^* < a$ and $(a + b) < x^* < (a + b + c)$ are free to vibrate.

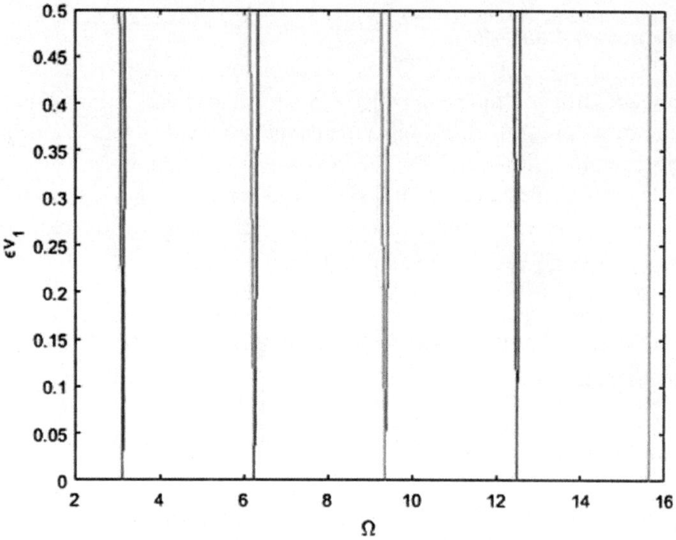

Fig. 6 Stability boundaries for the first five natural frequencies for a band-saw with pulley support parameter ($k = 0.22$) and dimensionless mean speed ($v_0 = 1.15011$)

For the sement in the spans $0 < x^* < a$ and $(a + b) < x^* < (a + b + c)$, the non-dimensional equation of motion is given as

$$\frac{\partial^2 u}{\partial t^2} + \frac{dc_v}{dt}\frac{\partial u}{\partial x} + 2c_v\frac{\partial^2 u}{\partial x \partial t} + (c_v^2 - 1)\frac{\partial^2 u}{\partial x^2} = 0 \tag{32}$$

for

$$\begin{aligned} u &= u^*/(a+b+c), \\ x &= x^*/(a+b+c), \\ t &= t^*\sqrt{T/\rho A(a+b+c)^2} \\ c_v(t) &= v\sqrt{\rho A/T}, \end{aligned} \tag{33}$$

where T is the tension in the string, ρ is the density and A is the cross-sectional area of the string.

Taking k as the stiffness coefficient per unit length of the foundation, the non-dimensional equation of motion for the span $a < x^* < (a + b)$ takes the form

$$\frac{\partial^2 u}{\partial t^2} + \frac{dc_v}{dt}\frac{\partial u}{\partial x} + 2c_v\frac{\partial^2 u}{\partial x \partial t} + (c_v^2 - 1)\frac{\partial^2 u}{\partial x^2} + \kappa^2 u = 0 \tag{34}$$

$$k = \sqrt{k(a+b+c)/T} \tag{35}$$

4.2.2 Method of Analysis

Using the method of multiple scales [33], mode shapes and natural frequencies of the system were determined. Stability analysis was carried out for principal and combination parametric resonance. The increase in the axial mean velocity tends to decrease the natural frequency of the system, whereas an increase in the stiffness factor and foundation length increases the natural frequency of the system.

4.3 Parametric Vibration of Travelling String Due to Tension Fluctuation

4.3.1 Governing Equation

Mahalingam [38] considered a travelling chain as a uniform axially travelling string.

The travelling chain is having m as the mass of chain per unit length, T as the chain tension, V as the speed of chain. The tension in the chain due to longitudinal excitation at any instant is given as $T(t) = T + \Delta T \cos \omega t$ where ω is the circular frequency of exciting force. The governing equation of motion for travelling chain becomes

$$\left(V_0^2 + \Delta V_0^2 \cos \omega t\right)\frac{\partial^2 y}{\partial x^2} = V^2 \frac{\partial^2 y}{\partial x^2} + 2V \frac{\partial^2 y}{\partial x \partial t} + \frac{\partial^2 y}{\partial t^2} \tag{36}$$

where $V_0 = \sqrt{(T/m)}$ is the velocity of propagation of transverse waves.

4.3.2 Method of Analysis

Substituting

$$y = e^{i\mu x} y_0(t) \text{ and } \omega t = 2z \tag{37}$$

We get

$$\ddot{y}_0 + 2iV\mu\dot{y}_0 + [\mu^2(V_0^2 - V^2 + \Delta V_0^2 \cos 2z)]y_0 = 0 \tag{38}$$

Let

$$y_0 = e^{-iV\mu z} u(z) \tag{39}$$

Substituting Eq. (38) in Eq. (37), we get Mathieu equation in the form

$$\ddot{u} + (\alpha + 2q \cos 2z)u = 0 \tag{40}$$

$$\alpha = \mu^2 V_0^2 \tag{41}$$

$$2q = \mu^2 \Delta V_0^2 \tag{42}$$

The solution of Mathieu equation leads to transition curves separating stability and instability regions.

Mote [3] studied the parametric excitation of an axially translating string with a periodic tension component. The Galerkin's method was used to reduce the equation of motion to a set of coupled Mathieu equation. The transition curves showing stability-instability zones were calculated by the application of numerical method.

Naguleswaran and Williams [39] examined the parametric excitation of axially translating string due to periodic variation in the band tension caused due to wheel eccentricity and joints. They used the Galerkin method up to four term approximation to study the stability analysis similar to the solution of Mathieu equation separating stability and instability zones.

Rhodes [40] investigated the parametric self-excitation of a belt due to periodic variations in belt tension. When the belt is in transverse vibration, the length of the vibrating span varies at twice the frequency of the belt vibration and the belt tension will have an alternating component at twice the natural frequency of transverse vibration. Varying tension on the belt is recorded as stress or strain pattern when the belt winds onto the pulley.

5 Non-linear Parametric Vibration

Non-linear parametric excitation in axially travelling string gives rise to complicated dynamic behaviours such as bifurcation and chaos which occur due to tension or velocity fluctuation.

5.1 Non-linear Parametric Vibration of Axially Travelling Viscoelastic String

Chen et al. [13] considered the transverse vibration of an axially accelerating viscoelastic string with geometric non-linearity. A uniform viscoelastic string having mass density ρ, cross-sectional area A, initial tension P and travelling with speed harmonically varying about an average speed is considered.

5.1.1 Governing Equation

The non-dimensional equation of motion governing the transverse vibration of viscoelastic string with boundary conditions is given as

$$\frac{\partial^2 u}{\partial \tau^2} + 2(\gamma + \gamma_1 \cos \omega\tau)\frac{\partial^2 u}{\partial \xi \partial \tau} + \left(\gamma^2 + \frac{\gamma_1^2}{2} + 2\gamma\gamma_1 \cos \omega\tau + \frac{\gamma_1^2}{2} \cos 2\omega\tau - 1\right)$$
$$\frac{\partial^2 u}{\partial \xi^2} - \omega\gamma_1 \sin \omega\tau \frac{\partial u}{\partial \xi} = \frac{\partial}{\partial \xi}\left(\varepsilon\varsigma(\xi, \tau)\frac{\partial u(\xi, \tau)}{\partial \xi}\right), \tag{43}$$

$$u(0, t) = 0, u(1, t) = 0 \tag{44}$$

$$\varsigma(\xi, \tau) = \frac{E_e}{2}\left(\frac{\partial v(\xi, \tau)}{\partial \xi}\right)^2 + \frac{E_v}{2}\frac{\partial}{\partial \tau}\left(\frac{\partial v(\xi, \tau)}{\partial \xi}\right)^2, \tag{45}$$

$$u = \frac{U}{l}, \xi = \frac{x}{l}, \tau = \frac{t}{l}\sqrt{\frac{P}{\rho A}}, \omega = \Omega l\sqrt{\frac{\rho A}{P}}, E_e = \frac{E_0 A}{P},$$
$$E_v = \frac{\eta b}{l}\sqrt{\frac{P}{\rho A}}, \gamma = c_0\sqrt{\frac{\rho A}{P}}, \gamma_1 = c_1\sqrt{\frac{\rho A}{P}}, \varepsilon\varsigma(\xi, \tau) = \frac{A\sigma(x, t)}{P}, \tag{46}$$

where $U(x, t)$ the transverse displacement, $\sigma(x, t)$ is the stress in the axial direction, $\varepsilon_L(x, t)$ is the Lagrangian strain, E_0 is the stiffness constant of the string and η is the dynamic viscosity.

5.1.2 Method of Analysis

Using the method of multiple scales [33], the amplitude of steady state response in two-to-one parametric resonance is investigated. Lyapunov's linearized stability theory is used to study the stability analysis of trivial and nontrivial solutions.

Yang et al. [41] used a novel procedure to analyse the non-linear vibration analysis of axially travelling string. Galerkin method and subsequently the invariant manifold method are applied to the differential equation of motion to derive descretised system. The method of multi-timescale is applied to analyse 1:3 internal resonance in the system.

Mockensturm et al. [42] analysed the weakly non-linear equation of motion derived by Thurman and Mote. Galerkin's discretization method was applied to evaluate the response near the principal parametric instability regions for nontrivial limit cycle motions. It is concluded that the amplitude of nontrivial limit cycles reduces with an increase in the axial speed and ultimately vanishes for large speeds.

5.2 Non-linear Parametric Vibration of Multi-supported Axially Travelling String

5.2.1 Governing Equation

Kesimli et al. [43] studied the non-linear parametric excitation of multi-supported axially travelling string considering geometric non-linearity. The axial velocity of the string is periodically varying about an average value (Fig. 7).

The dimensionless equation of motion governing transverse vibration between any two successive supports is given as

$$\left(\ddot{w}_{m+1} + 2\dot{w}'_{m+1}v + w'_{m+1}\dot{v}\right) + \left(v^2 - 1\right)w''_{m+1} = \frac{1}{2}v_b^2\left(\sum_{r=0}^{n}\int_{\eta_r}^{\eta_{r+1}} w'^2_{r+1}dx\right)w''_{m+1} \tag{47}$$

where the dimensionless parameters are

$$w_{m+1} = \frac{w^*_{m+1}}{L}, \eta = \frac{x^*_{m+1}}{L}, t = t^*\sqrt{\frac{p}{\rho AL^2}}, v = \frac{v*}{\sqrt{p/\rho A}}, v_b^2 = \frac{EA}{p}, v_k = \frac{kL}{p}$$

where $m = 0,1,2\ldots n$, and n is the number of supports.

x^*_{m+1} is the position of any support from the origin.

$x_0 = 0$, $x_{n+1} = L$ the total length of the string.

x_p is the location of multiple supports.

v_b represents longitudinal rigidity and v_k represents the effect of rigidity of spring coefficient. The axial velocity is made dimensionless by dividing with the critical velocity.

$$\eta_0 = 0, \eta_{n+1} = 1$$

With boundary conditions

$$w_1(0,t) = 0, w_{n+1}(1,t) = 0 \tag{48}$$

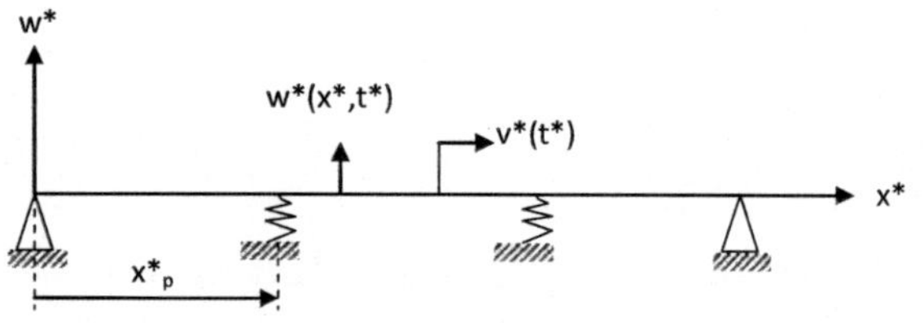

Fig. 7 Axially travelling string on multi-supported springs

$$w_p(\eta_p, t) = w_{p+1}(\eta_p, t), \tag{49}$$

$$v_{k_p} w_p(\eta_p, t) = \left[1 + \frac{1}{2}v_b^2\left(\sum_{r=0}^{n}\int_{\eta_r}^{\eta_{r+1}} w_{r+1}'^2 dx\right)\right] \times \left(w_{p+1}'(\eta_p, t) - w_p'(\eta_{p,} t)\right)(p = 0, 1, 2 \ldots n) \tag{50}$$

$\ddot{w}_{m+1}$ is the local acceleration, $2\dot{w}_{m+1}'v$ is the coriolis acceleration, $v_b^2 w_{m+1}''$ is the centripetal acceleration, η_p represents the locations of intermediate springs.

5.2.2 Method of Analysis

Approximate solutions were found using multiple scale method [33]. In case of parametric resonance, two solutions are obtained namely simple and complex. For the simple solution, amplitude is zero and for the complex solution the amplitude increases depending on the fluctuation frequency. Non-linear terms affect the natural frequency values and increase the frequency values. The increase varies depending on the spring constant, axial speed and the number of supports.

6 Conclusions

The researches activities presented in the paper show vast progress in the analysis of transverse vibration of axially travelling string. Different types of modelling, dynamic analysis, stability analysis based on linear and non-linear vibration have been discussed. Also modelling, dynamic analysis of linear and non-linear parametric vibration has been reviewed. The research activities in this area are important because axially travelling string is the simplest representation of the gyroscopic system. The future research requires an understanding of complicated dynamical behaviours such as bifurcation, chaos, patterns in transverse vibration of axially travelling string and experimental investigation on various devices that model axially travelling string as a mechanical system.

Acknowledgements The author expresses his deep gratitude to Professor Dr. L. N. Panda (C. E. T. Biju Pattnaik University of Technology) who guided him in the preparation of the paper.

References

1. C.D.J. Mote, A study of band-saw vibrations. J. Franklin Inst. **279**(6), 431–444 (1965)
2. C.D. Mote Jr., On the nonlinear oscillation of an axially moving string. J. Appl. Mech. **33**, 463–464 (1966)
3. C.D. Mote Jr., Parametric excitation of an axially moving string. J. Appl. Mech. **35**(1), 171–172 (1968)
4. A.L. Thurman, C.D. Mote Jr., free, periodic, nonlinear oscillation of an Axially moving strip. J. Appl. Mech. **36**(1), 83–91 (1969)
5. A.G. Ulsoy, C.D.J. Mote, Analysis of band-saw vibration. Wood Sci. **13**, 1–10 (1980)
6. J.A. Wickert, C.D. Mote Jr, Classical vibration analysis of axially moving continua. J. Appl. Mech. ASME **57**, 738–744 (1990)
7. J.A. Wickert, C.D. Mote Jr, Linear transverse vibration of an axially moving string particle system. J. Acoust. Soc. Am. **84**(3), 963–969 (1988b)
8. J.A. Wickert, Response Solutions for the vibration of a travelling string on an elastic foundation. J. Vibr. Acoust. **116**(1), 137–139 (1994)
9. M. Pakdemirli, A.G. Ulsoy, A. Ceranoglu, Transverse vibration of an axially accelerating string. J. Sound Vib. **169**(2), 179–196 (1994)
10. M. Pakdemirli, A.G. Ulsoy, Stability analysis of an axially accelerating string. J. Sound Vib. **203**, 815–832 (1997)
11. L.Q. Chen, N.H. Zhang, J.W. Zu, Bifurcation and chaos of an axially moving viscoelastic string. Mech. Res. Commun. **29**(2/3), 81–90 (2002)
12. L.Q. Chen, J.W. Zu, J. Wu, Principal resonance in transverse nonlinear parametric vibration of an axially accelerating viscoelastic string. Acta Mech. Sinica (English Ed) **20**(3), 307–316 (2004)
13. L.Q. Chen, J.W. Zu, J. Wu, Transverse vibrations of an axially accelerating viscoelastic string with geometric nonlinearity. J. Eng. Math. **48**(2), 171–182 (2004)
14. L.Q. Chen, Principal parametric resonance of axially accelerating viscoelastic strings constituted by the boltzmann superposition principle. Proc. Royal. Soc. Lond. (2005) (revised)
15. L.Q. Chen, N.H. Zhang, J.W. Zu, The regular and chaotic vibrations of an axially moving viscoelastic string based on 4-order galerkin truncation. J. Sound Vib. **261**(4), 764–773 (2003)
16. L. Zhang, J.W. Zu, Nonlinear vibration of viscoelastic moving belt, part 1: free vibration analysis. J. Sound Vib. **216**(1), 75–91 (1998)
17. L. Zhang, J.W. Zu, Nonlinear vibration of parametrically excited moving belt, part 1: dynamic response. J. Appl. Mech. **66**(2), 396–402 (1999)
18. L. Zhang, J.W. Zu, Nonlinear vibration of parametrically excited moving belt, part 2: stability analysis. J. Appl. Mech. **66**(2), 403–409 (1999)
19. L. Zhang, J.W. Zu, One to one auto-parametric resonance in serpentine belt drive systems. J. Sound Vib. **232**(4), 783–806 (2000)
20. H.R. Oz, M. Pakdermili, E. Ozkaya, Transition behavior from string to beam for an axially accelerating material. J. Sound Vib. **215**(3), 571–576 (1998)
21. H.R. Oz, M. Pakdermili, Vibrations of axially moving beam with time-dependent velocity. J. Sound Vib. **227**(2), 239–257 (1999)
22. H.R. Oz, M. Pakdermili, H. Boyaci, Non-linear vibrations of an axially moving beam with time-dependent velocity. Int. J. Non-linear Mech. **36**, 107–115 (2001)
23. L.N. Panda, R.C. Kar, Nonlinear dynamics of a pipe conveying pulsating fluid with parametric and internal resonances. Nonlinear Dyn. **49**, 9–30 (2007)
24. L.N. Panda, R.C. Kar, Nonlinear dynamics of a pipe conveying pulsating fluid with combination, principal parametric and internal resonances. J. Sound Vib. **309**, 375–406 (2008)
25. S. Zhou, T.-J. Yu, X.-D. Yang, W. Zhang, Global dynamics of pipes conveying pulsating fluid in the supercritical regime. Int. J. Appl. Mech. **9**, 29 (2017)
26. M. Javadi, M.A. Noorian, S. Irani, Stability analysis of pipes conveying fluid with fractional viscoelastic model. Mechanica 1–12 (2019)

27. R.D. Swope, W.F. Ames, Vibrations of a moving thread line. J. Franklin Inst. **275**, 36–55 (1963)
28. L. Meirovitch, *Analytical Methods in Vibration* (MacMillan, New York, 1967)
29. L. Meirovitch, *Principles and Techniques of Vibrations* (Prentice-Hall, New Jersey, 1997)
30. N.C. Perkins, Linear dynamics of a translating string on an elastic foundation. J. Vibr. Acoust. **112**(1), 2–7 (1990)
31. R.G. Parker, Supercritical speed stability of the trivial equilibrium of an axially moving string on an elastic foundation. J. Sound Vib. **221**(2), 205–219 (1999)
32. R.A. Malookani, S. Dehraj, S.H. Sandilo, Asymptotic approximations of the solution for a travelling string under boundary damping. J. Appl. Comput. Mech. **5**(5), 918–925 (2019)
33. A.H. Nayfeh, *Introduction to Perturbation Techniques* (Wiley, New York, 1981)
34. J.S. Chen, Natural frequencies and Stability of an axially travelling string in contact with stationary load system. J. Vibr. Acoust. **119**(2), 152–157 (1997)
35. F.Y. Huang, C.D. Mote Jr., On the translating damping caused by a thin viscous fluid layer between translating string and translating rigid surface. J. Sound Vib. **181**(2), 251–260 (1995)
36. L. Meirovitch, *Computational Methods in Structural Dynamics* (Sijthoff and Noordhoff, Groningen, The Netherlands, 1980)
37. M.H. Ghayesh, Nonlinear transversal vibration and stability of an axially moving string supported by a partial viscoelastic guide. J. Sound Vib. **314**, 757–774 (2008)
38. S. Mahalingam, Transverse vibration of power transmission chains. Br. J. Appl. Phys. **8**, 145–148 (1957)
39. S. Naguleswaran, C.J.H. Williams, Lateral vibration of band-saw, pulley belts and the like. Int. J. Mech. Sci. **10**, 239–250 (1968)
40. J.E. Rhodes, Parametric self-excitation of a belt into transverse vibration. J. Appl. Mech. **37**(4), 1055–1060 (1970)
41. X.D. Yang, H. Wu, Y.J. Qian, W. Zhang, C.W. Lim, Nonlinear vibration analysis of axially moving strings based on gyroscopic modes of decoupling. J. Sound Vib. **393**, 308–320 (2017)
42. E.M. Mochenstrum, N.C. Perkins, A.G. Ulsoy, Stability and limit cycles of parametrically excited, axially moving strings. J. Vibr. Acoust. **116**(3), 346–351 (1996)
43. A. Kesimili, E. Ozkaya, S.M. Bagdatli, Nonlinear vibrations of spring supported axially moving string. Nonlinear Dyn. **81**, 1523–1534 (2015)

A New Iterative Methods for a Nonlinear System of Equations with Third and Fifth-Order Convergence

Bijaya Mishra, Ambit Kumar Pany, and Salila Dutta

Abstract In this paper, we present a pair of iterative methods for solving a system of nonlinear equations. Both methods are constructed without using the second-order derivative. It is further shown that these iterative methods possess third and fifth order convergence respectively. Finally, some numerical experiments are given to confirm our theoretical findings.

Keywords Iterative method · Order of convergence · Nonlinear system of equations · Root-finding

1 Introduction

Solution of nonlinear equations and system of nonlinear equations using iterative methods are the most attractive fields in numerical analysis. There are good amounts of research contributions toward these areas but still there is some space for modification. The quadratically convergent Newton's approximation method [1, 2, 14] is a classic work in this regard.

Let the system of nonlinear equations be

$$F(x) = (f_1(x), f_2(x), \ldots, f_n(x))^T = 0. \tag{1}$$

B. Mishra
Department of Mathematics, Gandhi Institute for Technological Advancement, Bhubaneswar 752054, India
e-mail: bijayamishra.math@gmail.com

A. K. Pany (✉)
Center for Applied Mathematics and Computation, Siksha O Anusandhan University, Bhubaneswar 751030, Odisha, India
e-mail: ambit.pany@gmail.com

S. Dutta
Department of Mathematics, Utkal University, Bhubaneswar 751004, Odisha, India
e-mail: saliladutta516@gmail.com

S. R. Mishra et al. (eds.), *Recent Trends in Applied Mathematics*, Lecture Notes in Mechanical Engineering, https://doi.org/10.1007/978-981-15-9817-3_30

where each function f_i maps a vector $x = (x_1, x_2, \ldots, x_n)^T$ of n-dimensional space $\mathbb{R}^n$ into the real line $\mathbb{R}$. System (1) involves n nonlinear equations and n unknowns.

We need to find a vector $\alpha = (\alpha_1, \alpha_2, \ldots, \alpha_n)^T$ such that $F(\alpha) = 0$. Newton's method to solve (1) involves the scheme

$$x_{n+1} = x_n - \frac{F(x_n)}{F'(x_n)} \quad n = 0, 1, 2, \ldots, \tag{2}$$

where $F'(x_n)$ is the first Frechet derivative of $F(x_n)$.

Subsequently, many Newton type iterative methods have been developed by using different quadrature rules to solve the system of nonlinear equations with third-order convergence, see [3–10]. To discuss a few of them, we have

Noor and Waseem [4] have established two cubically convergent methods using open quadrature formula.

$$x_{n+1} = x_n - \left[F'(x_n) + 3F'\left(\frac{x_n + 2y_n}{3}\right)\right]^{-1} 4F(x_n), \quad n = 0, 1, 2, \ldots \tag{3}$$

and

$$x_{n+1} = x_n - \left[3F'\left(\frac{2x_n + y_n}{3}\right) + F'(y_n)\right]^{-1} 4F(x_n), \quad n = 0, 1, 2, \ldots \tag{4}$$

where $y_n = x_n - \frac{F(x_n)}{F'(x_n)}, \quad n = 0, 1, 2, \ldots$

Further, Liu et al. [3] have developed a cubically convergent method using two-point Gauss quadrature formula.

$$x_{n+1} = x_n - \frac{2F(x_n)}{\left[F'(\frac{x_n+y_n}{2} - \frac{y_n-x_n}{2\sqrt{3}}) + F'(\frac{x_n+y_n}{2} + \frac{y_n-x_n}{2\sqrt{3}})\right]}, \quad n = 0, 1, 2, \ldots \tag{5}$$

where $y_n = x_n - \frac{F(x_n)}{F'(x_n)}, \quad n = 0, 1, 2, \ldots$

Also they have constructed a fifth-order convergent method using two-point Gauss quadrature formula.

$$z_n = x_n - \frac{2F(x_n)}{\left[F'(\frac{x_n+y_n}{2} - \frac{y_n-x_n}{2\sqrt{3}}) + F'(\frac{x_n+y_n}{2} + \frac{y_n-x_n}{2\sqrt{3}})\right]}, \tag{6}$$

$$x_{n+1} = z_n - \frac{F(z_n)}{F'(y_n)}, \quad n = 0, 1, 2, \ldots \tag{7}$$

where $y_n = x_n - \frac{F(x_n)}{F'(x_n)}, \quad n = 0, 1, 2, \ldots.$

Some multi-step iterative methods with fith-order convegence has been developed by using quadrature rule. For detail see [11–13].

In this paper, we develop a couple of new iterative schemes to solve the system of nonlinear equations. These schemes posses third- and fifth-order convergence, respectively. Further, we estimate the Efficiency Index $E.I = p^{1/w}$, where p is the order of convergence and w is the sum of the number of functional evaluations and derivative evaluations per iteration. Additionally the computational order of convergence has been established.

This paper is organized as follows. Section 2 deals with development of new methods. In Sect. 3 the convergence of these methods are established. Finally in Sect. 4, several numerical examples are tested to find the solution of nonlinear equations and boundary value problems of nonlinear ODEs, affirming the consistency of the numerical results with the theoretical findings, and also a comparison is done corresponding to some existing results of some rules of the same class.

2 Development of the Methods

Modified Newton's method—(I)

we have proposed the new Newton type iterative method as

$$x_{n+1} = x_n - \frac{NF(x_n)}{\sum_{k=1}^{N} F'\left[x_n + \frac{(y_n - x_n)(k-0.5)}{N}\right]}, \quad n = 0, 1, 2, \ldots \tag{8}$$

where $y_n = x_n - \frac{F(x_n)}{F'(x_n)}$, $n = 0, 1, 2, \ldots$

In particular choosing $M = 1$, i.e., $N = 2$, we have

$$x_{n+1} = x_n - \frac{2F(x_n)}{\left[F'(\frac{3x_n + y_n}{4}) + F'(\frac{x_n + 3y_n}{4})\right]}, \quad n = 0, 1, 2, \ldots \tag{9}$$

where $y_n = x_n - \frac{F(x_n)}{F'(x_n)}$, $n = 0, 1, 2, \ldots$

The scheme (9) is named as NH-1 method.

Modified Newton's method—(II)

Let

$$y_n = x_n - \frac{F(x_n)}{F'(x_n)}, \quad n = 0, 1, 2, \ldots \tag{10}$$

be the nth iterate in the Newton's method. And let s_n be the (n+1)th iterate in generalized NH-1 method, i.e.,

$$s_n = x_n - \frac{NF(x_n)}{\sum_{k=1}^{N} F'\left[x_n + \frac{(y_n - x_n)(k-0.5)}{N}\right]}. \tag{11}$$

We can use the iterative values from (10) and (11) to construct a new method with higher order convergence.

$$\begin{cases} s_n = x_n - \dfrac{NF(x_n)}{\sum_{k=1}^{N} F'\left[x_n + \frac{(y_n - x_n)(k-0.5)}{N}\right]}; \\ x_{n+1} = s_n - \dfrac{F(s_n)}{F'(y_n)} \qquad n = 0, 1, 2, \ldots. \end{cases} \tag{12}$$

The scheme (12) is named as generalized NH-2 method.

In particular choosing $M = 1$, i.e., $N = 2$, we have

$$\begin{cases} s_n = x_n - \dfrac{2F(x_n)}{\left[F'(\frac{3x_n + y_n}{4}) + F'(\frac{x_n + 3y_n}{4})\right]}; \\ x_{n+1} = s_n - \dfrac{F(s_n)}{F'(y_n)} \qquad n = 0, 1, 2, \ldots. \end{cases} \tag{13}$$

The scheme (13) is named as NH-2 method.

3 Convergence Analysis

Lemma 3.1 *Let $F : A \subset \mathbb{R}^n \to \mathbb{R}^n$ be sufficiently Frechet differentiable in a convex set A. For any $x_0, t \in A$, the Taylor's expansion is as follows*

$$F(x_0 + t) = F(x_0) + tF'(x_0) + \frac{t^2}{2!}F''(x_0) + \frac{t^3}{3!}F'''(x_0) + \cdots \frac{t^{m-1}}{m-1!}F^{m-1}(x_0) + R_m, \tag{14}$$

where $\|R_m\| \leq \frac{1}{m!} \sup \|F^m(x_0 + \beta t)\| \|t\|^m$ and $0 \leq \beta \leq 1$.

Theorem 3.1 *Let $F : A \subset \mathbb{R}^n \to \mathbb{R}^n$ be sufficiently Frechet differentiable in a convex set A. Let $F(\alpha) = 0$ and $\alpha \in D$. If the initial guess x_0 is close to α then the iterative method NH-1 converges cubically to α and the error equation is*

$$e_{n+1} = ({c_2}^2 - \frac{c_3}{4N^2}){e_n}^3 + ({c_2}^3 + 3c_2c_3 + \frac{3c_2c_3}{4N^2} - \frac{2c_4}{N^2} - \frac{3c_4}{2N^3}){e_n}^4 + O(\|e_n\|^5), \tag{15}$$

where $e_n = x_n - \alpha$ and $c_k = (\frac{1}{k!}) \times \frac{F^{(k)}(\alpha)}{F'(\alpha)}$.

Proof Using Taylor's series expansion for $F(x)$ at $x = x_0$ and taking $F(\alpha) = 0$ and $e_n = x_n - \alpha$, we get

$$F(x_n) = F'(\alpha)(x_n-\alpha), F''(\alpha)(x_n-\alpha)^2, F'''(\alpha)(x_n-\alpha)^3, F^{iv}(\alpha)(x_n-\alpha)^4 + O(\|x_n-\alpha\|^5)$$
$$= F'(\alpha)\left[e_n + c_2e_n^2 + c_3e_n^3 + c_4e_n^4 + O(\|e_n\|^5)\right]. \tag{16}$$

Expanding $F'(x)$ at $x = x_n$, we get

$$F'(x_n) = F'(\alpha)\left[1 + 2c_2e_n + 3c_3e_n^2 + 4c_4e_n^3 + 5c_5e_n^4 + O(\|e_n\|^5)\right]. \tag{17}$$

From the above two equations

$$\frac{F(x_n)}{F'(x_n)} = e_n - c_2e_n^2 + 2(c_2^2 - c_3)e_n^3 + (7c_2c_3 - 3c_4 - 4c_2^3)e_n^4 + O(\|e_n\|^5). \tag{18}$$

Again, taking $P_k = \frac{k-0.5}{N}$

$$\begin{aligned} x_n - \left[\frac{(k-0.5)}{N} \times \frac{F(x_n)}{F'(x_n)}\right] &= x_n - P_k[e_n - c_2en^2 + 2(c_2^2 - c_3)e_n^3 + (7c_2c_3 - 3c_4 - 4c_2^3)e_n^4 \\ &\quad + O(\|e_n\|^5] \\ &= x_n + (-1 + (1 - p_k))e_n + C_2P_ke_n^2 - 2C_2^2P_ke_n^3 + 2C_3P_ke_n^3 - \\ &\quad - (7c_2c_3 - 3c_4 - 4c_2^3)e_n^4 + O(\|e_n\|^5) \\ &= \alpha + (1 - P_k)e_n + C_2P_ke_n^2 - 2C_2^2P_ke_n^3 + 2C_3P_ke_n^3 \\ &\quad - (7c_2c_3 - 3c_4 - 4c_2^3)e_n^4 + O(\|e_n\|^5). \end{aligned} \tag{19}$$

Then

$$\begin{aligned} F'(x_n - \left[\frac{(k-0.5)}{N} \times \frac{F(x_n)}{F'(x_n)}\right] &= F'(\alpha)[1 + (2C_2 - 2C_2P_k)e_n + (2C_2^2P_k + 3c_3 + 3c_3P_k - 6c_3p_k)e_n^2 \\ &\quad + (-4c_2^3p_k + 10c_2c_3p_k - 6c_2c_3p_k^2 + 4c_4 - 12c_4p_k + 12c_4p_k^2 - 4c_4p_k^3)e_n^3 \\ &\quad + (-26c_2^2c_3p_k + 6C_2c_4p_k + 8c_2^4p_k + 18c_2^2c_3p_k^2 + 12c_2c_4p_k + 12c_2c_4p_k^3 \\ &\quad - 24c_2c_4p_k^2 + 5c_5 - 20c_5p_k + 30c_5p_k^2 - 20c_5p_k^3 + 5c_5p_k^4)e_n^4 \\ &\quad + O(\|e_n\|^5]. \end{aligned} \tag{20}$$

Now using (16) and (20) in (8), we get

$$\begin{aligned} x_{n+1} &= x_n - \frac{N \times F(x_n)}{\sum_{k=1}^{N} F'\left[x_n - \left[\frac{(k-0.5)}{N} \times \frac{F(x_n)}{F'(x_n)}\right]\right]} \\ &= x_n - [e_n + (-c_2^2 + \frac{c_3}{4N^2})e_n^3 + (-c_2^3 - 3c_2c_3 - \frac{3c_2c_3}{4n^2} + \frac{2c_4}{N^2} + \frac{3c_4}{2N^3})e_n^4 \\ &\quad + O(\|e_n\|^5)]. \end{aligned} \tag{21}$$

$$e_{n+1} = (c_2^2 - \frac{c_3}{4N^2})e_n^3 + (c_2^3 - 3c_2c_3 - \frac{3c_2c_3}{4n^2} + \frac{2c_4}{N^2} + \frac{3c_4}{2N^2})e_n^4 + O(\|e_n\|^5). \tag{22}$$

Thus, (22) shows that the iterative scheme NH-1 is cubically convergent. This completes the rest of the proof. □

Theorem 3.2 *Let the vector function $F(x) = 0$ satisfies all the conditions of theorem-1. Then the iterative scheme NH-2 posses fifth-order convergence. More over the error equation will be*

$$e_{n+1} = \left[2\left[c_2^4 - \frac{c_2^2 c_3}{4N^2}\right] e_n^5\right] + O(\|e_n\|^6), \tag{23}$$

where $e_n = x_n - \alpha$ and $c_k = (\frac{1}{k!}) \times \frac{F^{(k)}(\alpha)}{F'(\alpha)}$.

Proof Using third-order convergence formula

$$e_{n+1} = (c_2^2 - \frac{c_3}{4N^2})e_n^3 + (c_2^3 + 3c_2c_3 + \frac{3c_2c_3}{4n^2} - \frac{2c_4}{N^2} - \frac{3c_4}{2N^2})e_n^4 + O(\|e_n\|^5)$$

that is

$$x_{n+1} = \alpha + (c_2^2 - \frac{c_3}{4N^2})e_n^3 + (c_2^3 + 3c_2c_3 + \frac{3c_2c_3}{4n^2} - \frac{2c_4}{N^2} - \frac{3c_4}{2N^2})e_n^4 + O(\|e_n\|^5) \tag{24}$$

Replacing x_{n+1} by s_n in (24), we have

$$\begin{aligned} s_n &= \alpha + (c_2^2 - \frac{c_3}{4N^2})e_n^3 + (c_2^3 - 3c_2c_3 + \frac{3c_2c_3}{4n^2} - \frac{2c_4}{N^2} - \frac{3c_4}{2N^2})e_n^4 + O(\|e_n\|^5) \\ &= \alpha + h_n, \end{aligned} \tag{25}$$

where $h_n = (c_2^2 - \frac{c_3}{4N^2})e_n^3 + (c_2^3 + 3c_2c_3 + \frac{3c_2c_3}{4n^2} - \frac{2c_4}{N^2} - \frac{3c_4}{2N^2})e_n^4 + O(\|e_n\|^5)$.

Now

$$\begin{aligned} F(s_n) &= F(\alpha) + F'(\alpha)h_n + \frac{F''(\alpha)}{2!}h_n^2 + \frac{F'''(\alpha)}{3!}h_n^3 + O(\|h_n\|^4) \\ &= F'(\alpha)[h_n + c_2h_n^2 + c_3h_n^3 + O(\|h_n\|^4)] \\ &= F'(\alpha)[(c_2^2 - \frac{c_3}{4N^2})e_n^3 + (c_2^3 - 3c_2c_3 + \frac{3c_2c_3}{4n^2} - \frac{2c_4}{N^2} - \frac{3c_4}{2N^2})e_n^4 + O(\|e_n\|^5)]. \end{aligned} \tag{26}$$

Again $y_n = x_n - \frac{F(x_n)}{F'(x_n)}$.

So by using (18), we obtain

$$\begin{aligned} y_n &= x_n - [e_n - c_2e_n^2 + 2(c_2^2 - c_3)e_n^3 + (7c_2c_3 - 3c_4 - 4c_2^3)e_n^4 + O(\|e_n\|^5)] \\ &= \alpha + c_2e_n^2 - 2(c_2^2 - c_3)e_n^3 - (7c_2c_3 - 3c_4 - 4c_2^3)e_n^4 + O(\|e_n\|^5). \end{aligned} \tag{27}$$

And

$$\begin{aligned} F'(y_n) &= F'[\alpha + c_2 e_n^2 - 2(c_2^2 - c_3)e_n^3 - (7c_2c_3 - 3c_4 - 4c_2^3)e_n^4 + O(\|e_n\|^5)] \\ &= F'(\alpha) + F''(\alpha)[c_2 e_n^2 - 2(c_2^2 - c_3)e_n^3 - (7c_2c_3 - 3c_4 - 4c_2^3)e_n^4 + O(\|e_n\|^5)] \\ &= F'(\alpha)[1 + 2c_2^2 + 4(c_2c_3 - c_2^3)e_n^3 + O(\|e_n\|^4)]. \end{aligned} \tag{28}$$

From (11), we get

$$(x_{n+1} - \alpha) = (s_n - \alpha) - \frac{F(s_n)}{F'(y_n)}$$

and

$$(x_{n+1} - \alpha)F'(y_n) = (s_n - \alpha)(F'(y_n)) - F(s_n). \tag{29}$$

Using (26) and (28) in (29) yields

$$KUL = KUV - KV, \tag{30}$$

where $K = F'(\alpha)$, $L = e_{n+1}$,
$U = [1 + 2c_2^2 + 4(c_2c_3 - c_2^3)e_n^3 + O(\|e_n\|^4)]$, and
$V = [(c_2^2 - \frac{c_3}{4N^2})e_n^3 + (c_2^3 - 3c_2c_3 + \frac{3c_2c_3}{4n^2} - \frac{2c_4}{N^2} - \frac{3c_4}{2N^2})e_n^4 + O(\|e_n\|^5)]$.
Simplifying (30), we obtain

$$UL = \left[2(c_2^4 - \frac{c_2^2c_3}{4N^2})e_n^5\right]$$

and

$$\begin{aligned} L &= \left[2(c_2^4 - \frac{c_2^2c_3}{4N^2})e_n^5\right][U]^{-1} \\ &= \left[2(c_2^4 - \frac{c_2^2c_3}{4N^2})e_n^5\right][1 - \{2c_2^2 + 4(c_2c_3 - c_2^3)e_n^3\} + \cdots]. \end{aligned} \tag{31}$$

Further

$$e_{n+1} = \left[2(c_2^4 - \frac{c_2^2c_3}{4N^2})e_n^5\right]O(\|e_n\|^6). \tag{32}$$

Hence, from (32) it is seen that the iterative scheme NH-2 posses fifth-order convergence. This completes the rest of the proof. □

4 Numerical Examples

Efficiency index We see that in each of the same class of methods, the functional values as well as derivatives are evaluated in each iteration. The efficiency index as $E.I = p^{1/w}$, where p is the order of convergence and w is the sum of the number of functional evaluations and derivative evaluations per iteration in the method. For a system of nonlinear equations with n equations and n unknowns, evaluating the function F is to calculate n functional values $f_i = 1, 2, \ldots, n$ and evaluating a derivative F' is to calculate n^2 derivative values $\frac{\partial f_i}{\partial x_j}, i = 1, 2, \ldots, n \; and \; j = 1, 2, \ldots, n$. It is seen that efficiency index of various convergent methods as follows:

NR-1: $E.I. = 3^{\frac{1}{n+2n^2}}$, NR-2: $E.I. = 3^{\frac{1}{n+3n^2}}$, NGM: $E.I. = 3^{\frac{1}{n+3n^2}}$ and M: $E.I. = 5^{\frac{1}{2n+4n^2}}$ and the efficiency index of the proposed methods are as follows:

NH-1: $E.I. = 3^{\frac{1}{n+3n^2}}$ and NH-2: $E.I. = 5^{\frac{1}{n+4n^2}}$

We have taken some examples to examine the convergence of our constructed methods. We have also compared this result with the same class of methods discussed earlier, namely, NR-1 [4], NR-2 [4], NGM [3] and M [3]. We have evaluated the number of iterations k, the error of the approximate solution $x_{(k)}$, the computational order of convergence (COC1/COC2), the computational asymptotic convergence constant (CACC), approximate value of the function $Fx_{(k)}$ for our constructed methods NH-1, NH-2 and are compared with some of the same class of methods. The comparison is shown in respective tables.

To compute the computational order of convergence (COC), we have used the following formulae:

For the nonlinear system with unknown exact solution (α)

$$COC1 = \frac{\log(\|x_{(k+2)} - x_{(k+1)}\| / \|x_{(k+1)} - x_{(k)}\|)}{\log(\|x_{(k+1)} - x_{(k)}\| / \|x_{(k)} - x_{(k-1)}\|)}.$$

For the nonlinear system with known exact solution (α)

$$COC2 = \frac{\log(\|x_{(k+2)} - \alpha\| / \|x_{(k+1)} - \alpha\|)}{\log(\|x_{(k+1)} - c\| / \|x_{(k)} - \alpha\|)}$$

For the computation of the above factor, we have used the last three approximations of the corresponding iterations.

Additionally, we have evaluated the computational asymptotic convergence constant (CACC) for pth order convergence as defined below

$$CACC = \frac{\|x_{(k+1)} - \alpha\|}{\|x_{(k)} - \alpha\|^p}.$$

Example 4.1 Consider the following system with five equations and five unknowns:

$$\begin{aligned}
&4(x_1 - x_2^2) + x_2 - x_3^2 = 0.\\
&x_2(x_2^2 - x_1) - 2(1 - x_2) + 4(x_2 - x_3^2) + x_3 - x_4^2 = 0.\\
&x_3(x_3^2 - x_2) - 2(1 - x_3) + 4(x_3 - x_4^2) + x_2^2 - x_1 + x_4 - x_5^2 = 0.\\
&x_4(x_4^2 - x_3) - 2(1 - x_4) + 4(x_4 - x_5^2) + x_3^2 - x_2 = 0.\\
&x_5(x_5^2 - x_4) - 2(1 - x_5) + x_4^2 - x_3 = 0.
\end{aligned} \tag{33}$$

where $x_{(0)} = (1.5, 1.5, 1.5, 1.5, 1.5)^T$ is the initial trial solution and $\alpha = (1, 1, 1, 1, 1)^T$ is the exact solution. The numerical comparison of the results of the system (33) is shown in Table 1.

Example 4.2 Consider the following system with two equations and two unknowns:

$$\begin{aligned}
&(x_1 - 1)^4 + e^{-x_2} - x_2^2 + 3x_2 + 1 = 0.\\
&4sin(x_1 - 1) - ln(x_1^2 - x_1 + 1) - x_2^2 = 0,
\end{aligned} \tag{34}$$

where $x_{(0)} = (0.5, -0.5)^T$ is the initial trial solution, and the approximate numerical solution of the above nonlinear system is $(1.271384307950132, -0.880819073102661)^T$. The numerical comparison of the results of the system (34) is shown in Table 2.

Table 1 Comparison of various methods for the system (33)

Method	NR1	NR2	NG-M	M	NH1	NH2
k	4	4	4	2	4	2
$\|x^{(k)} - \alpha\|_2$	7.0286e−15	6.9048e−15	6.9048e−15	2.0109e−12	4.6311e−15	1.8054e−12
COC1	2.7937	2.7959	2.7959	4.6774	2.7863	4.5104
CACC	0.0433	0.0436	0.0436	0.0056	0.0341	0.0048
$\|Fx^{(k)}\|_2$	3.3887e−14	3.3769e−14	3.3769e−14	1.0260e−11	2.4330e−14	8.9097e−12

Table 2 Comparison of various methods for the system (34)

Method	NR1	NR2	NG-M	M	NH1	NH2
k	3	3	3	2	3	2
$\|x^{(k)} - \alpha\|_2$	0.8200	0.8312	0.8249	0.8571	0.8293	0.8576
COC2	2.7711	2.7548	2.7571	4.7250	2.6761	4.7241
$\|Fx^{(k)}\|_2$	0.0814	0.0886	0.0847	0.0066	0.0863	0.0054

Table 3 Comparison of various methods for the system (35)

Method	NR1	NR2	NG-M	M	NH1	NH2
k	4	4	4	2	4	3
$\|x^{(k)} - \alpha\|_2$	3.5527e−15	4.4408e−16	2.2204e−15	3.6592e−13	6.6613e−15	7.1320e−13
COC1	2.9695	2.9590	2.9209	4.1680	2.9695	4.1613
CACC	0.0897	0.0886	0.0789	3.8162e−04	0.0949	4.3310e−04
$\|Fx^{(k)}\|_2$	1.0658e−14	3.5527e−15	7.1054e−15	1.0977e−12	1.7763e−14	2.1387e−12

Example 4.3 Consider the following nonlinear equation:

$$2e^{x-4} - 5x + 18 = 0, \tag{35}$$

where $x_{(0)} = 2$ is the initial trial solution and $\alpha = 4$ is the exact solution. The numerical comparison of the results of the Eq. (35) is shown in Table 3.

Example 4.4 Consider the following nonlinear equation:

$$\begin{aligned} &y''(t) + y^{1+p}(t) = 0, \ t \in [0, 1], (p > 0) \\ &y(0) = 0, y(1) = 1. \end{aligned} \tag{36}$$

Let the interval [0, 1] be partitioned as $t_0 = 0 < t_1 < t_2 < \cdots < t_{n-1} < t_n$, $t_i = t_0 + ih$ and $h = 1/n$.

Let $y_0 = y(t_0) = y(0) = 0$, $y_1 = y(t_1), \ldots y_{n-1} = y(t_{n-1})$ and $y_n = y(t_n) = y(1) = 1$. Now using numerical defferential formula, we have $y_i'' = \frac{y_{i-1} - 2y_i + y_{i+1}}{h^2}$, $i = 1, 2, \ldots (n-1)$. Choosing $p = \sqrt{3}$ and $n = 10$, we obtain the following system of nonlinear equation involving nine variables.

$$\begin{aligned} &2y_1 - h^2 y_1^{\sqrt{3}+1} - y_2 = 0. \\ &-y_{i-1} + 2y_i - h^2 y_i^{\sqrt{3}+1} - y_{i+1} = 0.i = 2, 3, \ldots, 8. \\ &-y_8 + 2y_9 - h^2 y_9^{\sqrt{3}+1} - 1 = 0, \end{aligned} \tag{37}$$

where $y_{(0)} = (1, 1, 1, 1, 1, 1, 1, 1, 1)$ is the initial trial solution and the approximate numerical solution of the above nonlinear system is $(0.10630974179490, 0.21259757834295, 0.31873991751981, 0.42444234737133, 0.52918276231997, 0.63216575267634, 0.73229207302272, 0.82814954506063, 0.91803296518217)^T$. The numerical comparison of the results of the system (37) is shown in Table 4.

Table 4 Comparison of various methods for the system (36)

Method	NR1	NR2	NG-M	M	NH1	NH2
k	3	3	3	2	3	2
$\|x^{(k)} - \alpha\|_2$	1.6182	1.6192	1.6186	1.6363	1.6156	1.6362
COC2	3.0651	3.0498	3.0589	4.9999	3.0580	4.9999
$\|Fx^{(k)}\|_2$	0.0814	0.0886	0.0847	0.0066	0.0863	0.0054

Table 5 Comparison of various methods for the system (37)

Method	NR1	NR2	NG-M	M	NH1	NH2
k	5	5	5	3	5	3
$\|x^{(k)} - \alpha\|_2$	2.5355e−04	2.5355e−04	2.5355e−04	0.0019	2.5355e−04	0.0019
COC1	2.9756	2.9756	2.9756	4.7451	2.9756	4.7451
CACC	0.1634	0.1634	0.1634	0.0177	0.1634	0.0177
$\|Fx^{(k)}\|_2$	0.0010	0.0010	0.0010	0.0078	0.0010	0.0078

Example 4.5 Consider the following system of nonlinear equations:

$$\sum_{j=1, j\neq i}^{n} x_j - (n-1){x_i}^2 = 0, \quad 1 \le i \le n, n = 5, \tag{38}$$

where $x_{(0)} = (3.5, 3.5, 3.5, 3.5, 3.5)^T$ is the initial trial solution and $\alpha = (1, 1, 1, 1, 1)^T$ is the exact solution. The numerical comparison of the results of the system (38) is shown in Table 5.

5 Conclusion

We have constructed a pair of iterative methods for the solution of nonlinear system of equations with third- and fifth-order convergence, respectively. The computational order of convergence (COC1/COC2), the computational asymptotic convergence constant (CACC), and the efficiency index (E.I) for these methods are equivalent to those of the same class of discussed methods. The numerical results also agree to the theoretical claim. Above all, these methods are very handy in solving the system of nonlinear equations as well as boundary value nonlinear ordinary differential equations. The proposed methods can be seen as alternatives to the existing methods of the same class.

References

1. S.D. Conte, C. De Boor, *Elementary Numerical Analysis* (Kogakusha Ltd., Mc Graw Hill, 1972)
2. A.M. Ostrowski, *Solution of Equations in Euclidean and Banach Space*, 3rd edn. (Academic Press, New York, 1973)
3. Z. Liu, Q. Zheng, C. Huang, Third- and fifth-order Newton-Gauss methods for solving nonlinear equations with n variables. Appl. Math. Comput. **290**, 250–257 (2016)
4. M.A. Noor, M. Waseem, Some iterative methods for solving a system of nonlinear equations. Comput. Math. Appl. **57**, 101–106 (2009)
5. M.A. Noor, Some applications of quadrature formulas for solving nonlinear equations. Nonlinear Anal. Forum **12**(1), 91–96 (2007)
6. M.T. Darvishi, A. Barati, A third-order Newton-type method to solve systems of nonlinear equations. Appl. Math. Comput. **187**, 630–635 (2007)
7. M. Frontini, E. Sormani, Third-order methods from quadrature formulae for solving systems of nonlinear equations. Appl. Math. Comput. **149**, 771–782 (2004)
8. H.H.H. Homeier, A modified Newton method with cubic convergence: the multivariate case. J. Comput. Appl. Math. **169**, 161–169 (2004)
9. M.Q. Khirallah, M.A. Hafiz, Novel three order methods for solving a system of nonlinear equations. Bull. Math. Sci. Appl. **2**, 01–14 (2012)
10. J. Kou, A third-order modification of Newton method for systems of nonlinear equations. Appl. Math. Comput. **191**, 117–121 (2007)
11. M.A. Noor, Fifth-order convergent iterative method for solving nonlinear equations using quadrature formula. J. Math. Control Sci. Appl. **1**, 241–249 (2007)
12. A. Golbabai, M. Javidi, A new family of iterative methods for solving system of nonlinear algebraic equations. Appl. Math. Comput. **190**, 1717–1722 (2007)
13. A. Cordero, J.R. Torregrosa, Variants of Newtons method using fifth-order quadrature formula. Appl. Math. Comput. **190**, 686–698 (2007)
14. J.E. Dennis, R.B. Schnable, *Numerical Methods for Unconstrained Optimization and Nonlinear Equations* (Prentice Hall, 1983)

Analytical Solution of Trapped Burgers' Equation with *Tan-hyperbolic* Method

Apul N. Dev and Manoj Kr. Deka

Abstract The features of the well-known *tanh* method and its detailed mechanism is discussed which is basically used to derive shock and solitons solutions of different non-linear trapped Burgers', non-linear Burgers' equation and different non-linear trapped K-P equation, as well as non-linear K-P equation. All the necessary mechanisms along with their procedures are discussed. We expect this article to be suitable for the targeted audience.

1 Introduction

The appearance of the non-linear phenomenon in different regime of fluid dynamics [1], wave interaction in plasma physics [2, 3], kinetic theory of chemistry [4], as well as mathematical application of biological phenomenon [5], is inevitable. As a result, different methodologies such as Painlev analysis [6], Hirota's bilinear technique [7], inverse scattering transform [8] have been widely developed and tested. Even based on these tools, numerical methods have been developed. But owing to the difficulty in handling these analytical methods, one has to dig into the thorough knowledge of the properties and possibilities of these methodologies in order to apply them for one's interest.

Here, based on some of our research works, we aim to develop an overview of the scope of the application of *Tan-Hyperbolic* method. We discuss basically a stationary solution of solitary and shock wave in both one and three dimensions using this technique. We expect this would be helpful to our targeted audience working in this domain of non-linear wave theories.

A. N. Dev (✉)
Center for Applied Mathematics and Computing, Siksha 'O' Anusandhan (Deemed To Be University), Bhubaneswar 751030, Odisha, India
e-mail: apulnarayan@gmail.com

M. Kr. Deka
Vill-Borka, PO-Borka, Kamrup, Guwahati 781101, Assam, India

S. R. Mishra et al. (eds.), *Recent Trends in Applied Mathematics*, Lecture Notes in Mechanical Engineering, https://doi.org/10.1007/978-981-15-9817-3_31

2 The *Tan-Hyperbolic* Method

A partial differential equation of the type

$$P(u,\ u_x,\ u_{xx},\ u_{xxx},\ \ldots\ldots..) = 0 \tag{1}$$

is normally converted to an ordinary differential equation

$$Q(u,\ u',\ u'',\ u''',\ \ldots\ldots..) = 0 \tag{2}$$

Now using a wave transformation $\chi = (\xi - U\tau)$, Eq. (2) is then integrated, provided all the terms contain derivatives and integration constants are considered as zeros. Introducing the new independent variable $z = \tanh(\chi)$,

then $\frac{dz}{d\chi} = \sec h^2(\chi) = 1 - z^2$,
$\frac{dW}{d\chi} = \frac{dz}{d\chi}\frac{dW}{dz} = \left(1 - z^2\right)\frac{dW}{dz}$,
$\frac{d^2W}{d\chi^2} = \frac{d}{d\chi}\frac{dW}{d\chi} = \frac{d}{d\chi}\left(1 - z^2\right)\frac{dW}{dz} = \left(1 - z^2\right)^2\frac{d^2W}{dz^2} - 2z\left(1 - z^2\right)\frac{dW}{dz}$.
We assume for tan-hyperbolic expansion,

$$W(z) = \sum_{r=0}^{m} a_r z^{\delta+r}. \tag{3}$$

Substituting (3) into the Ordinary differential Eq. (2), we find an algebraic equation in powers of *z*. To find *m*, we generally balance the highest order of linear terms and non-linear terms. As a result, we get a system of equations involving the parameters a_r (r = 0, 1, …m), U and c. From the resulting equation, all coefficients of powers of z are collected. On determining these parameters, we obtain an analytic solution *u(x, t)* in a closed form. In the following subsections, we describe the solution of different non-linear equations involving this *Tan-Hyperbolic* method.

3 Trapped Burgers' Equation with Its Solution

The Trapped Burgers' equation is

$$\frac{\partial}{\partial\xi}\left(\frac{\partial\phi^{(1)}}{\partial\tau} + A\left(\phi^{(1)}\right)^{1/2}\frac{\partial}{\partial\xi}\phi^{(1)} - B\frac{\partial^2\phi^{(1)}}{\partial\xi^2}\right) + C\left(\frac{\partial^2\phi^{(1)}}{\partial\eta^2} + \frac{\partial^2\phi^{(1)}}{\partial\zeta^2}\right) = 0 \tag{4}$$

where the non-linear coefficient *A*, the dispersion coefficient *B* and the transverse coefficient *C*, respectively. Now using the transformation $\chi =$

$c(\xi l + \eta m + \zeta n - U\tau)$, χ is reduced to a single parameter, where l, m, n represents the direction cosines along the x, y, z-axes and let $\left(\phi^{(1)}\right)^{\frac{1}{2}}(\xi, \eta, \zeta, \tau) = \psi(\chi)$ i.e. $\phi^{(1)}(\xi, \eta, \zeta, \tau) = \psi^2(\chi)$ Thus, from Eq. (20), one obtains

$$-Bcl^3\frac{d\psi^2}{d\chi} + Al^2\frac{2\left(\psi^2\right)^{3/2}}{3} - Ul\psi^2 + C\left(m^2 + n^2\right)\psi^2 = 0$$

$$-2cBl^3\psi\frac{d\psi}{d\chi} + Al^2\frac{2\psi^3}{3} - Ul\psi^2 + C\left(m^2 + n^2\right)\psi^2 = 0$$

$$-2cBl^3\frac{d\psi}{d\chi} + Al^2\frac{2\psi^2}{3} - Ul\psi + C\left(m^2 + n^2\right)\psi = 0. \tag{5}$$

To find the solution of modified 3D Burgers' Eq., we now apply the *tanh* method where we define $z = \tanh(\chi)$, $\psi(\chi) = W(z)$ where

$$\frac{dz}{d\chi} = \sec h^2(\chi) = 1 - z^2$$
$$\frac{dW}{d\chi} = \frac{dz}{d\chi}\frac{dW}{dz} = \left(1 - z^2\right)\frac{dW}{dz},$$

Then Eq. (5) becomes

$$\frac{2Al^2}{3}W^2 - 2Bcl^3\left(1 - z^2\right)\frac{dW}{dZ} + C\left(m^2 + n^2 - Ul\right)W = 0 \tag{6}$$

We assume $W(x) = \sum_{r=0}^{m} a_r x^{\delta + r}$ to obtain the series solution of Eq. (6) and then the leading order analysis of finite terms gives $r = 1$ and $\delta = 0$ because $2 + r - 1 = 2r \Rightarrow r = 1$ so that $W(x)$ becomes $W(x) = a_0 + a_1 z$. Now putting the value of $W(x)$ and $\frac{dW}{dZ}$ in Eq. (3)

$$\left\{C\left(m^2 + n^2 - Ul\right)(a_0 + a_1 z) + \frac{2A}{3}(a_0 + a_1 z)^2 - Bc2a_1\left(1 - z^2\right)\right\} = 0$$
$$\left\{C\left(m^2 + n^2 - Ul\right)(a_0 + a_1 z) + \frac{2A}{3}\left(a_0^2 + 2a_0 a_1 z + a_1^2 z^2\right) - Bc2a_1\left(1 - z^2\right)\right\} = 0$$
$$C\left(m^2 + n^2 - Ul\right)a_0 + \frac{2A}{3}a_0^2 - Bc2a_1 = 0 \tag{i}$$
$$C\left(m^2 + n^2 - Ul\right)a_1 + \frac{4A}{3}a_0 a_1 = 0 \tag{ii}$$
$$\frac{2A}{3}a_1^2 + Bc2a_1 = 0 \tag{iii}$$

$$
\begin{aligned}
&(iii) \Rightarrow a_1 = -\frac{3Bc}{A}, \\
&(ii) \Rightarrow a_0 = \frac{3C\left(m^2+n^2-Ul\right)}{4A}, \\
&(i) \Rightarrow -C\left(m^2+n^2-Ul\right)a_0 + \frac{2A}{3}a_0^2 - B2a_1\,c = 0 \\
&\Rightarrow -3C\left(m^2+n^2-Ul\right)a_0 + 2Aa_0^2 - B6a_1c = 0 \\
&\Rightarrow -4Aa_0^2 + 2Aa_0^2 - Bc6a_1 = 0, \\
&\Rightarrow -2Aa_0^2 + B6c\frac{3B}{A}c = 0 \\
&\Rightarrow -\frac{9}{8A}\left(C\left(m^2+n^2-Ul\right)\right)^2 + B^2\frac{18}{A}\,c^2 = 0 \\
&\Rightarrow -\frac{1}{8}\left(C\left(m^2+n^2-Ul\right)\right)^2 + 2B^2c^2 = 0 \\
&\Rightarrow \frac{1}{16B^2}\left(C\left(m^2+n^2-Ul\right)\right)^2 = c^2 \\
&\Rightarrow \frac{C\left(m^2+n^2-Ul\right)}{4B} = c \\
&\therefore\ a_1 = -\frac{3C\left(m^2+n^2-Ul\right)}{4A} = -a_0 \\
&\therefore\ W(\chi) = a_0 - a_0z = a_0(1-z)
\end{aligned}
$$

we can obtain the values of $a_0 = \phi_m$ and $W(z) = \phi_m\left\{1 - \tanh\left(\frac{\chi}{\omega_1}\right)\right\}$.

The required stationary solution of Trapped Burgers' Eq. (4) is

$$
\phi^{(1)} = \phi_m^2\left\{1 - \tanh\left(\frac{\chi}{\omega}\right)\right\}^2 \tag{7}
$$

where $\phi_m = \frac{3}{4}\left[\left\{U\,l - C\left(m^2+n^2\right)\right\}/A_1\,l^2\right]$ and $\omega = 4Bl^3/\left\{U\,l - C\left(m^2+n^2\right)\right\}$.

4 Burgers' equation and its solution:

The Burgers' equation

$$
\frac{\partial}{\partial\xi}\left(\frac{\partial\phi^{(1)}}{\partial\tau} + A\,\phi^{(1)}\frac{\partial\phi^{(1)}}{\partial\xi} - B\frac{\partial^2\phi^{(1)}}{\partial\xi^2}\right) + C\left(\frac{\partial^2\phi^{(1)}}{\partial\eta^2} + \frac{\partial^2\phi^{(1)}}{\partial\zeta^2}\right) = 0 \tag{8}
$$

where A, B and C are Non-linear, dissipative and transverse coefficients, respectively. The non-linear coefficient A can be positive, negative or zero. Now to solve

Eq. (8), the following transformation is used $\chi = (\xi l + \eta m + \zeta n - U\tau)$ to a single parameter χ and considering $\phi^{(1)}(\xi, \eta, \zeta, \tau) = \psi(\chi)$ which gives

$$-B_2 c l^3 \frac{d\psi}{d\chi} + A_2 l^2 \frac{\psi^2}{2} + \left\{C_2\left(m^2 + n^2\right) - Ul\right\}\psi = 0 \tag{9}$$

To study the solution Eq. (8), tanh method is used and for that the transformation $z = \tanh(\chi)$ and $\psi(\chi) = W(z)$ is used, where

$$\frac{dz}{d\chi} = \sec h^2(\chi) = 1 - z^2$$
$$\frac{dW}{d\chi} = \frac{dz}{d\chi}\frac{dW}{dz} = \left(1 - z^2\right)\frac{dW}{dz},$$

Then the Eq. (9) becomes

$$\frac{A_1}{2} W^2 l^2 - Bcl^3\left(1 - z^2\right)\frac{dW}{dZ} + \left\{C\left(m^2 + n^2\right) - Ul\right\}W = 0 \tag{10}$$

Now substituting $W(z) = \sum_{r=0}^{\infty} a_r z^{\rho + r}$ the series solution of Eq. (10) is sought and for leading order analysis of finite terms gives $r = 1$ and $\rho = 0$ because $2 + r - 1 = 2r \Rightarrow r = 1$ and then the $W(z)$ becomes $W(z) = a_0 + a_1 z$ and $\frac{dW}{dZ} = a_1$ Now putting the value of $W(z)$ and $\frac{dW}{dZ}$ in Eq. (10), we get the stationary solution of 3D Burgers Eq. (8) as

$$\phi^{(1)} = \phi_m\left\{1 - tanh\left(\frac{\chi}{w}\right)\right\} \tag{11}$$

where $\phi_m = \left\{U l - C\left(m^2 + n^2\right)\right\}/A l^2$ and $w = 2Bl^3/\left\{U l - C\left(m^2 + n^2\right)\right\}$ are the height and thickness of the shock wave, respectively, with l, m, n, being the direction cosine along x-axis, y-axis and z-axis, respectively, and U is the velocity. The solution describes a negative (positive) shock wave for $A < 0\,(A > 0)$, respectively. When $A = 0$ then $\phi_m \to \infty$, from Eq. (11), the formation of shock wave is not possible, then we move to another equation with different values of n.

$$\frac{\partial}{\partial\xi}\left(\frac{\partial\phi^{(1)}}{\partial\tau} + A_n\left(\phi^{(1)}\right)^{\frac{n}{2}}\frac{\partial\phi^{(1)}}{\partial\xi} - B\frac{\partial^2\phi^{(1)}}{\partial\xi^2}\right) + C\left(\frac{\partial^2\phi^{(1)}}{\partial\eta^2} + \frac{\partial^2\phi^{(1)}}{\partial\zeta^2}\right) = 0 \tag{12}$$

where n = 1, 2, 3, …, and the non-linear coefficient A_n have different values for different equations and the coefficient dissipative B, transverse coefficient C are as before. Now to solve the nth- 3D Burgers Eq. (12), we used the transformation $\chi = (\xi l + \eta m + \zeta n - U\tau)$ to a single parameter χ, then the above equation becomes

$$\left\{-Ul\phi^{(1)} + \frac{2A_n}{n+2}\left(\phi^{(1)}\right)^{\frac{n+2}{2}} l^2 - Bcl^3 \frac{d\phi^{(1)}}{d\chi}\right\} + C\left(m^2 + n^2\right)\phi^{(1)} = 0 \quad (13)$$

We considering $\left(\phi^{(1)}\right)^{\frac{n}{2}}(\xi, \eta, \zeta, \tau) = \psi(\chi)$, i.e. $\phi^{(1)}(\xi, \eta, \zeta, \tau) = \psi^{\frac{2}{n}}(\chi)$ which gives

$$\left\{-Ul\psi^{\frac{2}{n}} + \frac{2A_n}{n+2}\left(\psi^{\frac{2+n}{n}}\right)l^2 - Bcl^3 \frac{2}{n}\psi^{\frac{2-n}{n}} \frac{d}{dx}\psi\right\} + C\left(m^2 + n^2\right)\psi^{\frac{2}{n}} = 0$$

$$\left\{-Ul + \frac{2A_n}{n+2}\psi l^2 - Bcl^3 \frac{2}{n}\psi^{-1} \frac{d}{dx}\psi\right\} + C\left(m^2 + n^2\right) = 0$$

$$-Bcl^3 \frac{1}{n}\frac{d\psi}{d\chi} + A_n l^2 \frac{\psi^2}{n+2} + \left\{C\left(m^2 + n^2\right) - Ul\right\}\psi = 0 \quad (14)$$

Again, after using the transformation $z = \tanh(\chi)$ and $\left(\phi^{(1)}\right)^{\frac{n}{2}}(\xi, \eta, \zeta, \tau) = \psi(\chi) = W(z)$ the Eq. (14) becomes

$$\frac{dz}{d\chi} = \sec h^2(\chi) = 1 - z^2$$

$$\frac{dW}{d\chi} = \frac{dz}{d\chi}\frac{dW}{dz} = \left(1 - z^2\right)\frac{dW}{dz},$$

$$\frac{A_n}{n+1}W^2 l^2 - \frac{Bcl^3\left(1 - z^2\right)}{n}\frac{dW}{dZ} + \left\{C\left(m^2 + n^2\right) - Ul\right\}W = 0 \quad (15)$$

Now we substitute $W(z) = \sum_{r=0}^{\infty} a_r z^{\rho+r}$ to find the series solution of Eq. (15) and for leading order analysis of finite terms gives $r = 1$ and $\rho = 0$ and we obtain $W(z)$ as equal to $W(z) = (a_0 + a_1 z) \Rightarrow \left(\phi^{(1)}\right)(\xi, \eta, \zeta, \tau) = \{a_0(1 - z)\}^{\frac{2}{n}}$. Now introducing the value of $W(z)$, the stationary shock solution of nth degree modified 3D Burgers Eq. (12) as

$$\phi^{(1)} = \left[\phi_{mn}\left\{1 - \tanh\left(\frac{\chi}{w_n}\right)\right\}\right]^{\frac{2}{n}} \quad (16)$$

where $\phi_{mn} = \left(\frac{n+2}{4}\right)\left\{Ul - C\left(m^2 + n^2\right)\right\}/A_n l^2$ and $w_2 = 2Bl^3/n\left\{Ul - C\left(m^2 + n^2\right)\right\}$ are the amplitude and width of the shock wave, respectively, with l,m, n, representing the direction cosines for x-axis, y-axis and z-axis, respectively, and U is the velocity.

5 Conclusion

With the help of some well-established examples, the effectiveness, as well as applicability of the *Tan Hyperbolic* method, is discussed in detail. We believe, based on the analytical analysis of the solution, the exact nature of solitary and shock wave can be revealed, which is really very important from the viewpoint of non-linear plasma physics theory.

References

1. G. Whitham, *Linear and Nonlinear Waves* (Wiley, New York, 1974).
2. M. Kr. Deka, A.N. Dev, Plasma Phys. Rep. **44**(10), 1 (2018)
3. A.N. Dev, M. Kr. Deka, Phys. Plasmas **25**, 072117 (2018)
4. P. Gray, S. Scott, *Chemical Oscillations and Instabilities* (Clarendon Press, Oxford, 1990).
5. J. Murray, *Mathematical Biology* (Springer, Berlin, 1989).
6. R. Conte (ed.), *The Painleve Property* (Springer, Berlin, 1999)
7. R. Hirota, Direct method of 2nding exact solutions of nonlinear evolution equations, in *Backlund Transformations* ed. by R. Bullough, P. Caudrey, vol. 1157 (Springer, Berlin, 1980)
8. M. Ablowitz, D. Kaup, A. Newell, H. Segur, The inverse scattering transform-Fourier analysis for nonlinear problems. Stud. Appl. Math. **53**, 249 (1974)